Fifth Edition

PRINCIPLES OF MEAT *Science*

Elton D. Aberle
University of Wisconsin-Madison

John C. Forrest
Purdue University

David E. Gerrard
Virginia Tech University

Edward W. Mills
Pennsylvannia State University

Kendall Hunt
publishing company
4050 Westmark Drive • P O Box 1840 • Dubuque IA 52004-1840

Book Team

Chairman and Chief Executive Officer   Mark C. Falb
President and Chief Operating Officer   Chad M. Chandlee
Vice President, Higher Education   David L. Tart
Senior Managing Editor   Ray Wood
Assistant Editor   Bob Largent
Editorial Manager   Georgia Botsford
Senior Editor   Lynnette M. Rogers
Vice President, Operations   Timothy J. Beitzel
Assistant Vice President, Production Services   Christine E. O'Brien
Senior Production Editor   Carrie Maro
Permissions Editor   Caroline Kieler
Cover Designer   Jeni Fensterman

Cover image © 2012 Shutterstock, Inc.

Kendall Hunt
publishing company

www.kendallhunt.com
Send all inquiries to:
4050 Westmark Drive
Dubuque, IA 52004-1840

Copyright © 1975 by W.H. Feeman and Company

Copyright © 1989, 1994, 2001, 2012 by Kendall Hunt Publishing Company

ISBN 978-0-7575-9995-8

All rights reserved. No part of this publication may be reproduced, stored in a retrieval system, or transmitted, in any form or by any means, electronic, mechanical, photocopying, recording, or otherwise, without the prior written permission of the copyright owner.

Printed in the United States of America
10 9 8 7 6 5 4 3 2 1

BRIEF CONTENTS

CHAPTER 1   Meat as a Food, 1

CHAPTER 2   Structure, Composition, and Development of Animal Tissues, 7

CHAPTER 3   Muscle Contraction and Energy Metabolism, 61

CHAPTER 4   Principles of Animal Handling and Harvesting, 75

CHAPTER 5   Conversion of Muscle to Meat: Biochemistry of Meat Quality Development, 97

CHAPTER 6   Properties of Fresh Meat, 131

CHAPTER 7   Meat Merchandising, 141

CHAPTER 8   Principles of Meat Processing, 175

CHAPTER 9   Microbiology of Meat, 213

CHAPTER 10   Deterioration, Preservation, and Storage of Meat, 235

CHAPTER 11   Palatability and Cookery of Meat, 259

CHAPTER 12   Nutritive Value of Meat, 277

CHAPTER 13   Meat Inspection and Food Safety, 291

CHAPTER 14   Meat Grading and Evaluation, 317

CHAPTER 15   Electronic Assessment of Carcasses and Fresh Meat, 337

CHAPTER 16   By-Products of the Meat Industry, 351

CONTENTS

PREFACE, ix
ABOUT THE AUTHORS, xi
LIST OF TABLES, xiii
LIST OF FIGURES, xv

1
MEAT AS A FOOD, 1
What Is Meat?, 1
Muscle as Meat, 1
What Is Meat Science?, 2
Meat Consumption and the Economy of a Nation, 3
The Meat Industry in the United States, 5
Science and Meat, 6

2
STRUCTURE, COMPOSITION, AND DEVELOPMENT OF ANIMAL TISSUES, 7
Objectives, 7
Key Terms, 7
Muscle Tissue, 9
Epithelial Tissues, 24
Nervous Tissue, 26
Connective Tissue, 28
Muscle Organization, 37
Chemical Composition of the Animal Body, 46
Chemical Composition of Skeletal Muscle, 51
Carcass Composition, 53

3
MUSCLE CONTRACTION AND ENERGY METABOLISM, 61
Objectives, 61
Key Terms, 61
Nerves and the Nature of Stimuli, 62
Contraction of Skeletal Muscle, 66
Relaxation of Skeletal Muscle, 68
Sources of Energy for Muscle Function, 70

4
PRINCIPLES OF ANIMAL HANDLING AND HARVESTING, 75
Objectives, 75
Key Terms, 75
Introduction, 76
Personnel, 76
Transportation, 76
Inspectors, 78
Animal Physiology and Homeostasis, 78
Flight Zone and Point of Balance, 80
Animal Movement and Facilities, 80
Temperature and Humidity, 82
Animal Size and Densities, 83
Feed Withdrawal, 83
Lairage, 84
Harvesting, 84
Non-ambulatory Animals, 85
Stunning and Immobilization, 86
Exsanguination (Bleeding), 90
Ritual Slaughter, 92
Scalding and Skinning, 93
Evisceration, 94
Carcass Manipulation, 95

5
CONVERSION OF MUSCLE TO MEAT: BIOCHEMISTRY OF MEAT QUALITY DEVELOPMENT, 97
Objectives, 97
Key Terms, 97
Introduction, 98
Postmortem Muscle Metabolism, 98
Energy Metabolism, H^+ Production and Muscle pH Decline, 106
Factors Affecting Postmortem Changes and Meat Quality, 114

6
PROPERTIES OF FRESH MEAT, 131
Objectives, 131
Key Terms, 131
Meat Quality, 132
Water-Holding Capacity, 132
Chemical Basis of Water-Holding Capacity, 133
Color, 136
Structure, Firmness, and Texture, 139

7
MEAT MERCHANDISING, 141
Objectives, 141
Key Terms, 141
Retail Meat Distribution Systems, 142
Retail Store Types, 143
Fresh Meat Displays, 144
Frozen Meat Displays, 167
Processed Meat Displays, 168
Processing and Distribution for Food Service, 169
Types of Food Service, 169
Meat Acquisition for Food Service, 169

8
PRINCIPLES OF MEAT PROCESSING, 175
Objectives, 175
Key Terms, 175
Types of Processed Meat Products, 176
History of Meat Processing, 178
Basic Processing Procedures, 178

9
MICROBIOLOGY OF MEAT, 213
Objectives, 213
Key Terms, 213
Microbiological Principles, 215
Microbial Contamination of Meat Animal Carcasses and Products, 232

10
DETERIORATION, PRESERVATION, AND STORAGE OF MEAT, 235
Objectives, 235
Key Terms, 235
Deterioration of Meat, 236
Refrigerated Storage, 241
Thermal Processing, 252
Dehydration, 254
Irradiation, 255
Preservation by Chemical Ingredients, 257
Preservation by Packaging, 258

11
PALATABILITY AND COOKERY OF MEAT, 259
Objectives, 259
Key Terms, 259
Palatability Characteristics, 260
Factors Associated with Meat Palatability, 265
Cookery, 267
Serving Cooked Meat, 275

12
NUTRITIVE VALUE OF MEAT, 277
Objectives, 277
Key Terms, 277
Proteins, 279
Lipids and Calories, 282
Carbohydrates, 285
Minerals, 286
Vitamins, 287
Variety and Processed Meats, 287
Nutrient Retention during Heating, 288
Nutrient Labeling, 288

13
MEAT INSPECTION AND FOOD SAFETY, 291
Objectives, 291
Key Terms, 291
History, 292
Application and Enforcement of Inspection Laws, 294
Requirements for Granting Inspection Service, 296
Elements of Inspection, 297
Religious Inspection and Certification, 316
Seafood Inspection, 316

14
MEAT GRADING AND EVALUATION, 317
Objectives, 317
Key Terms, 317
Types of Grading and Evaluation, 318
Functions of Grades, 319
Federal Quality Grades, 320
Carcass Cutability and Federal Yield Grades, 330
Services Offered by the USDA Grading Service, 336

15
ELECTRONIC ASSESSMENT OF CARCASSES AND FRESH MEAT, 337
Objectives, 337
Key Terms, 337
Requisites for Successful Technology Adoption, 338
Principles of Electronic Assessment, 339

16
BY-PRODUCTS OF THE MEAT INDUSTRY, 351
Objectives, 351
Key Terms, 351

Edible Meat By-Products, 352
Inedible Meat By-Products, 358
Hides, Skins, and Pelts, 358
Tallows and Greases, 362
Animal Feeds and Fertilizers, 362
Glue, 363
Pharmaceuticals and Biologicals, 363

REFERENCES, 367
INDEX, 375

PREFACE

Meat science courses at many universities serve students having a broad range of interests and backgrounds. The goal in preparing this book is to provide a text for college students enrolled in meat science courses. When the first edition was prepared thirty-five years ago, many excellent reference texts were available for introductory meat science courses, but no single up-to-date text covered the bulk of information we believed was needed in those courses. *Principles of Meat Science* was generated to fill that void. Our objective in subsequent revised editions is to continue providing a comprehensive text, which includes the latest discoveries and principles that are necessary to understand the science of meat and the utilization of meat as food.

We believe this text is suitable for students enrolled in their first university-level meat science course if they have taken appropriate preparatory courses. *Principles of Meat Science* is written with the intent that students using this text would have the level of science knowledge provided by the completion of introductory biology, chemistry, and mathematics courses, and also have completed an introductory animal science, food science, or animal industry course.

Muscle is the primary component of meat and an understanding of muscle biology is basic to meat science and meat industry practices. To that end, muscle structure, composition, development, growth and function are presented as they influence the use of muscle as meat. Further, emphasis is placed on basic principles that govern meat industry practices—principles of postmortem muscle chemistry, meat processing and preservation, sanitation and food safety, meat inspection, meat grading and evaluation, meat cookery, meat distribution through marketing channels and use of meat animal by-products.

In our quest for completeness, we have attempted to provide information on meat produced by cattle, swine, sheep, poultry, seafood, and other species. Because many of the basic principles of meat science apply to all meat-producing species, specific references to species usually were considered unnecessary. Properties of meat that are unique to an individual species have been included where appropriate. This text is also oriented toward the meat industry as it exists in the United States; there is wide variation in much of the information given here when consideration is given to the industry in other areas of the world. This is particularly true with regard to meat inspection, regulations, standards and grades, and the nomenclature of meat cuts.

In this the fifth edition, we have incorporated new knowledge of muscle structure and biology, advances in meat processing, packaging and distribution technologies, and improvements in sanitation and food safety practices. Awareness and knowledge of animal welfare has led to increased emphasis on humane handling and harvesting practices; thus, we have included a new chapter on principles of animal handling and harvesting. Similarly, the increased use of electronic and vision-based meat evaluation tools stimulated inclusion of a separate chapter on principles of electronic assessment of carcasses and fresh meat. Many universities have instituted courses on animal growth and development. We believe this subject requires more comprehensive coverage than could be presented in this text. Several text books have been published on this topic. Thus, a separate chapter on growth and development has been deleted from this edition. However, the principles of muscle, adipose tissue, bone development and growth, and of carcass composition that are important to understanding the science and utilization of meat are included in the chapter on structure, composition, and development of animal tissues. Finally, retail and institutional meat merchandizing is discussed in a single meat merchandizing chapter. On the advice of instructors who use the text in their courses, the order of chapters is rearranged to more closely reflect the sequence in which topics are covered.

Several individuals provided suggestions, figures, illustrations, and assistance with parts of this revision. We express our appreciation to Dennis Burson, University of Nebraska-Lincoln; Stephen Ingham, University of Wisconsin-Madison and Wisconsin Department of Agriculture, Trade and Consumer Protection; Ronald Jenkins, Budenheim USA; Amanda King, University of Wisconsin-Madison; Mark Morgan, Purdue University; Chad A. Stahl, Meat Animal Consultation and Testing Services; Kurt Vogel, University of Wisconsin-River Falls; and Gang Yao, University of Missouri for their insight and originality to help bring this edition to fruition.

Finally, we recognize with gratitude the contributions of three co-authors of the first three editions of *Principles of Meat Science*—Harold Hedrick, University of Missouri (deceased); Max Judge, Purdue University (retired); and Robert Merkel, Michigan State University (retired). Although the content has been updated and reorganized in many cases, their contributions are evident in nearly every chapter. Most importantly, we continue the goals and objectives they helped establish in writing the first edition, specifically to present a comprehensive text for undergraduate meat science courses that emphasizes the scientific principles on which the study of meat science is based. Professors Hedrick, Judge, and Merkel were our teachers, mentors, colleagues, and friends. We are indebted to them.

ABOUT THE AUTHORS

Elton D. Aberle

Elton D. Aberle is Emeritus Dean and Emeritus Professor of Animal Sciences at the College of Agricultural and Life Sciences, University of Wisconsin-Madison. He has taught, conducted research and published in the areas of muscle biology, meat animal growth and development and food chemistry. Prior to the University of Wisconsin, he was a professor of animal sciences at Purdue University and professor and head in the Department of Animal Science, University of Nebraska-Lincoln. He has received teaching and research awards from the American Meat Science Association and the American Society of Animal Science and is a Fellow of the American Association for the Advancement of Science.

John C. Forrest

John C. Forrest is Emeritus Professor of Animal Sciences at Purdue University. His teaching responsibilities included the introductory course in Meat Science along with a hands-on lab. He also lectured in food science, foods and nutrition, and hospitality and tourism management courses. Research activities centered around meat quality and on the development of electronic technology to assess quality and composition of carcasses and cuts on-line. He was technical advisor to the Indiana Meat Packers and Processors Association and was responsible for meat science extension programs in Indiana. He has received research and extension awards from the American Meat Science Association and the American Society of Animal Sciences.

David E. Gerrard

David E. Gerrard is Professor and Head, Department of Animal and Poultry Science, Virginia Tech University. Prior to Virginia Tech, he held faculty positions in Animal Sciences at Purdue University and in Food Science at the University of Missouri-Columbia. Research activities focus on pre- and post-natal development and maturation of muscle fibers and the role of post mortem metabolism in modulating meat quality. He has taught courses in meat animal growth and development and muscle biology. He was a University Faculty Scholar at Purdue University and has received research awards from the American Meat Science Association and the American Society of Animal Science.

Edward W. Mills

Edward W. Mills is Associate Professor of Dairy and Animal Sciences at Pennsylvania State University. He has a courtesy appointment in Food Science. He was manager of technical services at Wilson Foods and on the faculty at Purdue University before his appointment at Penn State in 1988. He teaches courses in animal products technology and meat processing and advises undergraduate and graduate students. His research in meat science currently is focused on low-tech meat preservation, natural curing technology and control of lipid oxidation in fresh and precooked meat products. Dr. Mills has been recognized for outstanding teaching by the American Meat Science Association. He is helping to initiate community meat processing and preservation enterprises in Kenya and Namibia.

TABLES

Chapter 1

TABLE 1.1
Per Capita Meat Consumption in Selected Countries, 2010 (kg), 4

Chapter 2

TABLE 2.1
Characteristics of Muscle Fibers in Domestic Meat Animals and Birds, 45

TABLE 2.2
Elemental Composition of the Animal Body, 47

TABLE 2.3
Fatty Acid and Triglyceride Composition of Some Animal Fat Depots (Percent by Weight), 48

TABLE 2.4
Approximate Composition of Mammalian Skeletal Muscle (Percent Fresh Weight Basis), 50

TABLE 2.5
Distribution of Body Components of Cattle During Growth and Development, 52

TABLE 2.6
Heritability Estimates of Carcass Characteristics of Cattle, Poultry, Sheep, and Swine, 53

TABLE 2.7
Priority and Partition of Nutrients among Body Systems and Tissues, 58

Chapter 4

TABLE 4.1
Times and Temperatures for Maximal Feather Removal in Birds, 93

Chapter 5

TABLE 5.1
Products, Reactants and Enzymes, Including ATP And H+, Produced during the Anaerobic Metabolism of Glycogen or Glucose to Pyruvate or Lactate in Postmortem Muscle, 104

TABLE 5.2
Delay Time Before Onset of Rigor Mortis, 110

TABLE 5.3
Summary of Changes That Occur in Skeletal Muscle During Postmortem Aging, 113

TABLE 5.4
Heritability Estimates for Physical Properties of Meat, 118

TABLE 5.5
Resistance to Shear, and the Tenderness Rating of Selected Muscles of Cooked Beef, 121

TABLE 5.6
Effects of Electrical Stimulation on Meat properties, 129

Chapter 7

TABLE 7.1
Anatomical and Common Terms Used for Bones from the Beef, Pork, and Lamb Carcass, 145

TABLE 7.2
Approximate Yields of Wholesale Cuts, Parts, or Edible Portions, 150

TABLE 7.3
Institutional Meat Purchase Specification Identifying Numbers for Various Species, 171

Chapter 8

TABLE 8.1
Binding Classification of Meat Raw Materials, 196

TABLE 8.2
Types and Origins of Spices Commonly Used in Processed Meats, 200

Chapter 9

TABLE 9.1
Characteristics of Some Common Food Poisonings and Infections, 221

TABLE 9.2
The Storage Time Required, at Various Freezer Temperatures, to Destroy Trichinella Larvae in Various Thicknesses of Pork, 230

Chapter 10

TABLE 10.1
Maximum Recommended Length of Storage of Certain Meat Items at Various Temperatures for the Preservation of Optimum Quality, 250

Chapter 12

TABLE 12.1
Suggested Servings of Food from the Basic Food Groups, 280

TABLE 12.2
Contribution of Animal Products to the Diet, 281

TABLE 12.3
Proximate Composition and Caloric Content, 283

TABLE 12.4
Cholesterol Content of Common Measures of Selected Foods, 285

TABLE 12.5
Mineral Content of Meat and Poultry Retail Cuts, 287

TABLE 12.6
Vitamin Content of Meat and Poultry Retail Cuts, 287

TABLE 12.7
Protein and Fat Composition Percentages for Cooked Variety Meats, 289

TABLE 12.8
Proximate Composition and Caloric Content of Selected Sausages, Cooked and Smoked Meats, 289

TABLE 12.9
Mineral Content of Selected Sausages, Cooked and Smoked Meats, 290

Chapter 13

TABLE 13.1
A HACCP Plan Checklist, 300

TABLE 13.2
Generic Product Description for a HACCP Plan, 304

TABLE 13.3
Generic Example of Hazard Analysis for a HACCP Plan, 307

TABLE 13.4
Generic Example of a HACCP Plan, 309

TABLE 13.5
Generic Example of a Room Temperature Log for a HACCP Plan, 313

TABLE 13.6
Generic Example of a Thermometer Calibration Log for a HACCP Plan, 313

TABLE 13.7
Generic Example of a Metal Detection Log for a HACCP Plan, 314

TABLE 13.8
Generic Example of a Corrective Actions Log for a HACCP Plan, 314

TABLE 13.9
Generic Example of a Pre-Shipment Review Log for a HACCP Plan, 314

Chapter 14

TABLE 14.1
US Department of Agriculture Grades for Meat, 326

TABLE 14.2
Factors Used to Establish US Department of Agriculture Grades for Meat, 327

TABLE 14.3
Expected Retail Yield for Beef and Lamb Yield Grades and Expected Percent Lean Cuts for Pork Carcass Grades, 331

TABLE 14.4
Comparison of Yields of Retail Cuts and Retail Values, 333

TABLE 14.5
Dressing Percentages of Various Kinds of Livestock by Grades, 334

Chapter 16

TABLE 16.1
Estimated By-Product Value, 353

TABLE 16.2
Annual By-Product Raw Materials Available, 353

TABLE 16.3
Approximate Yield of Various Items Obtained from Meat Animals, 354

TABLE 16.4
Approximate Yield of Items Obtained from Broiler Chickens and Turkeys, 354

TABLE 16.5
Approximate Yield of Items Obtained from Cod Fish, 355

TABLE 16.6
Edible Usage of By-Products Obtained from Meat Animals, 355

TABLE 16.7
Inedible By-Products Obtained from Meat Animals, and their Major Usage, 359

TABLE 16.8
Classification of Cattle Hides, 360

TABLE 16.9
Classification of Sheep Pelts, 360

FIGURES

Chapter 2

FIGURE 2.1
Photomicrograph of skeletal muscle fibers, 9

FIGURE 2.2
Drawing of a skeletal muscle showing the structural relationships among tendons, blood vessels, and nerves, 10

FIGURE 2.3
Macroscopic and microscopic muscle structure, 10

FIGURE 2.4
Skeletal muscle fibers showing structural features and their longitudinal orientation, 11

FIGURE 2.5
Cross-section through a muscle fiber at the A band-I band junction, 12

FIGURE 2.6
Cross-section of skeletal muscle fibers of the pig, 12

FIGURE 2.7
Organization of skeletal muscle from the gross structure to the molecular level, 13

FIGURE 2.8
Portions of two myofibrils and a sarcomere, 14

FIGURE 2.9
Electron photomicrograph of a cross-section of a myofibril, 15

FIGURE 2.10
Representation of Z filaments and their attachment to actin filaments, 16

FIGURE 2.11
Representation of the construction and fine structure of an actin filament, showing the relation of tropomyosin and troponin, 17

FIGURE 2.12
Representation of the construction and fine structure of the myosin filament, 18

FIGURE 2.13
The location of titin molecules in the sarcomere with the N-terminus of each titin polypeptide at the Z disk and the C-terminus at the M line, 19

FIGURE 2.14
Representation of the sarcoplasmic reticulum and T tubules, and their relation to the myofibrils of mammalian skeletal muscle, 20

FIGURE 2.15
Smooth muscle fibers from the longitudinal muscle layer of a pregnant rabbit's uterus at term, 22

FIGURE 2.16
Photomicrographs of cardiac muscle, 23

FIGURE 2.17
Embryonic and fetal development of muscle fibers, 25

FIGURE 2.18
Illustration of squamous, cuboidal, and columnar epithelial tissue, 26

FIGURE 2.19
Illustration of a neuron and the motor end plates associated with it, 27

FIGURE 2.20
Illustration of the fibroblast showing the synthesis of tropocollagen and extracellular assembly of tropocollagen molecules into collagen fibrils and fibers, 29

FIGURE 2.21
Illustration of the amino acid sequence and molecular structure of collagen and tropocollagen molecules, 30

FIGURE 2.22
Illustration of the chicken wire-like network of Type IV collagen, 31

FIGURE 2.23
Representation of adipocyte development, 32

FIGURE 2.24
Representation of fibrocartilage at the insertion of tendon into bone, 33

FIGURE 2.25
Long bone showing the longitudinal structure, 35

FIGURE 2.26
Midshaft section of a compact bone, 35

FIGURE 2.27
Progressive ossification and growth of a long bone, 36

FIGURE 2.28
Illustration of the structure of a primary muscle bundle, 38

FIGURE 2.29
Photomicrograph of a bovine semitendinosus muscle, 39

FIGURE 2.30
Photomicrograph of a human sartorius muscle in cross section, 40

FIGURE 2.31
Illustration of the vascular supply to, and network associated with, skeletal muscle fibers, 40

FIGURE 2.32
Diagram and photomicrographs of fat cells, 41

FIGURE 2.33
Photomicrographs of the muscle-tendon junction, 43

FIGURE 2.34
Change in swine longissimus muscle fiber diameter with increasing age, 44

FIGURE 2.35
Photomicrograph of cross-section of porcine longissimus muscle, 45

FIGURE 2.36
Changes in the percentage of bone, muscle, and fat in beef carcasses during growth, 53

FIGURE 2.37
Bull and carcass that exhibit double muscling, 55

FIGURE 2.38
Comparison of callipyge and normal genotype lambs, 56

Chapter 3

FIGURE 3.1
Establishment of a membrane potential in the normal resting nerve fiber, 63

FIGURE 3.2
Membrane resting and action potentials, 64

FIGURE 3.3
Schematic representation of a motor endplate, 65

FIGURE 3.4
Diagram of a segment of actin filament showing the location of tropomysosin and troponin subunits, 67

FIGURE 3.5
One sarcomere is shown at various stages of shortening during contraction, 67

FIGURE 3.6
Representation of the cross-bridge cycle, 69

FIGURE 3.7
Representation of the conformation change in the lever arm of the myosin cross-bridge during the power stroke of a cross-bridge cycle, 69

FIGURE 3.8
Events during a complete muscle contraction-relaxation cycle, 71

FIGURE 3.9
Illustration of the pathways that supply energy for muscle function, 72

FIGURE 3.10
Cyclic nature of the pathways that provide energy for muscle contraction and heat production, 74

Chapter 4

FIGURE 4.1
Home slaughtering showing pork carcasses hung outdoors, 77

FIGURE 4.2
Typical store-front of the local meat shop, 77

FIGURE 4.3
Modern hog slaughtering facilities have the capacity to process in excess of 10,000 pigs per day, 77

FIGURE 4.4
Homeostasis, 79

FIGURE 4.5
Flight zones in farm animals, 79

FIGURE 4.6
Use of nylon flags and shaker/rattle paddles to encourage animal movement, 81

FIGURE 4.7
Livestock Weather Safety Index, 82

FIGURE 4.8
V-belt design for moving cattle in a processing plant, 84

FIGURE 4.9
V-design for moving pigs in to stunning position, 85

FIGURE 4.10
A downer (non-ambulatory) cow, 85

FIGURE 4.11
General design of a captive bolt stunning apparatus, 86

FIGURE 4.12
Schematic showing the approximate location of captive bolt placement for an effective stun of various farm animals, 87

FIGURE 4.13
General design of equipment used to restrain cattle for stunning, 88

FIGURE 4.14
Illustration of proper electric stunning in pigs, 89

FIGURE 4.15
Schematic of a gondola-type CO_2 stunning apparatus for pigs, 91

FIGURE 4.16
Illustration of placement of the knife for the normal sticking process in pigs, 91

FIGURES   xvii

FIGURE 4.17
Defects caused by improper electrical stunning of pigs, Color section

Chapter 5

FIGURE 5.1
Molecular structure of adenosine triphosphate (ATP), 99

FIGURE 5.2
Glycogenolysis, glycolysis, and the Tricarboxylic Acid (TCA) cycle pathways, 101

FIGURE 5.3
General structure of the branched glycogen molecule and molecular linkages of glucose residues in the glycogen molecule, 102

FIGURE 5.4
Reactions used to regenerate NAD+ and ATP in times of oxygen deprivation by the reduction of pyruvate to lactate, 105

FIGURE 5.5
Various postmortem pH decline curves, 106

FIGURE 5.6
Postmortem temperature decline curves, 109

FIGURE 5.7
Physiograph output showing the change in extensibility of muscle with time postmortem, 109

FIGURE 5.8
Isometric tension development in muscle during phases of rigor mortis, 110

FIGURE 5.9
Degradation of Z disks in beef longissimus muscle during postmortem storage at 2° C, 112

FIGURE 5.10
Changes in postmortem muscle, 115

FIGURE 5.11
Sites of production, and the general actions of the major hormones related to stress reactions, 116

FIGURE 5.12
Glycogen levels in chicken breast muscles immediately after death, 123

FIGURE 5.13
Thaw rigor shortening, 123

FIGURE 5.14
Micrographs of relaxed and cold-shortened beef muscle, 125

FIGURE 5.15
Effect of temperature on the degree of shortening, 125

FIGURE 5.16
Lipid oxidation, 126

FIGURE 5.17
Differences in rate of pH decline in electrically stimulated and nonstimulated beef carcasses, 127

FIGURE 5.18
Micrographs of nonstimulated and electrically stimulated beef muscle, 128

FIGURE 5.19
Longissimus muscle of electrically stimulated (right) and unstimulated (left) beef carcasses at 24 hours postmortem, Color section

FIGURE 5.20
Beef carcass electrical stimulation, 129

FIGURE 5.21
Effects of carcass electrical stimulation, 130

Chapter 6

FIGURE 6.1
Charged hydrophilic groups on the muscle proteins attract water, 134

FIGURE 6.2
Effect of pH on the amount of immobilized water present in meat, 135

FIGURE 6.3
Schematic representation of the heme complex of myoglobin, 137

FIGURE 6.4
Oxidatin state (valence) of the central iron atom, 138

FIGURE 6.5
Relation of oxygen pressure to pigment chemical state and color, Color section

FIGURE 6.6
An example of the varying firmness of pork muscle; butt end views of two fresh hams, 139

Chapter 7

FIGURE 7.1
Beef, pork, and lamb skeletal diagrams, 146

FIGURE 7.2
Skeleton of the chicken with bones and parts identified, 147

FIGURE 7.3
Diagrams of beef, veal, pork, and lamb carcasses, 148

FIGURE 7.4
Retail cuts from the breast, brisket, spare ribs, bacon side, and short plate, 149

FIGURE 7.5
Retail cuts from the shoulder (chuck) arm, 152

FIGURE 7.6
Retail cuts from the shoulder blade, 153

FIGURE 7.7
Retail cuts from the rib, 154

FIGURE 7.8
Retail cuts from the loin (short loin), 155

FIGURE 7.9
Retail cuts from the sirloin, 156

FIGURE 7.10
Retail cuts from the leg, round, or ham, 157

FIGURE 7.11
Chicken broiler parts, 158

FIGURE 7.12
Turkey parts, 159

FIGURE 7.13
Market forms of fish, 160

FIGURE 7.14
Shellfish and edible portions, 161

FIGURE 7.15
Retail cutting test record form, 165

FIGURE 7.16
Examples of IMPS beef cuts, Color section

FIGURE 7.17
Examples of IMPS cut specifications, Color section

FIGURE 7.18
Chicken breast cuts for institutional food service, Color section

FIGURE 7.19
Chicken leg cuts for institutional food service, Color section

Chapter 8

FIGURE 8.1
Multiple needle injection of curing solution into a bone-in ham, 180

FIGURE 8.2
Basic design of tumbling and massaging machines, 181

FIGURE 8.3
Chemical reaction pathways for myoglobin in meat products, 182

FIGURE 8.4
Designs of various comminuting devices, 187

FIGURE 8.5
Micrograph of batter-type sausage, 188

FIGURE 8.6
Drawing of an oil-in-water emulsion, 189

FIGURE 8.7
Emulsifying agent between the lip and water phases, 190

FIGURE 8.8
Raw meat batter, 190

FIGURE 8.9
Emulsion break in frankfurters, 192

FIGURE 8.10
Equipment for deboning meat, 194

FIGURE 8.11
Sausages stuffed in natural and artificial casings, 204

Chapter 9

FIGURE 9.1
Growth curve for a pure culture of bacteria under ideal growth conditions, 216

FIGURE 9.2
Generation intervals for one species of psychrophilic bacterium at different storage temperatures, 217

FIGURE 9.3
Time required at different storage temperatures for the spoilage of hamburger initially contaminated with 1 million psychrophilic bacteria/gram, 217

FIGURE 9.4
The life cycle of Toxoplasma gondii, 227

FIGURE 9.5
Life cycle of *Trichinella spiralis*, 228

FIGURE 9.6
Photomicrograph showing an encysted *Trichinella spiralis* larva in skeletal muscle, 228

FIGURE 9.7
Time-temperature relationship for *Trichinella spiralis* control in country hams, 229

FIGURE 9.8
Typical culture plate, showing bacterial and mold colonies, 231

Chapter 10

FIGURE 10.1
Typical cooling curves for beef and pork carcasses, 242

FIGURE 10.2
Freezing curves, showing the relative rates of freezing at various freezer temperatures, 245

FIGURE 10.3
The ultrastructural appearance of an unfrozen muscle fiber, 246

FIGURE 10.4
Freezing curves for 6-inch meat cubes, 247

FIGURE 10.5
Freezing curve of a thin section of beef, 249

FIGURE 10.6
Average temperature at the geometric center of cylindrical specimens, 251

FIGURE 10.7
Radura symbol, 256

Chapter 11

FIGURE 11.1
Degrees of cooked meat doneness, Color section
FIGURE 11.2
Tender and tough muscle fiber, 262
FIGURE 11.3
Relationship of marbling to overall palatability, 264
FIGURE 11.4
Influence of time and temperature of heating on the water-holding capacity of beef muscle, 267
FIGURE 11.5
Electron micrographs of beef muscle, 268–269
FIGURE 11.6
Effect of heating on toughening of beef, 271
FIGURE 11.7
Effect of heating temperature and time, 272

Chapter 12

FIGURE 12.1
Nutrition facts label, 278
FIGURE 12.2
MyPlate, 279

Chapter 13

FIGURE 13.1
Federal inspection stamp similar to those used on inspected fresh meat cuts, 295
FIGURE 13.2
Federal inspection stamp similar to those used on inspected meat products, 295
FIGURE 13.3
Imported meat stamp, 295
FIGURE 13.4
Federal inspection stamp and label similar to those used on fresh horse meat, 297
FIGURE 13.5
Decision tree to identify critical control points, 299
FIGURE 13.6
All animals are examined by inspectors before slaughter, 303
FIGURE 13.7
All parts of the carcass receive a thorough postmortem examination by an inspector, 303
FIGURE 13.8
Inspection is conducted at all processing stages, 304
FIGURE 13.9
Diagram for raw ground beef for a HACCP plan, 305
FIGURE 13.10
Sample of a US Department of Agriculture approved label, 315

Chapter 14

FIGURE 14.1
Brands used by a meat processor, 320
FIGURE 14.2
Identifying marks used for US Department of Agriculture grades for meat, 321
FIGURE 14.3
Indices of maturity in meat animal carcasses, 322
FIGURE 14.4
Marbling in beef, 323
FIGURE 14.5
Color-texture-exudation standards, Color section
FIGURE 14.6
Relationships among marbling, maturity, and quality, 325
FIGURE 14.7
USDA choice and standard grade beef carcasses, 328
FIGURE 14.8
Relationships among flank fat streaking, maturity, and quality, 329
FIGURE 14.9
Relationships among feathering and flank fat streaking, maturity, and quality, 329
FIGURE 14.10
Front view of stewing chickens, 330
FIGURE 14.11
Degrees of muscle development for pork carcasses, 332
FIGURE 14.12
Relationships between backfat thickness and muscling score, 332
FIGURE 14.13
Beef carcass yield grades, 335

Chapter 15

FIGURE 15.1
Tristimulus method of color description, Color section
FIGURE 15.2
Components of most fiber optic-based spectroscopy technologies, 340
FIGURE 15.3
Placement of an optical probe, 341
FIGURE 15.4
Application of an optical probe to determine pork carcass composition and quality, 342

FIGURE 15.5
Example of pH probe placement in a pork carcass, 345

FIGURE 15.6
Use of a hand-held pH/impedence probe in determining meat quality, 345

FIGURE 15.7
Approximate placement of electrodes on a carcass, 346

FIGURE 15.8
Real-time ultrasound image of a pig carcass, 347

FIGURE 4.17 | Defects caused by improper electrical stunning of pigs. (Upper) Hemorrhage on the surface of a pork loin. (Lower) Fractured thoracic vertebrae and hemorrhaging.

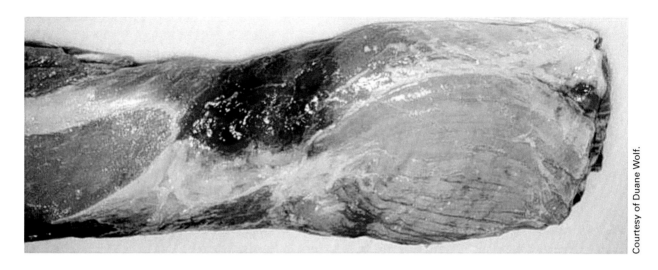

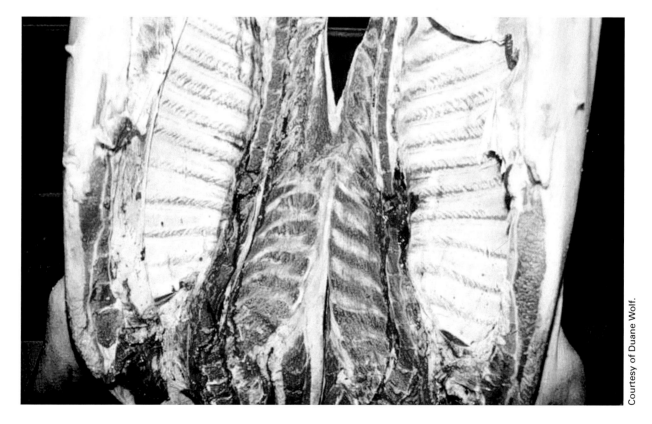

FIGURE 5.19 | Longissimus muscle of electrically stimulated (right) and unstimulated (left) beef carcasses at 24 hours postmortem.

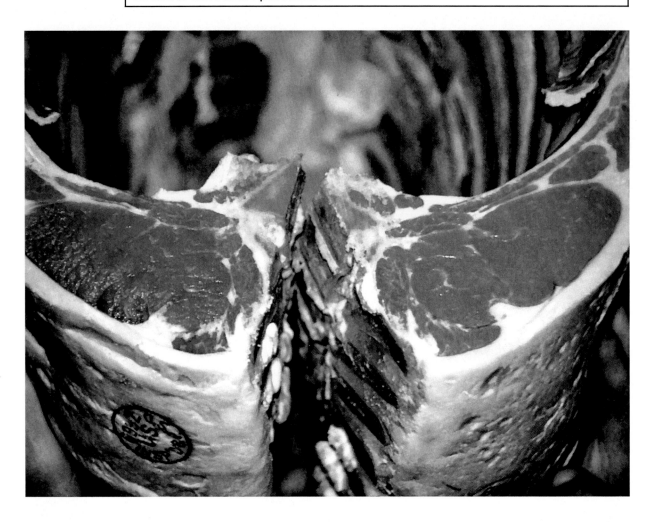

FIGURE 6.5 | Relation of oxygen pressure to pigment chemical state and color.

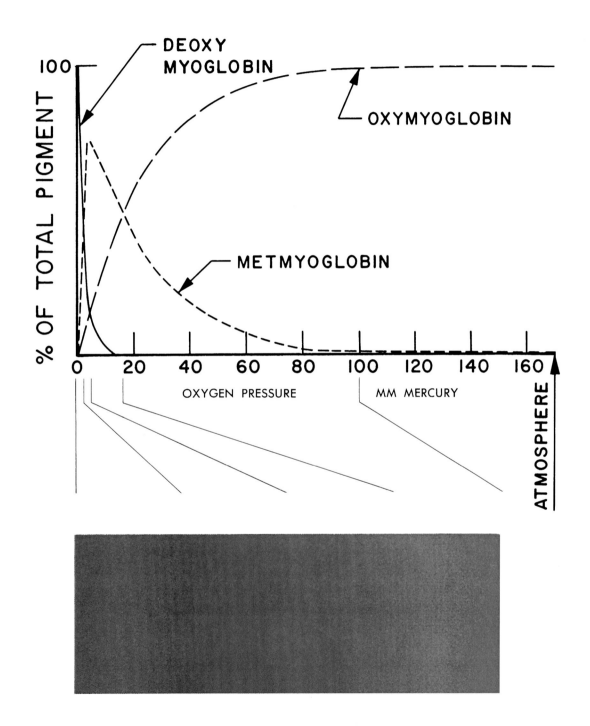

FIGURE 7.16 | Examples of IMPS beef cuts. (A) Chuck End and (B) Loin End views of 107 Beef Rib, Oven Prepared; (C) Rib End and (D) Sirloin End views of 180 Beef Loin, Strip Loin, Boneless. From *Meat Evaluation Handbook* by the American Meat Science Association. Copyright © American Meat Science Association. Reprinted by permission.

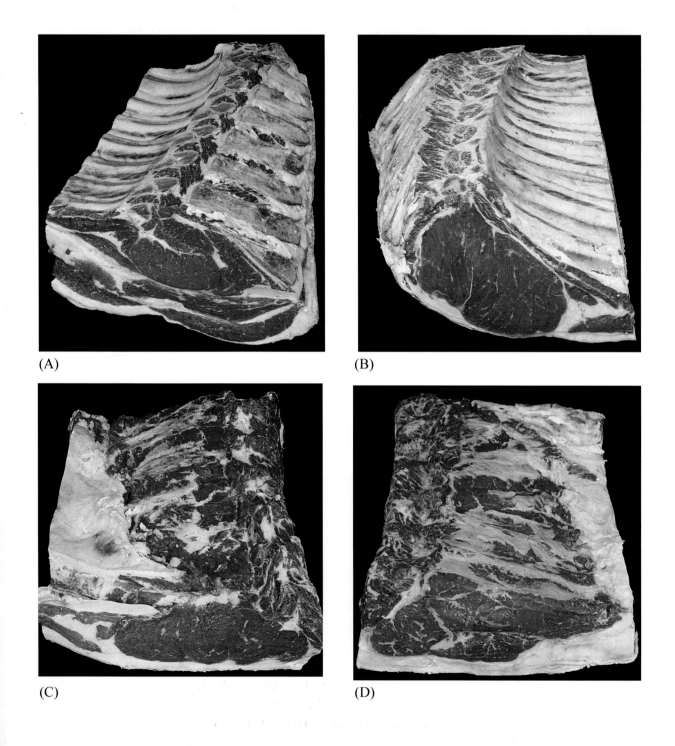

FIGURE 7.17 | Examples of IMPS cut specifications of (A) 1179 Beef Loin, Strip Loin Steaks, Bone In; (B) Purchaser Specified Options [PSO] for length of tail of Strip Loin Steaks. Left to right: PSO2, PSO3, PSO4, and PSO5; (C) 1180 Beef Loin, Strip Loin Steaks, Boneless. As illustrated in The Meat Buyers Guide, published by the North American Meat Processors Association, Reston, VA.

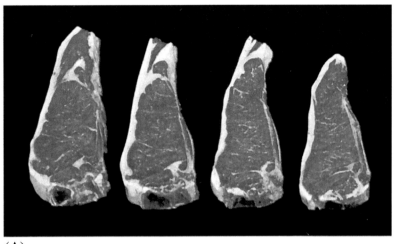

(A)

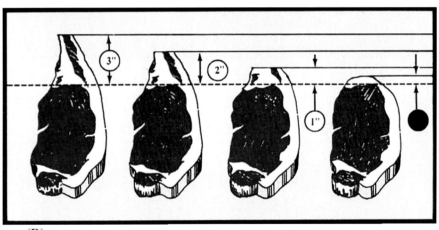

(B)

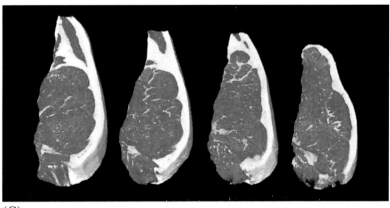

(C)

FIGURE 7.18 | Chicken breast cuts for institutional food service. Boneless, skinless whole breast (top), boneless, skinless breast halves (lower, far left and far right) and boneless skinless tenderloins (lower center). From AMS-628 Quality Poultry for Volume Buyers, US Department of Agriculture, Agricultural Marketing Service, 1997.

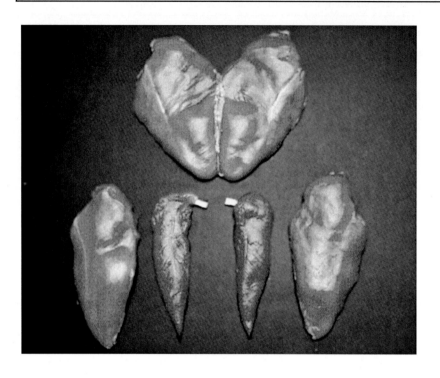

FIGURE 7.19 | Chicken leg cuts for institutional food service. Boneless, skinless drumstick (left) and boneless, skinless thigh (right). From AMS-628 Quality Poultry for Volume Buyers. US Department of Agriculture, Agricultural Marketing Service, 1997.

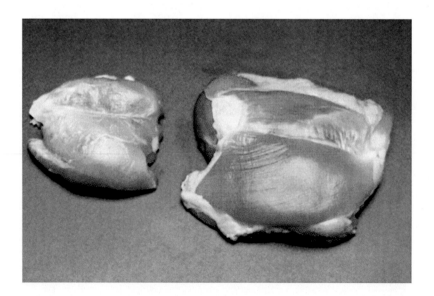

FIGURE 11.1 | Degrees of cooked meat doneness.

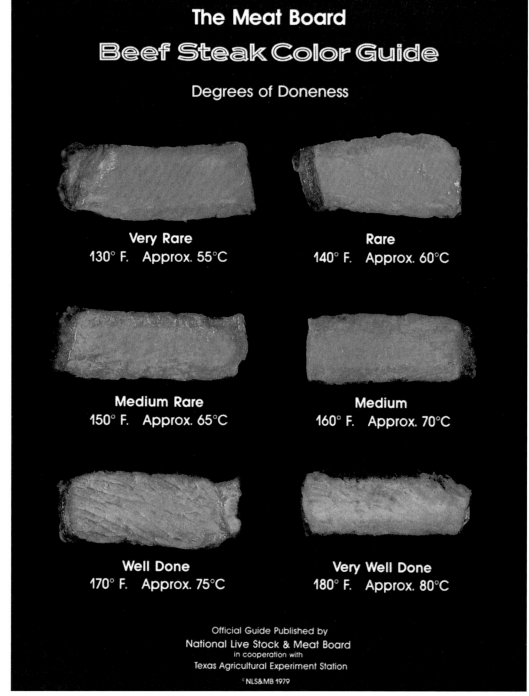

FIGURE 14.5 | Color-texture-exudation standards for ham muscles (top), color standards for *longissimus* muscle (center), and marbling scores for *longissimus* muscle (bottom) for pork carcass evaluation. Courtesy of the National Pork Producers Council, Des Moines, IA. Copyright © by National Pork Producers Council. Reprinted by permission.

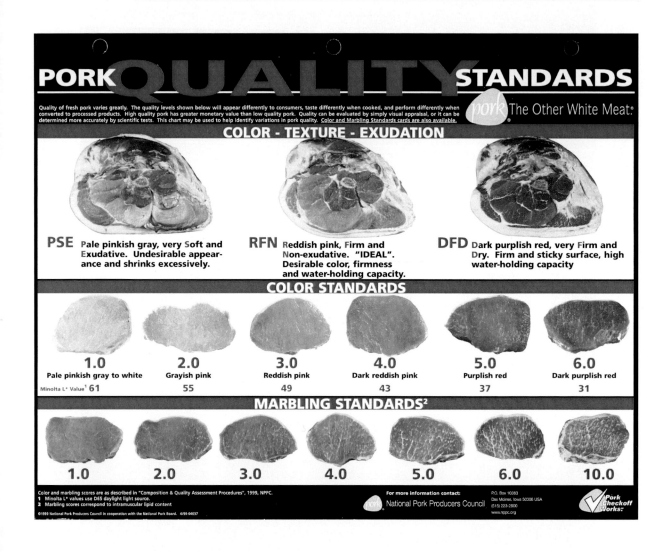

FIGURE 15.1 | Tristimulus method of color description. Note that X, Y and Z each correspond to a range of wavelengths in the visible spectrum that are perceived as red, green and blue, respectively.

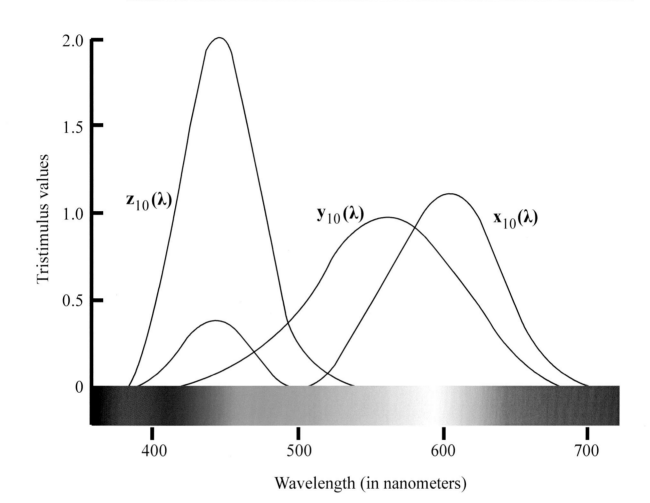

1

MEAT AS A FOOD

Eating is a process essential to the maintenance of life itself and few foods can quiet the pangs of hunger and satisfy the appetite so quickly and completely as meat. In Western society, so great is the pleasure derived from consumption of sizzling steaks, succulent roasts, spicy sausages, crisp fried chicken, juicy lobster tail, or any of the hundreds of other meat dishes available as the main attraction for a meal, that other nutritious foods are hard pressed to compete. A complex industry has developed to provide the nutritious, high-quality meat supply sought after by much of the world's human population. You are beginning a study of that industry and the unique food products it prepares for the world's human population.

WHAT IS MEAT?

Were you surprised to see lobster tail and fried chicken listed along with steaks and roasts? If so, it is time to broaden your concept of meat. Meat is defined as those animal tissues that are suitable for use as food. All processed or manufactured products that might be prepared from these tissues are included in this definition. While tissues from nearly every species of animal can be used as meat, most meat consumed by humans comes from domestic animals and aquatic organisms.

Meat as an entity can be subdivided into several general categories. The largest category, in terms of volume of consumption, is "red" meat. Beef, pork, lamb or mutton, and veal are commonly identified as "red" meats. (However, some muscles of pork and those of veal are rather pale and do not fit intuitively into the category of "red" meats.) Horse, goat, eland, llama, camel, water buffalo, and rabbit meats also are used for human consumption in many countries. Poultry meat is the flesh of domestic birds, and includes that of chickens, turkeys, ducks, geese, and guinea fowl. Sea foods are the flesh of aquatic organisms, of which the bulk is fish. The flesh of clams, lobsters, oysters, crabs, and many other species also is included in this category. A fourth category is game meat, which consists of the flesh of any non-domesticated animal.

MUSCLE AS MEAT

It should be emphasized that any discussion of meat must be directed primarily toward muscle. Although meat is composed of numerous tissues, such as adipose, epithelial, connective and nervous tissues, the major component of meat is muscle. Yet, muscle tissue did not evolve to serve the needs of meat consumers, but rather, to perform the functions of locomotion and heat production in living animals.

Because muscles are organs whose unique structure primarily serves biological functions, they possess several characteristics that influence their value as food. For example, certain muscles contain relatively large quantities of connective tissue, which is associated with meat toughness.

However, these muscles can be as palatable as any other, provided a cooking method is used that degrades the connective tissues.

Had muscle been engineered and designed to function primarily as a food product, it undoubtedly would have differed in several characteristics. With advances in technology, many foods are being engineered to specifications, and attempts are being made to substitute other protein sources for the muscle in meat. Surprisingly, it is extremely difficult to imitate the texture and flavor imparted to meat by muscle and other animal tissues. Attempts to prepare meat-like products without using animal tissue have met with limited success. However, the partial replacement of muscle with other protein sources in various meat products has great potential for extending the meat supply to more of the world's people.

Meat is consumed largely for the muscle it contains. Consumers usually leave the adhering or separable fat of meat servings on the plate. The important characteristics of manufactured meat products such as moisture-binding ability, fat-emulsifying capacity, and color are provided by the proteins of muscle fibers. Consequently, the focus of the next two chapters of this book is on muscle and associated tissues, and the principal objective is to bridge the gap between muscle biology and meat science. Understanding the structure and biological functions of muscle and its associated tissues is critical to understanding the utilization of meat as food, the palatability properties of meat, and the successful conversion of meat into a multitude of processed products.

WHAT IS MEAT SCIENCE?

Chapters 4 through 16 of this book address the broad field of study known as meat science. An important part of this field is the basic study of unique characteristics of muscle and other animal tissues as they are transformed into meat. A complete understanding of the basic properties of the tissues present in meat may lead to improved utilization and better meat products.

Meat science is not limited to the study of tissues. It is a component of all facets of the meat industry, beginning with animal production and ending with final preparation of meat for consumption. Animal breeding, feeding, and management are extremely important parts of the food chain because meat quality control actually starts on the farm or ranch or in the feedlot. Moreover, animal welfare is a critical consideration during animal transport and harvesting and has implications for meat quality characteristics.

The market system through which animals move from production units to packing plants encompasses economic and other dimensions of meat science. Market signals represent the primary line of communication between consumers and producers. The language is money, and producers adjust their production to meet market demands, which ideally reflect desires of consumers. Whether or not the market system accurately communicates consumer wishes to producers, it is a major determinant of the type of livestock and poultry being produced.

Meat science encompasses the activities of packers, processors, and purveyors, or that segment of industry that converts live animals into food products and then distributes such products to merchandisers. Meat technology is applied to maintain product quality and wholesomeness and to develop new and different products.

Retail meat markets, hotels, restaurants, and institutions are important components of the marketing system. Retailers and food service operators are the meat industry's representatives to consumers. Meat retailers prepare many fresh meat cuts, display all meat products in an attractive manner, and maintain product quality and wholesomeness. The hotel, restaurant, and institutional management group carries meat processing to its ultimate end, and places cooked meat before the consumer. The final cooking and serving of meat is just as important as any segment of the complex industry that brings meat from grazing lands, feedlots, and housing units to consumers.

One reason for the increasing complexity of the livestock and meat business is that new competitive food products are being developed continually. These new competitors seek to entice consumers with modifications in convenience, price, quality, uniformity, nutritional value, or even with novelty. If the meat industry is to maintain its present position of importance in the food production chain and maintain a dynamic and growing market, it must produce the highest quality products with the greatest possible efficiency, develop innovative new products, and employ sophisticated advertising and promotion programs. Such developments require the input of students trained in meat science.

Students who plan to be associated with the meat industry during their working careers must not be satisfied with learning only the status of the industry today, for this knowledge will soon be out of date. Instead they must learn basic concepts and be prepared to apply these to changing situations. Indeed, they should prepare to initiate change.

In this text, emphasis is placed on basic principles that govern meat industry operations. It is not the purpose to describe the intricacies of all phases of meat science as they affect the complex meat industry. This would provide only a historical background and would not prepare students for involvement in a dynamic industry. Rather, characteristics of animals that produce tissues used for meat are considered, with emphasis on growth mechanisms responsible for differences in body composition. Structure, composition, and function of muscle tissue components are examined in detail, and contributions that each make to characteristics of final products are stressed. Principles for minimizing stress during animal transport and harvesting and techniques for humane harvesting procedures are described. Meat processing and preservation are described, as are principles from which processes have been developed to prepare and preserve hundreds of different meat products. (Recipes and detailed instructions for manufacture of specific products are available from other sources, if they are needed for reference.) Many other topics, such as inspection, food safety, grading and standardization, measurement of palatability and physical properties, and by-products are included, providing an in-depth coverage of many aspects of the total meat industry.

MEAT CONSUMPTION AND THE ECONOMY OF A NATION

Throughout recorded history, consumption of meat has indicated a position of social and economic prestige among people and nations. As nations industrialize and improve their economic position, their meat consumption increases. Moreover, as persons raise their social or economic status, they tend to demand a greater quantity and higher quality of meat products.

Meat is one of the most nutritious foods used for human consumption. It is a particularly rich source of high-quality protein, iron, essential B vitamins, and vitamin A (liver). In most industrialized nations, no single food surpasses the contribution of meat to dietary protein, iron, or B vitamins. Meat provides virtually all of the vitamin B12 consumed.

Adequate protein nutrition is essential. A sufficient quantity of protein having all essential amino acids is necessary for physical well-being and for proper mental and intellectual development. The social and economic impact of adequate dietary protein on individuals and nations is striking. The health, vigor, and longevity of meat-eating people enable them to contribute effectively to science, industry, and the arts. If the people of the less-developed nations of the world could be supplied adequately with meat or other high-quality protein foods, it is likely that their capability for rapid industrial, social, political, and intellectual development would increase many times.

Per capita meat consumption among countries of the world is influenced by amount and type of livestock production in a particular country and by the resources available to purchase meat from other countries (Table 1.1). Quantities consumed of various species also are influenced by social custom, religious belief, and personal preference. But throughout the world, meat is a prized commodity, the supply of which falls far short of satisfying the appetites or meeting the needs of the world's population.

TABLE 1.1 Per Capita Meat Consumption in Selected Countries, 2010 (kg)

Country	Beef and Veal*	Pork*	Broiler Meat**
Argentina	55.8	6.7	33.7
Australia	35.3	22.5	35.5
Brazil	37.8	NA	45.4
Canada	29.6	23.9	29.7
China	4.1	37.9	9.2
European Union—27	16.6	43.2	17.8
Hong Kong	23.8	65.7	43.0
Japan	9.7	19.6	16.3
Mexico	17.2	15.8	29.7
New Zealand	29.2	21.2	NA
Russia	16.6	19.9	21.0
Saudi Arabia	NA	NA	42.6
South Africa	14.0	3.6	30.8
South Korea	12.5	31.6	15.3
United States	38.8	27.9	43.4
Venezuela	18.2	4.9	30.0
Vietnam	NA	21.0	7.2

* Carcass weight equivalent basis
** Ready-to-cook equivalent basis
NA Not available
Source: US Department of Agriculture, Foreign Agricultural Service, Office of Global Analysis. "Livestock and Poultry: World Markets and Trade." April 2011.

The breeding and rearing of livestock and poultry for meat production constitutes a significant proportion of total agricultural production in many countries. If we also consider production of forages and grains used as feed for livestock and poultry, the proportion of total agricultural resources devoted to meat production is very great (more than 60 percent in the United States). In the US, more than one-half of total farm income is derived from the sale of livestock (including poultry) and livestock products, and approximately 40 percent of total farm income is derived from the sale of livestock and poultry for meat production.

The food industry, which includes food production, processing, and distribution, is the largest industry in the private sector of the US economy in terms of total assets, employment, and gross business receipts. If the meat packing and processing industry is compared to other segments of the food industry, it ranks either first or second in total assets, value added by processing, and total employment. More importantly, the meat industry easily ranks first in total business receipts, earning approximately 28 percent of total food industry receipts. The next largest is the dairy industry, with about 16 percent of total food receipts. These figures reflect the significant proportion of consumer food dollars spent for meat and meat products and emphasize the important position of meat in the US food economy.

THE MEAT INDUSTRY IN THE UNITED STATES

The meat industry has had a long and colorful history that includes a nostalgic era of development as well as rougher times marked with scandals and labor problems similar to those experienced by other industries. The unsanitary practices of a few unscrupulous meat-packing companies rocked the entire industry when they were given nationwide publicity at the beginning of the twentieth century. However, a highly sophisticated industry with modern business practices that applies scientific knowledge to food manufacturing has emerged from this background to provide this country with one of the safest, highest-quality meat supplies in the world.

The US meat industry began when colonial butchers, most of who had learned their trade in Europe, began slaughtering and dressing animals for others beyond their immediate family. Colonial meat shops became the first retail meat markets. The term "meat packing" arose with the first commercial attempts at meat preservation in this country, i.e., the salting and packing of pork in wooden barrels for storage or for shipping to Europe.

As cities grew, small packing plants were established. These plants generally expanded in size as urban populations grew. The major packing companies originated in large cities because of availability of the labor force needed to operate large plants. Animals often were driven on hoof from production areas to railheads, and then moved by rail to large terminal livestock markets that had developed in conjunction with the growing packing industry.

Refrigeration and transportation both played major roles in the development of the meat industry. In earlier times, slaughter plants were located in large cities so that highly perishable products could be moved quickly to consumers. With slow transportation and no refrigeration, meat could best be moved from grazing lands and feedlots on the hoof. Before the advent of mechanical refrigeration, slaughter and processing were limited to cooler seasons except in areas where ice could be obtained from rivers and lakes and stored for use during summer months.

Large meat packing plants capable of producing full lines of fresh and processed meats were established in all major cities in the US. Chicago, with its geographical location in the center of the cornbelt, became immortalized as the "hog butcher of the world" by poet Carl Sandburg. Many factors led to the rise of the Chicago meat packing industry, including availability of all necessary resources, raw material (livestock), labor, transportation (both water and rail), and ready markets for products. The meat industry in Chicago and other major cities reached its peak in the early 1950s. Large plants that had started in the 1890s or earlier and grown piece-meal into huge units then became obsolete. Automation reduced the size of the labor force needed in meat packing and processing. The development of the interstate highway system, refrigerated truck transportation, improved sanitation, and improved packaging materials made possible the movement of perishable foods over long distances. In the 1960s, cropland irrigation changed the livestock industry in the Great Plains and the Southwest, and large cattle feedlots were established in areas formerly useful only for grazing. Integrated poultry production developed in the Southeast from Maryland to Texas and pork production developed in the Southeast, particularly in North Carolina. Consequently, growing and finishing of livestock and poultry for market is no longer concentrated in the cornbelt, but has spread over most of the US wherever feed supplies, labor availability, market access and environment factors make livestock and poultry production economically viable.

When new meat packing plants were constructed to replace obsolete ones, they were not built in large cities but rather, were placed close to areas of livestock production. It became more economical to ship meat and dressed carcasses than live animals. These modern plants are very different in character from the old ones. They usually are attractive in appearance and often are located in open country, rather than in crowded industrial areas. They are well lighted and easy to clean and provide comfort and safety to employees. Modern meat packing plants are very specialized, usually producing only dressed carcasses from one species or manufacturing a line of processed meat items from carcasses, wholesale cuts, or boneless meat received from outside sources.

Changes in meat merchandising practices have accompanied other changes in the meat industry. Many retail chains have established large central facilities where vacuum packaged, trimmed, wholesale cuts are received from slaughterers and further processed into retail cuts that are distributed daily to local supermarkets. This centralized processing reduces the labor required to operate meat counters in supermarkets. Similar efficiencies are realized when the processor is the final fabricator of fresh meat for specialized needs of hotel, restaurant, and institutional food service units. In fact, recent trends encourage packers and processors to perform final fabrication and packaging of fresh meat handled by all merchandisers including retail chains.

This is just a brief account of the recent history of a dynamic industry. Technological advances in meat processing, packaging, preservation, and transportation did not cause changes in the meat industry, but they made change possible. The industry is still in an evolutionary stage, and great opportunity exists for further innovations to better serve consumers of meat.

SCIENCE AND MEAT

In earlier times, manufacture of meat products was strictly an art handed down from generation to generation. Today, development of new products and improvement of old ones is a science. Many major meat companies have extensive research facilities and employ highly trained meat scientists to solve problems and develop new products. Major research programs in meat science are being conducted at universities, research institutes, and government research facilities in this country and abroad. Because of this research effort there has been a broad expansion of the science of meat as a food. Much of the information in this book is the result of this research. Yet, many basic questions remain unanswered and new technical problems frequently arise; consequently, great opportunity exists for the imaginative scientist.

2

STRUCTURE, COMPOSITION, AND DEVELOPMENT OF ANIMAL TISSUES

OBJECTIVES: *Understand the gross, cellular and subcellular organization of skeletal, cardiac, and smooth muscle and recognize the contribution of non-muscle cell types and extracellular components to their structure. Gain an appreciation of how muscle, fat, and bone develop and influence carcass composition.*

Key Terms

- Skeletal muscle
- Smooth muscle
- Cardiac muscle
- Striated muscle
- Muscle fiber
- Myofiber
- Muscle cell
- Sarcolemma
- Transverse tubules, T-system, T-tubules
- Myoneural junction
- Motor end plate
- Sarcoplasm
- Myofibril
- Myofilaments
- I band
- A band
- Z disk
- Sarcomere
- H zone
- Pseudo H zone
- M line
- Myosin
- Myosin filament
- Actin
- Actin filament
- Z filaments
- Titin
- Tropomyosin
- Troponin
- Nebulin
- C protein
- H protein
- Myomesin
- M protein
- Skelemin
- Creatine kinase
- α-actinin
- Cap Z
- Desmin
- Filamin
- Paranemin
- Synemin
- Dystrophin
- Talin
- Vinculin
- Costameres
- Sarcoplasmic reticulum
- Cisternae
- Fenestrated collar
- Terminal cisternae
- Longitudinal tubules
- Cathepsins
- Intercalcated disks
- Myogenesis
- Myoblasts
- Myotubes
- Central nervous system
- Peripheral nervous system
- Neuron
- Axon
- Neuroplasm
- Dendrites
- Nerve fibers
- Nerve trunks
- Synapse
- Schwann cells
- Myelin sheath

CHAPTER 2 STRUCTURE, COMPOSITION, AND DEVELOPMENT OF ANIMAL TISSUES

- Adipose tissue
- Extracellular substance
- Connective tissue proper
- Supportive connective tissue
- Proteoglycans
- Glucosaminoglycans
- Tropocollagen
- Tropoelastin
- Hyaluronic acid
- Chondroitin sulfates
- Dense connective tissue
- Loose connective tissue
- Dense irregular connective tissue
- Dense regular connective tissue
- Aponeuroses
- Collagen
- Elastin
- Collagen fibril
- Collagen fiber
- Intermolecular cross-linkages
- Cervical ligament (*ligamentum nuchae*)
- Fibroblasts
- Adipoblasts
- Adipocytes
- White fat (or adipose tissue)
- Brown fat (or adipose tissue)
- Subcutaneous adipose tissue
- Intermuscular adipose tissue
- Intramuscular adipose tissue
- Chondrocytes
- Hyaline cartilage
- Elastic cartilage
- Fibrocartilage
- Diaphysis
- Epiphyses
- Periosteum
- Compact bone
- Spongy bone
- Articular cartilage
- Synovial fluid
- Epiphyseal plate or cartilage
- Metaphysis
- Osteocytes
- Osteoblasts
- Canaliculi
- Chrondroblasts
- Endrochrondral ossification
- Intramembranous ossification
- Medullary cavity
- Osteoclasts
- Red marrow
- Yellow marrow
- Platelets
- Red cells (erythrocytes)
- White cells (leucocytes)
- Phagocytic
- Hemoglobin
- Fasciculi or muscle fiber bundles
- Primary bundles
- Endomysium
- Perimysium
- Epimysium
- Motor end plate
- Marbling
- Seam fat
- Myotendinal structure
- Aponeuroses
- Hyperplasia
- Hypertrophy
- Satellite cell
- Red fibers
- White fibers
- ATPase activity
- Myosin isoforms
- Type I fibers
- Type II fibers
- Oxidative metabolism
- Glycolytic metabolism
- Myoglobin
- Neutral lipids
- Triglycerides
- Glycogen
- Sarcoplasmic protein
- Myofibrillar protein
- Stromal protein
- Dressing percentage
- Phenotype
- Genotype
- Heritability estimate
- Double muscling
- Myostatin
- Callipyge
- Polar overdominance
- Growth hormone (somatotropin)
- Epinephrine
- Norepinephrine
- Beta-adrenergic agonists
- Paylean®
- Optaflexx®
- Zilmax®
- Androgens
- Estrogens
- Progesterone
- Synthetic estrogens
- Estradiol-17ß
- Zeranol
- Estradiol benzoate

Key Terms (continued)

CHAPTER 2 STRUCTURE, COMPOSITION, AND DEVELOPMENT OF ANIMAL TISSUES

The animal carcass is the product of the slaughtering process, which removes blood, viscera, head, hair, skin, and other tissues from a living animal. It is composed primarily of muscle, variable quantities of connective tissues, and some epithelial and nervous tissues. Although all connective tissue types are present in the carcass, adipose tissue (fat), bone, cartilage, and connective tissue proper predominate. The three primary components of the carcass: muscle, bone, and fat each contain multiple tissues. The properties and relative proportion of each tissue are responsible for the quality and leanness of meat. Because muscle and connective tissues are the most abundant tissues in meat, their structure and composition will be discussed in considerable detail. A brief discussion, however, of epithelial and nervous tissues will be included.

MUSCLE TISSUE

Skeletal muscle represents the bulk (35 to 65 percent) of carcass weight of meat animals. In addition to skeletal muscle, meat contains **smooth muscle**, primarily as a component of blood vessels. Another specialized form of muscle tissue, called **cardiac muscle**, is confined solely to the heart. Skeletal and cardiac muscles are referred to as **striated** muscles because of the transverse banding pattern observed microscopically, as shown in Figure 2.1.

Skeletal Muscle

Skeletal muscles are organs of the muscular system that are attached directly or indirectly to bones via ligaments, fascia, cartilage, or skin (Figure 2.2). There are more than 600 muscles in the animal body, and they vary widely in shape, size, and action. Specific characteristics of a given muscle are dictated by its function. Each muscle is covered with a connective tissue sheath, which is continuous with connective tissue that extends into the interior most aspects of the muscle. Nerve fibers and blood vessels enter and exit the muscle along these connective tissue networks. Many of the above features of muscle are shown in Figure 2.3.

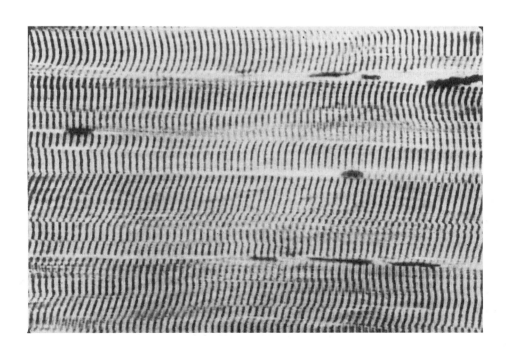

FIGURE 2.1 | Skeletal muscle is occasionally referred to as voluntary muscle while smooth and cardiac muscles are called involuntary muscles.

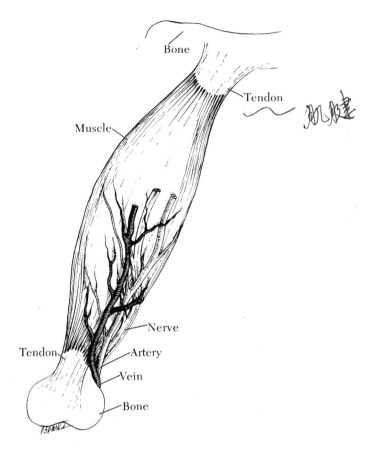

FIGURE 2.2 | Drawing of a skeletal muscle showing the structural relationships among tendons, blood vessels, and nerves. Courtesy of W.H. Freeman and Company.

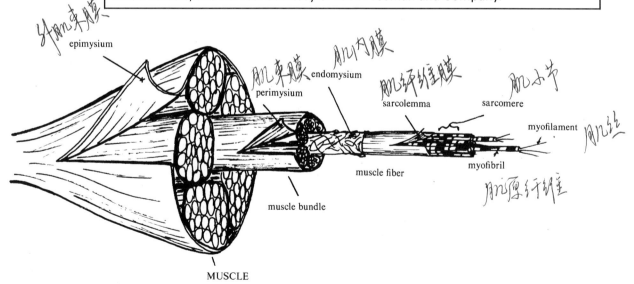

FIGURE 2.3 | Diagrammatic representation of macroscopic and microscopic muscle structure. From B.B. Chrystall, "Macroscopic, microscopic and physiochemical studies of the influence of heating on muscle tissues and proteins," Ph.D. Dissertation, Univ. Of Missouri (1970).

Skeletal Muscle Fiber

The structural unit of skeletal muscle tissue is the highly specialized cell known as a **muscle fiber**, **myofiber** or **muscle cell**. Muscle fibers constitute 75 to 92 percent of total muscle volume. The remaining volume consists of connective tissues, blood vessels, nerve fibers, and extracellular fluid. Mammalian and avian skeletal muscle fibers are long, multinucleated, unbranched, threadlike cells that taper slightly at both ends, as shown in Figure 2.4. Although fibers may attain a length of many centimeters, they generally do not extend the length of the entire muscle. Muscle fibers vary considerably in diameter, ranging from 10 to more than 100 μm (1 μm {micrometer} equals one millionth of a meter) within the same species, and even within the same muscle.

Sarcolemma

The membrane surrounding a muscle fiber is called the **sarcolemma** (Figure 2.4). The prefix *sarco-* is derived from the Greek word *sarx* or *sarkos*, which means flesh, and the suffix *-lemma*, is the Greek word for husk. The sarcolemma is composed of protein and lipid material and is relatively elastic, a property that enables it to endure great distortion during contraction, relaxation, and stretching. The structure, composition, and properties of the sarcolemma are similar to the plasmalemma (unit membrane) of other cells of the body. Periodically, along the length of the fiber and around its entire circumference, invaginations of the sarcolemma form a network of tubules called **transverse tubules** (Figure 2.5). These tubules are sometimes referred to as the **T-system** or **T-tubules**.

Motor nerve fiber endings terminate on the sarcolemma at the **myoneural junction**, as shown in Figure 2.19. The prefix *myo-* is derived from the Greek word *mys*, meaning muscle. The myoneural junction is the site where the endings of motor nerve fibers are implanted in small invaginations of the sarcolemma, as shown in Figure 3.3. Structures present at the myoneural junction form a small, raised mound on the surface of the muscle fiber called the **motor end plate** (Figure 2.19). A more detailed description of the structure of the myoneural junction is presented in Chapter 3.

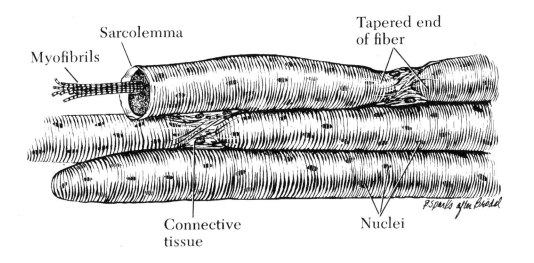

FIGURE 2.4 | Drawing of skeletal muscle fibers showing structural features and their longitudinal orientation. After M. Brodel, Johns Hopkins Hosp. Bull. 61:295, 1937; © The Johns Hopkins University Press.

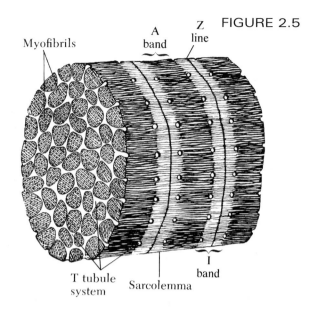

FIGURE 2.5 | Drawing of a cross-section through a muscle fiber at the A band-I junction, showing the transverse tubule system in mammalian skeletal muscle. Invaginations of the sarcolemma are shown along the fiber and around it's entire circumference. From L.D. Peachey, "Form of the Sarcoplasmic Reticulum and T System of Striated Muscle," in E.J. Briskey, R.G. Cassens, and B.B. Marsh, eds. *The Physiology and Biochemistry of Muscle as a Food*, 2. Madison: The University of Wisconsin Press; 1970 by the Board of Regents of the University of Wisconsin System, 307.

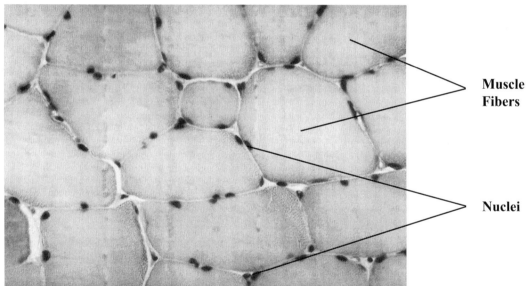

FIGURE 2.6 | Photomicrograph of a cross-section of skeletal muscle fibers of the pig, showing their polygonal shape and the peripheral position of nuclei (x 630).

Sarcoplasm
The cytoplasm of muscle fibers is called **sarcoplasm**. This intracellular colloidal substance contains all organelles and inclusions. Water constitutes about 75 to 80 percent of the sarcoplasm, but, in addition, sarcoplasm of skeletal muscle contains lipid droplets, glycogen granules, ribosomes, numerous proteins, nonprotein nitrogenous compounds, and a number of inorganic constituents.

Nuclei
Skeletal muscle fibers are multinucleated but, because of tremendous variation in their length, the number of nuclei per fiber is not constant. A fiber

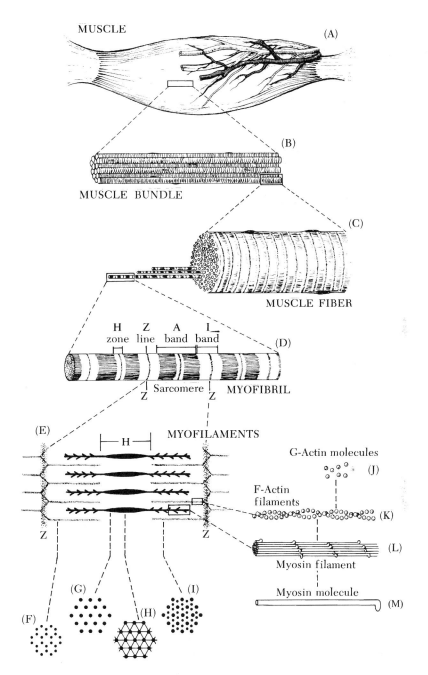

FIGURE 2.7 | Diagram of the organization of skeletal muscle from the gross structure to the molecular level. (A) skeletal muscle, (B) a bundle of muscle fibers, (C) a muscle fiber, showing the myofibrils, (D) a myofibril, showing the sarcomere and it's various bands and lines, (E) a sarcomere, showing the position of the myofilaments in the myofibril, (F-I) cross-sections showing the arrangement of the myofilaments at various locations in the sarcomere, (J) G-actin molecules, (K) an actin filament, composed of two-F-actin chains coiled about each other, (L) a myosin filament, showing the relationship of the heads to the filament, (M) a myosin molecule showing the head and tail regions. Modified from Bloom and Fawcett, *A Textbook of Histology*, 9th ed., W. B. Saunders Company, Philadelphia, 1968, 273.

several centimeters long may have several hundred nuclei with a regular distribution about every 5 μm along its length, except at tendinous attachments where the nuclei are more concentrated and more irregularly distributed. The concentration of nuclei also increases in the vicinity of the motor end plate. In mammalian and avian muscle, nuclei are located at the periphery of the fiber, just beneath the sarcolemma, as shown in Figure 2.6.

In contrast, the nuclei of fish skeletal muscle are usually centrally located within the fiber. Nuclei are ellipsoidal in shape, with th eir long axis oriented parallel to the long axis of the fiber. These features of muscle fiber nuclei can best be visualized by examining the transverse and longitudinal position of nuclei illustrated in Figures 2.4 and 2.6.

Myofibrils

The **myofibril** is an organelle unique to muscle tissue, as implied by the prefix. Myofibrils are long, thin, rods, usually measuring 1 to 2 μm in diameter. In most mammalian skeletal muscle, the long axis of a myofibril is parallel to the long axis of the fiber. Myofibrils are bathed in sarcoplasm and extend the entire length of the muscle fiber. Diagrams and electron photomicrographs of the myofibril are shown in Figure 2.7d and e and in Figure 2.8 respectively. Muscle fibers of meat animals with diameters of 50 μm have at least 1000 and could have as many as 2000 (or more) myofibrils.

Photomicrographs of myofibrillar cross-sections reveal a highly organized array of dots that have two distinct sizes (Figure 2.9). These dots represent the two types of **myofilaments** within myofibrils (shown in longitudinal section in Figure 2.7d and e and in Figure 2.8). Myofilaments are commonly referred to as thick and thin filaments of the myofibril. In longitudinal section, thick filaments are aligned parallel to each other and arranged in exact alignment across the entire

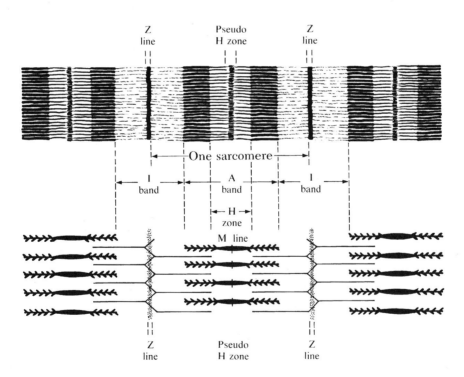

FIGURE 2.8   A drawing adapted from an electron photomicrograph, showing portions of two myofibrils and a sarcomere (× 15,333) and a diagram corresponding to the sarcomere, identifying it's various bands, zones, disks, and lines. Modified from H. E. Huxley, "The Mechanism of Muscular Contraction." Copyright © 1965 by Scientific American, Inc. All rights reserved.

myofibril. Likewise, thin filaments are aligned precisely across the myofibril, parallel to each other and to the thick filaments. This arrangement of myofilaments, and the fact that thick and thin filaments overlap in certain regions along their longitudinal axes, account for the characteristic banding or striated appearance of the myofibril (Figure 2.8). Furthermore, bands of each myofibril are aligned across the entire muscle fiber giving the fiber a striated appearance (Figure 2.1). This banding effect, which takes the form of alternating light and dark areas, explains the term striated muscle.

Areas of different density are visible within light and dark bands of myofibrils. Because the light band is singly refractive when viewed with polarized light, it is described as being isotropic and is called the **I band**; whereas, the broad dark band is doubly refractive, or anisotropic in polarized light, and is designated the **A band**. The A band is much denser than the I band but both bands are bisected by relatively thin, dense lines. The I band is bisected by a dark thin band called the **Z disk**. The unit of the myofibril spanning two adjacent Z disks is referred to as a **sarcomere**. The sarcomere includes both an A band and the two half I bands located on either side of the A band. The sarcomere is the repeating structural unit of the myofibril, and it is also the basic unit where muscle contraction and relaxation occur. Sarcomere length is not constant, and its dimensions, as well as those of the I band, are dependent on state of contraction at the time a muscle is examined. In resting mammalian muscle, a sarcomere length of 2.5 µm is fairly typical.

Other features of the myofibril exist as zones, lines, or bands with different densities. The **H zone**, **pseudo H zone** and **M line** (Figures 2.7d, e, and 2.8) are structures whose appearances change with contraction state.

Myofilaments

Thick and thin filaments of the myofibril differ not only in dimensions, but also in their chemical composition, properties, and position within the sarcomere. Thick filaments of vertebrate muscles are approximately 14 to 16 nm (nanometers) in diameter (1 nm equals one-billionth of a meter) and 1.5 µm in length. Thick filaments constitute the A band of the sarcomere, as illustrated in Figures 2.7e and 2.8. Since the predominant protein in thick filaments is **myosin**, they are referred to as **myosin filaments**. Myosin filaments are held in position by other proteins, some of which are located in the M line.

Thin filaments are about 6 to 8 nm in diameter and extend approximately 1.0 µm on either side of the Z disk. These filaments constitute the I band of the sarcomere and extend beyond the I band into the A band, lying alongside and in close proximity to the thick myosin filaments

FIGURE 2.9 | Electron photomicrograph of a cross-section of a myofibril, showing the thick and thin filaments and their hexagonal arrangement. Adapted from D.S. Smith, *Muscle* Academic Press, Inc., New York, 1972, 14.

(Figure 2.7e). Because **actin** is the most abundant protein in the thin filament, they are referred to as the **actin filaments**.

Distribution of myofilaments in cross-section (Figures 2.7 and 2.9) shows the orderly arrangement of thick (myosin) and thin (actin) filaments in a sarcomere. Only thick filaments are present in the H zone, but a cross-section through the A band where actin and myosin filaments overlap shows six thin filaments surrounding each thick filament. The I band contains only thin filaments.

Z Disk Ultrastructure

In longitudinal section, an actin filament on one side of the Z disk lies between two actin filaments on the opposite side of the Z disk (Figures 2.7e and 2.8). This arrangement indicates that the actin filaments per se do not pass through the Z disk but rather terminate at the Z disk in a very organized manner. Ultra-thin filaments, called **Z filaments**, constitute the material of the Z disk, and they connect with actin filaments on either side of the disk. Near the Z disk, each actin filament connects to four Z filaments that pass obliquely through the Z disk. Each of the four Z filaments then connects with an actin filament in the adjacent sarcomere, as shown diagrammatically in Figure 2.10.

Proteins of the Myofibril

There are more than 20 different proteins associated with the myofibril. Six of these proteins account for approximately 90 percent of the total myofibrillar protein. These six in decreasing order of their relative abundance are **myosin**, **actin**, **titin**, **tropomyosin**, **troponin**, and **nebulin**. Myofibrillar proteins are classified by their function as either contractile, regulatory, or cytoskeletal proteins (Table 2.4). Actin and myosin constitute the major contractile proteins. The major regulatory proteins include tropomyosin and troponin. As regulatory proteins, their role is to regulate actin-myosin interactions during muscle contraction. Cytoskeletal proteins, such as titin and nebulin, serve as the template and/or provide the scaffold for the alignment of myofilaments during myofibril and sarcomere formation. Cytoskeletal proteins are integral to the structure of the Z disk. In mature muscle, cytoskeletal proteins maintain overall longitudinal and lateral myofibril alignment and linkage with the cell membrane as well as myofilament structural integrity.

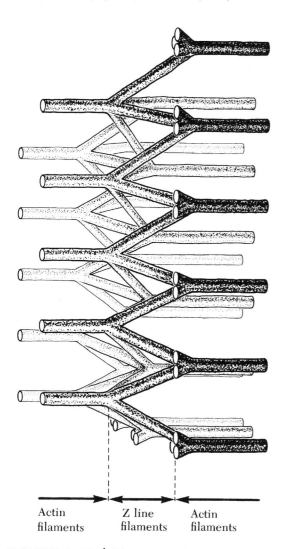

FIGURE 2.10 | Diagrammatic representation of Z filaments and their attachment to actin filaments. Reproduced from *The Journal of Cell Biology*, 1962, volume 13:332, by copyright permission of the Rockefeller University Press.

CHAPTER 2 STRUCTURE, COMPOSITION, AND DEVELOPMENT OF ANIMAL TISSUES

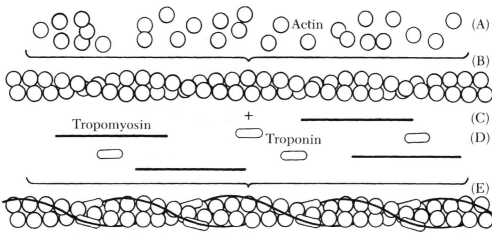

FIGURE 2.11 | Diagrammatic representation of the construction and fine structure of an actin filament, showing the relation of tropomyosin and troponin to its (A) G-actin molecules, (B) two chains of polymerized actin (F-actin) coiled about each other, (C) long thin molecules of tropomyosin, (D) pellet-shaped troponin molecules, and (E) an actin filament, showing how two coiled chains of F-actin and polymerized tropomyosin molecules form two long strands of tropomyosin, one strand being associated with each of the F-actin chains. The periodic location of troponin along the actin filament every seven G-actin molecules, and its association with the tropomyosin strand, are also illustrated. Modified from J. M. Murray and A. Weber, "The Cooperative Action of Muscle Proteins". © George V. Kelvin/Scientific American, Inc. with permission.

Major Contractile Proteins

Actin constitutes approximately 20 percent of the myofibrillar proteins. It is a globular (spherical) shaped molecule approximately 5.5 nm in diameter. This molecule is referred to as G-actin (for globular actin) and is the monomeric (single molecule) form of actin. The fibrous nature of the actin filament occurs when G-actin monomers polymerize to form F-actin (fibrous actin), as shown in Figure 2.11b and Figure 2.7j and k. In F-actin, the G-actin monomers are linked together in strands, much like beads on a string of pearls. Two strands of F-actin are spirally coiled around one another, to form a "super helix" that is characteristic of the actin filament.

Myosin is a fibrous protein that constitutes approximately 45 percent of myofibrillar protein. The myosin molecule is an elongated rod shape, with a thickened portion at one end. The thickened end of the molecule is usually referred to as the head region, and the long, thin portion that forms the backbone of the thick filaments is called the rod or tail portion. The portion of the molecule between the head and the rod regions is called the neck region. The head region of the molecule is double-headed, and it projects laterally from the long axis of the myosin filament (Figures 2.7m and 2.12b).

Myosin filaments are assembled with the myosin molecules oriented in opposite directions on either side of the M line. In the center of the A band, the myosin filament contains only the rod portion of myosin molecules without any of the globular heads. This region within the H zone, on

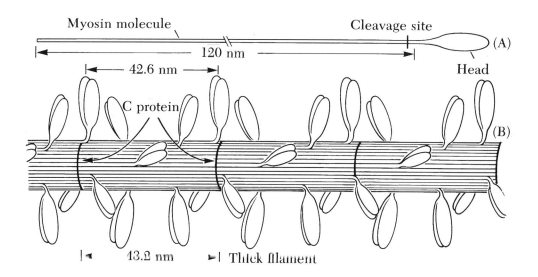

FIGURE 2.12 | Diagrammatic representation of the construction and fine structure of the myosin filament. (A) The myosin molecule, showing the head and tail, and the cleavage site. (B) A myosin filament, showing the projecting double heads and periodic (every 43.2 nm) encircling band of C protein. Modified from George V. Kelvin in "The Cooperative Action of Muscle Proteins" by J. M. Murray and A. Weber. Scientific American, February 1974.

either side of the M line, is called the pseudo H zone. In the myosin filament, the polarity of the myosin molecules is such that the heads on either side of the bare central region of the A band are oriented at an oblique angle away from the M line (Figures 2.7e and 2.8). The myosin heads contain the functionally active sites of the thick filaments, which function during muscle contraction by forming cross-bridges with actin filaments. During muscle contractions each myosin head attaches to a G-actin molecule of the actin filament. Formation of cross-bridges through this interaction of actin and myosin filaments is discussed in Chapter 3.

Major Regulatory Proteins

Tropomyosin constitutes approximately 5 percent of myofibrillar protein and lies in close contact with the actin filament. A strand of tropomyosin lies alongside, within each groove of the actin superhelix, and a single molecule extends the length of seven G-actin molecules in the actin filament. The relationship of tropomyosin to the actin filament is illustrated in Figure 2.11e.

Troponin also constitutes approximately 5 percent of myofibrillar protein and, like tropomyosin, is present at very well defined intervals in the grooves of the actin filament, where it lies astride the tropomyosin strands. Troponin units show a periodicity (repetitiveness) along the length of the actin filament, as shown in Figure 2.11e. There is one molecule of troponin for every seven or eight G-actin molecules along the actin filament.

Cytoskeletal Proteins

Titin is the most abundant cytoskeletal protein in muscle, composing about 10 percent of myofibrillar proteins. Titin molecules are among the largest polypeptides known. Single titin polypeptides extend longitudinally in each half sarcomere from the M line to the Z disk and constitute a third filament of the myofibril as illustrated in Figure 2.13. The portion of titin in the A band is inelastic, while that in the I band is elastic. Titin binds to the outside shaft of the thick filament and the protein (**C protein**) that encircles and stablizes the thick filament. Titin is believed to provide the scaffold for alignment of the filaments during myofibril

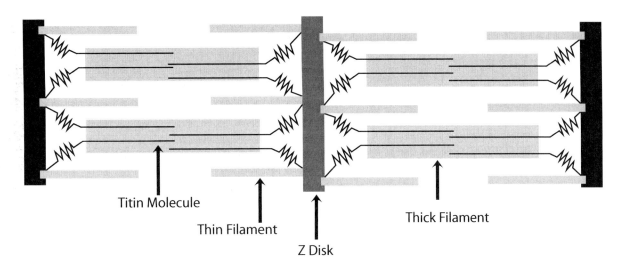

FIGURE 2.13 | Diagram of the location of titin molecules in the sarcomere with the N-terminus of each titin polypeptide at the Z disk and the C-terminus at the M line. The elastic domains of the titin molecules are located in the I-band. Modified from John M. Squire, Hind A. Al-Khayat, Carlo Knupp and Pradeep K. Luther. Molecular architecture in muscle contractile assemblies. In *Advances in Protein Chemistry, Vol. 71, Fibrous Proteins: Muscle and Molecular Motors.* John M. Squire and David A. D. Parry, eds. Elseview Academic Press, San Diego, 2005, 28.

and sarcomere formation. In mature myofibrils, it maintains the ordered structure and integrity of myofibrils within the sarcomere and is thought to be responsible for the resting tension associated with each sarcomere.

Nebulin composes approximately 4 percent of myofibrillar proteins. It is located close and parallel to the thin (actin) filament. Nebulin extends longitudinally along the entire length of the thin filament from the A band to the Z disk. In developing muscle, it is believed to play a role in the organization of thin filaments during myofibril formation. In mature muscle, it serves as a template for assembly and/or a scaffold for stability of thin filaments. It also may function in anchoring thin filaments to Z disks.

C protein constitutes approximately 2 percent of myofibrillar proteins. Seven bands of C protein encircle each thick filament on both sides of the H zone. **H protein**, approximately 1 percent of myofibrillar protein, also is present in this region. **Myomesin, M protein** and **skelemin** constitute 2 percent, 1 percent, and less than 1 percent of myofibrillar proteins, respectively, and are present in the structural components of the M line, where they stabilize the rod portion of myosin molecules of the thick filament. **Creatine kinase** also is present at the M line but does not have a role in this structure. **Alpha-actinin** and **Cap Z**, which represent 2 percent and less than 1 percent, respectively, of myofibrillar proteins, are integral components of the Z disk. All of the other cytoskeletal proteins presented in Table 2.4 are associated with the exterior of the Z disk or with intermediate filaments and each comprises less than 1 percent of total myofibrillar proteins. The proteins, **desmin, filamin, paranemin,** and **synemin,** are components of thin or intermediate filaments peripheral to the disk itself. These filaments arising at the Z disk from one myofibril interconnect with those of adjacent myofibrils. They are believed to play a role in the lateral alignment of adjacent myofibrils because in mature muscle all Z disks are aligned laterally across the entire muscle fiber. **Dystrophin, talin,**

and **vinculin**, along with filamin are components of **costameres** that lie beneath the cell membrane.

Sarcoplasmic Reticulum and T Tubules

In embryological origin, the **sarcoplasmic reticulum** (SR) corresponds to the endoplasmic reticulum of other cell types. The SR is a membranous system of tubules and **cisternae** (flattened reservoirs for calcium storage) that forms a closely meshed network around each myofibril. The SR and transverse tubules (T tubules), although usually discussed together, are two separate and distinct membrane structures. The T tubules are associated with the sarcolemma, whereas the SR is an intracellular membrane structure.

The SR consists of several distinct elements, structural features of which are discussed with reference to a single sarcomere. Relatively thin tubules, oriented parallel to the direction of the myofibrillar axis, constitute the longitudinal tubules of the reticulum. In the H zone region of the sarcomere, the longitudinal tubules converge, forming a perforated sheet that is called a **fenestrated** (window like) **collar**. These features of the SR are shown in Figure 2.14. At the junction of the A and I bands the longitudinal tubules converge and join with a pair of larger, transversely oriented tubular elements called **terminal cisternae**. **Longitudinal tubules** extend in both directions from the fenestrated collar to the terminal cisternae. Two tubular elements comprising the terminal cisternae lie parallel to each other with one tubule of the pair transversing the A band, and the other transversing the adjacent I band of the sarcomere. A T tubule also runs transversely across the sarcomere at the A-I band junction and lies between the two tubular elements of the terminal cisternae pair. Connecting elements exist between the terminal cisternae and T tubules in spite of their independent origin and structural integrity. The structure and orientation of the SR and T tubules, with reference to the sarcomere, are illustrated in Figure 2.14. This

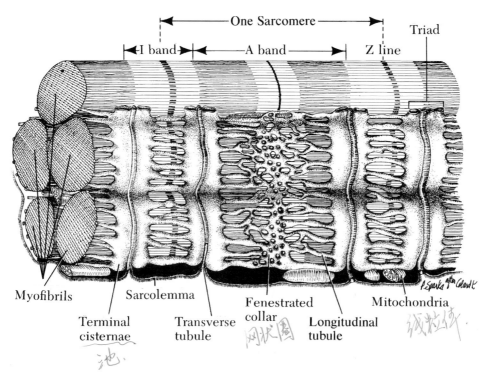

FIGURE 2.14 | Diagrammatic representation of the sacroplasmic reticulum and T tubules, and their relation to the myofibrils of mammalian skeletal muscle. Modified from L. D. Peachey, J. Cell Biol. 25:209, 1965, from Bloom and Fawcett, *A Textbook of Histology*, 9th ed. W. B. Saunders Company, Philadelphia, 1968, 281.

sarcotubular system encircles each myofibril at the A-I band junction of mammals, birds, and some fish but in some species it is located at the Z disks.

The extensive nature of the SR can be appreciated if one considers the fact that these structures are associated with each sarcomere along the entire length of the myofibril, and that the muscle fiber contains at least a thousand myofibrils. SR volume varies from one muscle fiber to another, but it is estimated to make up approximately 13 percent of total fiber volume. The T tubules, on the other hand, comprise only approximately 0.3 percent of fiber volume.

Mitochondria

Mitochondria are oblong organelles (Figure 2.14) located in the sarcoplasm. They frequently are referred to as the "powerhouse of the cell" because they capture energy derived from carbohydrate, lipid, and protein metabolism and then provide the cell with a source of chemical energy. They contain enzymes that the cell uses in oxidative metabolism. Great variation exists in mitochondrial numbers and size in muscle fibers. Skeletal muscle mitochondria are relatively abundant at the periphery of the fiber near the poles of the nuclei and are especially abundant at motor end plates. Additionally, a number of mitochondria are located between the myofibrils, adjacent to Z disks, I bands, or A-I band junctions (Figure 2.14).

Lysosomes

Lysosomes are small vesicles located in the sarcoplasm that contain a number of enzymes collectively capable of digesting the cell and its contents. Included among these lysosomal enzymes are a group of proteolytic enzymes known as **cathepsins**.

Golgi Complex

Structures called the Golgi complex are located in the sarcoplasm near the nuclei. The muscle fiber, being multinucleated, has numerous Golgi complexes located near nuclei and, compared to those of secretory cells such as in the liver, pancreas, and mammary gland, they are quite small. Vesicles of the Golgi complex resemble membranes of the sarcoplasmic reticulum.

Smooth Muscle

Although smooth muscle comprises only a small proportion of meat, a brief description of its ultrastructure must be included to exploit the principal differences between smooth and skeletal muscle. Smooth muscle is mainly located in the walls of arteries and lymph vessels, and in the gastrointestinal and reproductive tracts.

Smooth muscle fibers vary in size and shape, depending on their location. They are not always spindle shaped, as usually depicted, but may be quite uneven in contour along their length. In cross-section, smooth muscle fiber shapes vary from extremely flattened ellipsoids to triangles and polyhedrons. Photomicrographs of smooth muscle fibers are shown in Figure 2.15.

The sarcolemma of smooth muscle fibers forms tight membrane-to-membrane junctions that allow for communication between neighboring fibers. A smooth muscle fiber has only one nucleus and is usually centrally located within the cell, although in some larger fibers the nucleus may be displaced slightly from the center. The sarcoplasmic reticulum in smooth muscle fibers is much less developed than in skeletal muscle. Myofilaments of smooth muscle fibers are also less ordered than in skeletal muscle fibers. Rather myofilaments are arranged in pairs that run parallel to the longitudinal axis of the fiber. In contrast to skeletal muscle, a large number of filaments are found concentrated at each end of smooth muscle fibers. Among these filaments are actin-like molecules that resemble skeletal muscle F-actin. Myosin filaments, on the other hand, are not visible, even though myosin is very abundant in smooth muscle. Additionally, myofilaments of smooth muscle are attached to dark zones on the sarcolemma. These dark zones are evident along the surface of the sarcolemma, but are less abundant than those located at the ends of fibers. Dark zones are described as being analogous to the Z disks of skeletal muscle, since they move with contractile action. These structures most likely transmit the myofilamental contraction force to the sarcolemma.

Smooth muscle fibers occur either singly or in bundles. Regardless of the arrangement, each fiber or bundle is surrounded by a delicate network

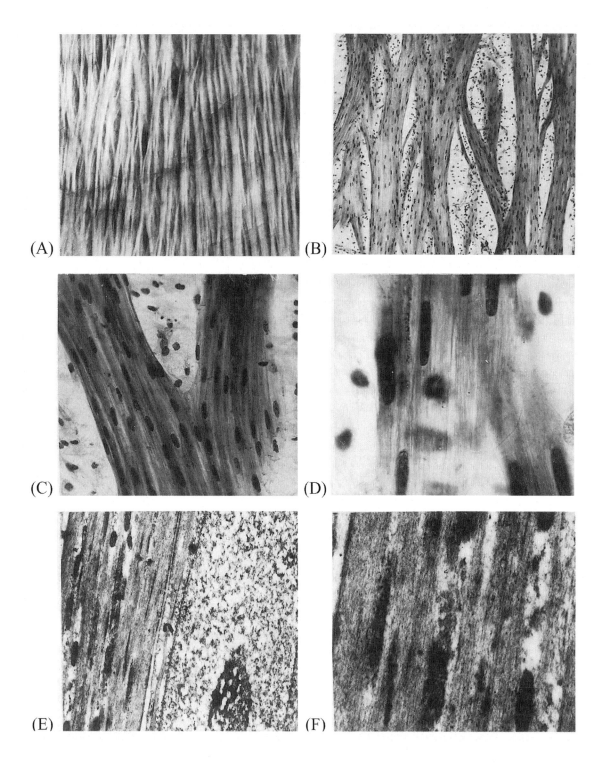

FIGURE 2.15 | Photomicrographs of smooth muscle fibers from the longitudinal muscle layer of a pregnant rabbit's uterus at term. Photomicrographs (A-D) light microscopy (x 12, x 75, x 300, and x 1000, respectively) and, (E, F) electron photomicrographs (x 10,000 and x 20,000 respectively). Courtesy of Butterworths Scientific Ltd.

of fine connective tissue that aids in transmitting force during contraction. Scattered throughout this connective tissue matrix, blood vessels and nerve fibers occupy the space between the fibers. Even so, smooth muscle is poorly supplied with blood compared to skeletal muscle.

Cardiac Muscle

Cardiac muscle possesses the unique property of rhythmic contractility, which continues ceaselessly from early embryonic life until death. Cardiac muscle has properties that resemble both skeletal and smooth muscle. Like smooth muscle, each fiber has a single centrally located nucleus. Fibers of cardiac muscle are branched, with the main trunk being larger in diameter than those of its branches, yet still smaller in diameter and shorter than the fibers of skeletal muscle. Sarcoplasm of cardiac muscle contains numerous glycogen granules, and the mitochondria are especially large and numerous. These and other features of cardiac muscle are evident in Figure 2.16. Myofilaments of cardiac muscle are not well organized into discrete myofibrils as in skeletal muscle. Instead, aggregates of myofilaments form fibrils of extremely variable size, and dimensions vary along the longitudinal axis of the fibrils. However, thick (myosin) and thin (actin) filaments are readily apparent under the microscope, and are aligned to give a striated appearance that is similar to that of skeletal muscle (Figure 2.16).

The T tubules of cardiac muscle are larger in diameter and occur at the Z disks, in contrast to their occurrence at the A-I band junction in mammalian and avian skeletal muscle. Elements of the SR are not only less developed in cardiac muscle, but structures comparable to the terminal cisternae are absent.

At regular intervals along the longitudinal axis of cardiac muscle, dense lines called **intercalated disks** transect the fiber. In mammalian cardiac muscle, this occurs at offset intervals across I bands of the fiber, giving a stepwise appearance to the disks (Figure 2.16). The intercalated disks are continuous across the entire fiber, and their paired membranes represent the cell membranes of

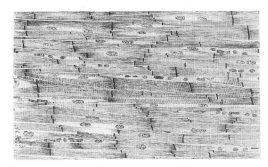

(A)

(B)

FIGURE 2.16 | Photomicrographs of cardiac muscle. A longitudinal section (A), showing fiber arrangements and nuclei locations and, (B) an electron photomicrograph showing the banding pattern, intercalated disks, mitochondria, and glycogen granules. From D. W. Fawcett, *A Textbook of Histology*, 9th ed. W. B. Saunders Company, Philadelphia, 1968, 285, 289.

adjoining muscle fibers. Thus, these disks provide a cohesive link among the fibers of the myocardium (heart muscle) as well as facilitate the transmission of contractile force in the direction of the fiber axis from one fiber to another.

The myocardium is the contractile layer of the heart and contains the bulk of the cardiac muscle. Fibers of the myocardium are held in place by connective tissue fibers, which are continuous with connective tissue sheaths that group them into fascicles (bundles) of fibers. Blood and lymph vessels and nerve fibers enter and exit the myocardium via the connective tissue between muscle bundles. Cardiac muscle is endowed principally with a capacity for oxidative metabolism; consequently, it has a more extensive capillary network than either smooth or skeletal muscles.

Development Of Skeletal Muscle Fibers

Myogenesis is the term used to describe the molecular, biochemical, and morphological events that occur during the formation of muscle fibers. Normally, myogenesis is thought to occur solely in the prenatal period of development, however, the full complexity (development) of muscle is not attained until puberty. Thus, events of myogenesis occur throughout the lifetime of most domestic meat animals. But, in meat animals, most muscle fiber formation occurs during the embryonic and fetal phases of prenatal development. Although alterations in growth and development of muscle can affect meat quality, they will not be covered in great detail in this text. Rather an understanding of growth processes important to meat scientists will be covered.

Formation of muscle fibers is quite unique compared to other cell types because muscle cells are multinucleated. Myogenic cells are differentiated early in the embryonic phase. These cells undergo proliferative mitosis (daughter cells are identical to parent cells) until further differentiation occurs into **myoblasts**. Myoblasts are cells that have the ability to fuse with one another to form immature muscle fibers. Thus, groups of myoblasts (muscle precursor cells) fuse and donate their nuclei to a single immature muscle fiber. In early myogenesis, these immature muscle fibers are known as **myotubes** because the nuclei are located in the center of the fiber and myofibrils are located peripherally beneath the sarcolemma when viewed transversely. The first myotubes formed in developing embryos and fetuses are known as primary myotubes and the muscle fibers they produce are primary muscle fibers (Figure 2.17). Secondary myotubes, also the result of myoblast fusion, form later in fetal development around the primary myotubes and produce secondary muscle fibers. As both types of fibers mature and begin to accumulate contractile protein, myofibrils are formed centrally in the muscle fiber, forcing the nuclei to the periphery. At this point, both primary and secondary muscle fibers structurally resemble mature muscle fibers.

EPITHELIAL TISSUES

Of the four tissue types present in the animal body, quantitatively, epithelial tissue contributes the least to meat. However, the characteristic flavor and crispness of fried chicken, for example, is partly due to properties of epithelial tissue and its underlying connective tissue. Epithelial tissue forms the linings that cover the surfaces of the body and organs and is usually removed during slaughter and processing. Most of the epithelial tissue remaining in the carcass is associated with blood and lymphatic vessels, but this tissue also is present in many edible organs, such as the kidney and liver, as well as in hides and skins.

Epithelial tissue is characterized as having little extracellular material and is classified according to cell shape and the number of layers forming the epithelium. Cell shapes vary from thin, flat cells to tall columnar cells (Figure 2.18). Some cells form single layers, while others form stratified layers. In either case, they maintain extensive intercellular contacts, thus forming cohesive cellular sheets that cover surfaces and line cavities. Additionally, the surface of the epithelium may be modified to increase its efficiency in performing such functions as protection, secretion, excretion, transport, absorption, and sense perception.

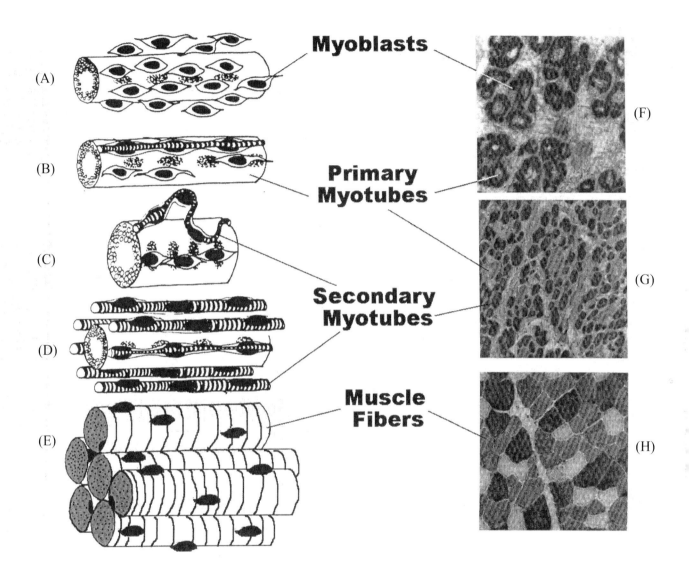

FIGURE 2.17 | Illustration of embryonic and fetal development of muscle fibers. (A) Myoblasts surrounding a primary myotube. (B and C) Secondary myotube forming on and detaching from the surface of a primary myotube. (D) Secondary myotubes surrounding a primary myotube. (E) Several fully formed muscle fibers. (F) Cross-section of early embryonic skeletal muscle showing the predominance of primary myotubes. (G) Cross-section of early fetal skeletal muscle showing secondary myotubes surrounding primary myotubes. (H) Cross-section of fully formed skeletal muscle fibers. From *Structure and Development of Meat Animals and Poultry* by Swatland. Copyright © 1994 by Technomic Publishing Company, Inc. Reprinted by permission.

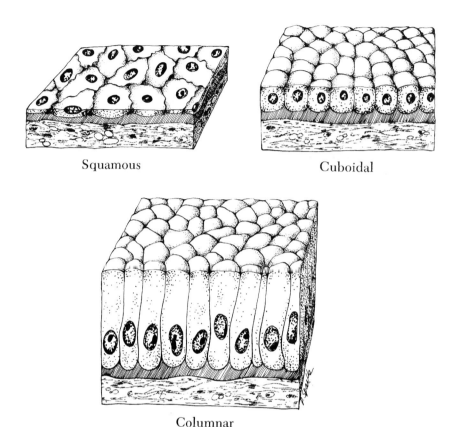

FIGURE 2.18 | Diagrammatic illustration of squamous, cuboidal, and columnar epithelial tissue. Modified from J. E. Crouch, *Functional Human Anatomy*, 2nd ed. Lea & Febiger, Philadelphia, 1985, 50.

NERVOUS TISSUE

Nervous tissue constitutes a small proportion of meat (less than 1 percent), but functions in the period immediately prior to and during slaughter and influences meat quality. The interrelationship of nervous and muscle tissues is described in chapters 3 and 5.

Nervous tissue is generally categorized structurally as being part of either the peripheral or central nervous systems. The **central nervous system** consists of the brain and spinal cord, and the **peripheral nervous system** consists primarily of nerve fibers in other parts of the body. The **neuron** comprises the bulk of nervous tissue. It consists of a polyhedrally shaped cell body and a long cylindrical structure called the **axon**. Encased within the neuron is **neuroplasm** (cytoplasm).

Neuroplasm in the axon of the neuron is frequently called axoplasm. Embedded in the neuroplasm of the cell body is a centrally placed nucleus, and radiating from the cell body are several short branched structures called **dendrites**. The axon terminates in one or more fine multi-branched structures called the axon ending. These features of the neuron are shown in Figure 2.19.

Nerve fibers are composed of groups of neuronal axons, and assembly of groups of fibers into fascicles results in formation of **nerve trunks**. In nerve fibers, neurons are oriented longitudinally in relation to each other; the axon ending of one cell interdigitates with the dendrite of another cell. Interdigitating areas of these neurons are called **synapses**. Neuron structures are not actually joined at synapses, rather they are close enough to permit chemical substances released by one neuron to

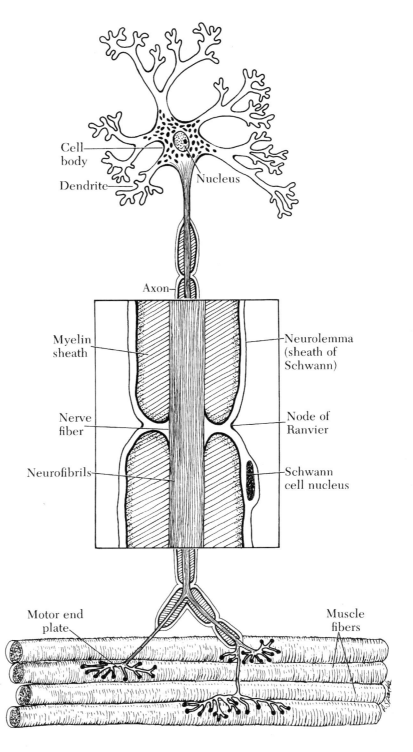

FIGURE 2.19 | Diagrammatic illustrations of a neuron and the motor end plates associated with it. The magnification of the axon (center) shows details of the nerve fiber, neurofibrils, myelin sheath, Schwann cell nucleus, and node of Ranview. Modified from J. E. Crouch, *Functional Human Anatomy*, 2nd ed., Lea & Febiger, Philadelphia, 1985, 480.

influence the adjacent neuron which, in turn, influences the next, and so on. Fascicles of nerve fibers are held together by sheaths of connective tissue, and the nerve trunk itself is ensheathed in a connective tissue covering. All peripheral nerve fibers are ensheathed by **Schwann cells** and, in addition, the larger fibers are enveloped in a **myelin sheath** within the sheath of Schwann cells (Figure 2.19). While small peripheral nerve fibers contain a Schwann cell sheath, they are devoid of a myelin sheath. Thus, nerve fibers are frequently referred to as either myelinated or unmyelinated.

CONNECTIVE TISSUE

As the name implies, connective tissue literally connects and holds various parts of the body together. Connective tissues are distributed throughout the body as components of the skeleton, in the framework of organs and blood and lymph vessels and in sheaths that surround structures such as tendons, nerve trunks, and muscles. The skin or hide is attached to the body by connective tissues. Connective tissues provide the body with a barrier against infective agents, and are critically important in wound healing. Fat storage cells are located within a specialized type of connective tissue known as **adipose tissue.** Connective tissue envelops muscles, muscle bundles, and finally, muscle fibers themselves.

Connective tissues are characterized generally as having relatively few cells and considerable **extracellular substance**. This extracellular substance varies from a soft jelly to a tough fibrous mass, but almost always contains embedded fibers that provide structural integrity to the connective tissues. In cartilage, extracellular substance has a rubbery consistency, whereas the extracellular substance in bone is much stronger and is impregnated with calcium salts. Both cartilage and bone contain fibers, but they are associated with other extracellular substances. Adipose tissue is composed mainly of cells with a limited number of extracellular fibers, while blood and lymph have extracellular fluid with no fibers.

Connective tissue surrounding muscles, muscle bundles, and muscle fibers is fibrous and is known as **connective tissue proper**. Bone and cartilage, on the other hand, are **supportive connective tissues** because other tissues are attached to them, providing the body with structural support. Although preceding generalizations about connective tissues might indicate that they are quite different from each other, they have many similarities in composition and function. Fibers of connective tissue proper are sometimes continuous with bone and cartilage, and some chemical compounds are common to a number of connective tissues.

Connective Tissue Proper

Connective tissue proper consists of a structureless mass called ground substance plus embedded cells and extracellular fibers. Extracellular fibers include those of collagen and elastin. A detailed structure of each is described later in this section. Connective tissue proper contains two different populations of cells, which are referred to as fixed or wandering cells. Fixed cells include fibroblasts, mesenchymal cells, and specialized adipose (fat storage) cells. Wandering cells are primarily involved in injury reactions; they include eosinophils, plasma cells, mast cells, lymph cells, and free macrophages. Extracellular components (ground substance and fibers) are produced and maintained by the cellular component, the fibroblast (Figure 2.20) and other cell types as defined in the next two sections.

Ground Substance

Ground substance is a viscous solution containing soluble glycoproteins (carbohydrate-containing proteins). These glycoproteins are usually referred to as **proteoglycans**. Proteoglycans are large molecules composed of a core protein to which are attached various combinations of **glycosaminoglycans** (Figure 2.20). Ground substance also contains substrates and end products of connective tissue metabolism. Included among these precursors of collagen and elastin synthesis, which include **tropocollagen** and **tropoelastin**, respectively. Notable among the glycosaminoglycans are **hyaluronic acid** and the **chondroitin sulfates**. Hyaluronic acid is a very viscous substance found

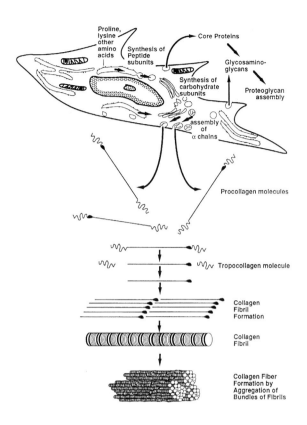

FIGURE 2.20 | Diagrammatic illustration of the fibroblast showing the synthesis of tropocollagen and extracellular assembly of tropocollagen molecules into collagen fibrils and fibers. Also illustrated is the synthesis of carbohydrates and proteins for the assembly of glycoaminoglycans and proteoglycans.

in joints (synovial fluid) and between connective tissue fibers. Chondroitin sulfates are found in cartilage, tendons, and adult bone. These two glycosaminoglycans and associated proteins function as lubricants, intercellular cementing substances, and structural matter in cartilage and bone. They also provide a barrier against infectious agents.

Extracellular Fibers

Extracellular fibers in densely packed structures are referred to as **dense connective tissue,** and those forming a loosely woven network are called **loose connective tissue**. Dense connective tissue may be further characterized by fiber arrangement. In **dense irregular connective tissue**, fibers are densely interwoven, but in a random arrangement. However, in **dense regular connective tissue**, fibers are arranged in bundles lying parallel to each other, such as in tendons, or in flat sheets, such as those present in **aponeuroses** (tendinous extensions of connective tissues surrounding muscle) (Figure 2.33). Extracellular fibers include **collagen** and **elastin**.

Collagen is the most abundant protein in the animal body and significantly influences meat tenderness. In most mammalian species it constitutes 20 to 25 percent of total body protein. Collagen is the principal structural protein of connective tissue. It is a major component of tendons and ligaments, and to a lesser extent, of bones and cartilage. Additionally, networks of collagen fibers are present in essentially all tissues and organs, including muscle. Distribution of collagen is not uniform among skeletal muscles, but the amount present generally parallels physical activity. Muscles of limbs contain more collagen than those around the spinal column and, consequently, the former are less tender than the latter.

Glycine is the most abundant amino acid in collagen, and comprises approximately one-third of the total amino acids. Hydroxyproline and proline account for another one-third of its amino acid content as shown in Figure 2.21. Hydroxyproline is a relatively constant component of collagen (13 to 14 percent) and does not occur to a significant extent in other animal proteins. Thus, chemical assays for hydroxyproline are commonly used to determine the amount of collagen in tissues. Also, collagen is a glycoprotein in that it contains small quantities of the sugars galactose and glucose.

The **tropocollagen** molecule is the structural unit of the **collagen fibril**. Tropocollagen molecules are composed of three α-chains that form a triple helix (Figure 2.21). There are at least 19 different α-chains, which are combined in various combinations of three to form the tropocollagen molecules characteristic of more than 12 different types of collagen. The three α-chains of tropocollagen may be all alike, all different, or two alike

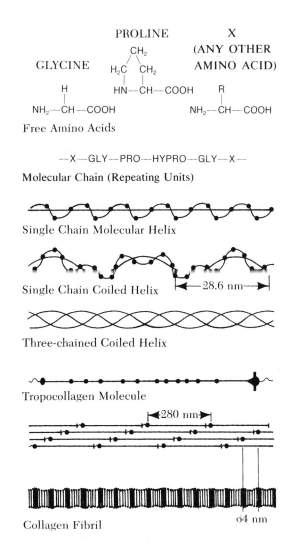

FIGURE 2.21 Diagrammatic illustration of the amino acid sequence and molecular structure of collagen and tropocollagen molecules, and of collagen fibril formation. Adapted from J. Gross "Collagen." Copyright © 1961 by Scientific American, Inc. All rights reserved.

and one different depending on the specific type of collagen. Of the more than 12 types of collagen, Types I, III, IV, V, and VII are present in the connective tissues associated with skeletal muscle.

Following synthesis of tropocollagen molecules in fibroblasts, they are secreted into the intercellular matrix where they are assembled into collagen fibrils (Figure 2.20). During fibril assembly, tropocollagen molecules are aligned longitudinally end to end and laterally in a slightly less than one-fourth overlapping stagger (Figures 2.20 and 2.21). The overlapping stagger gives rise to the unique striated appearance of collagen fibrils, which differs from the striations of skeletal muscle. Not all types of collagen form fibers. Of the collagen types present in muscle, only in Types I and III are collagen fibrils assembled to form fibers. Type I collagen forms large fibers, while Type III collagen forms fine fibers. Type IV collagen, which is present in basement membranes, is non-fibrous. Tropocollagen molecules in Type IV collagen form a chicken wire-like sheath that surrounds individual muscle fibers (Figure 2.22). In Type V and VIII collagens, tropocollagen molecules are aligned longitudinally to form microfibrils. Microfibrils of Type VII collagen are very long and thin. Microfibrils of type V collagen may form thin fibers.

In the fibrous collagens of Types I and III, a variable number of fibrils is grouped together to form a **collagen fiber** (Figure 2.20). The number of fibrils is dependent on the load and stress that the collagen fibers must endure. Likewise, the number of collagen fibers present is dependent on the load they must bear in the tissue. Collagen fibers are almost completely inextensible. Individual fibers are colorless, but aggregations, such as those in muscle sheaths or in tendons, are white.

The relative insolubility and high tensile strength of collagen fibers results from **intermolecular cross-linkages**. These cross-linkages are few in number, and more easily broken, in young animals. As the animal grows older, the number of cross-linkages increases and the linkages are converted to stable linkages. Consequently, collagen is more soluble in young animals and becomes

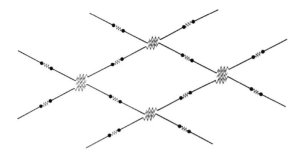

FIGURE 2.22 | Diagrammatic illustration of the chicken wire-like network of Type IV collagen.

Connective Tissue Cells

Of the several kinds of cells found in connective tissue, only fibroblasts, mesenchymal cells, and adipose cells are discussed in this chapter because of their immediate relevance to the properties of meat.

Fibroblasts

Fibroblasts vary greatly in shape but generally possess a long and slender morphology. Fibroblasts synthesize precursors of extracellular components of connective tissue, namely tropocollagen, tropoelastin, and ground substance. These connective tissue proteins are synthesized within the fibroblast and released into the extracellular matrix. Construction of collagen and elastin fibers from tropocollagen and tropoelastin subunits, respectively, occurs extracellularly in the connective tissue matrix. Synthesis of tropocollagen within the fibroblast and extracellular formation of fibrils are illustrated in Figure 2.20. Construction of elastin probably follows a sequence similar to that of collagen.

less soluble as the animal ages. This characteristic, in particular, is an important structural feature of muscle and becomes important when considering meat tenderness.

Elastin is a much less abundant connective tissue protein than collagen, and its ultrastructural characteristics are not as well known. Elastin is a rather rubbery protein that is present throughout the body in ligaments and arterial walls, as well as in the framework of a number of organs including muscle. The **cervical ligament (*ligamentum nuchae*)** supporting the neck of ruminants is a large cable-like strand of elastin fibers arranged in a parallel fashion. Aggregations of elastin fibers, such as in the *ligamentum nuchae,* have a characteristic yellow color. Elastin fibers are easily stretched and, when the tension is released, return to their original length.

Although the amino acid composition of elastin differs from that of collagen, glycine is still present in the greatest quantity. Elastin contains about eight amino acid residues of each of two unique amino acids, desmosine and isodesmosine, per 1000 residues. The extreme insolubility of elastin is due largely to its high content (about 90 percent) of nonpolar amino acids and desmosine cross-links. Elastin is highly resistant to digestive enzymes, and cooking has little solubilizing effect on it; thus this protein contributes little to the nutritive value of meat.

Mesenchymal Cells

Mesenchymal cells are spindle-shaped cells that are slightly smaller than fibroblasts. This relatively nondescript population of cells can differentiate into any one of several different types of cells, depending on the specific environmental stimuli. They are precursors of fibroblasts as well as the cells that accumulate lipids, known as **adipoblasts**. It should be emphasized that fibroblasts per se do not directly give rise to adipose cells. Rather, mesenchymal cells give rise to adipoblasts, which are immediate precursors of fat cells. Adipoblasts are located in the loose connective tissue matrix close to blood vessels. As these primitive cells accumulate fat, they eventually become adipose cells (**adipocytes**).

Adipose Tissue

When a tissue accumulates adipocytes to the extent that they are the predominant cell type, **adipose tissue** or fat occurs. Many such locations in the body, known as adipose tissue depots, are present in the animal body. The composition of several adipose tissue depots of meat animals is presented in Table 2.3.

In many animal species, the body contains two types of adipose tissue; **white fat** and **brown fat**. The predominate type of adipose tissue in meat animals is white, however, appreciable amounts of brown fat are present in mammals, especially around birth but this tissue is normally converted to white fat shortly after birth even though some brown fat persists into the adult stage in some species. Brown adipocytes are smaller in size than those of white fat, and their color is mainly due to a high content of cytochrome contained in the mitochondria.

When the primitive adipose cell begins to accumulate lipids it is known as an adipoblast; and, when it is filled with lipid, it is a mature adipocyte. As the adipoblast begins to accumulate lipid in the form of droplets near the center of the cell, the droplets fuse and form one large lipid globule. The cytoplasm, nucleus, mitochondria, and other organelles are pushed to one side of the cell as it becomes filled with lipid (Figures 2.23 and 2.32a). The mature adipocyte is large, having a diameter of up to 120 um, whereas the diameter of the primitive cell is only 1 to 2 um. Adipose tissue gradually develops into lobes and lobules that are enclosed in a delicate sheath of collagenous fibers and supplied with a network of blood capillaries (Figure 2.32b,c,d).

Adipoblasts develop at widely varying rates in different parts of the body, which results in marked variations in proportions of fat present in different body locations. In young animals, deposits of fat usually appear first in visceral areas. Then, if nutrient intake is adequate, fat is deposited beneath the skin (**subcutaneous**), between muscles (**intermuscular**), and between muscle bundles (**intramuscular**). Intramuscular fat deposits are the latest to develop and occur in fat cells associated with perimysial connective tissues. These fat cells accumulate and grow between the bundles of muscle fibers in close proximity to the arterioles that permeate the muscle. Adipose tissues continue to develop in growing and adult animals as

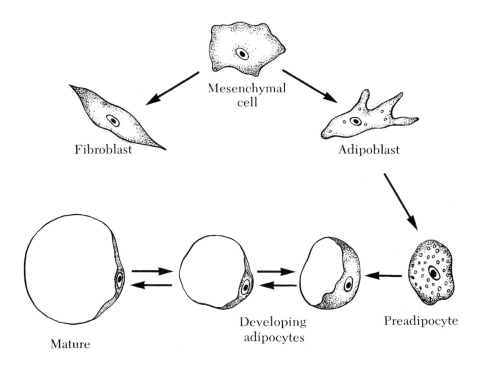

FIGURE 2.23 | Diagrammatic representation of adipocyte development. Courtesy of W. H. Freeman and Company.

a smaller proportion of nutrient intake is required for growth of other tissues and a greater proportion is available for energy storage. Therefore, fat cells are numerous and widespread in connective tissues of normal, healthy, adult animals. In very obese animals, hyperplasia to form new adipoblasts may occur repeatedly in phases.

During the fattening phase of growth, the percentage of lipid in adipose tissue increases and the percentages of water, protein, and other constituents decrease. However, lipid content of all adipose tissues does not increase at the same rate. For example, the pig has three distinct layers of backfat, the middle of which increases in lipid content most rapidly, while the inner layer shows the slowest rate of increase.

Adipose tissue is dynamic, which means that lipid is being constantly stored and mobilized. Extensive metabolic activity requires a rich supply of blood and enzyme systems. Most adipose tissues represent storage of energy that may be drawn on in response to needs of other tissues. During periods of poor nutrition and, consequently, of limited energy intake, adipocytes release their stored lipid into the vascular system. If energy intake is restricted over a prolonged period of time, adipocytes revert to aggregations of irregularly shaped cells within supporting connective tissues of the depot. These cells retain their ability to again accumulate lipid if and when the animal's energy intake is restored to a level above that needed for maintenance. This series of events may occur many times throughout the life of the animal.

Cartilage

Cartilage and bone are specialized connective tissues and constitute the supportive elements of the animal body. During embryonic development, most of the skeleton is formed as cartilage and is later converted to bone. However, not all cartilage is converted to bone, and some, such as the articular surfaces of joints, persist throughout the life of the animal.

Cartilage is composed of cells and extracellular fibers that are embedded in a gel-like matrix. The predominant cell type found in cartilage is **chondrocytes** that exist as clusters of cells isolated in small cavities within the extracellular matrix. These features of cartilage are shown in Figure 2.24. Collagen fibrils form a fine network in the cartilage matrix. These fibrils constitute as

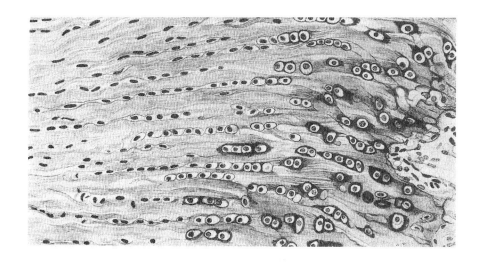

FIGURE 2.24 | Diagrammatic representation of fibrocartilage at the insertion of tendon into bone. From D.W. Fawcett, *A Textbook of Histology*, 9th ed. W. B. Saunders Company, Philadelphia, 1968, 219.

much as 40 percent of the cartilage, but normally are not visible since they are completely surrounded by the matrix.

Cartilage differs in the relative amount of collagen and elastin fibers and in the abundance of the matrix in which the fibers are embedded. These differences allow cartilage to be categorized as **hyaline**, **elastic**, or **fibrocartilage**. Hyaline cartilage is found on joint surfaces, on the ventral ends of ribs (costal cartilages), on the dorsal tips of vertebrae, between individual vertebrae and in other locations. Hyaline cartilage is relatively inelastic, semitransparent with a bluish-white tint. Elastic cartilage, on the other hand, is less transparent, more elastic (more flexible) and has a more yellowish tint than hyaline cartilage. Its matrix has a number of branched elastin fibers. Elastic cartilage is a component of the epiglottis as well as the internal and external portions of the ear. Fibrocartilage is characterized by the presence of numerous collagenous fibers and is found at the attachment of tendons to bone and within the capsules and ligaments of joints.

Bone

Bone, like other connective tissues, contains cells, fibrous elements, and the extracellular matrix (ground substance). Unlike other connective tissues, the extracellular matrix is calcified. This provides rigidity and protective properties characteristic of the skeleton. Bones are storage sites for calcium, magnesium, sodium, and other ions.

In long bones, such as the femur, the long central shaft, called the **diaphysis**, is a hollow cylinder of compact bone. On both ends of the diaphysis are enlargements of the bone, called the **epiphyses**. The entire bone is covered with a thin membrane of specialized connective tissue called the **periosteum**. The periphery of the epiphysis is covered with a thin layer of **compact bone**, but the inside is composed of **spongy bone**. These features of bone are illustrated in Figure 2.25. Articulating surfaces of the epiphyses are covered with a thin layer of hyaline cartilage, called the **articular cartilage**. The joint formed by the articular cartilage portions of adjacent bones contains the specialized connective tissue substance known as **synovial fluid**. The cartilaginous region separating the diaphysis and epiphysis of growing animals is called the **epiphyseal plate**. Cartilage of the epiphyseal plate lies adjacent to a transitional region of spongy bone in the diaphysis called the **metaphysis** (Figure 2.25). Spongy bone of the metaphysis and adjacent epiphyseal cartilage constitute the region of the bone where growth in length occurs. The central interior of the diaphysis contains marrow that, in the young animal, is primarily red marrow. As the animal grows older, however, red marrow gradually is converted to yellow marrow.

In compact bone, the calcified bone matrix is deposited in layers (Figure 2.26). Small cavities are distributed throughout the matrix and contain cells known as **osteocytes** and **osteoblasts**. Tubular elements called **canaliculi** radiate in all directions forming networks of canals between the cell cavities. This network of canaliculi is important in bone cell nutrition. Osteocytes are the principal cells of fully formed bone, and they are located in the cell cavities within the calcified matrix. Osteocytes are primarily involved with bone formation and osteoclasts with the process of bone resorption.

Two principal components of bone matrix are the organic matrix and inorganic salts. The organic matrix consists of the ground substance that contains proteoglycan complexes. Embedded in the ground substance are numerous collagenous fibers. The inorganic component of bone matrix consists primarily of calcium phosphate salts. Crystals of these calcium salts are deposited in the organic matrix between the collagen fibers.

Bone is formed, both in prenatal and postnatal life, by a transformation of connective tissue. It is formed directly or indirectly by cells known as osteoblasts, fibroblasts, and chondroblasts. **Osteoblasts** are directly active in production of bone. **Fibroblasts** are primarily responsible for synthesis of fibers and the ground substance of tendons, ligaments, and connective tissue proper. **Chondroblasts** form cartilage. Both prenatal and postnatal formation of bone may be preceded by formation of a cartilage "model," or it may occur by direct transformation of connective tissue. When a bone is formed by replacement of cartilage, the process

CHAPTER 2 STRUCTURE, COMPOSITION, AND DEVELOPMENT OF ANIMAL TISSUES

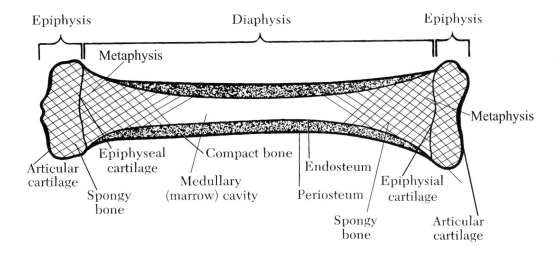

FIGURE 2.25 | Diagram of a long bone showing the longitudinal structure. Courtesy of W. H. Freeman and Company.

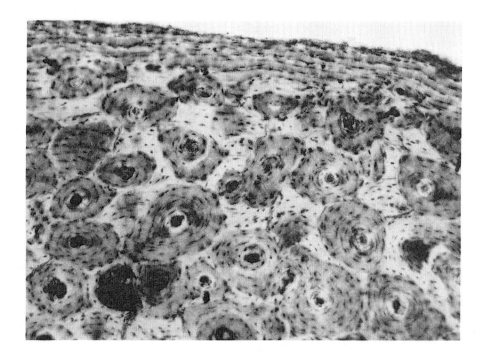

FIGURE 2.26 | Photomicrograph of midshaft section of compact bone. Dark circular structures are Haversian systems. Courtesy of D. C. Van Sickle, Department of Veterinary Anatomy, Prudue University.

is known as **endochondral ossification**. Bones formed by this process include those of the limbs. Growth in length of bones occurs by this process. When bone formation occurs directly, in the absence of cartilage, the process is known as **intramembranous ossification**. Bones of the skull are formed by this process. In intramembranous bone formation, ossification occurs in and replaces connective tissue. An example is the widening of bone that occurs within or beneath the periosteum. The series of changes in growth of bone is illustrated in Figure 2.27.

At birth, a long bone such as the femur or tibia has a diaphysis with a **medullary cavity** and cartilaginous extremities. Ossification centers appear in the cartilaginous ends to begin formation of the bony epiphyses. The cartilage between the diaphysis and epiphysis does not ossify immediately; it becomes the epiphyseal cartilage. Postnatal growth in length of a typical long bone results from growth of cartilage on the epiphyseal side of the epiphyseal cartilage and subsequent ossification of cartilage on the diaphyseal side. Growth in thickness occurs by continuous deposition beneath the periosteum. Osteoblasts are responsible for bone deposition in the various sites of bone development, and osteocytes help to maintain it.

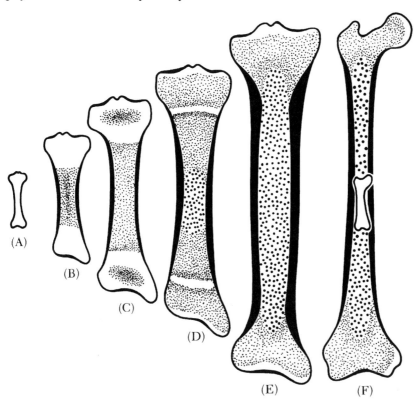

FIGURE 2.27 | Progressive ossification and growth of a long bone. (A) The cartilaginous stage. (B) Deposition of spongy, endochondral bone (stipple), and the compact, perichondral bone (black). (C) An epiphysis appears at each end. (D) The marrow cavity (sparse stipple) appears as endochondral bone is resorbed. (E) Each epiphysis ossifies, leaving articular cartilage at both ends. Notice that the enlargement of the marrow cavity continues by the resorption of bone centrally as deposition continues on the periphery. (F) A bone at birth superimposed over the same bone of an adult to show their relative sizes, and the amount of deposition and internal resorption that occurs during growth. Courtesy of W. H. Freeman and Company.

Concurrently with bone growth in length and diameter, the marrow cavity becomes enlarged due to the continuous resorption of the medullary wall by activity of **osteoclasts**. This cavity is occupied by marrow, a soft tissue composed of networks of delicate reticular connective tissue filled with various cells. Two types of marrow are recognized: red and yellow. **Red marrow** is the chief blood cell forming organ of the adult body and is found in spongy bone of the epiphyses of long bones, and the main bodies of the sternum, ribs, and vertebrae. **Yellow marrow** consists mainly of adipose tissue and is found in the medullary cavity of bones. In the adult animal, the site of the original diaphysis has been replaced by the medullary cavity, and the bone shaft has been derived from intramembranous growth of bone beneath the periosteum.

It is by control of the two processes, deposition and resorption, that the size and shape of the skeleton is changed during growth. These opposing balanced processes continue during adult life and are responsible for a continuous remodeling of bone tissue. When the bone has attained its full size, growth ceases. The epiphyseal cartilage stops growing and becomes completely ossified, fusing the epiphyses and diaphysis into a single bone. Deposition of additional layers of compact bone beneath the periosteum ceases. Termination of growth does not occur simultaneously in all bones, i.e., the sacral vertebrae reach maturity before the lumbar, and the lumbar before the thoracic vertebrae. The full extent of postnatal bone growth may be seen by comparing a bone at birth to that of an adult animal, as shown in Figure 2.27f.

Blood And Lymph

Blood and lymph, and their respective vessels, are derived from connective tissues. It already has been noted that connective tissues are present in walls of blood and lymph vessels. Blood provides a medium, both for nutrient (including oxygen) transport to cells and for removal of waste byproducts. Blood also serves as a defense mechanism against infectious agents and as a regulator of pH and fluid balance.

Blood consists of a fluid medium, plasma, in which various cellular components are suspended. Blood cells are generally classified as **platelets**, **red cells (erythrocytes)**, and **white cells (leukocytes)**. These cells constitute approximately 40 percent of total blood volume in meat animals. The primary function of erythrocytes is that of carrying oxygen. The function of leukocytes within the bloodstream is largely unknown. Outside the vascular system, some leukocytes are **phagocytic** (digest foreign matter) and others function in immune reactions. Platelets participate in the blood clotting mechanism. Erythrocytes contain **hemoglobin** that is directly responsible for oxygen transport to cells and gives blood a characteristic red color. Under normal physiological conditions, blood constitutes roughly 7 percent of the body weight of mammals.

Lymph is fluid found within lymphatic vessels. Cells in lymph are predominately lymphocytes, although a few erythrocytes and leukocytes may be present. Lymph circulates continually and passes through all tissues, organs, and lymph nodes of the body, into the thoracic duct, and ultimately into the bloodstream.

MUSCLE ORGANIZATION

Thus far, discussion of skeletal muscle has dealt with the muscle fiber, its organelles, and their relationship to overall functional properties. In skeletal muscles, groups of muscle fibers are bound together by connective tissues (Figure 2.3). A small amount of epithelial and nervous tissue is also associated with skeletal muscles. The organization, role, and function of each of these tissues in muscle are the topics of the following section.

Muscle Bundles and Associated Connective Tissues

In structural terms, muscle is composed of many individual fibers that are grouped together into **bundles** or **fasciculi**. The process of grouping of individual fibers into bundles begins during prenatal muscle development. The number of fibers varies from one bundle to another; consequently, bundle size is variable. A number of bundles, in turn, are grouped in various patterns to form a muscle. Cross-sectional inspection of muscle reveals that bundles appear as irregularly shaped polygons (Figure 2.3). They assume these shapes largely because of mutual deformation, but pressure of connective tissue septa (partitions) between adjacent bundles also affects their shape. The size of the bundles, and the thickness of their connective tissue septa, determine the texture of muscle; those with small bundles and thin septa have fine texture, and muscles with large bundles and thicker connective tissue septa have coarse texture. The finer the texture, the greater is the precision of movement in the living muscle.

The sarcolemma (outer cell membrane) of individual muscle fibers is surrounded by the **endomysium**, a delicate connective tissue covering shown in Figures 2.3 and 2.28. It should be emphasized that the sarcolemma and endomysium are two separate and distinct structures, even though both encase the muscle fiber. The endomysium is composed of fine collagen fibers.

Approximately 20 to 40 muscle fibers, and the associated endomysium, are grouped into structures called **primary bundles**. A variable number of primary bundles are grouped together to form larger bundles known as secondary bundles. Both primary and secondary bundles are ensheathed in the **perimysium**, a sheath of collagenous connective tissue (Figures 2.3 and 2.29). Finally, a variable number of secondary bundles are grouped together to form a muscle, which is itself contained in a connective tissue sheath called the **epimysium**. The collagen fibers of the epimysium are continuous with those of the perimysial septa surrounding individual bundles that, in turn, are continuous with collagen fibers of the endomysial septa, which surround individual muscle fibers. These connective tissue septa bind the fibers and bundles together and they support the associated nerves and blood vessels. A few fine elastin fibers frequently are found associated with collagen fibers in these connective tissue sheaths. These features of muscle are illustrated in Figures 2.3 and 2.30. Individual muscles in a group, such as in limbs, may have interconnections in the form of loose connective tissue septa between the epimysial sheaths of adjacent muscles.

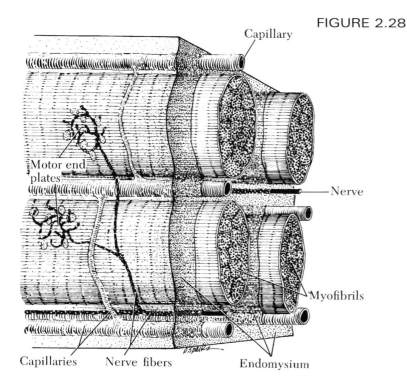

FIGURE 2.28 | Diagrammatic illustration of the structure of a primary muscle bundle, showing the relationship of fibers, endomysium, blood vessels, and nerves. Modified from J. E. Crouch, *Functional Human Anatomy*, 2nd ed. Lea & Febiger, Philadelphia, 1985, 183.

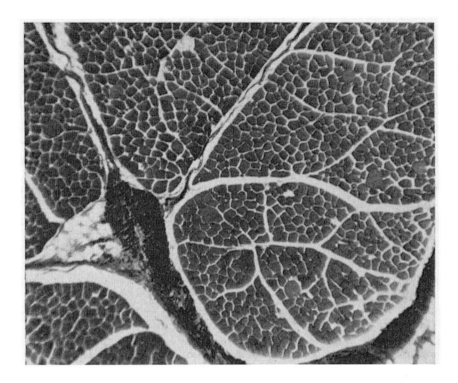

FIGURE 2.29 | Photomicrograph of a bovine semitendinosus muscle showing muscle fibers, primary and secondary bundles, and the associated endomysial and perimysial connective tissues (x 50). From American Meat Institute, *The Science of Meat and Meat Products*. W. H. Freeman and Company.

Nerve and Vascular Supply

Each muscle is supplied with at least one artery and vein, as well as with nerve fibers from a nerve trunk (Figure 2.2). The latter contains both sensory and motor fibers. Nerve fibers and blood vessels supplying the muscle are associated with the epimysium and they enter its interior via the perimysium. In turn, branches that arise from these nerves and blood vessels supply individual muscle fibers and are supported by the endomysium (Figure 2.28). The finer arterioles and venules are oriented transversely to the muscle fibers, and the bulk of the capillaries are arranged parallel to the long axis of the fibers. Each fiber is supplied with a rather extensive vascular bed composed principally of these longitudinally oriented capillaries, lying in the endomysium, with occasional transverse branches encircling them as shown in Figure 2.31. This arrangement of blood vessels provides extensive coverage of the muscle fiber surface for exchange of nutrients and waste products of metabolism.

Like blood vessels, lymph vessels enter and exit the muscle via connective tissue septa. Similarly, nerve fibers and their branches enter and leave the muscle along avenues of connective tissue septa. As a nerve fiber approaches the interior of a muscle, its myelin sheath is lost. The nerve axon and its associated Schwann cell and connective tissue sheaths are found in, and become continuous with, the endomysium surrounding muscle fibers. Axon endings make contact with muscle fibers at the **motor end plate** (Figure 2.28). Each nerve fiber may branch and innervate numerous muscle fibers.

Intramuscular fat, called **marbling** in meat, is deposited within the muscle in loose networks of perimysial connective tissue in close proximity to blood vessels, as shown in Figure 2.32. Likewise, epimysial connective tissue between individual muscles contains variable quantities of adipose tissue. This latter type of fat deposit is referred to as **intermuscular fat** or **seam fat**.

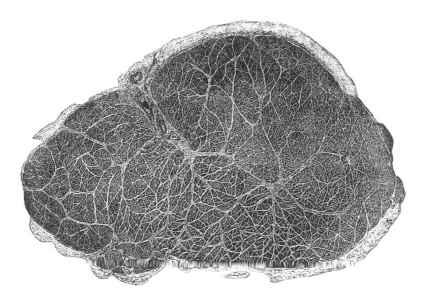

FIGURE 2.30 | A photomicrograph of a human sartorius muscle in cross section, showing its subdivision into bundles of various sizes, and the endomysial, perimysial, and epimysial connective tissues (x 3). From D. W. Fawcett, *A Textbook of Histology*, 9th ed. W. B. Saunders Company, Philadelphia, 968, 271.

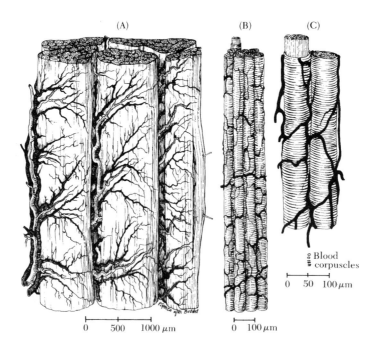

FIGURE 2.31 | A diagrammatic illustration of the vascular supply to, and network associated with, skeletal muscle fibers. (A) The vascular network that surrounds the muscle fiber bundles. (B) The capillary network supplying muscle fibers. (C) A close-up view of the capillary network and two muscle fibers. From M. Brodel, *Johns Hopkins Hosp. Bull.* 61: 295, 1937. © The Johns Hopkins University Press.

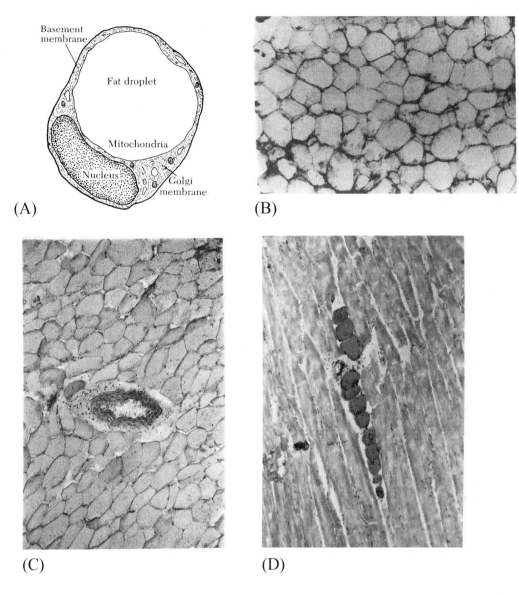

FIGURE 2.32 | Diagram and photomicrographs of fat cells. (A) Diagram of a typical fat cell. (B) Photomicrograph of bovine subcutaneous fat cells. (C) Intermuscular bovine fat cells surrounding an artery. (D) Intramuscular bovine fat cells arranged longitudinally between muscle bundles. Photomicrographs courtesy W. G. Moody, University of Kentucky.

Muscle-Tendon Junction

Structural details of how the force of contracting myofibrils is transmitted to the muscle tendon are unresolved. However, it is believed that the ends of myofibrils are connected to tendon fibrils. As the myofibrils approach the tapered ends of the muscle fiber, the characteristic banding pattern is lost and the myofibrils become continuous with strands of non-contractile fibers (Figure 2.33). Tendon fibrils are inserted in the sarcolemmal membrane, opposite the point at which non-contractile connecting fibers of each myofibril are attached. This entire **myotendinal structure** is continuous with the tendon. The epi-, peri-, and endomysial network of connective tissue sheaths encases the muscle; in effect, its bundles and fibers form a harness that transmits contraction force to the tendon and, ultimately, to the bone. Tendinal attachments, as well as myotendinal junctions, are referred to as **aponeuroses**. However, some muscles are attached by connective tissue *fasciae* to bones and to other structures such as skin, ligaments, and other muscles. Such fasciae are composed primarily of white collagenous fibers with a few elastin fibers and form a sheet of connective tissue that performs the function of the tendon in aponeuroses.

Muscle Fiber Growth

During embryonic and fetal development, fibers in a muscle increase in number and become grouped into distinct bundles. Individual fibers increase only slightly in diameter during the first two-thirds of the prenatal period. Most of the increase in muscle weight during this period is due to **hyperplasia** (increase in number of fibers). During the last one-third of the prenatal period, **hypertrophy** (increase in size of existing fibers) contributes comparatively more to muscle growth than it does during the initial two-thirds. Developing muscle fibers grow in length by the addition of complete sarcomere units to the ends of existing myofibrils.

The greatest increase in muscle size occurs after birth, but the rate of increase declines as animals approach maturity. Muscle fibers grow by increasing both their diameter and length. After birth, the number of muscle fibers does not increase appreciably, even though exceptions to this rule have been shown to exist in some mammals. Therefore, postnatal muscle growth is accomplished primarily through hypertrophy of existing muscle fibers. Diameter of individual muscle fibers is increased by both increases in number and size of myofibrils. The number of myofibrils within a single muscle fiber may increase by 10 to 15 times during the lifetime of the animal. This proliferation of myofibrils occurs through longitudinal splitting of large myofibrils into two small daughter myofibrils, which may subsequently enlarge through addition of myofilaments. By repeating this process, a single myofibril may split four or more times.

Maximum muscle fiber diameter of the pig is attained after approximately 150 days of postnatal growth (Figure 2.34). Obviously, the age at which maximum fiber diameter is attained varies among individual animals and species, but all follow similar growth patterns. Animals that mature at an early age attain maximum muscle fiber diameter sooner than animals that mature at later ages. On the other hand, animals that attain large adult size at maturity do not have larger diameter muscle fibers than relatively small animals. Their large muscle mass is caused by greater numbers of fibers rather than the diameter of individual fibers. Therefore, muscle fiber size is not proportional to body size among or within species.

Growth in length of a muscle fiber generally precedes growth in diameter. Increases in muscle fiber length are accomplished postnatally both by increases in length of existing sarcomeres and by addition of new sarcomere units. However, the major portion is due to formation of new sarcomeres. Sarcomere number may increase more than threefold from birth to maturity, and new sarcomeres are added at the tendon-ends of myofibrils in skeletal muscle. There is good evidence that, after addition of new sarcomeres is complete, the muscle cannot increase its length by more than 20 percent through sarcomere lengthening.

In spite of large amounts of myofibrillar protein added to muscle fibers during growth, the protein-DNA ratio undergoes increases that are far less than would be expected from the amount of protein deposited. This indicates that nuclei are increasing in numbers in existing muscle fibers, even though myonuclei are no longer able to divide. The source of new nuclei in myofibers is a specialized muscle cell known as the **satellite cell**. These cells are located between the basement membrane and the sarcolemma of myofibers. They are capable of proliferating, differentiating, and fusing with muscle fibers. In this way, satellite cells provide a growing muscle fiber with added DNA, which ultimately increases the capacity of the muscle fiber to synthesize protein and grow. More than 95 percent of the total nuclei found in mature myofibers are derived from satellite cells. Satellite cells are also responsible for the formation of new myofibers during postnatal muscle growth.

Muscle Size

Diameter and length of individual muscles increase during growth as consequences of the growth of muscle fibers discussed above. In addition, some of the dimensional changes in muscles may occur as the result of shifts in fiber-to-fiber relationships. If some fibers terminate within muscle bundles rather than on tendinous structures, their lengthening increases muscle diameter and cross-sectional area. Invasion of muscle connective tissues by fat cells (Figure 2.32d) also increases the gross diameter of muscles.

Connective tissue increases in quantity in epimysial, perimysial, and endomysial locations during muscle growth. Collagen and elastin fibers are interspersed. Their relative and total quantities depend on specific muscle location and function. Even though the quantity of connective tissue increases during growth, the mass of muscle fibers increases more rapidly, resulting in dilution or reduction of percentage of connective tissue in muscle. Nevertheless, the tensile strength of connective tissue increases because of chemical crosslinking, which occurs in collagen with advancing age. When muscle fiber degeneration occurs during senescence, the percentage of connective tissue again increases.

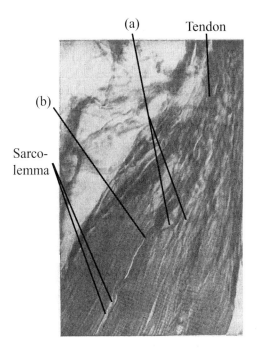 

FIGURE 2.33 | Photomicrographs of the muscle-tendon junction. (A) Visible light photomicrograph showing a muscle-tendon connection (x 500). (a) and (b) indicate tendon fibrils that connect to the sarcolemma. (B) An electron photomicrograph of the myotendinal junction showing the relation of the sarcomere to tendon collagen fibers. (A) From Maximow and Bloom, *Textbook of Histology*, 6th ed. W. B. Saunders Company, Philadelphia, 1952, 149. (B) D. W. Smith, Muscle Academic Press, Inc., New York, 1972, 56. Courtesy of Dr. D. S. Smith.

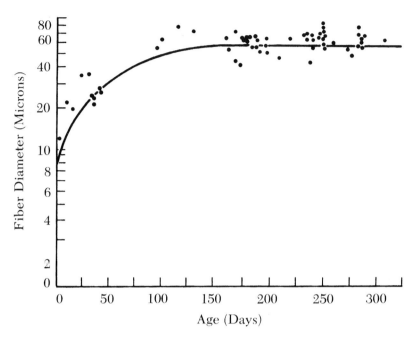

FIGURE 2.34 | Change in swine longissimus muscle fiber diameter with increasing age. Data from B. B. Chrystall, University of Missouri. Courtesy of W. H. Freeman and Company.

After animals are fully developed, muscle and fiber size may increase or decrease due to work-induced hypertrophy or atrophy (decrease in size) due to inactivity. As animals grow older and enter senescence, the total number of muscle fibers decreases through atrophy while remaining fibers become larger. Hypertrophy probably compensates for loss of function originally performed by the atrophied fibers. Thus, skeletal muscles in old age contain fewer fibers that are larger in diameter.

Muscle Fiber Types

As stated earlier, skeletal muscle consists of a heterogeneous population of myofibers. Classically, muscles are usually classified as red or white, based principally on their fresh color intensity, which is attributed to the proportion of **red** and **white fibers** they contain. It should be emphasized that few muscles are composed solely of red or white fibers, but rather most contain a mixture of red and white fibers. Additionally, most muscles of meat animals contain a higher proportion of white fibers than red fibers, even in muscles that are visibly red. Thus, red muscles are those with a higher proportion of red fibers than those found in white muscles; or alternatively, white muscles have fewer red fibers than do red muscles. Leg muscles of chickens and turkeys appear red, whereas breast muscles appear almost white because of the relatively high proportion of white fibers. A few muscles, such as the semitendinosus in pigs, exhibit regions that are predominately red while other portions of the same muscle are distinctly white.

In addition to classification as red or white, muscle fibers are classified based on their inherent **ATPase activity**, which is indicative of the predominant **myosin isoform** of the cell. A protein isoform simply refers to a protein that is closely related in structure and function to other proteins in the same family. However, subtle differences in amino acid sequence create differences in optimal protein functionality. In the case of myosin, four adult myosin isoforms have been identified in skeletal muscle: **type I**, **IIa**, **IIx(d)** and **IIb**. Therefore, each fiber can be classified according to the predominate type of myosin contained within a fiber. Type I and type IIa fibers are considered

CHAPTER 2 STRUCTURE, COMPOSITION, AND DEVELOPMENT OF ANIMAL TISSUES

TABLE 2.1 Characteristics of Muscle Fibers in Domestic Meat Animals and Birds*

Characteristic	Type I	Type IIa	Type IIx(d)	Type IIb
Reddness	++++	+++	+	+
Myoglobin content	++++	+++	+	+
Fiber diameter	+	+	+++	+++
Contraction speed	+	+++	+++	++++
Fatigue resistance	++++	+++	+	+
Contractile action	tonic	tonic	phasic	phasic
Number of mitochondria	++++	+++	+	+
Mitochondria size	++++	+++	+	+
Capillary density	++++	+++	+	+
Oxidative metabolism	++++	++++	+	+
Glycolytic metabolism	+	+	+++	++++
Lipid content	++++	+++	+	+
Glycogen content	+	+	++++	++++
Z disk width	++++	+++	+	+

* The characteristics are relative to the other fiber types.

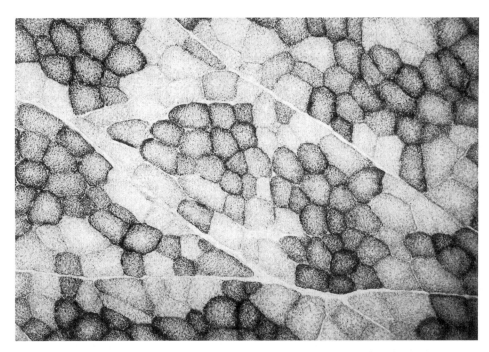

FIGURE 2.35 Photomicrograph of cross-section of porcine longissimus muscle, showing red and white fibers (dark and light, respectively) (× 175).

red muscle fibers whereas type IIx(d) and type IIb fibers are considered white muscle fibers. Another property commonly used to characterize muscle fibers is related to how fast the fibers contract when stimulated. Obviously all muscles contract rapidly, but differences between fast and slow contracting fibers exist, albeit only by fractions of a second. In accordance with the aforementioned fiber types, type I fibers are slower contracting fibers whereas type II fibers are fast contracting.

The structural, functional, and metabolic characteristics of the four adult muscle fiber types are different, and some of these features are presented in Table 2.1. However, these differences are relative, and considerable variation exists for each characteristic within each fiber type; thus, some overlap between types occurs. A photomicrograph illustrating different fiber types is shown in Figure 2.35. In most mammalian species, red and white fibers and type I and II fibers are more or less randomly intermingled within muscle bundles. However, in pig muscle red fibers are located near the center of the bundles in discrete groups that are surrounded by white fibers that form the periphery of the bundles.

The higher **myoglobin** content of red fibers, as compared to white fibers, accounts for their red color. Oxygen storage by myoglobin is consistent with the high proportion of enzymes involved in **oxidative metabolism** (requires oxygen for energy production) and the low levels of **glycolytic** enzymes found in red fibers. White fibers, on the other hand, have a high content of glycolytic enzymes and a low oxidative enzyme activity. Consistent with metabolic activity, the mitochondria (which contain oxidative enzymes) of red fibers are more numerous and larger in size than in white muscle fibers. Red fibers also have greater capillary density, which facilitates transfer of "metabolic wastes" and nutrients (particularly oxygen) to and from the vascular bed. Their smaller fiber diameter reduces the diffusion distances these materials must travel in red as compared to white fibers. Red fibers also have greater lipid content, some of which presumably serves as a source of metabolic fuel, and a lower content of glycogen than white fibers. **Glycolytic metabolism**, which predominates in white fibers, may occur in either the presence (aerobic) or absence of oxygen (anaerobic). Thus, white fibers have a lower capillary density than red fibers. White fibers have a more extensively developed sarcoplasmic reticulum (SR) and T tubule system, both of which are consistent with their more rapid contraction speed. White fibers also have narrower Z disks than red fibers.

White fibers have a phasic mode of action; they contract rapidly in short bursts and they are relatively easily fatigued. Red fibers contract more slowly, but for a longer time. This slower but sustained mode of action is generally referred to as tonic (continual tension) contraction. Red fibers are functionally important in posture and, because of their metabolism, they are less easily fatigued, so long as an oxygen supply is available.

All muscle fibers appear to be red at birth, but shortly thereafter some of them differentiate into white and intermediate fiber types. Evidence suggests that the nerve which innervates a given muscle fiber determines, in part, whether it differentiates into a white or intermediate fiber. Other evidence suggests that relative time of embryological development and sequence of morphological changes experienced by developing fibers are associated with its ability to undergo postnatal changes in fiber type.

CHEMICAL COMPOSITION OF THE ANIMAL BODY

The animal body normally contains about one-third of the approximately 100 chemical elements. About 20 of these are essential to life. An additional 30 to 35 elements have been detected in mammalian and avian tissues, but their presence is probably coincidental with their presence in the animal's environment during its lifetime. The most abundant elements (by weight) in the animal body are those present in water and in the organic compounds, such as proteins, lipids, and carbohydrates: oxygen, carbon, hydrogen, and nitrogen. These four elements account for approximately 96 percent of total body chemical composition, as shown in Table 2.2. Because water constitutes 60 to 90 percent of the soft tissues in the animal body, the reason for the high percentage of oxygen

TABLE 2.2 — Elemental Composition of the Animal Body

Major elements	Percent of body weight	Trace elements	
Oxygen	65.0	Aluminum	Lithium
Carbon	18.0	Arsenic	Manganese*
Hydrogen	10.0	Barium**	Molybdenum*
Nitrogen	3.0	Boron	Nickel
Calcium	1.5	Bromine**	Rubidium
Phosphorus	1.0	Cadmium**	Selenium*
Potassium	0.35	Chromium**	Silicon
Sulfur	0.25	Cobalt*	Silver
Sodium	0.15	Copper*	Strontium**
Chlorine	0.15	Fluorine**	Titanium
Magnesium	0.05	Iodine*	Vanadium
		Iron*	Zinc*
		Lead	

*Essential microelements.
**Microelements present in the animal body but essential functions are not known. (The microelements without asterisks are present in measurable quantities, but they have no known metabolic function.)

and hydrogen is obvious. Lipid and carbohydrate materials are primarily combinations of carbon, hydrogen, and oxygen atoms. Protein molecules contain these three elements plus nitrogen, sulfur, phosphorus, and small quantities of several other elements. Numerous other elements are present in the body as inorganic constituents (Table 2.2). The abundance of these inorganic elements varies from a high of 1.5 percent of the total body weight, as in the case of calcium, to barely detectable quantities. However, the total amount of an element in the body is not necessarily an accurate indication of its functional importance.

Water

Water is the fluid medium of the body, and some also is associated with cellular structures, in particular, with colloidal protein molecules. Water serves as a medium for transporting nutrients, metabolites, hormones, and waste products throughout the body. It also is the medium in which most chemical reactions and metabolic processes of the body occur.

Proteins

Proteins constitute a very important class of chemical compounds in the body. Some are necessary for its structure and others function in vital metabolic reactions. Except in fat animals, protein ranks second only to water in abundance (by weight) in the animal body. Most of the protein is located in muscle and connective tissues. Proteins of the body vary in size and shape; some are globular while others are fibrous. Structural differences in protein molecules contribute to their functional properties. For example, fibrous proteins form structural units and globular proteins include numerous enzymes that catalyze metabolic reactions.

Lipids

The animal body contains several classes of lipids, but the **neutral lipids** (fatty acids and glycerol) predominate. Of the various lipids in the body, some serve as sources of energy for the cell; others contribute to cell membrane structure and function;

TABLE 2.3 Fatty Acid and Triglyceride Composition of Some Animal Fat Depots (Percent by Weight)

Component	Chicken*	Pig SQ** Outer	Pig SQ** Inner	Pig PR***	Cattle SQ	Cattle PF	Sheep SQ	Sheep PR	Fish***** Herring oil
Fatty acids									
Lauric	—	trace	trace	trace	0.1	0.2	0.1	0.1	—
Myristic	0.1	1.3	0.1	4.0	4.5	2.7	3.2	2.6	6.7
Palmitic	25.6	28.3	30.1	28.0	27.4	27.8	28.0	28.0	11.5
Stearic	7.0	11.9	16.2	17.0	21.1	23.8	24.8	26.8	1.4
Arachidic	—	trace	trace	trace	0.6	0.6	1.6	2.6	—
Total saturated	32.7	41.5	47.3	49.0	53.7	55.1	57.7	59.5	19.6
Palmitoleic	7.0	2.7	2.7	2.0	2.0	2.2	1.3	1.9	8.6
Oleic	20.4	47.5	40.9	36.0	41.6	40.1	36.4	34.2	14.4
Linoleic	21.3	6.0	7.1	11.8	1.8	1.8	3.5	4.0	1.2
Linolenic	—	—	0.3	0.2	0.5	0.6	0.5	0.6	0.9
Parinaric	—	—	—	—	—	—	—	—	1.45
Gadoleic	—	—	—	—	—	—	—	—	14.3
Arachidonic plus chupandonic	0.6	2.1	1.7	1.0	0.4	0.2	0.6	0.8	1.94
Cetoleic	—	—	—	—	—	—	—	—	22.0
Total unsaturated	67.3	58.5	52.7	51.0	46.3	44.9	42.3	41.5	73.76
Approximate triglyceride composition fully saturated									
Tripalmitin		1			3		trace		
Dipalmitostearin		2			8		3		
Palmitodistearin		2			6		2		
Tristearin		—			—		—		

Mono-oleo-disaturated			
Oleodipalmitin	5	15	13
Oleopalmitostearin	27	32	28
Oleodistearin	—	2	1
Dioleo-monosaturated			
Palmitodiolein	53	23	46
Stearodiolein	7	11	7
Tri- unsaturated Triclyride			
Triolein	3	0	0

*Fat associated with the skin.
**Subcutaneous (SQ) fat from the dorsal, thoracic, and lumbar regions.
***Perirenal (PR) fat around the kidneys.
****From Pearson et al. Adv. Food Res. 23: 1(1977).

TABLE 2.4 — Approximate Composition of Mammalian Skeletal Muscle (Percent Fresh Weight Basis)

Component (range)	Percent	Component	Percent
Water (65 to 80)	75.0	**Non-protein nitrogenous substances**	**1.5**
Protein (16 to 22)	18.5	Creatine and creatine phosphate	
Major contractile proteins		Nucleotides	
myosin		(adenosine triphosphate (ATP),	
actin		adenosine diphosphate (ADP),	
Regulatory proteins		(etc.)	
tropomyosin		Free amino acids	
troponin (complex)		Peptides	
tropomodulin		(anserine, carnosine, etc.)	
Cytoskeletal proteins		Other nonprotein substances	
titin		(creatinine, urea, inosine	
nebulin		monophosphate (IMP), nicotinamide	
C-protein		adenine dinucleotide (NAD),	
myomesin		nicotinamide adenine dinucleotide	
M-protein		phosphate (NADP))	
desmin		**Carbohydrates and non-nitrogenous**	
filamin		**substances** (0.5 to 1.5)	1.0
vinculin		Glycogen	
synemin		Glucose	
creatine kinase		Intermediates and products of	
α-actinin		cell metabolism	
H-protein		(hexose and triose phosphates,	
skelemin		lactic acid, citric acid, fumaric	
Cap Z		acid, succinic acid, acetoacetic acid,	
paranemin		etc.)	
dystrophin		**Inorganic constituents**	
talin		Potassium	1.0
Sarcoplasmic proteins		Total phoshorous	
soluble sarcoplasmic and		(phosphates and inorganic phosphorous)	
mitchondrial enzymes		Sulfur (including sulfate)	
myoglobin		Chlorine	
hemoglobin		Sodium	
cytochromes and flavo-		Others	
proteins		(including magnesium, calcium,	
Stromal proteins		iron, cobalt, copper, zinc,	
collagen		nickel, manganeze, etc.)	
elastin			
other insoluble proteins			

Lipids (range 1.5 to 13.0)	3.0
Neutral lipids	
Phospholipids	
Cerebrosides	
Cholesterol	

still others, such as some of the hormones and vitamins, are involved in metabolic functions. Most lipids in the body are found in various fat depots in the form of **triglycerides**, glycerol esters of long chain fatty acids. The fatty acid and triglyceride composition of some fat depots in several species of meat animals is presented in Table 2.3. Except for milk fat, fatty acids having chains of 10 or fewer carbon atoms are rarely found in animal fats. On the other hand, C16 and C18 (molecules with chains of 16 and 18 carbon atoms, respectively) fatty acids predominate, with C12, C14, and C20 acids present only in small quantities. Of the saturated fatty acids, palmitic and stearic (C16 and C18, respectively) predominate in animal fats. The predominant unsaturated fatty acids in animal fats are palmitoleic, oleic (Cl6 and C18, respectively, with one double bond), linoleic (C18, two double bonds), and linolenic (C18, three double bonds). Oleic acid is the most abundant fatty acid in the animal body. Of the triglycerides in animal body fats, those that contain one palmitic fatty acid and two oleic fatty acid molecules in each triglyceride molecule are most abundant, followed by triglycerides containing one molecule each of oleic, palmitic, and stearic acid.

Carbohydrates

The animal body is a poor source of carbohydrates, but most of those present are found in muscles and liver. **Glycogen**, the most abundant carbohydrate, is present in the liver (2 to 8 percent of the fresh liver weight), but muscle generally contains only very small amounts. Other carbohydrate compounds found in the animal body include intermediates of carbohydrate metabolism and glycosaminoglycans present in connective tissues. Although carbohydrates make up a small proportion of body weight, they have extremely important functions in energy metabolism and in structural tissues.

CHEMICAL COMPOSITION OF SKELETAL MUSCLE

Since muscle is the principal component of meat, a brief discussion of its composition is necessary. Like the animal body, muscle contains water, protein, fat, carbohydrate, and inorganic constituents (Table 2.4). Muscle contains approximately 75 percent water (range: 65 to 80 percent) by weight. Water is the principal constituent of extracellular fluid and numerous chemical constituents are dissolved or suspended in it. Because of this, it serves as the medium for transport of substances between the vascular bed and muscle fibers.

Proteins constitute 16 to 22 percent of the muscle mass (Table 2.4) and they are the principal component of solid matter. Muscle proteins generally are categorized as **sarcoplasmic**, **myofibrillar**, or **stromal**, based primarily on their solubility. Sarcoplasmic proteins are readily extracted in aqueous solutions of low (0.15 or less) ionic strength. The more fibrous proteins of myofibrils are not soluble in low ionic strength solutions but require higher (0.3 or greater) ionic strength solutions of sodium or potassium salts for their extraction. Because the myofibrillar proteins are extracted by salt solutions, they are called salt-soluble proteins. Stroma proteins include the proteins of the connective tissues. Because of their very fibrous nature, they are not soluble even in high ionic strength salt solutions and are the insoluble protein fraction of muscle. Sarcoplasmic proteins include myoglobin and enzymes associated with glycolysis, the tricarboxylic acid cycle, and the electron transport

TABLE 2.5   Distribution of Body Components of Cattle During Growth and Development*

Component	Liveweight (kg)						
	111	204	313	517	853	882	
	Age (months)						
	3	5	10	19	44	47	
Percentage of empty weight** to live weight	88.0	84.3	87.6	88.8	90.4	92.2	
Percentage of carcass weight to live weight	54.2	53.7	56.5	60.5	65.3	69.0	
Percentage of carcass weight to empty weight		61.6	63.7	64.5	68.1	72.6	74.8
Percentage of hide and hair weight to empty weight	10.5	8.2	8.4	7.4	5.9	6.2	
Percentage of blood weight to empty weight		6.2	5.2	5.0	4.1	3.3	3.5
Percent of stomach and intestines weight to empty weight	4.8	4.6	5.0	4.7	2.5	2.3	
Percentage of liver weight to empty weight		1.8	1.8	1.5	1.2	0.8	0.8
Percentage of offal fat weight to empty weight		1.4	3.9	4.7	5.2	7.0	4.7

*Cattle were fed ad libitum.
**Empty weight is the liveweight minus the intestinal tract contents.
Source: Moulton, C.R., P.F. Towbridge and L.D. High, "Studies in Animal Nutrition. II. Changes in Proportion of Carcass and Offal on Different Planes of Nutrition," Mo. Agr. Exp. Sta. Res. Bul. 54 (1922).

chain. The latter two enzyme groups are contained within the mitochondria but are readily extractable. Myofibrillar or salt-soluble proteins include all the proteins associated with the myofibril. In addition to proteins, other nitrogenous compounds are present in muscle. They are categorized as nonprotein nitrogen (NPN) and include a host of chemical compounds. Notable among these are amino acids, simple peptides, creatine, creatine phosphate, creatinine, some vitamins, nucleosides, and nucleotides including adenosine triphosphate (ATP) (Table 2.4).

Lipid content of muscle is extremely variable, ranging from approximately 1.5 to 13 percent (Table 2.4). It consists primarily of neutral lipids (triglycerides) and phospholipids. While some lipid is found intracellularly in muscle fibers, the bulk of it is present in adipose tissue depots associated with loose connective tissue between bundles (Figure 2.32d).

Carbohydrate content of muscle tissue is generally quite small. Glycogen, the most abundant carbohydrate in the muscle, comprises approximately 0.5 to 1.3 percent of muscle weight. The bulk of the remaining carbohydrate consists of glycosaminoglycans (associated with connective tissues), glucose and other mono- or disaccharides, and the intermediates of glycolytic metabolism.

Finally, muscle contains numerous inorganic constituents, notable among which are cations and anions of physiological significance such as calcium, magnesium, potassium, sodium, iron, phosphorus, sulfur, and chlorine. Many other inorganic constituents found in the animal body (Table 2.2) also are present in muscle.

CHAPTER 2 STRUCTURE, COMPOSITION, AND DEVELOPMENT OF ANIMAL TISSUES

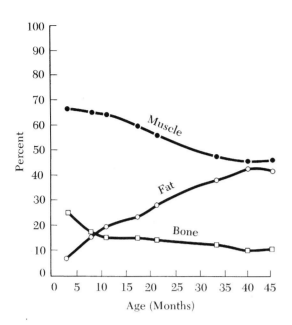

FIGURE 2.36 | Changes in the percentage of bone, muscle, and fat in beef carcasses during growth. Data from Moulton, C. R., P.F. Trowbridge and L. D. Haigh, Studies in Animal Nutrition; "III. Changes in Chemical Composition on Different Planes of Nutrition," Mo. Agr. Exp. Sta. Res. Bul. 55 (1922).

CARCASS COMPOSITION

Body Components

Although the industry is changing the way it markets livestock, many still pay for animals on a liveweight basis. Because the industry readily realized that the yield of carcass from live animals affected profitability, the proportion of the live animal that yielded a carcass was considered one of the first fundamental concepts of interest to those pioneers studying growth. Data presented in Table 2.5 show that the percentage of carcass to liveweight (**dressing percentage**) increases with liveweight. Concurrent with this increase, the percentages of non-carcass components such as hide, blood, stomach, intestines, and liver decline. Since animals become older as they become larger, one can deduce that a higher proportion of the body of young animals is composed of slaughter by-products than is the case in older or larger animals. This concept is of particular interest to those individuals buying livestock for slaughter, as those more mature animals generally yield a greater proportion of their weight as carcass.

TABLE 2.6	Heritability Estimates of Carcass Characteristics of Cattle, Poultry, Sheep, and Swine	
Characteristic	**Species**	**Heritability estimate* approximate average**
Fat thickness	Cattle	38
	Sheep	20
	Pigs	49
Longissimus muscle cross-sectional area	Cattle	70
	Sheep	48
	Pigs	48
Percent carcass fat	Broilers	30

*Estimates were compiled from various sources.

Proportion Of Muscle, Fat, and Bone

Carcass composition is generally of greater interest to animal and meat scientists than is total body composition. Muscle, fat, and bone are the major tissues considered when evaluating carcass composition and are of the greatest importance with regard to evaluation of livestock production practices. As the amount of one tissue increases, a concurrent decrease in the proportion of one or both of the remaining tissues must occur. The relative proportions of these three tissues in carcasses determine carcass value, to a large extent if weight is constant. These general principles are illustrated by the data from beef animals presented in Figure 2.36. Note that the decline in percentage of bone is greater than muscle, up to 10 months of age. After 10 months, the decline in percentage of muscle is greater than that of bone. However, the decline in percentage of both muscle and bone reflect the marked increase in percentage of fat. These changes in proportions of bone, muscle, and fat tend to parallel the order of inherent growth of these tissues: first bone, then muscle, and finally fat.

Factors That Influence Carcass Composition

Meat animal carcasses vary in composition through genetic, nutritional, and environmental effects. The following discussion addresses several factors within control of livestock producers that may be manipulated to achieve desirable effects.

Genetics

Animals of given breeds grow and develop in characteristic manners and produce carcasses with distinctive characteristics that are peculiar to the breed. For example, Duroc pigs and Angus cattle are known for their tendency to deposit intramuscular fat. A major difference between dairy and beef breeds is the distribution of various fat depots: dairy-type animals tend to have higher proportions of kidney and pelvic fat and smaller proportions of subcutaneous fat than beef-type animals. Mature size also is a breed characteristic, e.g., Southdown sheep are smaller than Suffolk sheep.

In meat animals, **phenotypic variations**, the outward visible expression of the genetic potential of an individual are due to genotype, environment, or to an interaction of both. Both genotype and environment are of great importance in determining the characteristics of any animal. **Genotype** or "genetic make-up" of an animal provides the necessary potential for growth and development, and the environment will tend to maximize or minimize the realization of this potential. Interaction of genetics and environment means that animals with a certain genotype might perform better in one environment than in another.

The genetic portion of phenotypic variance, expressed as a percentage, is called the **heritability estimate**. The percentage of heritability subtracted from 100 then gives an estimate of the variation due to environment. Heritability estimates in Table 2.6 show the extent to which muscle size and carcass fatness are heritable. Furthermore, these data indicate that livestock producers may make improvements in animals by selecting breeding animals that possess desirable carcass traits.

Major Genes

Occasionally, mutations occur in genes that encode for specific factors that control animal or muscle growth traits. As a result, growth is exaggerated in animals possessing these mutations. These mutations can occur naturally, and when a perceived beneficial attribute is detected in these animals, they are likely propagated unless it is lethal. Not until recently have advances in DNA technology allowed scientists to identify and study some very important genes that control muscle growth.

Double muscling is one such genotype in cattle that results in thick, bulging muscles, as illustrated in Figure 2.37. The term double muscling is misleading because it implies that animals have twice as many muscles as normal animals. In fact, double muscled animals have the same number of muscles as normal animals but rather have nearly twice as many muscle fibers, which results in an exaggerated muscle hypertrophy. Coupled with

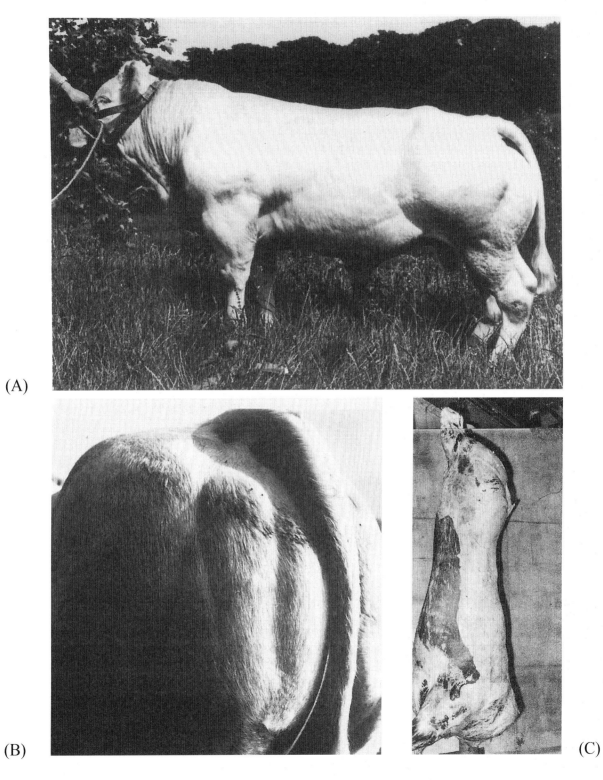

FIGURE 2.37 | Bull (A) and (B) and carcass (C) that exhibit double muscling. Note the two depressions in the muscling of the round (B) and the fore quarter (A) of the live animal and the bulging round of the carcass (C).

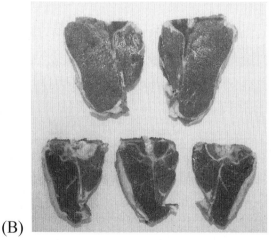

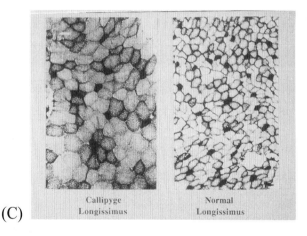

FIGURE 2.38 | Comparison of callipyge and normal genotype lambs. (A) Callipyge lambs (first and third, left to right) contrasted to normal lambs (second and fourth) of the same breeds. Photo courtesy of Sam Jackson, Texas Tech University. (B) Loin chops from callipyge (top) and normal (bottom) lambs. (C) Cross-section of *longissimus* muscle from callipyge (left) and normal (right) lambs showing muscle fiber hypertrophy of the callipyge genotype. Photo/illustration courtesy of C. E. Carpenter, Utah State University.

nearly twice the muscle fibers, double muscle animals have a lower proportion of red fibers and a higher proportion of white fibers than normal animals. Recall that white fibers are larger than red fibers and likely contribute to the muscle hypertrophy observed in animals with this genotype.

When double-muscled cattle are compared to normal cattle, growth is more rapid, development of muscle is greater, and fat deposition is less. In fact, muscle-to-bone ratios can approach 6:1 for double muscled animals compared to only 4:1 and 5:1 for traditional dairy and beef breeds, respectively.

The gene responsible for the double muscled condition has been identified as **myostatin.** Myostatin belongs to a family of growth factors that regulate myogenesis. In those cattle exhibiting the double muscled phenotype, researchers found that the myostatin gene has been disrupted, meaning the sequence was either mutated or portions of the sequence were lost over countless years of selection. As a result, the function of the translated protein is impaired. Although not completely understood at present, during development of double muscled fetuses, myogenesis is not controlled properly and animals with greater number of muscle fibers are developed. Because the economic benefits of controlling such a gene could have dramatic ramifications on meat production, many growth biologists are extremely interested in this and other genes that control muscle development.

Conditions similar to double muscling exist in other meat animal species. Pigs, sheep, and turkeys sometimes exhibit muscle development that is equal to the extremes of double muscling, however these conditions at present, have not been attributed to changes in the myostatin gene. Unfortunately, as is the case with double muscled cattle, extremes in muscle size often are accompanied by physiological conditions that interfere with animal reproduction, stress resistance, and locomotion.

Callipyge, illustrated in Figure 2.38, is another genotype that results in enlarged muscle development in various breeds of sheep. Callipyge is a Greek word meaning "beautiful buttocks", likely because of the excessive muscling observed in the hindlegs of lambs. Lambs possessing this genotype have greatly exaggerated hypertrophy (as much as 40% more than normal) in specific muscles of the loin and hindleg and less hypertrophy of muscles in the forelimb. They have less fat than those lambs lacking this genotype and their meat is significantly tougher. At the cellular level, muscle of callipyge lambs has greater protein-to-DNA ratios, a higher percentage of white fibers, and greater fiber diameters. In contrast to the double muscling genotype, however, muscle fiber number seems to be unaltered. The callipyge phenotype is caused by a point mutation in the telomeric end of ovine chromosome 18. The phenotype shows an unusual non-Mendelian inheritance pattern, **polar overdominance**, in which the phenotype is only expressed in heterozygotes that inherit the mutation from their paternal parent. Expression patterns of a number of the imprinted genes that surround the callipyge mutation are dysregulated in the affected muscles. Type IIb fibers seem to be differentially affected.

Physiological Age

All animals within species or breeds or among sexes do not grow, develop, fatten, or mature at the same chronological age. The term physiological age refers to the stage of development of an animal that can be described by identifiable stages of body development or function, such as body height and weight, body composition, or onset of puberty. Animals may attain these stages of physiological age at different chronological ages and may be described as being early or late maturing. Thus, at a given chronological age, animals from an early maturing breed would be physiologically older than animals from a late-maturing breed.

Nutrition

Although genotype dictates the maximum amount of growth and development that is possible, nutrition along with other environmental factors govern actual rate of growth and extent to which development is attained. Utilization of ingested nutrients is partitioned among various tissues and organs according to their metabolic rate and physiological importance (Table 2.7). Maintenance and function of vital physiological systems such as the nervous, circulatory, digestive, and excretory systems, take precedence over muscle growth and fat deposition. The order of precedence is as follows:

- Tissues that constitute vital organs and physiological processes
- Bone
- Muscle
- Fat deposition.

During pregnancy, the developing fetus holds a priority similar to that of vital tissues and organs of the dam.

TABLE 2.7	Priority and Partition of Nutrients among Body Systems and Tissues	
	Systems	**Tissues**
Highest	Circulatory	Muscle
	Respiratory	Adipose
	Digestive	Mesenteric
	Reproductive	Perirenal
		Subcutaneous*
		Intermuscular*
		Intramuscular
Lowest		

*Subcutaneous and intermuscular adipose tissue may be interchangeable, depending on species.

Plane of Nutrition

When food is plentiful, all tissues of the body receive sufficient nutrients for maintenance, normal growth, and fattening. However, if the food supply is limited, tissues are affected in reverse order of physiological importance. Vital tissues are maintained at the expense of others, so that a severe restriction of nutrients results in body tissues of less importance, for example, adipose tissue and in more severe instances, muscle, being utilized to maintain those of more vital importance.

It is possible to control the rate at which different tissues and parts of the body grow and develop by altering the nutritional level of animals at critical times. For example, when pigs are maintained on a high level of nutrition during the first several weeks of postnatal growth, the growth rate of bone, muscle, and fat is greater than that of pigs on a low level of nutrition. When animals are changed from a low to a high level of nutrition mid-growth, there is a marked recovery in growth rate, known as compensatory growth, but recovery is greater for fat than for muscle and bone and the carcasses are excessively fat. If pigs are started on a high level of nutrition and switched to a low level, they produce carcasses with more muscle and less fat than those on continuous high levels of nutrition.

The stage of postnatal growth over which nutritional deprivation is imposed affects the nature of response. If undernutrition occurs early in the postnatal growth period, long-lasting effects are greatest on the earliest maturing tissue (bone) and least on the latest maturing tissue (fat). Subsequent feeding at a high nutritional level may occur when growth intensity of bone and muscle is declining and that of fat is increasing. Yet, various tissues and organs in the bodies of growth-retarded animals exhibit remarkable compensatory growth when high levels of nutrients are available. If the animal has not been subjected to severe malnutrition for a long period, and if sufficient time is allowed, underdeveloped organs or tissues may completely recover from retarding effects sustained earlier. However, if the undernutrition is severe enough and of sufficient duration, irreversible damage may occur.

The efficiency of meat animals in converting feed into meat is generally related to the level of feed intake, but the relationship is rather complex. Highest efficiency in converting feed energy into body weight gain is achieved when animals are fed ad libitum. But, if feed energy intake exceeds the amount needed for lean tissue growth, the excess is used for fat deposition. Thus, animals full-fed high-concentrate diets usually produce more carcass fat, and consequently, are less efficient in converting feed to lean meat than are animals fed slightly below ad libitum energy intake, even though the ad libitum-fed animals would be more efficient in total feed energy retention. This is particularly evident in the later growth stages, as muscle and bone approach their mature sizes.

Slight to moderate feed restriction is an effective procedure to modify body or carcass composition. Abdominal fat in broilers can be reduced by a 3- to 5-day period of feed restriction, beginning at 5 to 7 days of age. Slight feed restriction of monogastric animals, particularly in the later growth stages, will produce leaner animals at slaughter, but limit feeding requires more labor and housing and animals reach acceptable slaughter weight more slowly. The additional costs associated with restricted feeding may surpass the additional value of leaner carcasses, depending on the premiums associated with leaner carcasses or the penalties deducted for fatter carcasses. Dietary energy of ruminant animals may be restricted conveniently by including variable amounts of fiber in the diet.

Protein

An adequate and continuous supply of protein is required in animal diets for growth and maintenance of tissues. Proteins are composed of varied amounts and kinds of amino acids, some of which cannot be synthesized in the animal body. These are called essential amino acids, and must be present in the diet of the animal.

Every animal has a daily need for dietary protein. Although animals cannot synthesize tissue proteins beyond their genetic potential by consuming excess protein, the rate of tissue accretion (growth) is readily reduced by inadequate dietary protein. Growth rates in monogastric animals are reduced by an inadequate total amount of protein, a deficiency of any one of the essential amino acids, or an imbalance of amino acids in the diet. In ruminant animals, amount and quality of dietary protein are less critical than in monogastric animals. If animals consume a surplus of protein, the excess is broken down and used as energy, or stored as fat.

Fat

Dietary fats are used by the animal for energy, and certain fatty acids are essential for growth. They also may be assimilated and deposited as body fat. Composition of deposited fat varies among species. All meat animals are able to synthesize fatty acids in the liver and/or adipose tissue from carbohydrates and proteins, and the fat that is deposited is characteristic of the species. Fats in the diet of monogastric animals may be assimilated and deposited in relatively unchanged form; whereas, dietary fats consumed by ruminants undergo degradation and resynthesis of more saturated fat by rumen bacteria before assimilation and deposition. Carcasses from monogastric animals fed a diet containing a specific type of fat will have fat deposits of similar chemical composition to the dietary fat. For example, pigs or chickens fed a diet high in unsaturated fat will have soft, oily carcass fat.

Hormones and Hormone-like Materials

Hormones are substances secreted into body fluids by ductless endocrine glands or other tissues such as the intestinal tract. Hormones act as regulators of chemical reactions involved with growth of tissues, maintenance of tissues, and other physiological processes. Combinations of hormones are involved in the growth process, and their interactive effects result in "normal" carcass composition. Only those hormones and hormone-like substances known to have major effects on carcass composition are included in the following discussion.

Growth hormone or **somatotropin** produces lean tissue growth throughout the animal. It is produced by the anterior pituitary gland and promotes the release of a class of hormones from the liver and other tissues known as insulin-like growth factors (IGFs). The IGFs are named as such because of their insulin-like activity and are responsible for protein synthesis associated with somatotropin. Such hormone-induced protein accretion partitions the utilization of nutrients toward lean tissue growth and away from fat deposition.

The dramatic effects of somatotropin have been demonstrated in meat animals by the administration of synthetic somatotropin produced by genetically engineered microorganisms. Some data suggest that carcass fat may be reduced by one-third or more by such treatments. Availability of plentiful supplies of growth hormone for administration to meat animals offers great promise for improvements in carcass leanness and meat production efficiency.

Hormones of the adrenal medulla, **epinephrine** and **norepinephrine**, exert widespread effects in the body. With reference to muscle tissue, these hormones assist in the mobilization of glycogen to provide energy. However, their effects influence muscle protein and lipid metabolism as well. These effects may relate to the ability of epinephrine to activate certain tissue receptors known as ß-receptors.

Hormone-like substances have been developed that have chemical structures very much like epinephrine and norepinephrine. These materials are known as **beta-adrenergic agonists** because they are effective in activating the beta-receptors. In addition, they are effective repartitioning agents: they shift available nutrients away from fat deposition and toward muscle accretion. Animals receiving such commercially available repartitioning agents (for example, **Paylean®**, **Optaflexx®**, or **Zilmax®**) produce carcasses that have increased muscle size and decreased quantities of fat. However, decreased meat tenderness is a generalized effect of beta-adrenergic agonist administration and some compounds have a greater effect on tenderness than do others.

The hormones of the testes and ovaries play an important role in growth and development of the body. Differential rate of growth and development as well as tissue composition are associated with the sex of animals. Males usually grow faster, mature later, and have carcasses that are more muscular and less fat than females. Meat animal producers frequently castrate male animals, a practice that modifies behavior and sometimes improves meat quality, but causes deposition of more fat and less muscle during growth.

The principal hormones produced by the testes are **androgens**; those produced by the ovaries are **estrogens** and **progesterone**. Androgens stimulate growth in muscles by increasing protein synthesis, an action that is accompanied by decreased fat deposition. Certain muscles are more sensitive to androgens, depending on their role in reproduction. In particular, muscles of the forequarter of the male, especially those in the neck and crest region, show greater development than in females or castrates. Androgens also stimulate deposition of bone salts causing increased bone growth as compared to females and castrates. However, when androgen levels become sufficiently high, they cause closure of epiphyseal plates and consequent skeletal maturation that precedes that of castrates.

Estrogens generally have little or no effect on skeletal muscle protein synthesis, but they are effective in promoting deposition of body fat. Specific effects depend on proximity to puberty and estrogen concentration. Females generally fatten at younger ages and lighter weights than males. With the exception of gilts, they also fatten more quickly than male castrates. Gilts tend to fatten and mature later than barrows. Estrogens, like androgens, stimulate bone salt deposition and are even more effective in causing epiphyseal plate closure. Therefore, females mature earlier than males.

Synthetic estrogens (estradiol-17ß, zeranol, or **estradiol benzoate**), singly or in combination with synthetic progesterone, are effective in increasing carcass leanness of growing wethers and steers by stimulating muscle growth and suppressing fat deposition. The protein content of carcasses from hormone-implanted animals is usually increased by 10 percent or more. The mechanism of action for this effect is unknown, but some evidence suggests that these exogenous hormones increase somatotropin secretion. Their effects are variable among species, however. In chickens, they increase the rate of fattening; in barrows and gilts, they depress body weight gain; and in bulls, they increase the rate of fattening.

3

MUSCLE CONTRACTION AND ENERGY METABOLISM

OBJECTIVES: *Identify the steps involved in muscle contraction and relaxation including the mechanism of stimulation to initiate contraction. Identify sources of energy for muscle contraction, relaxation, and function.*

Motor nerves
Membrane potential
Nodes of Ranvier
Sodium-potassium pump
Action potential
Depolarization
Electrochemical process
Myoneural (neuromuscular) junction
Motor endplate
Acetylcholine
Cholinesterase
Transverse (T-tubule) system
Sarcoplasmic reticulum
Triad junction
Dihydropyridine receptor
Ryanodine receptor
Contractile proteins
Cross-bridge
Regulatory proteins
Actomyosin
ATP (adenosine triphosphate)
ADP (adenosine diphosphate)
Power stroke
Lever arm
Repolarization
Calcium sequestering
Calcium pump
Phosphocreatine
Phosphorylation
Aerobic metabolism
Glycogenolysis
Glycolysis
Tricarboxylic acid (TCA) cycle
Pyruvate
Electronic transport chain
Anaerobic metabolism
Lactate

Key Terms

In the living body, muscle is a highly specialized tissue capable of converting chemical energy into mechanical energy primarily as a means for locomotion. It becomes a highly nutritious and palatable food when converted to meat after the harvest of an animal. As elaborated in Chapter 2, the overall structure of muscle is designed for contraction and relaxation, which leads to movement and locomotion. This ability to contract and relax is lost during the transformation of muscle to meat. Yet, the events surrounding this whole process of muscle contraction and relaxation, and subsequent loss of these abilities, dramatically affect the palatability of meat. The biochemical processes that provide energy for muscle function in living animals are similar to those processes that cause an accumulation of metabolites and subsequent loss of water-holding capacity during the postmortem period. Thus, an understanding of how muscles function normally to produce movement and generate heat forms the basis for our understanding why they are important in determining the ultimate functional characteristics of meat as a source of food.

In this chapter, muscle function is discussed from the point where a nerve stimulates muscle to contract to a point where the entire contractile structure and nerve return to a resting or "relaxed" state. In addition, the chemical reactions and processes that furnish energy to muscle are discussed in some detail. Before studying the mechanism whereby muscle contracts, generates force, and performs work, a clear understanding of the structure of the muscle fiber and myofibril (see Chapter 2) is necessary. Particular emphasis should be given to understanding and visualizing the transverse tubules, sarcoplasmic reticulum, the three-dimensional positions of the sarcoplasmic reticulum and myofibrils, as well as the structure, banding pattern, and proteins of myofilaments. It cannot be overemphasized that muscle is a highly specialized tissue whose contractile mechanisms are related directly to its unique structure.

NERVES AND THE NATURE OF STIMULI

Although muscle can be stimulated to contract through several means, most contractions are initiated by stimuli that arrive at the surface of the muscle fiber (the sarcolemma). In skeletal muscle, contraction is usually initiated by an electrical stimulus that starts in the brain or spinal cord, and is transmitted to the muscle via nerves (structure of neurons, their cell body, dendrites, and axons was described in Chapter 2). Nerve fibers that transmit stimuli to skeletal muscles are called **motor nerves**. At regular intervals along the motor nerve fiber, Schwann cells wrap around the axon of the nerve and secrete a myelin sheath that encapsulates the fiber. Between adjacent Schwann cells, the myelin sheath comes in close contact with the axon and these structures are referred to as the **nodes of Ranvier**. Because Schwann cells insulate portions of the axon against ion movements, the stimulus is forced to jump from node to node, which allows for a faster rate of stimulus conduction along the fiber. As a result, myelinated nerve fibers have thirty- to forty-fold greater conduction velocities than do unmyelinated fibers.

Transmembrane Potentials

Under normal resting conditions, an electrical potential (difference in net charge) exists between the inside and outside of most cells. These potentials vary from 10 to 100 millivolts, depending on the type of cell, but in resting (unstimulated) nerve and muscle fibers the **(trans)membrane potential** is about 90 millivolts. Fluids residing inside and outside of these fibers contain positive and negative ions. Generally, an excess of negative ions accumulates in the intracellular fluid and resides along the inner surface of the membrane, whereas the reverse is true outside the cell, where an excess of positive ions assembles in opposition along the extracellular surface of the membrane. As a result, an electrical potential (charge difference) or **resting membrane potential** is developed across the cell membrane (Figure 3.1), positive on the outer surface and negative inside the cell. Development of these transmembrane or resting membrane potentials in nerve and muscle result from:

- active transport of ions through the membrane
- selective membrane permeability to ions and small molecules, and
- the unique ionic composition of the intracellular and extracellular fluids.

Extracellular fluid contains high concentrations of sodium (Na+) and chloride (Cl-) ions (Figure 3.1). In contrast, concentrations of potassium K+ and nondiffusible negative ions (e.g., proteins and peptides) are very high in the intracellular fluid while Na+ and Cl- concentrations are quite low. Concentration gradients of Na+ and K+ across the plasma membrane are maintained by the active transport of Na+ out of the cell, and K+ into the cell. The system that accomplishes the active transport of Na+ and K+ is located in the plasma membrane, and is commonly referred to as the **sodium-potassium pump**. Energy required to pump both types of ions across the membrane against a concentration gradient is furnished by adenosine triphosphate (ATP). Permeability of a plasma membrane is 50 to 100 times greater for the diffusion of K+ than Na+. Thus K+ passes through the membrane with relative ease compared to Na+. As a result, K+ diffuses out of the cell more rapidly than Na+ diffuses in. Of course, nondiffusible negative ions in the intracellular fluid pass through the membrane only with extreme difficulty. Therefore, the net flow of electrical charges across the membrane is due to the diffusion of positively charged potassium ions into extracellular spaces, leaving fewer positive charges inside the cell. As a result, oppositely charged ions line the membrane, with the positive ions on the outer surface attracting negative ions on the inner surface of the membrane, thus establishing the resting membrane potential.

Net diffusion of positive charges out of the cell does not continue indefinitely, however. Once the resting membrane potential is established, further flow of K+ out of the cell is impeded. As the membrane potential increases, the forces opposing K+ diffusion increase. Thus, equilibrium is established between outward diffusion of K+ (along its concentration gradient) and the force opposing outward diffusion: the positive potential established by the presence of cations on the outer surface of the plasma membrane.

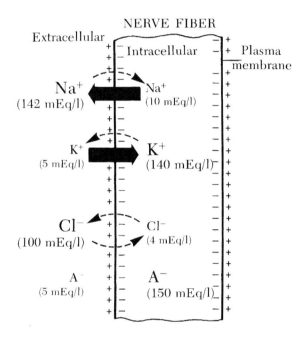

FIGURE 3.1 Establishment of a membrane potential in the normal resting nerve fiber, and the development of concentration differences of sodium (Na+), potassium (K+), chloride (Cl-), and nondiffusible negative ions (A-), between the two sides of the membrane. Dashed arrows represent diffusion, and solid arrows represent active transport ("pumps"). The concentration of each ion is indicated in parentheses in units of milliequivalents/liter. From Guyton, A. C., *Textbook of Medical Physiology*, 4th ed. W. B. Saunders Co. Philadelphia, 1971.

Action Potentials

As stated above, nerve and muscle fibers exhibit membrane potentials, as do other cells, but they have a unique capability not shared by any other cell type. They are able to transmit an electrical impulse, called an **action potential**, along their membrane surfaces. When an action potential is transferred from a motor nerve to muscle fibers, it initiates muscle contractions. An action potential travels along the membrane surface of the nerve fiber, and is actually a wave of reversing electrical charges or **"depolarization"** that results from chemical changes in the membrane. Thus, an action potential is often referred to as an **electrochemical process**. As described in the previous section, in the resting state, the membrane is positive on its outer surface and negative on the inside. The action potential is initiated by a sudden and dramatic increase in the permeability of the membrane to Na+. As a result, Na+ rushes into the cell to establish equilibrium between the concentrations inside and outside of the cell. At the same time, permeability of the membrane to K+ remains the same and, as a result, an excess of positive charges enters the cell, reversing the resting membrane potential (Figure 3.2a). An increased permeability to Na+ lasts only for a small fraction of a millisecond. The newly reversed membrane potential reduces Na+ permeability to its former low level. The outward flow of K+ continues and re-establishes the resting membrane potential (Figure 3.2b). The sodium-potassium pump in the cell membrane redistributes Na+ to the outside and K+ to the inside without disrupting the membrane potential. The entire sequence of events from initiation of an action potential to re-establishment of a resting membrane potential on the nerve fiber requires about 0.5 to 1 millisecond.

Myoneural Junction

The stimulus (action potential) that initiates muscle contraction is transferred from the nerve fiber to the muscle fiber at the **myoneural (neuromuscular) junction**. At this junction, the motor nerve branches into several terminal endings that come in close contact with the muscle fiber through a series of small invaginations in the sarcolemma (Figure 3.3). These terminal endings adhere tightly to, but do not penetrate, the sarcolemma. The combined structures of the myoneural junction form a small mound on the surface of the muscle fiber, called the **motor end plate**.

Although muscle fibers may be stimulated to contract by a strong synthetic electrical impulse, the electrical impulse of the action potential from a nerve is not strong enough to elicit a response alone. Therefore, a mechanism is required to amplify neuronal signals to the muscle fiber. When an action potential arrives at the motor end plate, it causes a chemical "transmitter," **acetylcholine**, to be released. Acetylcholine is stored in small vesicles found in the terminal nerve branches,

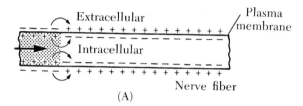

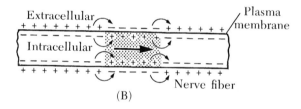

FIGURE 3.2 Membrane resting and action potentials. (A) The beginning of an action potential in a nerve fiber, showing the reversed membrane potential. (B) The action potential, as it progresses along a nerve fiber, showing the depolarization at the front and repolarization at the rear. The large solid arrows indicate the direction of the action potential's movement. The curved arrows indicate points of reversing potential in the membrane. Courtesy of W. H. Freeman and Company.

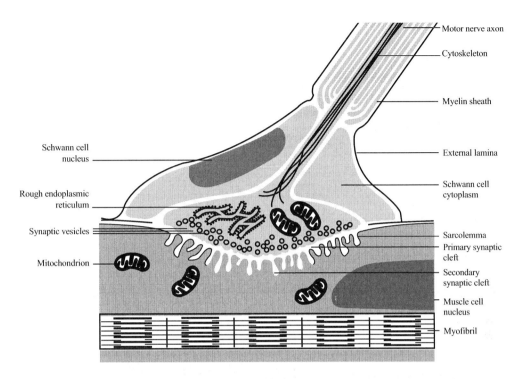

FIGURE 3.3 | Schematic representation of a motor endplate, as seen in a longitudinal section along the longitudinal axis of a muscle fiber. This image was published in Wheater's Functional Histology: *A Text and Colour Atlas*, 3rd Edition, H. George Burkitt, Barbara Young and John W. Heath, 124. Copyright © 1993 Churchill Livingstone. Reprinted by permission.

and the contents of a few vesicles are released with each action potential. When the sarcolemma comes in contact with acetylcholine, it becomes more permeable to Na+, the membrane depolarizes, and an action potential (similar to that in a nerve fiber) is propagated along the fiber in all directions. Again, the effects of acetylcholine on the sarcolemma occur for only a few milliseconds, as the enzyme **cholinesterase** is quickly released by the motor end plate and neutralizes the stimulant. Practical implications surrounding this mechanism are that a small signal from the nerve is amplified and spreads over the whole muscle fiber. Some extremely potent poisons act at the myoneural junction. Toxin produced by the bacterium *Clostridium botulinum* (responsible for botulism food poisoning) and curare (of poison-arrow fame) both prevent the transmission of impulses from nerves to muscles by interfering with the release or action of acetylcholine.

Muscle Action Potentials

The action potential in a muscle fiber triggered by acetylcholine is nearly identical to that which occurs in nerve fibers. The major difference is the duration of the action potential: 5 to 10 milliseconds in skeletal muscle fibers, as compared to 0.5 to 1 millisecond in nerve fibers.

Most muscle fibers have only one myoneural junction from which the stimulus is transmitted to all parts of the fiber. An action potential begins at the myoneural junction and progresses in all directions along the sarcolemma, stimulating the entire fiber. If you recall, a muscle fiber consists of numerous myofibrils. Many of these myofibrils are located deep in the muscle fiber, relatively remote from the sarcolemma. Therefore, the question arises as to how the action potential is propagated deep into an individual muscle fiber. This whole process of communicating to the inner most myofibrils is accomplished by the **transverse tubule (T tubule) system**. T tubules are a continuation of the muscle cell membrane and resemble small invaginations in the sarcolemma when viewed with a microscope. These membranes course inward from the outer surface of the muscle fiber (Chapter 2) and provide an avenue for an action potential to penetrate the inner most aspects of the fiber.

Calcium Release

As an action potential transverses a muscle fiber via the T tubules, it is ultimately responsible for calcium release from the **sarcoplasmic reticulum** that surrounds each myofibril. The sarcoplasmic reticulum is a membranous cellular organelle responsible for regulating the amount of calcium ions in the cytoplasm of the muscle fiber. Recall that the T tubule system and the sarcoplasmic reticulum are separate structures, but they are in extremely close proximity at the **triad junction**. The mechanism by which depolarization of the T tubule is communicated to the sarcoplasmic reticulum involves two membrane-bound calcium channel proteins, the **dihydropyridine receptor** and the **ryanodine receptor** (the latter named for its ability to bind the plant alkaloid ryanodine) located in the T-tubules and the sarcoplasmic reticulum, respectively. These proteins are located in the triad junction where two sarcoplasmic reticuli meet with one T tubule. Simply stated, the dihydropyridine receptor responds to depolarization of the T tubule and releases calcium ions into the cytoplasm of a muscle cell. As soon as the ryanodine receptor (calcium channel) detects this calcium release, it responds by allowing the calcium sequestered (held) in the sarcoplasmic reticulum to flood into the cytoplasm of the muscle cell, which initiates contraction as described in the next section of this chapter. When muscle is in a relaxed state, the concentration of calcium ions in the sarcoplasmic fluid is less than 10^{-7} moles/liter. However, the total concentration of calcium in skeletal muscle is more than 1000 times this level (greater than 10^{-4} moles/liter). This illustrates how much calcium is sequestered in the sarcoplasmic reticulum.

CONTRACTION OF SKELETAL MUSCLE

In simplest terms, skeletal muscle contraction involves four myofibrillar proteins: actin, myosin, tropomyosin, and troponin. Actin and myosin are the major **contractile proteins** and form the actin and myosin filaments of the myofibril. **Cross-bridges** formed directly between the thin (actin) and thick (myosin) filaments generate contractile force during contraction. In contrast to actin and myosin, tropomyosin and troponin play the role of **regulatory proteins** and assist in turning the contractile process "on" and "off" depending on the concentration of calcium ions present in the sarcoplasm of the fiber.

For muscle to remain in the relaxed state, calcium must be sequestered in the sacroplasmic reticulum and there is a relatively high concentration of ATP, which is complexed with magnesium ($Mg2+$) ions. This complex prevents interaction of actin and myosin (cross-bridge formation). Thus, when sarcoplasmic calcium concentrations are low (less than 10^{-7} moles/liter), and the Mg-ATP concentrations are high, cross-bridge formation between the actin and myosin filaments is inhibited. It is necessary to increase the concentration of free calcium ions to about 10^{-6} or 10^{-5} moles/liter (a 10- to 100-fold increase) to initiate a contraction.

In skeletal muscle, troponin functions as the $Ca2+$ sensitive "latch" that fixes tropomyosin's location on the thin filament in the "off" state of contraction. Troponin is a complex of three subunits, designated T, I and C. Troponin T always maintains a tight linkage to tropomyosin. Troponin I has a variable linkage to actin and inhibits actin-myosin interaction. Troponin C is the $Ca2+$ sensor of the troponin complex, and of the myofibril itself. When $Ca2+$ are released into the sarcoplasm, they preferentially bind to the troponin C subunit and induce a conformational change in the entire troponin complex. Collectively, these conformational changes result in movement of tropomyosin and troponin I toward the cleft of the actin filament and away from their blocking location on the filament. This exposes the myosin binding sites on actin. Figure 3.4A is a diagram of an actin filament in

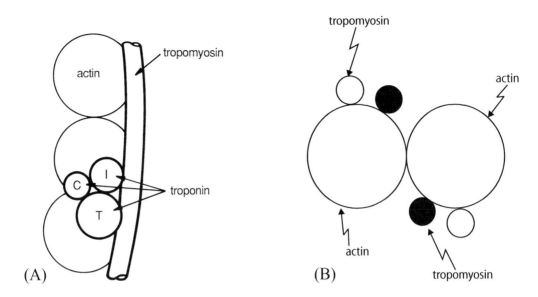

FIGURE 3.4 | Diagram of a segment of actin filament showing the location of tropomyosin and troponin subunits. (A) Longitudinal view of the location of tropomyosin and troponin. (B) Cross-sectional view of the actin filament showing the locations of tropomyosin in the resting state (open circles, myosin binding site blocked) and in the activated or "on" state (solid circles, myosin binding site exposed).

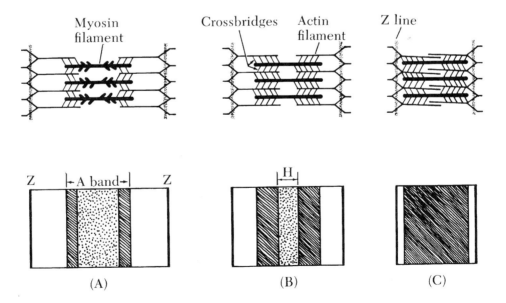

FIGURE 3.5 | One sarcomere is shown at various stages of shortening during contraction (top). The banding pattern in a muscle fiber at corresponding degrees of contraction (bottom). Contraction is shown for the following stages: (A) in a muscle that is extended, (B) at rest length, and (C) severely shortened. Modified from W. Bloom and D. W. Fawcett, *A Textbook of Histology*, 9th ed. W. B. Saunders Co., Philadelphia, 1968.

longitudinal view, showing the position of tropomyosin and troponin. Figure 3.4B illustrates the actin filament in cross-section and shows how the myosin binding site is exposed when tropomyosin moves deeper into the cleft formed by the two F-actin chains.

The shift of tropomyosin's position to the "on" state of contraction allows the myosin heads to form cross-bridges between the thick (myosin) and thin (actin) filaments. These cross-bridges develop a contractile force, and the actin filaments in each half of the sarcomere are pulled toward the center of the sarcomere (Figure 3.5). The protein complex formed when actin and myosin interact at the cross-bridge is called **actomyosin**. During contraction, the length of actin and myosin filaments does not change. Rather, the filaments slide along each other, pulling the Z disks closer to the myosin filaments, thereby decreasing sarcomere length.

The A band width is constant during all phases of muscle contraction, but the I band and H zone widths change. These widths are greatest when the muscle is stretched (Figure 3.5a), and decrease as the muscle shortens (Figure 3.5b). In severely contracted muscle, actin filaments meet, or even overlap, in the center of the A band, and the Z disks may abut on the ends of the myosin filaments (Figure 3.5c). Under these conditions, the H zone and I band are not discernible in electron micrographs.

Muscle contraction requires energy in addition to that normally consumed by resting muscle. This energy is derived from **ATP (adenosine triphosphate)** that is hydrolyzed in the following reaction.

$$ATP + H_2O \rightarrow ADP + H^+ + Pi + (energy)$$

The cross-bridge is responsible for splitting ATP into **ADP (adenosine diphosphate)** and Pi (inorganic phosphate) and for converting part of the energy released from splitting of ATP into force that slides the filaments. The rest of the energy is released as heat. Each cross-bridge generates enough movement to slide the actin filament 10 nanometers (nm). During contraction, filament sliding requires a cyclic "making" and "breaking" of cross-bridges, with each cycle contributing a small amount to the total contraction.

One cycle of cross-bridge formation is illustrated in Figure 3.6. The cycle contains a **power stroke** portion and a recovery portion. Four positions of the cross-bridge during a contraction cycle are illustrated. Binding of ATP to the cross-bridge at the end of the power stroke causes rapid dissociation of the cross-bridge from the actin filament (position 1 $\rightarrow$ position 2). The cross-bridge then changes conformation and hydrolyzes ATP to ADP and Pi. This is the recovery process of the cycle (position 2 $\rightarrow$ position 3) and the cross-bridge is in the pre-power stroke form (position 3). The cross-bridge then complexes with the actin filament (position 3 $\rightarrow$ position 4) and undergoes a conformation change, the power stroke (position 4 $\rightarrow$ position 1). P^i and then ADP are released from the cross-bridge during the power stroke. Binding of ATP to the cross-bridge at the end of the power stroke prepares the cross-bridge for the next cycle. The conformation change during the power stroke occurs in the portion of the cross-bridge known as the **lever arm** (Figure 3.7). Most of the mass of the cross-bridge does not participate in this motion.

RELAXATION OF SKELETAL MUSCLE

Relaxation of skeletal muscle involves reestablishment of the resting state, and is measured by a decrease in muscle tension. In the relaxed state, muscle generates very little tension and can easily be stretched. This means that there are no cross-bridges between actin and myosin filaments, and the filaments of each sarcomere slide passively over one another. For relaxation to occur, conditions prevailing during the resting state must be reestablished: namely, calcium concentration within the sarcoplasm must be reduced to 10^{-7} moles/liter (or less), and ATP levels restored. Therefore, relaxation can proceed only when the events that activated the contractile process are reversed. Obviously, the first step in the relaxation process is **repolarization** of the sarcolemma and returning the membrane potential to its resting value. Subsequent steps in the relaxation process are then able to occur.

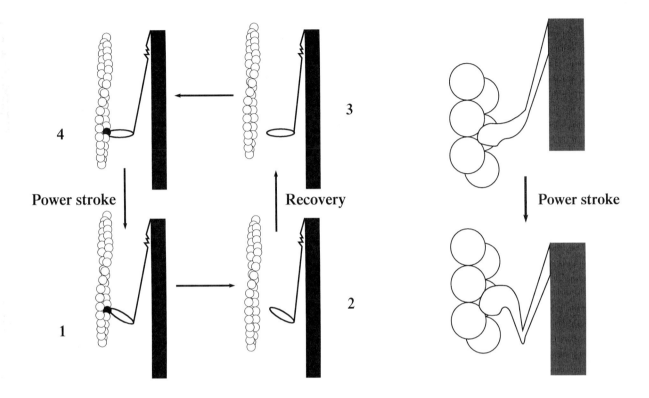

FIGURE 3.6 | Schematic representation of the cross-bridge cycle, showing one myosin head forming a cross-bridge (3→4), undergoing a power stroke (4→1), the cross-bridge dissociating from actin (1→2), and recovering to prepare for the next cycle (2→3). Modified from Michael A. Greeves and Kenneth C. Holmes. The molecular mechanism of muscle contraction. In *Advances in Protein Chemistry, Vol. 71, Fibrous Proteins: Muscle and Molecular Motors.* John M. Squire and David A. D. Parry, eds. Elseview Academic Press, San Diego, 2005, 164.

FIGURE 3.7 | Schematic representation of the conformation change in the lever arm of the myosin cross-bridge during the power stroke of a cross-bridge cycle. Modified from Michael A. Greeves and Kenneth C. Holmes. The molecular mechanism of muscle contraction. In *Advances in Protein Chemistry, Vol. 71, Fibrous Proteins: Muscle and Molecular Motors.* John M. Squire and David A. D. Parry, eds. Elseview Academic Press, San Diego, 2005, 164.

Intracellular free calcium concentrations in the sarcoplasm must be returned to original resting levels. **Calcium sequestering** by the sarcoplasmic reticulum is accomplished against a calcium concentration gradient. Therefore, an active "pumping" process is involved, similar to the process that establishes and maintains sodium and potassium gradients across nerve and muscle fiber membranes. This pump utilizes ATP as an energy source, i.e., hydrolyzes ATP to ADP, and pumps calcium back into the sarcoplasmic reticulum where it is stored in the terminal cisternae until release by a subsequent stimulus. The **calcium pump** is activated by increased cytosolic calcium. As free calcium concentrations in the sarcoplasm decrease, troponin senses the concentration change and allows tropomyosin and troponin I to return to the "off" location on the actin filament in which cross-bridge formation is inhibited. In the absence of cross-bridges, tension is not generated and the stretching imposed by elastic components in the muscles causes the filaments to slide passively over one another.

Thus, calcium has multiple effects on muscle when released into the sarcoplasm. Calcium activates troponin, which shifts the location of tropomyosin and troponin I on the actin filament, exposes the myosin binding sites on actin, and allows cross-bridge formation. Calcium also activates myosin ATPase, which releases energy for contraction. And calcium stimulates the ATPase of the calcium pump, which pumps calcium back into the terminal cisternae to end the contraction.

Figure 3.8 is a flow diagram of the events that occur during a complete contraction-relaxation cycle in skeletal muscle.

SOURCES OF ENERGY FOR MUSCLE FUNCTION

From the discussion presented to this point, it should be evident that ATP is the ultimate source of energy for:

- the contractile process
- the pumping of calcium back into the sarcoplasmic reticulum during relaxation
- maintaining the sodium/potassium ion gradients across the sarcolemma.

Of these three uses of energy, contraction is by far the most energetically demanding process. It has been estimated that, during a single muscle twitch, contraction uses a thousand times more energy than does reversal of membrane potential, and at least ten times more energy than does the calcium pump in the sarcoplasmic reticulum. Yet, the amount of ATP present in the muscle is sufficient to supply energy for only a few twitches. Therefore, a very rapid and efficient means must be available for re-synthesis of ATP within living muscle.

When an animal is slaughtered, the muscle does not instantaneously stop living and become meat. In fact, muscle does not 'die' or cease to function until its capacity to generate ATP stops. Pathways that provide for ATP synthesis by rephosphorylation or converting ADP to ATP in the living muscle continue sometime after respiration has ceased. Those biochemical reactions that occur as muscle attempts to maintain these homeostatic conditions (stable equilibrium) after harvest cause profound changes in muscle properties. These reactions constitute a major part of the processes described as the "conversion of muscle to meat" and are discussed in detail in Chapter 5. The following discussion of muscle energy metabolism is specific to understanding the function of living muscle. However, these same mechanisms also function in post mortem muscle as the tissue attempts to maintain homeostasis.

As stated previously, muscle contraction uses large amounts of ATP. Calcium pumping by the sarcoplasmic reticulum and maintenance of membrane potential at the sarcolemma require lesser amounts. In living muscle, ADP is very rapidly rephosphorylated to ATP. The most immediate source of energy for ATP re-synthesis is phosphocreatine. ATP re-synthesis is accomplished by the reaction:

$$H^+ + ADP + phosphocreatine \underset{}{\overset{creatine\ kinase}{\rightleftharpoons}} ATP + creatine + H_2O.$$

CHAPTER 3 MUSCLE CONTRACTION AND ENERGY METABOLISM

CONTRACTION PHASE

Resting state
↓
Motor nerve action potential arrives at motor end plate
↓
Acetylcholine released, sarcolemma and membranes depolarized
(Na^+ flux into fiber)
↓
Action potential transmitted via T-tubules to SR
↓
Ca^{2+} released from SR terminal cisternae into sarcoplasm
↓
Ca^{2+} binds to troponin C
↓
Tropomyosin moves to expose myosin binding site on actin
↓
Actin-myosin crossbridge formation
↓
Myosin ATPase activated & ATP hydrolyzed
↓
Repeated formation & breaking of crossbridges resulting in sliding
of filaments and sarcomere shortening

RELAXATION PHASE

Cholinesterase released and acetylcholine breakdown
↓
Sarcolemma & T-tubules repolarized
↓
SR Ca^{2+} pump activated & Ca^{2+} returned to SR terminal cisternae
↓
Tropomyosin blocks myosin binding site on actin
↓
Actin-myosin crossbridge formation terminated
↓
Mg^{2+} complex formed with ATP
↓
Passive sliding of filaments
↓
Sarcomeres return to resting state

FIGURE 3.8 | Flow diagram of the events during a complete muscle contraction-relaxation cycle.

This reaction occurs in the sarcoplasm and is catalyzed by the enzyme creatine kinase. Therefore, ATP broken down during a contraction is rapidly restored. Unless special precautions are taken to prevent ATP re-synthesis from **phosphocreatine,** the reaction occurs so rapidly that ATP break-down during a single muscle twitch cannot be measured; only phosphocreatine breakdown is seen. Concentration of phosphocreatine in resting muscle is about twice that of the resting level of ATP. Therefore, it also is subject to depletion during extended periods of contraction and must be replenished during rest periods by other mechanisms. **Rephosphorylation** of creatine occurs at the mitochondrial membrane.

The most efficient mechanism for ATP synthesis is a series of reactions collectively referred to as **aerobic (with oxygen) metabolism**, in which food nutrients, such as carbohydrates, proteins, and lipids, are degraded to carbon dioxide and water, while part of the energy released is used to form ATP. The degradation products or metabolites of these nutrients may all pass through some or all of this series of reactions. Although glucose constitutes only a fraction of the total nutrients used in the body, its metabolism when derived from glycogen, is used in this section to illustrate aerobic metabolism.

Recall, from reading Chapter 2, that glycogen (composed of glucose molecules) is stored in muscle as an energy reserve and constitutes about 1 percent of the muscle by weight. Glycogen is broken down into individual glucose molecules by a process known as **glycogenolysis**. The product of glycogenolysis is glucose 1-phosphate. The entire process occurs in the sarcoplasm. Each

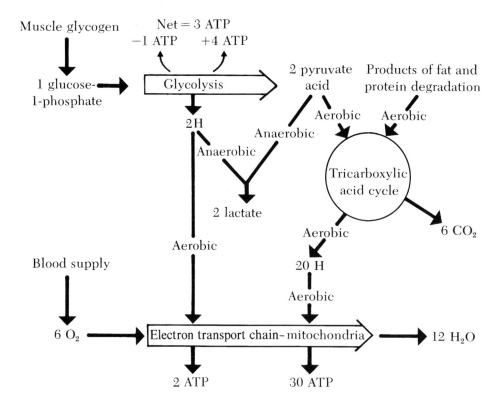

FIGURE 3.9 | A diagrammatic illustration of the pathways that supply energy for muscle function. One molecule of glucose 1-phosphate split from glycogen is degraded to CO_2 and H_2O in the glycolytic pathway, tricarboxylic acid cycle, and electron transport chain. The energy yield (in terms of molecules of ATP produced) is indicated at each step. When oxygen is limited, energy may be supplied by glycolysis and conversion of pyruvic acid to lactic acid. Courtesy of W. H. Freeman and Company.

glucose-1-phosphate is subsequently divided into two 3-carbon fragments and then into pyruvate through a series of reactions known as **glycolysis**. (See chapter 5 for a more complete discussion of glycolysis.)

The useful energy yield from glycolysis is three rephosphorylations (3 ADP converted to 3 ATP) and 2 hydrogen ions (H+) per glucose l-phosphate molecule taken from glycogen. These ions are accepted by a carrier compound, nicotinamide adenine dinucleotide (NAD+), to form NADH and transported to the mitochondria for use in other rephosphorylation events.

The second part of aerobic metabolism is a series of reactions collectively referred to as the **tricarboxylic acid cycle** (TCA cycle) or Krebs cycle, which occurs in the mitochondria. In this cycle, **pyruvate** is sequentially broken down into carbon dioxide, which readily diffuses out of the cell to the bloodstream and is removed as waste, and H+, which is accepted by the carrier NAD+, to form NADH. Degradation products of fatty acids and proteins also may enter this cycle and be converted into useful energy (Figure 3.9).

Most useful rephosphorylation occurs in the third element of aerobic metabolism, the **electron transport chain**. The electron transport chain is a group of iron-containing enzymes located in the mitochondria, along with the TCA cycle enzymes. In the electron transport chain, H+ from glycolysis and the TCA cycle are transferred from NAD+, and are combined with molecular oxygen to form water. A large part of the energy released is used to rephosphorylate ADP, and the remainder is lost as heat. For each pair of H+ ions from the TCA cycle, three ATPs are produced. In addition, two ATPs are produced for each pair of H+ ions that are released in glycolysis.

Under aerobic metabolism, when one molecule of glucose is split from glycogen and carried through this entire sequence of reactions the net energy yield would be 35 molecules of ATP. Three ATP molecules are obtained from glycolysis along with 2 H+, which will yield 2 ATP molecules in the electron transport chain. At the end of glycolysis, one glucose molecule from glycogen will yield two pyruvate molecules. Each pyruvate molecule will yield 10 H+, for a total of 20 H+ in the TCA cycle. These 20 H+ are converted, in turn, to 30 ATPs in the electron transport chain.

If a muscle is working slowly, and oxygen is supplied in adequate amounts, aerobic metabolism and phosphocreatine breakdown can adequately meet the energy needs of the tissue. However, when the muscle is contracting rapidly, the lack of oxygen may limit ATP re-synthesis via aerobic metabolism. Under limited oxygen supplies, **anaerobic metabolism** is able to supply energy to the muscle for a short period of time. A major feature of anaerobic metabolism is that **lactate** accumulates in the tissue. When the oxygen supply is inadequate, all of the H+ released from glycolysis and the TCA cycle cannot combine with oxygen. Also, as previously indicated, hydrolysis of ATP to ADP also produces H+. Thus, H+ ions accumulate in the muscle. Some of the excess H+ are then used to reduce pyuvate to lactate, which generates NAD+ that is required for one reaction in the glycolytic pathway. This permits glycolysis to proceed at a rapid rate. As mentioned above, each glucose yields three ATP molecules in glycolysis, so anaerobic metabolism can supply energy for muscle function. It is important to emphasize that the amount of energy available to muscle tissue using this anaerobic route is limited because as more and more lactate and H+ accumulate in the muscle, the pH is lowered. When pH values reach about 6.5, the rate of glycolysis is drastically reduced, with a proportional reduction in ATP resynthesis. Under these conditions, fatigue develops quite rapidly, and the muscle is no longer able to contract due to insufficient energy and excess acidity (low pH).

This whole process is quite typical in the muscle of most animals. As muscle recovers from fatigue, lactate is transported out of the muscle via the bloodstream, and is reconverted to glucose in the liver or metabolized to carbon dioxide and water by the heart (via a specialized enzyme system). Muscle ATP and phosphocreatine are replenished by normal aerobic metabolism. The recovery process occurs quite rapidly for a slight case of muscle fatigue, or may require extended periods if the fatigue is severe.

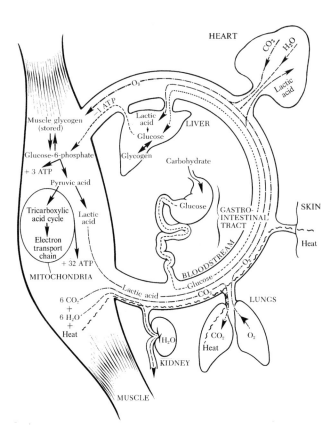

FIGURE 3.10 | Cyclic nature of the pathways that provide energy for muscle conteraction and heat production is summarized in this diagram. Courtesy of W. H. Freeman and Company.

The cyclic process of energy provision for muscle contraction and function is summarized in Figure 3.10. To follow this process, begin with the gastrointestinal tract where nutrients (potential energy-producing compounds) are absorbed by the body; in the case illustrated, glucose. Glucose is carried by the circulatory system either to the liver where it can converted to glycogen for storage or to the muscle, where it may be metabolized for energy immediately or stored as glycogen. Liver glycogen may be hydrolyzed to glucose and carried to the muscle as needed. In muscle, glycogen is metabolized to pyruvate via the glycolytic pathway yielding a small amount of ATP. Pyruvate may be then further metabolized in the TCA cycle and electron transport chain to produce water and carbon dioxide, yielding significantly greater amounts of ATP, or in the absence of oxygen, lactate. Lactate, water and carbon dioxide are removed from the muscle by the bloodstream. Carbon dioxide is expelled from the body through the lungs, water is excreted though the kidneys, and lactate may be re-synthesized to glucose in the liver or metabolized to water and carbon dioxide in the heart. Much of the energy liberated from metabolism is not captured for muscle contraction and is released as heat in the muscle to maintain body temperature. Excess heat is removed by the bloodstream and is dissipated through the skin and lungs. Thus, we can see that a dynamic system is available to provide energy to muscle. It is only during periods of very rapid muscle contraction that this system is unable to keep pace with demands. But when this occurs, fatigue develops rapidly and the muscle ceases to contract to allow time for recovery.

4

PRINCIPLES OF ANIMAL HANDLING AND HARVESTING

OBJECTIVES: *Understand biological, functional, and social principles for the harvest of meat animals, including antemortem and postmortem practices to ensure optimal food safety and quality. Describe the principles and procedures for animal transport and handling from production sites until harvest begins. Understand the impact of stressors on animal physiology and how facilities and practices can be modified to optimize animal well-being and reduce stress. Describe the various antemortem and early postmortem practices that affect the quality of meat.*

Key Terms

- Lairage
- Homeostasis
- Flight zone
- Point of balance
- Livestock Weather Safety Index
- Thermal neutral zone
- Shrink
- Fill
- Live hanging
- Downer
- Stun
- Exsanguination
- Captive bolt
- Stunning box
- Knock box
- Sticking
- Heart stick
- Shoulder stick
- Blood splash
- Blood spots
- Blood speckling
- Kosher
- Halal
- Scalding
- Singe
- Slack scald
- Hard or full scald
- Plucking
- Evisceration
- Bung
- Weasand
- Viscera
- Pluck
- Dressed
- Giblets
- Caul fat
- Split
- Steam vacuum
- Steam pasteurization

INTRODUCTION

Food Safety

For centuries people have harvested animals for food. Many traditional harvesting and processing practices inadvertently also contributed to safety of meat products. For example, widespread illness seldom resulted from meat or animals because small groups of animals were harvested on the farm for family consumption or in very small slaughter plants or butcher shops that sold their meat products in a limited geographical area (Figures 4.1 and 4.2) In contrast, today's modern processing plants harvest thousands of animals on a daily basis (Figure 4.3) in a single location and distribute the products nationally or in a multi-state area. To accomplish this task, these plants employ hundreds of workers, all focusing on single repetitive processes in the overall harvesting process. The potential for cross contamination from one animal or carcass to another combined with the wide distribution of products creates the potential for food safety threats to affect a large number of people. In response to these potential threats and because of the large costs incurred when a product re-call must be implemented, all segments of the meat industry have implemented programs to reduce the potential for food-borne illness caused by microbial contamination or other food safety issues. Such efforts are most effective if they begin during animal rearing and continue during animal transport, pre-harvest, and slaughter. Details of the types and kinds of contamination are covered in Chapter 9.

Animal Well-being

In addition to addressing food safety concerns, much emphasis is placed on the animal's well-being during all aspects of animal rearing, during transport and immediately prior to and during the harvesting process. Many of these practices are consistent with husbandry goals of the animal producer and many are mandated by law. Laws and regulations also apply to how processing companies provide for animal well-being during the harvesting process. Furthermore, reducing the stress of animals during harvesting is economically beneficial to the company as handling can greatly impact ultimate product quality. This chapter covers the principles and processes used prior to, during and immediately after the harvest of meat animals, reasons for implementing these processes and how they affect the animal or final product quality.

PERSONNEL

As is the case with virtually every process involving human intervention, employee commitment to the ideals and standards of optimal animal well-being, humane animal handling and meat safety is critical for success. Individuals engaged in the meat industry must understand that they can dramatically impact animal well-being and ultimately meat quality at virtually every point in the harvesting process. Individuals must also understand the "critical points" in the process that have the greatest impact on contamination and ultimate safety of products. Managers must very proactively provide educational programs and incentive-based activities that keep employees striving to maintain a high level of performance and high levels of quality and safety in the end product.

TRANSPORTATION

Most animals are reared in highly integrated facilities under intensive rearing systems. This is certainly the case for most poultry and pork produced in the US. Cattle, on the other hand, are raised in more extensive systems, where animals are less confined. Even so, the majority of cattle are fed in rather high volume feedlots for a period of time before harvest. These feedlots are considered intensive management schemes. Because the numbers of animals populating intensive production units are quite large, these operations may send multiple truck loads of animals to processing facilities on a daily basis. Many animal rearing facilities may be located in close proximity to the processing plant

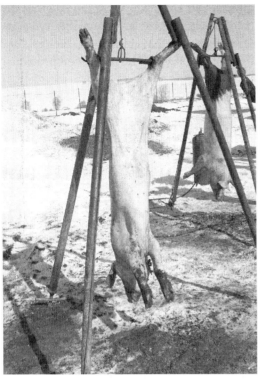

FIGURE 4.1 | Home slaughtering showing pork carcasses hung outdoors.

FIGURE 4.2 | Typical store-front of the local meat shop.

FIGURE 4.3 | Modern hog slaughtering facilities have the capacity to process in excess of 10,000 pigs per day.

and as a result, animals may to be in transit a short time before reaching their destination. At the processing plant, they remain in **lairage** or a place to rest until harvesting. In contrast, other production facilities may be located some distance from the processing plant and require extended travel time. By law (49 USC Sec 80502), "*...animals being transported may not be contained in a vehicle for more than 28 hours without being unloaded and given feed and water, and be allowed to rest.*" Obviously, animals transported long distances may require different lairage conditions than those transported short distances. The stress of transport can manifest itself in a physiological response by the animal. The whole process of loading animals into a truck, transporting them various distances and then unloading them at a processing plant can dramatically impact the animal's sense of well-being, and therefore, its physiology. Because an animal's perception of well-being can impact physiological responses, which in turn can affect ultimate meat quality development, the industry attempts to reduce these types of stress as much as possible.

INSPECTORS

Although inspection is covered in some detail in a later chapter, the harvesting process is understood more clearly if some of the principles of inspection are discussed here. Inspectors work for the government, either federal or state, to ensure that animals are harvested in a humane manner and under sanitary conditions. These individuals are positioned throughout the plant and are charged with monitoring all harvesting and ancillary processes that may affect the wholesomeness of the final product. Only healthy animals are eligible for harvesting. To that end, inspectors are charged with antemortem inspections to make sure that animals appear to be healthy. Although somewhat subjective in nature, inspectors are given free access to handling and holding facilities where animals are off-loaded from vehicles and allowed to rest prior to harvest. They observe animals for signs of unhealthiness, which are often manifested in some sort of deviation from a physical norm. For example, extremely thin animals are often highly scrutinized as are those with abnormal gaits or movements. Once inspectors are satisfied that animals are healthy, they are allowed to proceed to the harvesting process.

ANIMAL PHYSIOLOGY AND HOMEOSTASIS

One of the most detrimental states for an animal prior to harvesting is that of heightened anxiety, where the animal experiences or perceives danger. During these times, animals, like humans, release hormones that stimulate the well-known "fight or flight" response. Epinephrine and norepinephrine are particularly important because they have dramatic and immediate effects on the cardiovascular system and mobilize energy substrates needed to fuel a highly excited state. When animals get excited, their heart rate increases and they rapidly mobilize energy from reserves so that they can escape or combat their captor. Animal handlers attempt to eliminate, or more appropriately, try to reduce this physiological response prior to harvesting, especially immediately (< 1 hour) prior to slaughter. Prodding, poking, or other stimuli that result in a heightened level of awareness are not beneficial for the well-being of the animal, the quality of the meat harvested or the safety of workers.

Though an understanding of animal physiology is important for anyone involved in meat animal production, an exhaustive review of the area is beyond the scope of this textbook. However, one concept is particularly important: all living organisms function optimally within a relatively narrow range of physiological conditions. Maintenance of this well-orchestrated physiological state is termed **homeostasis**. Homeostatic regulation gives an organism the ability to survive under many different, sometimes adverse, environmental conditions that may include extreme variations in temperature, oxygen availability and psychological stresses. Homeostatic control is largely by the nervous and endocrine systems (Figure 4.4). These two systems serve as communication and triggering mechanisms that coordinate adjustments in the function of various physiological systems during periods of stress. To accomplish this collective

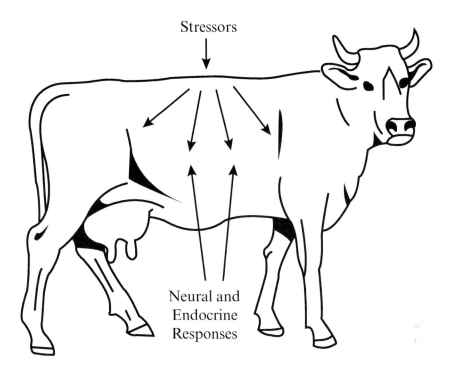

FIGURE 4.4 | Homeostasis. Environmental, physical or psychological stressors elicit neural and endocrine responses to help animals maintain balanced physiological conditions during a range of conditions.

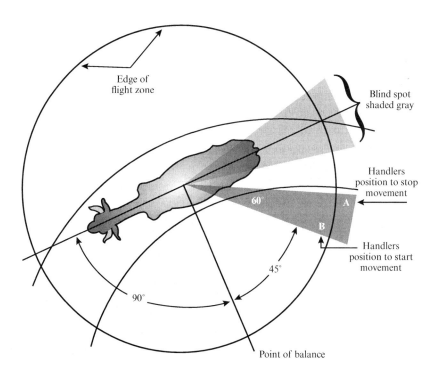

FIGURE 4.5 | Flight zones in farm animals.

task, individual organs, tissues and even cells possess individual homeostatic control mechanisms, yet they function and interact to maintain an environment under which each can perform its function efficiently. Most organs in the body, including muscle, function efficiently only within a narrow range of physiological conditions (pH, temperature, oxygen concentration, and energy supply). During the harvesting process, tissues of the body attempt to maintain homeostasis but fail because various organs and physiological systems are systematically disrupted. As a result, living tissue is transformed into meat (covered in detail in Chapter 5).

FLIGHT ZONE AND POINT OF BALANCE

The less complex nature of animal eyes and their placement on most meat animals makes an animal's use of visual cues very predictable. As shown in Figure 4.5, all animals have a **flight zone** or a virtual perimeter surrounding them where encroachment by humans or other perceived threats will cause the animal to react and move away from an intruder. This theoretical distance is different depending on the type and extent of lifetime interactions an animal has had with humans or other animals. Broaching this distance should be minimized whenever possible because in most handling facilities, there is always the possibility an animal may elect to "fight" if the path of "flight" seems less accommodating. Another important area for animal handlers is the virtual position roughly perpendicular to the animal's shoulder, the **point of balance**. Positioning ones self anterior or posterior to this location will normally cause the animal to reverse their path or to move forward, respectively. Animals are completely blind to anyone located immediately behind them unless vocal cues are used, which understandably can be startling to the unsuspecting animal. As a result, this location is quite risky simply because animals can not see individuals at that location. As a result, animals can not evaluate the threat and, in the case of cattle, kicks are justifiably targeted to anyone or anything in that particular area. As a result, the best position from which to move animals in a forward direction is located between 30 and 45 degrees forward from the longitudinal axis of the animal and opposite the most posterior part of the animal.

ANIMAL MOVEMENT AND FACILITIES

Moving animals from a location where they have resided most of their lives is always stressful. Moving animals onto trucks can be even more stressful. However, when care is taken, animal stress can be reduced. Slope of ramps, alley-way flooring and siding, corners, air movement, changes in flooring levels, moving objects, loading ramp design and even size of contemporary animals can affect stress level. Animals naturally are reluctant to move into unfamiliar areas. This reluctance is further exacerbated by poor lighting conditions and unnecessary shadows created by poor gating and chute designs. Modern, efficient facilities make the appropriate accommodations to reduce these types of animal stresses. Perhaps one of the most dramatic changes in livestock handling in the past ten years has been dramatic reduction in electric prod usage. In the past, this device was used to force animals through the maze of alleyways and pens in antiquated and poorly designed facilities. Today, prods are mandated by USDA to have no more than 50 volts. But, because of the obviously aggressive nature of prod usage and the animal's response to it, the practice has been virtually eliminated and replaced with less aggressive, animal-friendly means of persuading animals to move throughout a facility. Less aggressive means of moving livestock through facilities are physical or visual barriers such as sorting boards or a Matadors cape, respectively. Other aids consist of rattle/shaker paddles or nylon flags which respectively target audio and visual stimuli centers to stimulate movement without eliciting an undue level of anxiety (Figure 4.6). These simple, yet well-documented changes have resulted in much more animal friendly handling facilities and improved product quality.

Because of their smaller size, chickens, turkeys, and ducks are easier to handle physically, yet their responses to human interactions and being

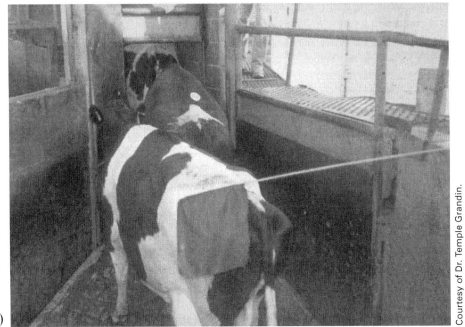

(A)

(B)

FIGURE 4.6 | Use of nylon flags and shaker/rattle paddles to encourage animal movement.

herded together, captured, and ultimately placed in smaller transportation containers are quite similar to the larger meat animals. Most broiler rearing houses contain about 100,000 birds. The process of capturing and crating birds and loading crates onto trucks is very labor intensive and subject to great variation in handling. Because nearly half the bruising in poultry occurs during the catching and crating process, automated devices have been developed to aid in gathering flocks from their housing. Even so, the process of placing live birds in transportation containers is arduous, time-consuming and physically demanding. Great care must be taken to educate handlers of the consequences of their actions during this process.

TEMPERATURE AND HUMIDITY

Ambient temperature and humidity are critical to animal comfort. Ventilation systems in on-farm facilities adjust rapidly to ever-changing ambient temperatures to insure optimal comfort and subsequently maximize animal performance. When animals are removed from their rearing facilities, as is the case when they are transported to processing plants, they are more vulnerable to stresses caused by adverse ambient temperatures. At one extreme, it is not uncommon for temperatures to reach well below freezing in some regions of the US during the winter. Under these extreme conditions, bedding material and ventilation of the vehicle must be adjusted in an attempt to meet the comfort level of the animals. For example, under conditions of less than -10° C (12° F), pigs require heavy bedding and minimal ventilation from side-slats on semi-trailers. In contrast, maximal ventilation and minimum bedding is needed during most other times when ambient temperatures are over 10° C (50° F).

Humidity likewise contributes significantly to animal comfort. Figure 4.7 shows the curvilinear

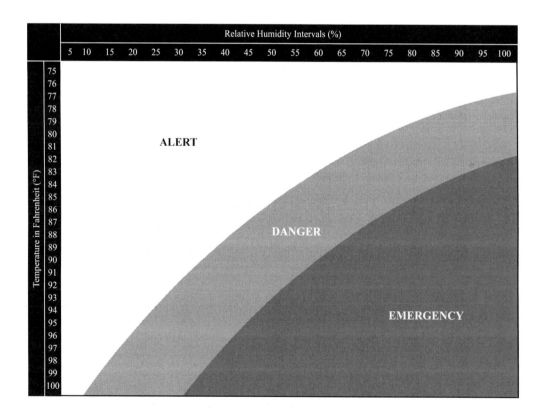

FIGURE 4.7 | Livestock Weather Safety Index

relationship between temperature, humidity, and livestock comfort. Extreme caution should be taken when the **Livestock Weather Safety Index** reaches the danger designation. In cases where animals must be moved during this designation, loading densities should be greatly reduced. As a rule of thumb, livestock should not be moved between 11:00 a.m. and 4:00 p.m. during the summer when the heat index, a combination of temperature and humidity, is greater than or equal to 100° F (38° C). Most often, animals are loaded in the early morning for delivery sometime late morning. If possible, providing moisture to pigs in the form of sprinklers or dampened bedding (shavings) can reduce heat-induced discomfort.

Great care must be exercised when moving poultry to processing plants, especially during hot weather. The thermal neutral zone is a range in ambient temperatures where the animal does not need to alter its basal metabolic rate or change its behavior to remain comfortable. For a normal broiler, the **thermal neutral zone** is 13°–24° C (55°–75° F). When ambient temperature is above the thermal neutral zone, movement of air must be facilitated during transportation and great care must be made to continue such air movement if birds are held on the truck prior to processing. At the other extreme, trucks usually are covered in the winter, but this often results in greater moisture condensation adjacent to the cover, which will accelerate heat loss in those birds.

Even though extreme care is taken during the loading and unloading process, some animals become stressed. If these animals are identified quickly, it is possible to keep them from getting worse and they should recover with ample rest. Indicators for stressed pigs are open mouth breathing, squealing, blotchy skin, stiffness, muscle tremors, and reluctance to move.

ANIMAL SIZE AND DENSITIES

As animals grow, they need more space. This increase in space is accommodated during the production phase by either moving fewer animals into similar spaces or by increasing the space per animal. Likewise, during transportation, great care must be taken when loading trucks. As a rule of thumb, typical market weight (~115 kg) pigs require approximately 0.4 square meters of trailer space for optimal comfort. Typical market steers (550 kg), without horns, need about 1.4 square meters of trailer space. As a result, one could expect approximately 170 market weight pigs or 50 market weight steers on a typical semi-trailer. Of course, this number depends on the distance traveled, size of the animal, and truck dimensions.

FEED WITHDRAWAL

Feed often is removed from animals 12-24 hours prior to transportation and harvesting. One rationale for this practice is to reduce the overall weight being transported. This practice reduces animal weight by reducing the weight of the gastrointestinal tract. In addition, feed consumed during this time is not fully utilized by the animal for growth or maintenance. Finally, feed withdrawal reduces ingesta movement through the digestive tract, which can reduce excreta during transport. This is particularly important as most microbial contamination of carcasses occurs from foreign matter located on the hide and hair of animals prior to harvesting. Thus, feed withdrawal is economically advantageous and can reduce chances of carcass contamination. But, even when feed is withdrawn, animals defecate during transportation. As a result of this and other weight loss, largely caused by dehydration, an animal's weight changes during transportation. This **shrink** is defined as that weight or fraction of weight (%) lost between the time of loading at the production site and unloading at the destination. Transportation shrink is a concern if animals are sold on a live weight basis because their live weight at the processing plant is less that that when they leave the farm. There is less concern when animals remain under the same ownership after leaving the production site until they are harvested. In this case, payment for animals is based on carcass weight. There is little shrinkage of carcass weight during transportation of animals. Another important feature is that shrunken animals (those containing minimal intestinal contents or feedstuffs) have less distended gastrointestinal tracts. Therefore, the risk of gastrointestinal tract lacerations during processing is reduced. This is particularly important because the gastrointestinal tract contains a large number of bacteria, which may be pathogenic. The reduction in **fill** will mathematically increase carcass yield and decrease the costs of handling manure in the slaughter facility. Finally, some data suggest that this practice improves meat quality but the evidence to support this thesis is rather scant.

LAIRAGE

As defined above, **lairage** is the time for animals to rest immediately prior to slaughter and after arrival at the slaughtering facility. This period of time can range from a very short to an overnight time frame. Lairage time should be no less than two hours. As discussed earlier, one of the most critical factors affecting meat quality is the animal's physiological state prior to slaughter. Though stress several hours prior to slaughter can have a dramatic impact on subsequent quality development (discussed later), the hour prior to slaughter is very critical, because hormones released at this time will have residual effects at a critical point when muscle becomes meat. Therefore, it is in the best interest of the processor to have enough lairage time for livestock to rest prior to slaughter. Usually, the density of animals in pens is closely monitored to ensure sufficient movement by all animals and there is free access to water at all times. Most commercial lairage facilities are equipped with misting units or drip sprinklers to keep livestock comfortable in the event of extreme heat.

HARVESTING

Animals are typically moved in groups from lairage to harvest. The herding nature of animals facilitates this activity, though with some difficulty. Depending on the design of the facility and the species harvested, animals may be rendered immobile by herding animals to conveyor systems, which may physically elevate animals off the floor or may simply move animals toward a particular location as a group. Either way, this eliminates their ability to resist movements toward a given destination. In the poultry industry, birds are shackled or hung upside down on a continuously moving rail. Once these animals are positioned on the conveyer system, the process can be continued with little resistance from the animal.

There are as many different handling systems as there are harvesting plants. Each plant is different and therefore automation differs. The most critical point is getting the animal loaded into the

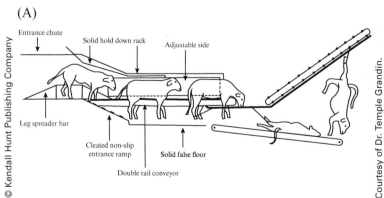
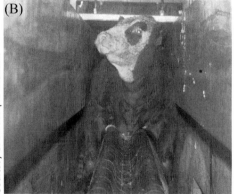

FIGURE 4.8 | V-belt design for moving cattle in a processing plant. (A) Illustration of the V-belt restrainer design. (B) Steer loaded into a V-belt restrainer.

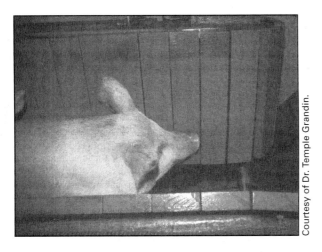

FIGURE 4.9 | V-design for moving pigs into stunning position.

FIGURE 4.10 | A downer (non-ambulatory) cow.

process. For cattle, they are often moved single file into an alley containing a V-shaped chain system (see Figure 4.8) where the floor is gradually sloped downward from beneath their feet. As a result, cattle are loaded on a conveyor, resting on their sternums and belly. This protects the cattle from hurting themselves or employees. Pigs, on the other hand, are gathered by a V-belt system that essentially performs the same task of removing the animals feet from the ground but does so by means of two apposed belts as shown in Figure 4.9. Once these animals are positioned, they are effortlessly transported to the initial process of harvesting, which is stunning.

As outlined above, poultry may be transported in small cages or crates, which can be conveniently loaded on a semi-trailer. Once at the processing plant, trucks are staged for entering the facility, which acts as the lairage period. Most often, birds are lairaged on the truck in cages. During this time, every attempt is made to maintain favorable environmental conditions to ensure optimal comfort and reduce antemortem stress. Once identified for processing, poultry can be removed from the cages and hung by their feet on shackles within an elaborate conveyor system that weaves throughout the plant. This procedure, often referred to as **live hanging**, is used when electricity is used as the stunning method. This has met with much resistance by a number of animal welfare-based organizations, especially in Europe, but the process is still quite popular in the US. When chemical immobilization (discussed in more detail below) is used, poultry may be immobilized prior to hanging.

NON-AMBULATORY ANIMALS

Animals not capable of moving on their own are considered non-ambulatory or **downers** (Figure 4.10). United States meat inspection regulations specify that all non-ambulatory cattle arriving at a processing plant should be condemned and destroyed. This is to prevent cattle infected with bovine spongiform encephalopathy (BSE) from entering the food supply because BSE cattle normally are downers. This does not mean that all downers have BSE, however. Other non-ambulatory animals (pigs, sheep and goats) can still be inspected, passed and used in the food supply, though the future of this practice is currently being debated aggressively. Given the nature of transportation crates and volume of birds handled on one vehicle, it is easy to imagine that dead or dying birds could be processed. This is strictly prohibited and can be monitored on the line, as birds dying in transport yield a carcass with much different characteristics.

STUNNING AND IMMOBILIZATION

The first step of the harvesting (slaughtering) process is to **stun** and immobilize animals, that is, to render them unconscious and insensible to pain. Methods for stunning are discussed below. Immobilization renders live animals and poultry unconscious and immobilized in preparation for **exsanguination** (bleeding); however death can occur from stunning and immobilization if the processes are not monitored frequently. The following physical signs are associated with a 95 percent effective stun.

- Head must be limp and floppy
- No voluntary blinking
- No rhythmic breathing
- No response to ear or nose pinching
- No arched backs
- No vocalization

Limb movement or kicking during this period is ignored and considered involuntary movement, though great care and experience is necessary to verify the involuntary nature of these movements. Extensive struggling and excessive movement, even though the animals are unconscious, can result in small hemorrhages, which allow blood

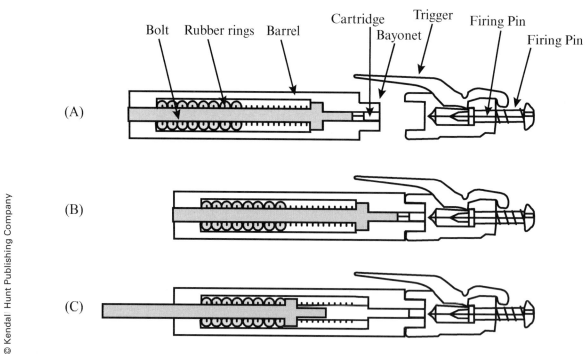

FIGURE 4.11 | General design of a captive bolt stunning apparatus. (A) The apparatus disassembled. (B) Unit assembled and ready to be discharged. (C) The unit after firing.

to leak from the capillaries into muscle or other tissues. In addition to these blemishes, improper stunning of birds results in broken wings, broken clavicles, red wing tips, and poor bleeding.

Reasons for stunning are twofold. First, in most developed countries, laws require humane immobilization methods that minimize needless suffering of animals. In the US, this is provided for in the Humane Slaughter Act. Moreover, it is respectful of animals to perform this act in a manner that is as pain-free as possible. Second, safety is of utmost importance for employees performing this step in the harvesting process. Stunning gives workers time to perform critical procedures that otherwise would place them in harm's way if animals were not rendered unconscious and immobile. Third, stunning improves meat quality development. As mentioned above, excessive movements can result in hemorrhages and subsequent bruising in the meat. If excessive hemorrhaging occurs, not only does it make muscle tissue unsightly to the consumer and more vulnerable to bacterial growth, it also inhibits maximal blood removal from the carcass. Therefore, effective stunning is a very crucial step in the harvesting process. Exsanguination (blood removal) occurs immediately after stunning as will be discussed in the next section.

Physical stunning

Physical stunning immobilizes animals by eliciting trauma to the head. Many physical methods are available such as sledge hammers and firearms. The instrument of choice in most harvesting operations, especially for stunning cattle, goats, and sheep, is known as the **captive bolt** (Figure 4.11). This device, resembling a firearm in some cases, causes physical trauma to the brain by propelling a retractable bolt or metal shaft into the skull of an animal, thereby rendering them unconscious and insensible to pain. In smaller facilities, propulsion of the bolt is by a powder charge, similar to blank cartridges used in starting pistols. In larger facilities, these devices are operated using compressed air. Some captive bolt devices have a large, blunt end attached to the bolt, which does not penetrate the skull instead delivering a concussive blow that stuns the animal. Placement of the captive bolt is critical in order to traumatize the brain. Correct placement can be visualized as the intersection of two lines, connecting the eyes and ears on opposite sides of the head (Figure 4.12). Placement of the captive bolt at this location will result in a contusion of the skull and the brain by the captive bolt. Recently, some stunning technicians have targeted a few centimeters above the intersection to guarantee striking the brain cavity and compensating for the angle of the stunning device.

Mechanical stunning is impractical in large poultry establishments. However, it is useful in emergencies or to immobilize small numbers of

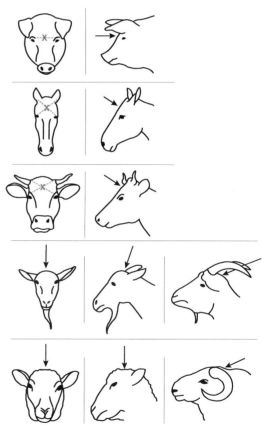

FIGURE 4.12 | Schematic showing the approximate location of captive bolt placement for an effective stun of various farm animals.

FIGURE 4.13 | General design of equipment used to restrain cattle for stunning.

Courtesy of Elton Aberle.

live poultry, which makes it a practical method in small and very small establishments. Decapitation, cervical dislocation, and blunt trauma to the head are the most common forms of mechanical immobilization in poultry operations.

Large animals such as cattle are often restrained for mechanical stunning in a **stunning box**, or **knock** box (Figure 4.13). (The latter term is historical stemming from a time when individuals positioned themselves above the cattle and "knocked" them in the head with long-handled sledge hammers to render them unconsciousness.) Stunning boxes are built from heavy-duty iron or stainless steel material to withstand the aggressiveness of a 600 kg-plus animal. The drawback of such a box is that it is labor intensive and slow, as the animal's head is free to move away from the stunner. To improve this process, head restraints may be used to minimize head movement and stunning may be incorporated on the line, as part of the conveyor system. Either way, cattle movement is minimized so that a more precise, efficient placement of the stunning gun is possible.

Electric stunning

Electric stunning had been a predominant stunning method in most swine and poultry processing facilities for several years. In the case of pigs, a current is passed through the animal and results in an epileptic-like seizure, which creates an immobilized state. Though many designs exist for the electric wands used to deliver electricity to the pig, two overarching approaches prevail: the head-only stun or the head-to-back stun (Figure 4.14). The former stun is partially reversible, while the latter design causes cardiac arrest. An effective stun is a function of amperage and voltage, and the electrical resistance of the animal, which is a function of animal size and its lean to fat composition. Moreover, the effectiveness of the electric stun can be affected by the positioning of the wand according to anatomical landmarks and the degree of contract between the wand and the animal. Often, pigs are wet from conditions in lairage, or have been subjected to a mist, which improves the electrical stun. Delivery of amperages of less than 1.25 amps to market weight animals may cause paralysis

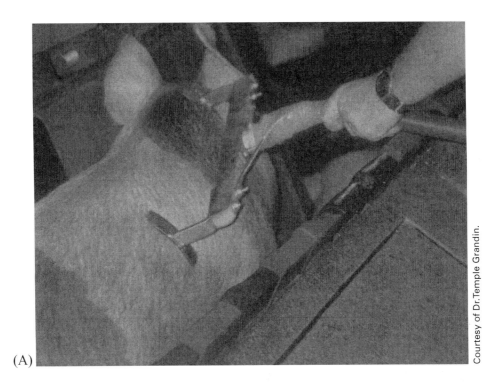

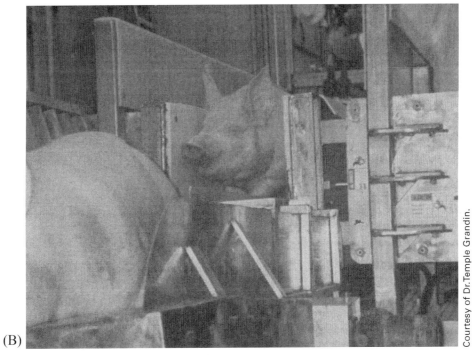

FIGURE 4.14 | Illustration of proper electric stunning in pigs. (A) Electrodes are applied to a pig for a head-to-back stun. (B) Stunned pig exiting the stunning restrainer.

instead of the desired grand mal seizure and therefore, should not be used. Voltages of less than 250 volts should not be used, though less critical than amperage.

Electrical stunning is the primary method to stun poultry. Birds are shackled live, hung on a moving chain and immediately transported to and partially submersed in a saline solution that is electrified. Saline is also sprayed on the feet and shackles to ensure good contact with the grounded rail. Saline is used because of its electrical conductivity properties.

Carbon dioxide (CO_2) stunning

Within the past ten years, CO_2 stunning has become popular, particularly for pigs. Most often cited as justification for adopting this stunning practice is improved ultimate product quality, especially color and reduced wateriness in fresh product. Also, it is suggested the process is more humane, though others have contested this notion. Carbon dioxide competes with oxygen for binding to hemoglobin and thus, animals placed in air containing 80 to 90 percent CO_2 lose consciousness from extended hypoxia, or reduced oxygen availability. During this time of unconsciousness, which normally occurs between 13 and 30 minutes after exposure to these conditions, animals are exsanguinated so consciousness in not regained. Poultry stunned using this manner, require slightly less CO_2 (65 percent). Other inert gases such as nitrogen and argon can be used, though effective concentrations may differ.

The organization of a pig processing line that uses CO_2-based stunning differs among facilities, but always involves a means to move pigs through some sort of pit that contains a high concentration of CO_2. Carbon dioxide is heavier than air and thus can be contained effectively in the enclosed pit. One system uses a batch-type approach in a Ferris Wheel-like design, where pigs are loaded at floor level into a gondola, which then descends into a pit or basement of CO_2 (Figure 4.15). The unconscious animals then ascend from the pit, are exsanguinated and shackled to a rail. Alternatively, other systems consist of a large belt-like conveyor on which pigs stand as they descend into a CO_2-containing pit. The speed of the process can be controlled to meet the conditions necessary for an optimized stunning protocol.

Relatively few controlled atmosphere systems (CAS) are operating in the poultry industry, though those in existence use CO_2. These systems allow stunning prior to shackling. Birds may remain in their transportation crates and the crates are moved through a CO_2 or other gas tunnel on a conveyor belt. After transition through the tunnel, crates are opened and stunned birds are shackled. Another equally effective strategy is to use on-truck stunning. In this case, an entire section of the truck is subjected to a gas-based stun. Either of these approaches is beneficial because stunned animals are relaxed and refractive to stressors and do not struggle as they are shackled and hung, which reduces potential quality defects. The disadvantage is a protracted time from stunning to exsanguination, which improves bleeding time but can negatively affect feather removal.

EXSANGUINATION (BLEEDING)

Removal of blood is known as **exsanguination** and is accomplished by severing the carotid arteries and jugular veins. In cattle and pigs, the preferred location is at the base of the neck where it joins the sternum (Figure 4.16) because the cut does not affect any muscles of the neck. It is important to sever the carotid arteries to accomplish maximum blood removal; severing only the jugular veins is not as effective. This method of severing the arteries and veins is also known as **sticking**. It also is known as a **heart stick** because the point of the knife may be inserted far enough to reach under the sternum. A **shoulder stick** occurs when the process is done incorrectly and results in blood spots and clots in some shoulder muscles. Lambs and birds are exsanguinated by severing the carotid arteries and jugular veins close to the head.

Animals should be exsanguinated as quickly as possible after stunning. Electrically stunned pigs should be bled within 10 seconds of the stun, while those CO_2-stunned should be bled within 30 seconds of stun. An optimum exanguination

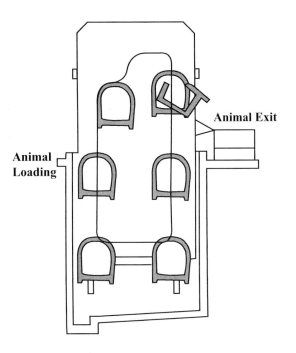

FIGURE 4.15 | Schematic of a gondola-type CO_2 stunning apparatus for pigs.

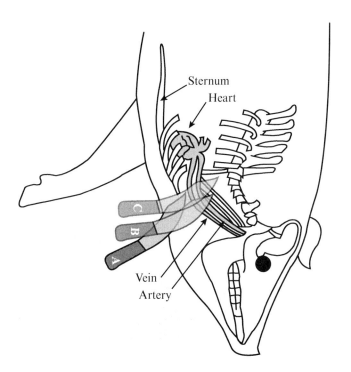

FIGURE 4.16 | Illustration of placement of the knife for the normal sticking process in pigs. (A) Placement of the knife at insertion. (B) and (C) Sequential movements of the knife as the carotid arteries and jugular veins are severed. The knife is then withdrawn.

process, at least for pigs, is close to 5 minutes long. Most animals are bled while hanging vertically from the overhead rail. However, some facilities elect to bleed pigs immediately after stun, while pigs are positioned on their side on a flat conveyor. Birds that are electrically stunned contort in such a manner, wings tucked and neck arched, that severing of the major blood vessels in the neck is facilitated. Low voltage (10-25V, 20 mA) stunning is considered partially reversible in birds, while high voltage (150 mA) stunning is considered irreversible and often mandated for fear of return to consciousness. However, if birds are bled within 7–10 seconds, birds don't recover because of failing blood pressure and circulatory failure. High voltage stunning essentially results in electrocution of the bird and cardiac arrest. Unfortunately, high voltage stunning causes significant carcass damages and reduced yields.

Immediately during exsanguination, the body's homeostatic control mechanisms perceive a drop in blood pressure, the heart begins pumping faster and peripheral vessels constrict in an attempt to maintain pressure and force blood to the vital organs. These responses produce greater blood removal than if the heart is not beating. But, only about 50 percent of total blood volume can be removed via this bleeding process. The remainder is pooled in the vital organs. Because blood is an excellent medium for growth of spoilage organisms and excess blood in meat cuts is unacceptable to consumers, a thorough bleeding is essential to the slaughtering process. Exsanguination quickly after stunning is important because stunning, particularly by electrical or mechanical means, causes an immediate increase in blood pressure. Also, the animal may exhibit involuntary movements. These factors can combine to cause product blemishes. Product blemishes are often a result of extended time between stun and exsanguination. Of particular significance are: **blood splash**, **blood spots**, **blood speckling** and **broken backs** (see Figure 4.17 in the color section). The first three blemishes often occur in whiter, larger, more glycolytic muscles and severely impact consumer acceptability and shelf life, as the properties of blood affect fresh meat color and potential bacterial growth. The latter, a fractured vertebral column, occurs when electrical stunning causes a severe contraction in the loin muscle and puts excessive pressure on the vertebrae in the middle of the back. Because there is massive hemorrhaging in this area this section of the loin must be trimmed extensively, which lowers the value of the loin and the entire carcass.

RITUAL SLAUGHTER

Some religions specifically prohibit stunning prior to exanguination in accordance with religious mandates and thus are exempted from legislation mandating stunning practices. **Kosher** refers to food processing accepted by the Jewish community as congruent with their beliefs. Accordingly, animals are not stunned and are exsanguinated by a single incision across the throat of an animal. The ritual of Kosher slaughter often involves a means of restraining cattle in a manner that exposes the throat. Once restrained with the throat fully exposed, the Shochet, or Jewish person conducting the slaughter, severes the veins, arteries, and trachea of an animal in the absence of any pausing, pressing, covering, tearing, or piercing action. Violation of this process results in meat that must be marketed as non-Kosher. Because consumption of blood is strictly forbidden by Jewish law, veins must be stripped from the carcass. Given the extensiveness of the capillary network in the beef hind quarter and the difficulty of performing a meticulous deveining, hindquarters are often sold in non-Kosher outlets.

The other frequent religious slaughter is termed **Halal**, which is Arabic meaning allowed or lawful. The process is termed Thabiha or Dhabiha and is similar to the Jewish equivalent of Kosher. Only healthy, blemish-free animals can be used for Halal product and must be slaughtered without stunning, using virtually an identical process to Kosher slaughters. No animals slaughtered in the name of anyone, other than Allah can be designated Halal meat.

SCALDING AND SKINNING

These processes occur immediately after exsanguination is complete. Cattle and sheep are almost always skinned to remove the hide and hair, while the hair usually is removed from pigs and feathers removed from birds, leaving the skin intact in each case.

The procedure that facilitates hair removal from pigs and feather removal from birds is called **scalding** and involves placing the animal or bird in hot water. The purpose is to loosen the hair in the hair follicle or the feather in the feather follicle so they it can be physically removed.

Though seemingly quite simple, all pig hair is not the same. Some is relatively fine and young, while others are older and coarser in appearance, and are positioned in larger, more robust follicles. Loosening of hair from the follicle is a function of both temperature and time. Placing pigs in water at 60° C (140° F) for 5 to 8 minutes satisfactorily releases hair from most follicles. The term "hard hair" is used to describe hair that is more difficult to remove using this approach. There is some evidence that hard hair occurs seasonally, predominately in the fall and winter. In these cases, the duration of the scald may be extended to 9 minutes. However, great care must be taken when increasing scald times as this process can affect the integrity of the skin and make it more subject to ripping or tearing. The skin may begin to cook if temperatures are too high which also damages the skin and may necessitate excess trimming of the carcass. Under normal scalding conditions, the skin is considered an impervious capsule to microbial entry so any increase in scalding times must be monitored carefully. Even more critical is the fact that scalding generally increases carcass temperatures by 1° C. While this appears nominal in the overall scheme of harvesting an animal, this change in temperature can easily impact the ultimate quality of the pork produced, especially given the muscle underlying the skin is actively metabolizing energy (discussed in detail in Chapter 5). Once the hair is loosened in the follicle, it is removed mechanically using an elaborate set of rubber paddles in the presence of water. Any hair remaining after this de-hairing process is singed from the carcass using quick bursts of flame.

The principles for feather removal are similar to those for hog hair removal. Immersion scalding is the most common technology in most poultry operations. Birds are mechanically moved through a tank of hot water. Immersion systems can be configured in a number of ways to combine various mechanical and physical actions. Scalding targets the muscles that hold the feathers in place. As is the case for hair removal in pigs, scalding temperatures vary with the type of poultry and the endpoint of the bird being processed (Table 4.1). A semi-scald or **slack scald** is performed in 50°–55° C (125°–130° F) water. The time and temperature combination leaves the epidermal layer intact and results in a more attractive, consumer acceptable carcasses. Water at slightly hotter temperatures—60° C (138°–140° F)—loosens the outer layer of skin and results in a loss of some of the yellow pigment in the skin but the feathers are easier to remove. **Hard-scalds** or **full scalds** require water temperatures close to 65° C (140°–150° F).

TABLE 4.1	Times and Temperatures for Maximal Feather Removal in Birds		
Species	**Time (s)**	**Temperature**	
		Celsius	**Fahrenheit**
Broilers (hard scald)	30–75	59–64	140–148
Broilers	60–120	60	140
Broilers (soft scald)	60–120	51–54	125–130
Turkeys	50–125	59–63	138–145
Quail	30	53	127
Waterfowl	30–60	68–82	154–180

This optimizes feather removal, especially small pin feathers, but results in carcasses that dry out quickly and have a less desirable appearance. Waterfowl and mature birds are often scalded at this temperature or higher. **Plucking** or removal of feathers, like hair removal, is accomplished using mechanical pluckers.

In the case of cattle and sheep, the hair or wool and skin are removed intact to yield a hide from cattle, which can weigh as much as 50 kg from large cattle with thick hides, or a pelt from sheep. Removal of the hide begins with removal of the feet usually using shank/hock (lower leg) cutters. Great care is taken at this point to ensure that contamination on the outside of the animal is not transferred to the carcass. This is particularly difficult as a fair amount of manual labor is needed to move the appendages. Once the feet and shanks are removed, the hide is marked-out, skinned or reflected (released) from the legs. In most modern processing plants, hides are pulled mechanically from the carcass using chains and pneumatic cylinders. Again, it is worth mentioning that removal of the hide or pelt is a critical step in hygienic dressing of cattle and sheep because of the potential introduction of contamination. A number of cattle slaughter plants have investigated hide washing stations in an attempt to reduce the chances of carcass contamination by the hide. However, results of such attempts have been mixed most likely because of variation in the amount of material on the surface of animals. Furthermore, adding water to already contaminated surfaces, unless properly managed, could exacerbate contamination due to the tendency of liquids to accumulate on parts of the carcass already at high risk for contamination, such as the brisket, flank and cod regions.

EVISCERATION

The process of removing internal organs, especially from the abdominal cavity is known as **evisceration**. This is one of the most critical steps to insure hygienic harvest, as the gastro-intestinal tract harbors a plethora of microorganisms that can cause undue and untimely spoilage of meat or in some cases cause serious food-borne illnesses. Most animals are eviscerated while hung on a rail by their back legs. Birds are also hung on a rail by their legs.

In cattle, pigs and sheep, typically the **bung** or rectum is loosened by cutting around the anus. At the anterior end of the alimentary canal, the esophagus of cattle, or **weasand**, is separated from the trachea before it is severed from the head in the area of the larynx. Often times in cattle, both the rectum and esophagus are "tied off" to minimize potential contamination from inadvertent loss of gastrointestinal tract contents. Feed restriction for 24 hours before slaughter greatly reduces the contents in the alimentary canal and helps facilitate a successful, hygienic removal of the **viscera**. Once the abdominal organs (stomach, rumen, liver, large and small intestines) have been released, they are still joined to the esophagus, which transects the diaphragm. Usually, the organs of the thoracic cavity, or **pluck**, are removed simultaneously with the viscera. The connective tissue portion of the diaphragm is pierced and loosened from muscle of the diaphragm. The lungs, heart, and trachea then are pulled from the carcass. At this point, the carcass has been **dressed**. It is observed for evidence of contamination, which if present is removed by trimming or washing, and then the carcass is released to the inspector for final disposition.

Evisceration of birds is preformed, in general, similar to that outlined above. Even so, the anatomy is quite different and thus the carcass is manipulated a bit differently in the process. Each lower leg (crus) is removed, the oil gland is removed and the vent, synonymous with the rectum in animals, is freed from its connection. The body cavity is opened, the viscera are removed and the **giblets** (heart, liver, and gizzard) are harvested. The intestines are removed, as are the lungs, trachea, crop, and air sacs. A number of these processes occur with little human labor because poultry plants are highly automated.

CARCASS MANIPULATION

Several other processes occur before the carcass proceeds to the cooler, or ice water in the case of most poultry operations, for chilling. The kidneys, which are the only remaining organs in the body cavity of animals, are freed from the surrounding fat and left exposed for visual and physical inspection. Depending on the plant, **caul fat**, (adipose tissue lining the abdominal cavity of pigs) is removed. Cattle and pig carcasses are **split** with a saw along the midline of the carcass through the vertebral column. Once split, any visually apparent contamination or tissues with excessive hemorrhages are trimmed from the carcass. On cattle carcasses, spinal cords are removed as a precautionary measure against spread of BSE.

Finally, carcasses are washed to remove bone fragments from the splitting process and extraneous blood and other undesirable components on the outer surfaces of the carcass. Washing with water can remove visible surface foreign matter but has little effect on the presence of surface microbes. The increased frequency of food-borne disease outbreaks traced to animals has created greater food safety expectations of meat products by consumers. In response to these concerns, the industry, scientific community, and governmental agencies have introduced antimicrobial interventions in the slaughter process to improve sanitation and food safety.

Steam vacuum

Vacuum systems equipped with steam are designed to remove small areas of visible contamination from carcass surfaces. This system is hand-held and uses a hot water spray (185° F) in a vacuum nozzle. In addition, steam is used to treat above and below the area. The hot water kills about 90 percent of the bacteria and loosens surface contamination so it can be vacuumed from the carcass. **Steam vacuum** systems are used at multiple points in the slaughter process. Fecal and ingesta contamination less than one inch in its greatest dimension may be removed by the steam vacuum system. Contamination of greater than one inch must be trimmed with a knife. Of course, temperatures and vacuum pressures need to be monitored to ensure proper operations.

Organic acid rinses

The entire carcass may be subjected to a water wash followed by an organic acid rinse. A wash immediately after hide removal and evisceration can serve to remove some bacteria before they attach to the carcass surface. The subsequent organic acid rinse provides a kill step for bacteria that remain on the carcass surface. Acetic, lactic, and citric acids are commonly used for this purpose. The concentration of the organic acid is between 1.5 percent and 2.5 percent and may be applied as a mist, fog, or a small droplet rinse.

Hot water rinses

High temperature water (>160° F, for beef carcass and 140° F for poultry) sprayed on the carcass as the last step prior to chilling are effective in substantially reducing the numbers of *E. coli* O157:H7, *Salmonella* and *Campylobacter* on carcasses.

Steam pasteurization

Steam pasteurization is accomplished by placing carcasses into slightly pressurized, enclosed chambers and injecting steam into the chamber that blankets and condenses over the entire carcass. This raises the surface temperature to ~70° C (185° F) and kills 95–99 percent of bacteria. Carcasses are then sprayed with cold water.

Antimicrobial chemicals

Other chemicals are used as antimicrobials for beef and poultry carcasses. They include:

- *Acidified Sodium Chlorite* (Sanova®)–effective in control of *E. coli* O157:H7, *Listeria*, *Campylobacter*, *Salmonella*, and other bacteria. Applied at 500–1200 ppm in a spray at ambient temperatures for beef carcasses. Applied at the same concentration for dips for birds.
- *FreshBloom®*—contains three different organic acids: citric acid, ascorbic acid, and erythorbic acid. The effective concentration is 6.25 percent sprayed on carcasses.
- *Lactoferrin*—Applied up to 2 percent of a water-based antimicrobial spray in the beef slaughter process, lactoferrin has been shown to be effective against more than 30 foodborne pathogens, including *E. coli* O157:H7, *Salmonella*, and *Listeria*.
- *Trisodium Phosphate*—Used in poultry establishments as a drench, spray, or dip. Effective in preventing bacteria attachment to the skin. Used in concentrations of 8–12 percent for reducing *Salmonella* incidence 27–47 percent.
- *Chlorine*—Used at 20 ppm in spray for poultry to produce a significant reduction in bacterial numbers.

5

CONVERSION OF MUSCLE TO MEAT: BIOCHEMISTRY OF MEAT QUALITY DEVELOPMENT

OBJECTIVES: *Understand how muscle transitions from a living tissue to meat. Describe how these changes affect the chemical, biochemical, and biophysical nature of the various tissues in the carcass, especially the lean tissue, and can impact fresh meat quality. Appreciate the biochemical pathways controlling this conversion and define those factors that are responsible for modulating changes in muscle temperature and pH declines postmortem.*

- ATP (adenosine triphosphate)
- ADP (adenosine diphosphate)
- Phosphagen system
- Glycolysis
- Mitochondrial respiration
- Phosphocreatine
- Creatine kinase
- Adenylate kinase (myokinase)
- AMP (adenosine monophosphate)
- Aerobic metabolism
- Tricarboxylic acid (TCA) cycle
- Glycogen
- Glycogenesis
- Glycogenolysis
- Glycogen phosphorylase
- Anaerobic metabolism
- Pyruvate
- Lactate
- Denaturation
- Exudative
- Rigor mortis
- Onset of rigor mortis
- Completion of rigor mortis
- Resolution of rigor
- Aging
- Conditioning
- Calpains
- Multicatalytic proteinase complex
- Cathepsins
- Calpastatin
- Water binding capacity
- Homeostasis
- Pale, soft, exudative (PSE) muscle
- Dark, firm, dry (DFD) muscle
- Dark cutting
- Myostatin gene
- Halothane gene
- Rendement napole gene
- Callipyge gene
- Double muscling
- Sex odor
- Blood splash
- Thaw rigor
- Cold shortening
- Accelerated or pre-rigor processing
- Electrical stimulation

INTRODUCTION

As addressed to some extent in previous chapters, a number of antemortem factors affect final meat quality. Most of these factors tend to manifest themselves in the biochemical changes that occur in tissues during the postmortem period, beginning immediately after the harvesting process starts until the carcass is fabricated into wholesale or retail cuts. The collective changes that occur in muscle through the postmortem period contribute immensely to the transformation of muscle to meat, and specifically impact meat quality development. The largest two contributors are deviations in tissue temperature and pH, which are mainly due to biochemical changes that occur postmortem in the muscle. These two factors singly or in combination affect virtually every chemical reaction that occurs in the field of meat science. As such, they are keys to solving many technical issues that often arise during subsequent processing events.

Changes in pH, though not independent of temperature occur largely from changes in the type and/or rate of biochemical reactions that occur when the entire organism is moved away from its homeostatic set points. Muscle pH values decrease from a physiological pH of near 7.4 in living animals to around 5.4, or lower, in fresh meat cuts of most animals. Living tissue and the underlying biological systems supporting its function are extremely sensitive to slight changes in pH. To illustrate the critical nature of a stable, neutral pH to living organisms and their tissues, consider that changes of just over 0.1 pH unit in blood are sufficient to cause acidosis that often results in death. Although muscle tissue is uniquely capable of functioning with periodic drops in pH to around 6.5, this simply illustrates the extreme sensitivity of organisms, their tissues and cellular components to subtle changes in pH.

Temperature also is extremely important in preserving the biological integrity of a number of tissues and systems in animals. Body temperature and that of the corresponding carcass, changes from a homeostatic set point of approximately 38°C in living animals to very close to 0°C by 24 hours after slaughter. Simply changing the temperature by this magnitude has tremendous effects on the physical nature of the carcass and on the mileu of reactions that are activated, enhanced, or arrested during the slaughtering process.

POSTMORTEM MUSCLE METABOLISM

As stated in an earlier chapter, most tissues do not cease functioning at death. Muscle tissue is no exception and continues to retain its capability to contract for sometime after exsanguination. In fact, under most circumstances, metabolism within the muscle increases substantially postmortem due to the disruption of central nervous system control or from other physiological impacts on the animal prior to harvesting. Muscle metabolism is so robust that it may retain biochemical activities for several hours after being placed in a cooler; in beef as much as 24 hours postmortem. This metabolism, though containing essentially all the elements of living muscle and attempting 'feverishly' to resist changes from the norm, quickly breeches the normal homeostatic set-points mandated by the whole animal and thus, muscle metabolism occurs much differently in dying muscle compared to that in muscle of an animal frolicking about a paddock. Even though quite different, it is helpful to understand energy metabolism in living muscle before a working knowledge of postmortem metabolism can be created.

ATP, The Energy Currency in Muscle Cells

Though repetitive of information covered in Chapter 3, it is critical to keep in mind that **ATP** is the predominant energy currency in most cells, even though other compounds are capable of harnessing energy as well. **Adenosine triphoshpate** (Figure 5.1), as the name implies contains, three phosphate residues attached to an adenosine base structure that consists of adenine and a ribose sugar. These phosphates are attached to ribose by means of high energy bonds. When these

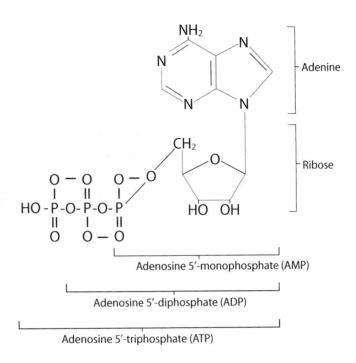

FIGURE 5.1 | Molecular structure of adenosine triphosphate (ATP). Removal of one phosphate residue produces adenosine diphosphate (ADP), while removal of two phosphate residues yields adenosine monophospate (AMP).

bonds are broken, sufficient energy is liberated to power a number of enzyme reactions that aide in a myriad of cellular functions. One reaction of particular significance to meat science and outlined in Chapter 3 is the interaction of myosin and actin in cross-bridge cycling that is mandatory for muscle contraction. Regardless of the reaction, when ATP is utilized, it is hydrolyzed according in the following reaction.

$$ATP + H_2O \rightarrow ADP + H^+ + Pi + (energy)$$

Of particular significance to understanding muscle energy metabolism is to recognize that the above hydrolyzation reaction of ATP yields ADP (**adenosine diphosphate**) and a hydrogen ion as well as energy liberated from breaking of the terminal or third phosphate bond of the ATP molecule. Because ATP is so critical to cell function, especially in muscle cells, cells must maintain a certain level of ATP in order to function properly or maintain homeostasis. This concept is critical to understand muscle energy metabolism, either in living or dying muscle, as all biochemical reactions are regulated to maintain a critical concentration of ATP in the cell. When ATP is consumed, such as during muscle contraction, cells are designed to respond or buffer against such a dramatic loss in ATP.

In muscle, there are three means to generate, or buffer against loss of ATP: 1) the **phosphagen system**, 2) **glycolysis** and 3) **mitochondrial respiration**. Each of these three systems plays a critical role in postmortem energy metabolism, albeit at different times during the transformation of muscle to meat.

Initially, the phosphagen system uses **phosphocreatine** and ADP to maintain cellular ATP concentrations. Creatine is capable of being phosporylated with a high energy bond similar to that of ATP. As such, when ATP is plentiful in the muscle cell, creatine is phosphorylated. In other words, when an animal is resting, the amount of phosphocreatine is at its maximal concentration in the muscle. This high energy compound is the first source to re-synthesize ATP in muscle cells when ATP levels drop below a given threshold. This is best illustrated by the following reaction:

$$H^+ + ADP + \text{phosphocreatine} \underset{\text{creatine kinase}}{\rightleftharpoons} ATP + \text{creatine} + H_2O.$$

This reaction is maintained at equilibrium in the cell. **Creatine kinase** catalyzes this reaction, in both directions, and exists as two forms in muscle, one located in the cytoplasm and one associated with the mitochondria. For the purposes of this chapter, we will consider the function of the cytoplasmic creatine kinase only. Because of the availability of phosphocreatine, ATP broken down during or after a contraction, or series of contractions is rapidly restored. In fact, this reaction occurs so rapidly that ATP breakdown during a single muscle twitch cannot be measured; only phosphocreatine breakdown is observed in such cases. The concentration of phosphocreatine in resting muscle is about twice that of the resting levels of ATP. Even so, extended periods of contraction can deplete phosphocreatine and therefore, phosphocreatine must be replenished during periods of rest by other mechanisms. During times of increased ATP consumption in a cell, the reaction is pushed to the right in an effort to keep ATP elevated or rephosphorylated. Correspondingly, during times of increased ATP production, greater amounts of creatine can be phosphorylated and the reaction is pushed to the left. This is but one aspect of the phosphogen system.

ATP stores within the muscle can be replenished easily by stored energy in the form of phosphocreatine in response to a single muscle twitch. However, in more severe cases of sustained contractions where huge amounts of ATP are hydrolyzed and the level of ADP in the cell increases rapidly, additional means are necessary to regenerate ATP. In response to this scenario, muscle has the ability to generate ATP from ADP using the enzyme **adenylate kinase**, or **myokinase**, according to the following reaction, the second element of the phosphogen system.

$$ADP + ADP \underset{\text{adenylate kinase}}{\rightleftharpoons} ATP + AMP$$

This reaction is also at equilibrium so the reaction will move to the left or right depending on the amount of each product on either side of the reaction. This reaction becomes important later in this chapter when we discuss 'acid meat' from pigs with a high influence of Hampshire breeding. When ATP is synthesized from ADP, AMP is formed along with ATP. The presence of **AMP** (**adenosine monophosphate**) is a critical point of control for the cell because increased levels of AMP in muscle cells signal that the energy charge (ATP) is low or decreasing rapidly, much like the fuel gauge in a car signals to a driver that fuel is needed to operate the car. The presence of AMP is very powerful in stimulating collateral biochemical pathways to produce energy in muscle. This makes AMP an attractive point of control for energy metabolism in the cell, but it must be tightly controlled in living muscle. In muscle that is dying, as is the case during the transformation of muscle to meat, the feedback mechanism to generate more energy is most likely lost as the entire biochemical means to generate ATP is fighting a losing battle by scrambling to maintain ATP levels. The system performs in whatever manner necessary to squeeze every molecule of ATP from the system. To that end, biochemical reactions have a means to pull reactions to a desirable endpoint, in this case ATP production, by reducing the products of ATP breakdown, as you can see from the following reaction,

$$AMP \underset{\text{deaminase}}{\rightleftharpoons} IMP + NH_3.$$

When coupled to the previous reaction, as AMP is deaminated, it is removed from the products side (right) of the reaction and therefore allows the adenylate kinase to proceed, which helps drive that reaction to the right, or to more ATP.

All of the aforementioned reactions occur rapidly in living muscle tissue, mainly because 1) AMP accumulation in living tissue is tightly controlled and 2) it can hamper attempts to make ATP with adenylate kinase. Therefore, detecting

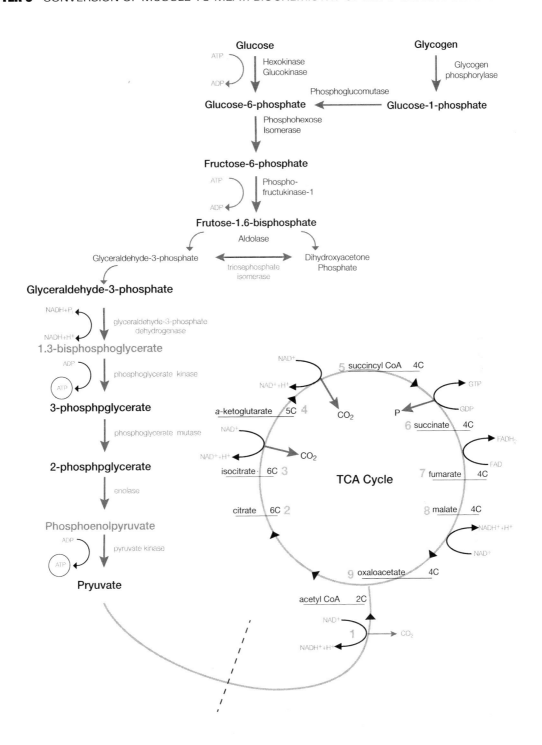

FIGURE 5.2 | Glycogenolysis, glycolysis and the Tricarboxylic Acid (TCA) cycle pathways. Dashed line between pyruvate and the TCA cycle signifies a metabolic blockage when oxygen is limiting. The product of glycogenolysis is glucose-1-phosphate. The product of glycolysis is pyruvate.

appreciable changes in these metabolites is quite difficult except with the aid of rather sophisticated instrumentation. In postmortem muscle, these reactions also occur rapidly after stunning but because the circulatory system has been destroyed, these metabolites in muscle can be readily observed and measured. Initially, ATP exists in muscle at basal levels. Within literally minutes of stunning, the phosphogen system via creatine kinase and adenylate kinase-based reactions attempts to rephosphorylate ADP to ATP. Though briefly successful, ADP continues to rise and though AMP rises, it is quickly deaminated in a futile attempt to maintain ATP production.

Glycolysis and Mitochondrial Respiration

When ATP consumption continues to a point where the phosphagen system is incapable of replenishing ATP levels, ATP regeneration must be accomplished by **glycolysis** (the breakdown or "lysis" of glucose, "glyc") and by the **aerobic metabolism** of the products of glucose breakdown, at least in living muscle. These processes collectively represent some of the most critical events that occur during the conversion of a living muscle to meat. Changing the rate of glycolysis in muscle, or the temperature at which it occurs will dramatically alter the quality of meat produced. Aerobic metabolism that occurs predominately in the mitochondria, though not important long-term in the muscle after exsanguination, clearly impacts the energy charge of the muscle prior to and immediately after stunning.

ATP production from glycolysis (Figure 5.2) occurs through a series of biochemical reactions (Table 1) predominately located in the sarcoplasm. ATP generated from glycolysis is by a distinctly different metabolic pathway from that generated in the **tricarboxcylic acid (TCA) cycle** (Figure 5.2). Under some physiological circumstances

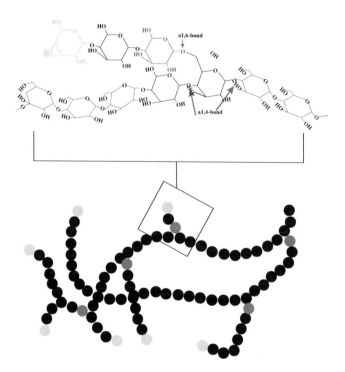

FIGURE 5.3 | General structure of the branched glycogen molecule (lower) and molecular linkages of glucose residues in the glycogen molecule (upper).

these two pathways can be linked (Figure 5.2) and the pyruvate produced by the glycolytic pathway can be further metabolized or broken down after entering the mitochondria. In this case, the mitochondrial- and the cytoplasmic (glycolytic)-based pathways work in tandem to produce massive amounts of energy (ATP) for cells. In other situations, the sequential nature of this energy production scheme is disrupted because of changes in physiological status and the capacity to generate ATP is reduced severely.

Glycogen Metabolism

Remember from earlier chapters and introductory biology courses that very little free glucose exists in muscle cells. In fact, glucose, which diffuses through glucose channels in the cell membrane, is converted immediately to glucose-6-phosphate. Though somewhat academic at this point, this is a critical step in ensuring that glucose stays in the cells. If glucose were not phosphorylated, it would simply diffuse from the cells back into the circulation or surrounding tissues when blood glucose levels plummet between meals. Once in the cell, and depending on the needs of the cells, glucose-6-phosphate may enter the glycolytic pathway, or in times of excess energy supply, glucose-6-phoshpate molecules may be converted to glucose-1-phosphate, linked together and stored as **glycogen** (Figure 5.3). Recall that glycogen constitutes about 1 percent of the muscle by weight. The process of building of energy reserves in the form of glycogen is known as **glycogenesis**. The breakdown of glycogen to glucose molecules is known as **glycogenolysis** (Figure 5.2). Glycolysis and glycogenolysis are different events but are generally closely linked because the products of glycogenolysis often enter glycolysis. The process of glycogenolysis is mediated enzymatically by the enzyme **glycogen phosphorylase**. This enzyme exists in two forms: a and b; the a form is generally considered active and the b form inactive. Depending on the relative proportion of these forms in the muscle of animals prior to slaughter, the rate of glycogenolysis may differ.

To summarize glycogenolysis and glycolysis under aerobic conditions, when one glucose residue is liberated from glycogen as glucose-1-phosphate and metabolized fully through the entire sequence of glycolytic reactions and TCA cycle reactions, the net energy yield is 35 molecules of ATP. This differs by 1 ATP from the ATP yield from one molecule of glucose diffusing into the cell, which is 34. The reason for this discrepancy is that free glucose must be phosphorylated by consuming one ATP, while the glucose residue released from glycogen (glucose-1-phosphate) is already phosphorylated during the process of removal from the larger glycogen polymer. To specify more precisely the origin of these ATP molecules, three ATP molecules are generated from glycolysis when glucose-1-phosphate is metabolized through glycolysis. In addition, two or three additional H+ are generated from the breakdown of glycogen or glucose (Table 5.1), respectively, before the products of glycolysis are presented to TCA cycle in the mitochondria. These H+ produce 2 ATPs in the mitochrondial electron transport chain during aerobic metabolism. Each of the two 3-carbon pyruvate molecules subsequently yields 10 H+, for a total of 20 H+ when metabolized through the TCA cycle. The resulting 20 H+ are converted, in turn, to 30 ATPs by the electron transport chain in the mitochrondia. Either 34 or 35 total molecules of ATP are created by the full metabolism of glucose or glucose-1-phosphate, respectfully.

Aerobic versus Anaerobic Metabolism

Muscle cells and their associated energy producing pathways function in conditions that vary from high oxygen availability (aerobic metabolism) to conditions where oxygen is limiting (**anaerobic metabolism**). Under either extreme, a six-carbon glucose molecule is metabolized or broken down into two, three-carbon pyruvate molecules (Figure 5.2). When oxygen is optimal in aerobic conditions, pyruvate can be taken up by the mitochondria and metabolized further in the TCA cycle (Figure 5.2). A simplification of this metabolism

TABLE 5.1 Products, Reactants and Enzymes, Including ATP And H^+, Produced during the Anaerobic Metabolism of Glycogen or Glucose to Pyruvate or Lactate in Postmortem Muscle

Step	Reaction	Enzyme	ATP from:		H^+ from:	
			Glucose	Glycogen	Glucose	Glycogen
1a	Glucose + **ATP** ⮕ Glucose 6-phosphate + ADP + **H^+**	Hexokinase	-1	0	1	0
1b	Glycogen$_n$ + Pi ⮕ Glucose 1-phosphate + Glycogen$_{n-1}$	Glycogen phosphorylase	0	0	0	0
	Glucose 1-phosphate ⮕ glucose 6-phosphate*	Phosphoglucomutase	0	0	0	0
2	Glucose 6-phosphate ⮕ Fructose 6-phosphate	Phosphoglucose isomerase	0	0	0	0
3	Fructose 6-phosphate + **ATP** ⮕ Fructose 1,6-bisphosphate + ADP + **H^+**	Phosphofructokinase	-1	-1	1	1
4	Fructose 1,6-bisphosphate ⮕ dihydroxyacetone phosphate + glycerolaldehyde 3-phosphate	Aldolase	0	0	0	0
					(X 2)**	
5	Dihydroxyacetone phosphate ⮕ glycerolaldehyde-3-phosphate	Triphosphate isomerase	0	0	0	0
6	Glycerolaldehyde-3-phosphate + Pi + NAD+ ⮕ 1,6-bisphosphoglycerate + NADH + **H^+**	Glycerolaldehyde 3-phosphate dehydrogenase	0	0	2	2
7	1,6-Bisphosphoglycerate + ADP ⮕ 3-phosphoglycerate + **ATP**	Phosphoglycerate kinase	2	2	0	0
8	3-Phosphoglycerate ⮕ 2-phosphoglycerate	Phosphoglyceratmutase	0	0	0	0
9	2-Phosphoglycerate ⮕ phosphoenopyruvate + H_2O	Enolase	0	0	0	0
10	Phosphoenopyruvate + ADP ⮕ pyruvate + **ATP**	Pyruvate kinase	2	2	0	0
	Total from glycolysis		2	3	4	3
	Pyruvate + NADH + **H^+** ⮕ Lactate + NAD+	Lactate dehydrogenase	0	0	-2	-2
	Total from anerobic glycolysis		2	3	2	1

*Occurs spontateously
**Multipled by 2 because initial substrate is split in two

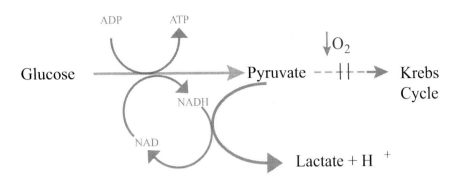

FIGURE 5.4 | Overview of reactions used to regenerate NAD+ and ATP in times of oxygen deprivation by the reduction of pyruvate to lactate.

is that under aerobic conditions a six-carbon glucose from glycogen is metabolized completely to carbon dioxide and water and yields 35 ATP.

$$\text{glucose} + \text{oxygen} \rightarrow CO_2 + \text{water} + 35\ \text{ATP (energy)}$$

A way to understand an aerobic type of metabolism is to keep in mind that aerobic exercise is usually at moderate levels with maximum breathing and occurs over a relatively extended period of time. The body consumes as much air (oxygen) as possible to facilitate this type of exercise regime. Once oxygenated, blood readily perfuses throughout the tissues, delivering maximal oxygen to the muscle cells, in particular the mitochondria, so glucose or a glycogen residue can be fully oxidized using glycolysis and mitochondrial respiration. It would be remise at this point not to point out that fats can also be oxidized by the mitochondria to produce several times more energy per unit weight than glucose. Fat is not metabolized in the glycolytic pathway and its metabolism requires even greater amounts of oxygen than for glucose. Fatty acid metabolism (oxidation) is only tangentially related to the subject of postmortem metabolism and is not a means to generate energy postmortem.

In contrast to aerobic metabolism, anaerobic metabolism, which occurs as a result of muscle working in the absence of oxygen, only occurs for a short duration, yet it can do so with rather powerful bursts. Continuing the exercise illustration, an anaerobic type of workout would be a 100M sprint, where the runner's muscles are forced to work very quickly at maximal capacity, yet the athlete may take only a few breaths during the entire event placing the muscles at an extreme oxygen debt. This latter example is fundamentally what happens in the muscles of carcasses after slaughter. They utilize energy in the absence of oxygen until the muscle is incapable of functioning any further.

A key to understanding the transformation of muscle to meat is to recognize that in the absence of oxygen, pyruvate, the three-carbon product of glucose metabolism in glycolysis, cannot be further metabolized in the mitochrondria because the electron transport chain in the mitochondria requires oxygen for ultimate function. But, though mitochondrial respiration stops, muscle tissue retains the capacity to function in the absence of oxygen and working mitochondria, much like a sprinting athlete. Though much less ATP (3 versus 35) is produced through anaerobic metabolism, the tissue is provided some energy so that the structural integrity of the cells can be maintained, at least for a short period of time. This is accomplished by the conversion of **pyruvate**, the final product of glycolysis, to **lactate** according to the following reaction:

$$\text{Pyruvate} + \text{NADH} + \text{H}^+ \rightarrow \text{Lactate} + \text{NAD}^+ + \text{H}^+$$

It is critical to understand that this one step allows glycolysis to continue producing ATP only because it allows the regeneration of nicotinamide adenine dinucleotide (NAD^+), which is necessary for several reactions up-stream in the glycolytic pathway (Figure 5.4). Under normal conditions, such as the sprinter that produces an incredible about of lactate prior to finishing the short race, at rest lactate is transported from the muscles to the liver, where it is re-synthesized into glucose and glycogen, or to the heart, where it is metabolized to carbon dioxide and water.

ENERGY METABOLISM, H+ PRODUCTION AND MUSCLE PH DECLINE

As a brief review, when living muscle is working and oxygen is supplied in adequate amounts, aerobic metabolism can meet all the energy needs of the tissue, and muscle is working at or very close to steady-state equilibrium. However, when consumption of ATP exceeds that produced by any of the above reactions, either due to elevated consumption rates or in the absence of oxygen, tissues are in a state of disequilibrium in terms of ATP production. ATP is consumed at a much greater rate than it can be produced. Under these conditions lactate accumulates in the tissue as pyruvate is converted to lactate to regenerate NAD^+. Also, each time an ATP is hydrolyzed a hydrogen ion (H^+) is released in the cell. Under these physiological

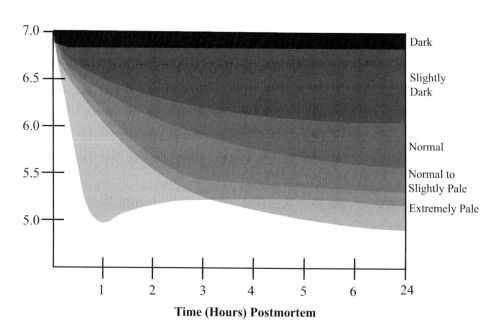

FIGURE 5.5 Various postmortem pH decline curves and resulting colors (shaded to reflect differences in fresh meat pinkness) observed in pork muscle during the transformation of muscle to meat.

conditions, glycolysis is being accelerated and 2 H^+ are being liberated with each glucose residue metabolized through glycolysis, or 1 H^+ in the case of a glucose residue from glycogen (Table 5.1). As a result, the concentration of H^+ increases. Hydrogen ions are quickly bound and buffered until their concentration exceeds the buffering capacity of the cells and the pH of the tissue declines. If allowed to proceed indefinitely, as in the case of dying muscle, glycolysis stops because no more ADP is available for rephosphorylation. Alternatively, in living muscle, fatigue occurs and rigor, or cramping sets in because there is insufficient ATP produced to maintain resting membrane potentials and low free Ca^+ levels and to break the myosin/actin cross-bridges that form.

Thus, muscle pH decreases with time postmortem mainly due to accumulation of H^+ that results from the hydrolysis of ATP to ADP and from glycolysis (Table 5.1). As mentioned previously, this decrease in muscle pH is one of the most significant postmortem changes in muscle postmortem. The rate and extent of muscle pH decline after an animal has been exsanguinated are highly variable. Factors that influence this decline are discussed later in this chapter.

A normal pH decline in porcine muscle, for example, is represented (Figure 5.5) by a gradual decrease from approximately pH 7.4 in living muscle to a pH of 5.6 to 5.7 within 6 to 8 hours postmortem, and then to an ultimate pH (reached at approximately 24 hours postmortem) of 5.3 to 5.7. In some animals the pH drops only slightly during the first hour after slaughter and then remains stable at a relatively high level, giving an ultimate pH in the range of 6.5 to 6.8. In other animals, muscle pH drops rapidly to around 5.4 to 5.5 during the first hour after exsanguination. Meat from these animals ultimately develops a pH in the range of 5.3 to 5.6.

Accumulation of H^+ and a correspondingly rapid pH decline early in the postmortem period can adversely affect meat quality. Development of a low pH (acidic) in muscle before natural body heat and heat of continuing metabolism have been dissipated through carcass chilling, causes **denaturation** of muscle proteins. Though carcass temperature will be discussed later, it is important to understand at this point that temperature and pH play key roles in contributing to protein denaturation, a process whereby proteins lose their native structure and in the process, release or lose water. Low pH and high (near normal body) temperatures is the critical combination of variables. Muscle may attain a low pH (5.2 to 5.4) after it has been thoroughly chilled without undergoing excessive protein denaturation. In addition, muscle proteins in some species are more sensitive to this type of denaturation than are those of other species or even breeds. For example, fish muscle proteins are more labile than most mammalian muscle proteins and are denatured at lower temperatures and higher pH value combinations.

Denaturation of proteins causes a loss of protein solubility, loss of water- and protein-binding capacity, and in more highly pigmented muscle, a loss of color intensity. All these changes are undesirable, whether the muscle is going to be used as fresh meat or subjected to further processing.

Muscles that have a very rapid and/or extensive pH decline (Figure 5.5) will be pale and have very low water-holding capacity, the latter of which causes cut surfaces to appear very wet. The term used to describe the wet surface condition is **exudative**. The exudate is liquid arising from the lean tissues. Though often described as "watery", the exudate contains a high concentration of proteins, Its loss represents a considerable economic loss to the meat industry. In severe cases, the exudate can literally drip, or ooze from the surface of the meat. In contrast, muscles that maintain a high pH during conversion of muscle to meat may be very dark in color, and very dry on the exposed cut surface, because naturally occurring water is tightly bound to proteins, as it is in living muscle.

Heat Production And Dissipation

Changes in temperature are the second of the two important factors regulating the transformation of muscle to meat. The following events lead to an increase in carcass temperature, mostly in the muscle tissues during the earliest times postmortem, within 20 minutes of stunning. Figure 5.6

illustrates some of the variation in heat production and dissipation encountered in pork muscle.

Animals experience heightened levels of anxiety associated with the entire harvesting process; strange surroundings, unrecognizable noises and altered lighting all contribute to this anxiousness. This nervousness coupled with some element of fear will cause a release of adrenaline into the circulatory system by the adrenal glands. Adrenaline targets a number of tissues, the most important being muscle. As a result, there is an increase in muscle activity, partly stimulated by the hormone and partly by increased movement, which in turn causes increased muscle contractions. As discussed in greater detail in an earlier chapter, it is important to recall that muscle contraction requires energy and the by-product of contraction is the release of heat and mechanical energy. In much the same way that athletes get nervous and warm-up prior to a competition, animals can experience increases in body temperature prior to the harvesting process.

Normally, the increase in body temperature would be dissipated rapidly through perspiration and respiration. However, these functions are immediately curtailed during the harvesting process, which leads to the second reason for carcass temperature increases immediately postmortem. The primary means of regulating body temperature has been eliminated at exsanguination and the heat generated accumulates.

Increased activity of many biochemical reactions in muscle after stunning and exsanguination also contribute to temperature increases in the carcass. As outlined above, the hydrolysis of ATP causes a decrease in the energy charge in the muscle (ADP/ATP), which essentially creates an 'organized chaos' as muscle cells attempt to re-establish energy levels. This heightened metabolism increases the overall heat produced in the carcass. In some cases, this can be directly attributed to genetic conditions, especially in pigs as will be discussed later.

The final contribution to increased carcass temperatures, or retarded declines in carcass temperature, are external factors associated with the harvesting process and reflect the procedures used to perform this function. During pig and poultry harvesting as outlined in Chapter 4, most carcasses are scalded in approximately 60°C water to loosen hair or feather follicles, respectfully. This is nearly 20 degrees above body temperature. After the scalding process, pig carcasses in particular may be subjected to flames that singe the remaining hair from the carcass. These factors also contribute to an increase in carcass temperature early postmortem, especially in pigs and birds. Other issues that may affect the duration and extent of this induced temperature increase include but are not limited to: 1) time to evisceration, 2) degree of animal fatness, and 3) size of the animal. Ambient temperature in the slaughter room, duration of the slaughter and dressing operation, and temperature of the initial chill cooler all have considerable influence on the rate of carcass temperature decline. Therefore, muscle denaturation is inevitable unless some means is employed to remove heat rapidly from muscle as it is produced.

Rigor Mortis

One of the most dramatic changes in physical properties of the carcass that occur postmortem during conversion of muscle to meat is the stiffening of muscles after death. The phenomenon of **rigor mortis** (Latin for "stiffness of death") has been recognized for hundreds of years but only during recent decades has its physiological basis been explained.

Stiffening observed in rigor mortis is due to formation of permanent cross-bridges that form between actin and myosin filaments directly from a lack of ATP necessary to break these bridges. This is the same chemical reaction that forms actomyosin during muscle contraction in living muscle. The difference between living and rigor muscles is that relaxation is impossible in the latter state because no energy (ATP) is available to break these actomyosin bonds. To illustrate this phenomenon consider Figure 5.7. Muscle is extremely extensible during the period immediately after exsanguination. If the muscle is loaded, or if a force is applied, it will passively stretch (trough); and, when the force is removed, the natural elasticity of the muscle will return it to its original length

CHAPTER 5 CONVERSION OF MUSCLE TO MEAT: BIOCHEMISTRY OF MEAT QUALITY DEVELOPMENT

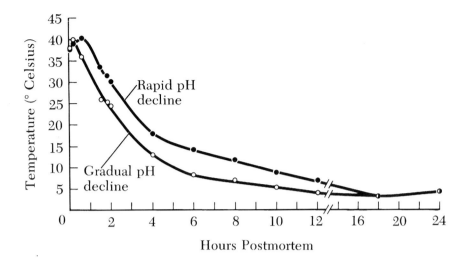

FIGURE 5.6 | Postmortem temperature decline curves. Modified from Briskey, E. J. and J. Wismer-Pedersen, *Biochemistry of Pork Muscle Structure*; "I. Rate of Anaerobic Glycolysis and Temperature Change versus Ultimate Muscle Structure," J. Food Sci. 26, 306 (1961).

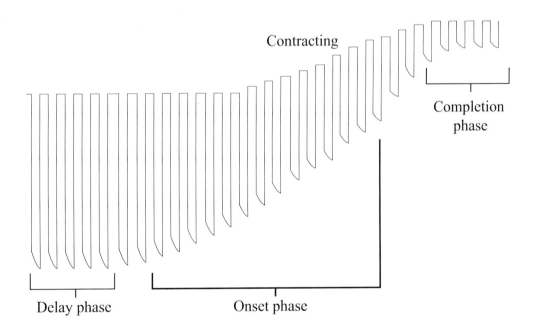

FIGURE 5.7 | Physiograph output showing the change in extensibility (y-axis) of muscle with time postmortem (x-axis). The troughs are when force is applied to stretch the muscle, the apex is when the force is removed. The delay, onset and completion of rigor mortis are indicated.

(apex). At this time (left, Figure 5.7), few if any actomyosin bridges are present to prevent force-induced extension. ATP complexed with Mg2+ (as noted in Chapter 3) is required for a muscle to remain in the relaxed state. During the earliest postmortem phases as pointed out previously, stores of creatine phosphate are used to re-phosphorylate ADP to ATP. The period of time during which the muscle is relatively extensible and elastic is called the delay phase of rigor mortis. However, as ATP falls, creatine phosphate is depleted, and phosphorylation of ADP becomes insufficient to maintain the tissue in a relaxed state. As a result, glycogen stores are metabolized so that ADP can be re-phosphorylated. The amount of ATP formed, however, is finite and a fraction of that necessary for homeostasis. Therefore as ATP becomes limiting, actomyosin bridges begin to form and the muscle gradually becomes less extensible when an external force is applied. This process signals the **onset phase of rigor mortis**, when muscle begins to lose extensibility. It lasts until the **completion** of **rigor mortis**. Once the muscle becomes relatively inextensible, rigor mortis is complete (right, Figure 5.7). Table 5.2 shows typical amounts of time required for muscle of different animal species to enter the onset phase of rigor mortis.

As cross-bridges form, muscles contract, if not restrained by various ligaments, tendons or other physical constraints. Though sarcomeres clearly shorten in unrestrained muscle experiencing rigor mortis, they can create tension even in the absence of shortening. These events are maximal when the ATP supply has been exhausted and all binding sites have formed actomyosin bridges. In the absence of ATP, the actomyosin bonds cannot be broken; consequently, tension is at its maximum. During postmortem storage, however, tension decreases but never returns to prerigor levels.

Figure 5.8 shows a typical pattern of tension developed in muscle when constant length is maintained during rigor mortis development and subsequent postmortem storage. The decrease in tension with time postmortem is referred to as "resolution" of rigor mortis. This is a misnomer because the actomyosin bonds of rigor mortis are not broken during postmortem storage as implied by the term **resolution of rigor**. Rather, the decrease in tension is due to events other than breaking actomyosin bonds. Specifically, proteolytic degradation of specific myofibrillar proteins results in dissolution of Z disks and loss of ultrastructural integrity. Rigor mortis development also affects meat tenderness. When rigor mortis is complete, meat tenderness is at its lowest value. Tenderness increases during postmortem storage and is related to the proteolytic changes during resolution of rigor, as will be described later.

Because actomyosin bonds formed during the development of rigor mortis are the same as those

TABLE 5.2	Delay Time Before Onset of Rigor Mortis
Species	**Hours**
Beef	6–12
Lamb	6–12
Pork	1/4–3
Turkey	< 1
Chicken	< 1/2
Fish	< 1

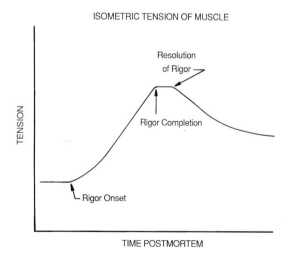

FIGURE 5.8 Isometric tension development in muscle during phases of rigor mortis.

formed during muscle contraction, rigor mortis may be thought of as an irreversible muscle contraction. Unrestrained muscles shorten as permanent bonds form during development of rigor mortis. Alternatively in muscles restrained from shortening by the skeleton, tension develops within the tissue that contributes to muscle stiffening. Rigor shortening differs from normal contraction because more cross-bridges are formed. During normal contraction, cross-bridges form at only about 20 percent of the possible binding sites, but in rigor, nearly all binding sites in the area of overlap between actin and myosin filaments contain cross-bridges.

The pattern of development of these physical changes varies from animal to animal, and even from muscle to muscle in the same animal. These variations are associated with properties of postmortem muscle that are important in the use of muscle as meat.

Rigor mortis and pH decline are closely correlated because both are related to energy metabolism, in particular the loss of ATP through hydrolysis. Muscle that undergoes pH changes shown in the top and bottom curves of Figure 5.5 will have rapid development of rigor mortis. In muscle with abbreviated pH decline (top curve), onset and completion of rigor mortis are rapid because the initial energy supply is limited. In the case of rapid early pH decline, (bottom curve), onset and completion of rigor also are rapid because energy supplies are rapidly metabolized. The intermediate pH decline curve is associated with a longer time of rigor mortis development than either of the two extremes. Likewise, an extended pH decline, likely is associated with a delay in rigor mortis development.

Failure Of Protective Mechanisms

With increasing time postmortem, all homeostatic mechanisms are eventually lost. As stored metabolites are depleted, muscle temperature declines because there are no operable mechanisms for generating heat. Nervous control from the central system is lost within 4 to 6 minutes after exsanguination. Uncontrolled nerve impulses may arise locally, causing muscle twitches for some time after exsanguination. These local impulses may affect postmortem metabolism.

In healthy living animals, muscle is protected from invasions of microorganisms by a series of defenses, the first line of which is the tissues that cover the body and surround many of the internal organs. Connective tissues and cell membranes also may offer some protection. The lymphatic system and circulating white blood cells are available to destroy organisms that enter the body. During the conversion of muscle to meat, membrane properties are altered and the tissue becomes susceptible to bacterial invasion. Because the circulatory and lymphatic systems are no longer operable, they cannot prevent the spread of microorganisms. Thus, most of the changes (other than pH lowering) that occur during conversion of muscle to meat favor microorganism proliferation, and care must be exercised to prevent contamination of meat during postmortem handling and storage.

Disruption Of Muscle Structure

During development and completion of rigor mortis, the microscopic structure of muscle in a cross-sectional view has much the same appearance as living muscle. Longitudinal sections, however, reveal shorter sarcomere lengths. It is almost certain that subtle degradative changes, such as loss of membrane integrity and function, begin soon after exsanguination. Carcasses and primal meat cuts are kept at refrigeration temperatures for various lengths of time after slaughter depending on their usage and merchandising schedules. During this postmortem period, many changes occur that alter meat properties. Most notable among the changes is an increase in tenderness. Holding carcasses or meat at refrigeration temperatures for extended periods after initial chilling is called **aging** in the United States and **conditioning** in many other countries of the world. This process is performed to improve meat tenderness.

The first observable changes in the ultrastructure of postmortem muscle occur in myofibrils where degradation of Z disks begins. This change

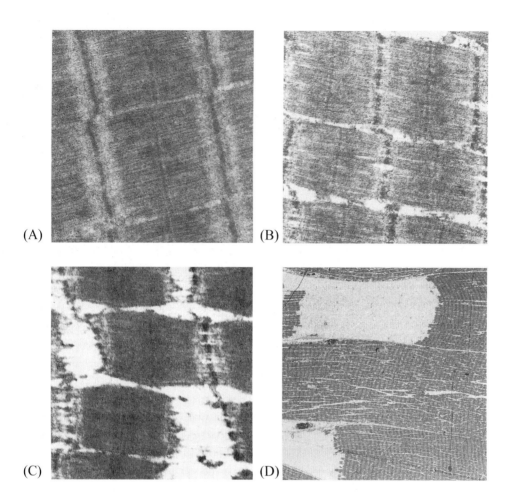

FIGURE 5.9 | Degradation of Z disks in beef longissimus muscle during postmortem storage at 2° C. (A) One hour after slaughter, Z disks are intact (x 12,500). (B) Twenty-four hours after slaughter, some Z disk degradation is apparent (x 12,500). (C) Forty-eight hours after slaughter, Z disk degradation and myofibril breaks are extensive (x 12,500). (D) Low magnification shows complete lateral breaks of the myofibrils in several muscle fibers 8 days after slaughter (x 650). From Gann, G. L. and R. A. Merkel, "Ultrastructural Changes in Bovine Longissimus Muscle During Postmortem Aging," *Meat Sci.* 2, 129 (1978).

causes muscle to become more extensive with increased storage time (Figure 5.9). Complete loss of these structures (Z disks) occurs due to proteolytic degradation of proteins associated with the disk, notably desmin, nebulin, and titin. While several other proteins of the myofibril are also degraded during postmortem storage, myosin, and actin are not. The contribution of proteolytic degradation of each of these myofibrillar proteins to meat tenderness is not completely understood. A summary of the postmortem proteolytic changes that occur in skeletal muscle is presented in Table 5.3.

As Z disks are degraded, the tension developed during the contraction of rigor causes myofibrils to rupture at the Z disk-I band junction, resulting in breaks that may extend to every myofibril across the entire lateral axis of muscle fibers (Figure 5.9). Breaks in the myofibrils release some of the tension developed during rigor contraction, and this is the mechanism whereby the "resolution" of rigor

TABLE 5.3 Summary of Changes That Occur in Skeletal Muscle During Postmortem Aging

1. Z disk degradation, which leads to weakening and fragmentation of myofibrils,
2. Degradation of desmin, which causes disruption of transverse cross-linking between myofibrils and leads to fragmentation of myofibrils.
3. Degradation and disappearance of troponin-T, Because of its location in myofibrils, the exact relationship between meat tenderness and troponin-T degradation is not yet understood.
4. Degradation of titin and nebulin, Because of their ability to maintain longitudinal stability of myofibrils, disruption of these structures would lead to fragmentation of myofibrils.
5. Degradation of these myofibrillar proteins results in appearance of new polypeptides of 95,000 and 28,000 to 32,000 molecular weight as seen by polyacrylamide gel electrophoresis.
6. Perhaps the most significant observation is that the major contractile proteins, myosin and actin, are not affected even after 56 days of postmortem aging.

Sources: Goll et al. (1983), Robson and Huiatt (1983), Koohmaraie (1992) and Robson et al. (1997).

occurs. These breaks produce fragments of myofibrils. In fact, fragmentation is often used as an index of meat tenderness because the two properties are highly correlated.

Several enzyme systems present in skeletal muscle have been implicated in the postmortem proteolytic degradation of myofibrillar proteins. These systems include the **calpains**, the **multicatalytic proteinase complex** (both of which are present in the sarcoplasm), and the **cathepsins**, present in the lysosomes of muscle fibers. To qualify as the active participant, the enzyme(s) must degrade the same myofibrillar proteins that are degraded during postmortem storage of meat as shown in Table 5.3. The multicatalytic proteinase complex does not cause postmortem proteolysis of any myofibrillar protein that plays a role in improving meat tenderness. Evidence indicates that the cathepsins are not released from lysosomes even after electrical stimulation and extended postmortem storage. Without their release, these enzymes cannot degrade myofibrillar proteins. The proteins would have to be endocytosed into lysosomes for proteolysis to occur as is the case in living muscle, but this is impossible in postmortem muscle because endocytosis is an active process requiring energy. Of these three enzyme systems, the calpains produce all of the proteolytic changes in myofibrillar proteins during postmortem storage presented in Table 5.3. The calpain system consists of two calcium-dependent enzymes, and a specific inhibitor, calpastatin. The two enzymes of this system differ in their calcium requirement for activation. One calpain requires millimolar calcium (m-calpain) and the other, micromolar calcium (u-calpain) concentrations for activation. Calcium released from mitochondria and the sarcoplasmic reticulum during postmortem storage activates the calpains.

Evidence is rather convincing that calpains are the enzymes accountable for the proteolytic changes in postmortem muscle. Infusion of calcium into carcasses or meat increases meat tenderness and produces all of the proteolytic changes presented in Table 5.3. In contrast, when a chelator of calcium is infused to tie up calcium ions, none of the proteolytic changes occur and tenderness is not improved. Compared with the calpain system, however, neither the multicatalytic proteinase complex nor the cathepsins are affected by calcium. Infusion of calcium into meat, especially beef, to improve tenderness is currently receiving much attention in the industry.

The third component of the calpain system, **calpastatin**, which inhibits calpain activity, varies in amount in beef. Cattle whose meat tends to be tough (bulls or those with Bos indicus breeding, such as Brahman cattle) typically have a greater content of calpastatin in their muscles. Thus, high calpastatin content in these cattle limits the extent of postmortem proteolysis by the calpains, and consequently the tenderization process is reduced. Selection of cattle for low calpastatin potentially offers a means of improving tenderness of beef. A limited number of such selection schemes, however, have been implemented at present.

Improvement in tenderness during postmortem storage is due almost entirely to proteolytic degradation of myofibrillar proteins. Yet, degradation of collagen would dramatically improve meat tenderness, especially if intermolecular cross-linkages were degraded. However, postmortem proteolysis of collagen is minimal, and intermolecular cross-linkages are not degraded. The small amount of proteolytic degradation of collagen that does occur appears to take place only during the early postmortem period.

Changes In Physical Properties of Muscles

In living animals, muscles with sufficient oxygen have a red appearance. If the tissue is in oxygen debt, it is dark red or purple in color. In postmortem muscle, as oxygen is used in metabolism, the color becomes dark purplish red. When fresh meat is cut, the exposed surface has this dark red appearance immediately after cutting, but, on exposure to the atmosphere for a few minutes, the myoglobin becomes oxygenated and changes to a brighter red color. If the muscle has been subjected to severe denaturation, color intensity will be greatly impacted. The complete chemistry of myoglobin and its color changes are discussed in Chapter 6.

Living muscles maintain a certain amount of tone because skeletal attachments impose a small degree of tension or stretch. As muscles enter full rigor they become even firmer to the point of stiffness. Later in the conversion process, as enzymatic degradation and protein denaturation proceed, they become less firm. If protein denaturation is severe, muscles can become very soft.

Water accounts for 65 to 80 percent of total muscle mass. In living muscle fibers, as in other living cells, water plays a major role in cellular function. Thus, much of the water in muscle cells is tightly bound to various proteins. If proteins are not denatured, they continue to bind water during conversion of muscle to meat, and to a great extent, even during cooking. This retained water contributes to juiciness and palatability of meat as a food.

Changes occurring in water-binding during the conversion of muscle to meat depend on the rate and extent of pH decline and on the amount of protein denaturation. When the pH of postmortem muscle remains very high (>6.5), the water-binding properties of meat are similar to those of living muscle. When the pH drops rapidly during the early postmortem period, low **water-binding capacity** results. Mechanisms by which water is held in meat are discussed in Chapter 6.

Discussion thus far in this chapter describes the complex of changes occurring when muscle is converted to meat. Figure 5.10 summarizes these events and shows their sequence. The following discussion describes many external factors that hasten, slow, enhance, or minimize these changes.

FACTORS AFFECTING POSTMORTEM CHANGES AND MEAT QUALITY

Tenderness, juiciness, color, and flavor may be influenced by changes occurring during conversion of muscle to meat. Important processing characteristics, such as emulsifying capacity, binding properties, cooking losses, and cooked meat color, also may be affected by these changes. From the discussion thus far, it is obvious that postmortem changes are highly variable and influence the way in which muscle can best be used as meat. These changes may be controlled, to a certain extent, to improve product quality. Therefore, factors affecting this conversion are extremely important in the overall effort to control and improve the quality of the meat supply.

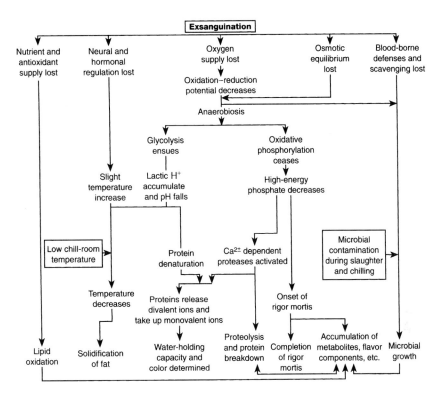

FIGURE 5.10 | Flow diagram of changes in postmortem muscle.

Antemortem Effects

Stress

As outlined in Chapter 4, when animals are moved to unfamiliar surroundings, they may become excited, fatigued, overheated, or chilled. All these conditions result from responses within the animal body that are caused by various factors in the new environment. When referring to the reactions of animals under these conditions, it often is noted that such animals are experiencing stress. The term stress is a general expression referring to physiological adjustments, such as changes in heart rate, respiration rate, body temperature, and blood pressure that occur during the exposure of the animal to adverse conditions. Such conditions, called stressors, occur when the environment becomes uncomfortable or hazardous to the animal's **homestasis**.

Adjustments in metabolism that occur during periods of stress are aided by the release of various hormones (Figure 5.11). Hormones especially important in a response to stressors are epinephrine and norepinephrine that arise from the adrenal medulla, adrenal steroids from the adrenal cortex, and thyroid hormones from the thyroid gland. These hormones influence many chemical reactions, some of which come into play in almost every type of stress-related condition. Epinephrine aids in the breakdown of glycogen that is stored in the liver and muscles, as well as fat stored in the body. Thus, release of epinephrine allows the body to mobilize an immediate source of energy for immediate use. In addition, epinephrine and norepinephrine help maintain adequate blood flow via their influence on the heart and blood vessels. Hormones from the adrenal cortex, on the other hand, are effective in reinforcing the ability of the tissues to respond during stress while thyroid hormones increase metabolic rate, and thereby provide additional energy to the animal.

The consequences of stress, and the associated metabolic adjustments, can be noted in the muscles. These adjustments usually place an increased demand on the muscles for contraction, so there is a need for an increased rate of blood flow to the muscles. Circulatory adjustments therefore are made on the basis of nutrient and oxygen demand, and on the need for temperature regulation. Many times the circulatory system is unable to provide the quantity of blood needed to maintain proper temperature and support increased activity (contractions) in the muscles. At such times, muscle temperature rises, and muscles become depleted to below normal levels of stored oxygen.

When the stress hormones are released and available to muscles, muscles become prepared to meet the demand for contractions that an emergency might require. Epinephrine, stimulates conversion of phosphorylase, the enzyme responsible for breaking down glycogen in muscle cells, from the inactive form to the active form (b → a). Muscle is now poised to liberate a great deal of energy quickly in the absence of oxygen (anaerobic glycolysis) leading to the creation of a highly acid environment because of the tremendous release of H+ with ATP hydrolysis and glycolysis. Consequently, the shift in metabolism that occurs spontaneously in the living animal is similar to that described for rapidly contracting muscle (Chapter 3) and postmortem energy metabolism in muscle.

When the demands on muscle tissue exceed the capacity of energy produced by aerobic metabolism, hydrogen ions buildup in muscles, particularly in muscles where white fibers predominate. Because lactate tends to buffer the H+ produced by binding them, lactic acid is formed but it cannot be utilized in skeletal muscle. Therefore, in the living animal it is removed and transported to the liver for the conversion to glucose or glycogen, or to the heart, where it can be utilized directly for energy. Animals that accumulate excess H+ in their muscles require a cessation of (or adaptation to) the conditions causing the stress, and adequate time to rid their muscles of these ions. If the quantities of lactic acid entering the bloodstream are too great for the liver and heart to neutralize, a generalized acidosis develops, and death may occur.

There is considerable variation in the tendencies of animals to accumulate and remove hydrogen ions from the muscle. If the circulatory system is removing lactic acid from the muscle, muscle tissue simply utilizes the bulk of stored glycogen as a result of the stress. With adequate time and nutrition, however, glycogen can be restored to normal levels.

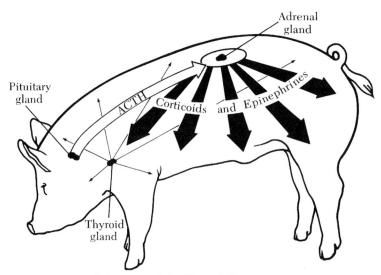

Influences of the "Stress" Hormones

FIGURE 5.11 | A simplified diagram of the sites of production, and the general actions of the major hormones related to stress reactions.

Muscle glycogen stores may be reduced when animals are handled, even though the stress hormones are not acutely released. When animals are fasted or exercised, muscle glycogen is used for energy. If adequate feed and rest are not provided for the animal, reduced muscle glycogen levels may exist at slaughter.

Many elements of the environment can become stressful to animals under certain conditions. Extremes in temperature, humidity, light, sound, and space are effective in producing the changes described in this chapter. However, they differ in their effects because the response that any one environmental condition produces depends on the species, weight, age, sex, inherent stress resistance, and emotional state of the animal. Some animals become frightened in unfamiliar surroundings, while others become hostile. These differences in reaction are probably associated with many factors in the animal, such as hormone balance (epinephrine/norepinephrine ratio), fatigue, or previous handling experience. Because unpredictable emotional responses are elicited, highly variable responses are produced in the muscles.

Stress and Muscle Characteristics

It is important to recognize that any of the environmental stress factors discussed can result in changes in the metabolites of muscle. These changes, in turn, are responsible for differences in the ultimate properties of meat. The nature of the changes depends on such factors as the duration or severity of the stress, and the level of the animal's stress resistance at the time of death.

Animals may be characterized as having, in variable degrees, stress susceptibility or stress resistance. If an animal is highly stress susceptible, it is likely to experience heatstroke, shock, and circulatory collapse when exposed to a stressor. Even a mild stress, not associated with high temperatures, may produce changes that result in death. In swine, susceptible animals exhibit a stress syndrome characterized by external signs such as extreme muscularity, anxious behavior, muscle tremors, and reddening of the skin.

In general, stress-susceptible animals have unusually high temperatures, rapid glycolysis (pH drop), and early postmortem onset of rigor mortis in their muscles. Although the postmortem changes are rapid, some degree of antemortem muscle temperature rise, H+ increase, and depletion of ATP also occurs. This combination of conditions results in an exaggeration of the muscle-to-meat transformation (rapid pH decline and an elevated carcass temperature resulting in protein denaturation). Muscles from stress-susceptible animals usually become pale, soft, and moist or exudative after a normal 18- to 24-hour chilling period. This **pale**, **soft**, **exudative** (**PSE**) condition most often results in lowered processing yields, increased cooking losses, and reduced juiciness. It usually is restricted to the muscles of the pork loin and ham, but it occasionally is seen in the dark muscles of the pork shoulder, as well as in some muscles of beef, lamb, and poultry. A **dark**, **firm**, **dry** (**DFD**) **muscle** condition also can be produced in meat from animals with a degree of stress susceptibility, if they have survived a stress sufficient enough to deplete their glycogen reserves.

Stress-resistant animals are able to maintain normal temperature and homeostatic conditions in their muscles in spite of relatively severe stress. However, they may accomplish this at the expense of muscle glycogen stores. Glycogen deficiency usually occurs when animals survive stress, such as that associated with fatigue, exercise, fasting, excitement, fighting, restraint, or electrical shock, but are slaughtered before they have sufficient time to replenish their muscle glycogen stores. Muscle glycogen depletion in these animals results in limited glycolysis in the muscle after death and results in a high ultimate pH. As a consequence of a high ultimate pH, changes in muscle color that otherwise occur during the postmortem transformation of muscle to meat, do not occur. Specifically, the resulting muscle reflects less total light due to the light-absorptive properties of the muscle and the pigments themselves reflect dark-red or purplish light. Tissue of this type is very dry or sticky in texture because of its excellent water-binding capacity. (Fresh meat color and water-holding properties are discussed more fully in Chapter 6). **Dark cutting meat** occurs frequently in beef (approximately 3 percent) and lamb but less often in

TABLE 5.4 — Heritability Estimates for Physical Properties of Meat

Characteristic	Species	Heritability estimate, approximate average percentage
Color score or color-firmnessstructure score	Swine	30
	Cattle	30
Marbling score (intramuscular fat)	Swine	25
Resistance to shear or tenderness score	Swine	30
	Cattle	60

pork. The major disadvantages of this situation are an unattractive dark, firm, dry appearance and a favorable pH for bacterial growth. This condition results in monetary loss, particularly in the merchandising of fresh beef.

There are many variations between the extremes of PSE and dark cutting meat. In general, the conditions can be described by the rate and extent of pH decline coupled with high carcass temperatures immediately postmortem. Figure 5.5 illustrates some of the possible pH patterns, along with the resulting conditions of the meat. Thus, one can visualize the biological variation in animal stress resistance in terms of variation in the ultimate quality of the muscle as food.

Genetics and Major Genes

Table 5.4 shows heritability estimates for some physical properties of meat. These estimates indicate that meat quality attributes are at least moderately heritable. Livestock producers, therefore could make major improvements in the acceptability of muscle as food by selecting breeding animals whose close relatives have muscles with normal color, moderate firmness, optimum or minimal intramuscular fat, and low resistance to shearing force (high degree of tenderness). The importance of these physical properties of meat is described in Chapter 6.

It is rare that any single gene controls an economically important trait in animals or quantitative or qualitative properties of meat. Occasionally, however, due to naturally occurring mutations or long-term selection pressure, a gene surfaces that controls an important trait. An example of such a major gene is that for **myostatin**. A defect in this gene results in cattle that are extremely muscled (see Chapter 3). Regarding meat quality, two genes have been identified that have a major influence on fresh pork quality, the **halothane (Hal)** and **Rendement Napole (RN;** or **acid meat)** genes. The **Hal gene** was discovered as a result of screening for stress susceptibility by subjecting pigs to halothane gas anesthesia. Animals possessing the Hal gene immediately responded by extreme rigidity in the hind legs. Once rigidity was observed, halothane gas was removed and most animals recovered normally. Pigs homozygous for this allele are stress-susceptible. The Hal gene encodes for a mutated calcium channel protein that resides in the sarcoplasmic reticulum. Because the protein is mutated, the functionality of the calcium channel is altered to a point where greater amounts of calcium leak from the sarcoplasmic reticulum in affected muscle tissue. As mentioned earlier, stress susceptible pigs often develop PSE meat. Because Hal-positive pigs possess a defective calcium channel, greater amounts of cytoplasmic calcium are released from the sarcoplasmic reticulum. This increased level of calcium, in turn, hastens metabolism in the muscle and results in a more rapid pH decline early postmortem while carcass temperature is still relatively high (Figure 5.5). The combination of low pH and high muscle temperature causes greater protein denaturation than normal and results in PSE pork. It should be emphasized that halothane-hastened postmortem metabolism can be mimicked through other postmortem handling treatments that are not related to the Hal gene. They will be discussed later.

The **RN gene** also causes development of PSE pork, but by another mechanism. Pigs possessing the RN gene have an altered adenosine monophosphate kinase (AMPK). AMPK is a rather novel enzyme because it is a very powerful regulator of energy charge in the cell. As the name implies, AMPK binds AMP. As discussed earlier, when AMP is plentiful in the cell, ATP levels have been compromised or severely depleted. Once AMP binds to AMPK, AMPK is capable of increasing the activity of pathways needed to generate energy, such as glycolysis, glucose uptake, fatty acid and carbohydrate oxidation, and inhibits metabolic pathways that store energy, such as glycogen synthase. Because feedback inhibition by the altered AMPK is minimal, RN pigs synthesize greater and greater amounts of glycogen. As a result, RN pigs often have higher amounts of muscle glycogen at the time of slaughter. During the transformation of muscle to meat, rates of postmortem glycolysis and pH decline are normal. But because there is a greater energy charge in the muscle prior to slaughter, more hydrogens are produced resulting in a lower than normal ultimate muscle pH. Although not completely understood at present time, the lower ultimate pH in acid meat causes protein denaturation similar to that which occurs under conditions of rapid pH decline and high muscle temperature.

A genetic condition identified in sheep over twenty years ago has been mapped to chromosome 18 (also see Chapter 3). Sheep possessing this genetic condition have over 32 percent more muscle, yet have normal birth weights and grow similar to those animals lacking the mutation. This mutation is commonly known as the **callipyge gene**, which is Greek meaning "beautiful buttocks". As captured in the name, these animals display a very muscular phenotype especially in the hind legs, but also noticeably in the loin, rack and shoulder muscles. Obviously, a genetic mutation with such desirable traits offers the meat industry great opportunities to increase production efficiency. However, in the case of this muscle-altering mutation the resulting meat is very unappealing from a tenderness standpoint. As mentioned above, meat tenderizes with postmortem age largely due to proteases that become active after harvesting. These same proteases can be inhibited by other proteins, which is the case with sheep possessing the callipyge gene. Calpastatin, the inhibitor of calpain proteases, is present in extremely high levels in the muscle of callipyge sheep, even many days postmortem. Thus, little aging-related changes in tenderness occur in lambs possessing this mutation. The use of these genetics has been limited in the sheep industry.

Though the **double muscling** genetic condition in cattle results in similar increases in muscling as the callipyge gene mutation in sheep, these conditions are quite different in the manner in which the mutations manifest themselves in the muscle tissue of animals. The double muscled condition is caused by a mutation in the **myostatin gene**, which largely controls muscle cell replication and formation. However, these animals do not have twice the number of muscles but rather possess nearly double the number of muscle fibers per muscle, thus giving them a distinct advantage in overall muscularity. These animals differ from callipyge sheep also in that double muscled beef is not less tender because of greater calpastatin content. Rather, meat from double muscled cattle tends to be less tender than that from normal cattle because of a combination of issues related to the overall lack of fatness, cold-shortening, age, and cookery method used; issues that are more related with management of the animal or processing and cooking of the resulting product. In fact, Texel sheep, a breed developed in the Netherlands, possesses a mutation in the myostatin gene similar to double muscled cattle and these sheep populate the bulk of lambs produced in the United Kingdom. Even so, use of double muscled cattle has been limited in the industry, not because of toughness, per se, but because characteristics like birth weight, growth rate, and quality grading characteristics are less desireable in this type of cattle.

Age

Animal aging is accompanied by darkening of muscle color due to increasing myoglobin concentration and other proteins of the muscle. High myoglobin muscles are desirable when used in processed meats for their contribution to product color, but otherwise, dark color is useful only as a guide to animal age. Of greater consequence is the nature of connective tissue in muscles of old animals. Decreases in tenderness that accompany animal age result largely from changes in connective tissue properties. As animals age, stable intermolecular cross-links within the collagen fibrils increase. This results in decreased solubility of collagen during the cooking process and increased resistance to shearing or chewing action. This whole process is likely associated with the additional exercise old animals have experienced that necessitates or accommodates strengthening of connective tissue fiber structure.

Although it is well-accepted that meat from very young animals is more tender than that of aged animals, changes occurring with age are not linear. During the rapid phase of growth, tenderness increases with time because rapid development of muscle fiber size dilutes existing connective tissue. Thus, market weight beef animals (12 to 18 months of age) often have more tender meat than growing calves (6 months of age). Substantial muscle toughening in beef animals is evident at about 30 months of age. Beyond this time there is further gradual toughening, but the rate of change becomes progressively slower. Similar patterns of meat toughening occur in animals of other species during growth and development, but comparable changes occur at different ages.

Although the above discussion of age-associated changes in tenderness has included reference to chronological age, it should be emphasized that age-related changes are simply a reflection of physiological age, which may or may not be equivalent to chronological age. Since animals within a species and muscles within an animal age at different rates, indicators of the physiological age of an animal are generally better predictors of tenderness than chronological age. These indicators, located principally in various components of the skeleton, are described in Chapter 15.

Meat flavor intensity increases with animal age. The likely cause of this flavor change is increased concentration of nucleotides in muscle, which degrade to inosinic acid and hypoxanthine postmortem. Flavor intensity may become so great that it is objectionable to some consumers, an example being the strong mutton flavor of mature sheep or game animals.

Muscle Location

Certain muscles that are free to shorten during rigor mortis onset often lack tenderness. Degree of tension placed on individual muscles by the skeleton and temperature achieved pre-rigor influence sarcomere length and ultimate tenderness. In addition, some of the differences in tenderness among muscles result from different quantities of connective tissue. Table 5.5 shows the tenderness of some beef muscles as determined both by resistance to mechanical shearing, and by palatability tests. The values are representative of muscle-to-muscle differences expected and are the basis for variations in tenderness often noted within single cuts of meat. Consequently, first impressions of meat cuts by consumers may depend largely on the particular muscle tasted.

Sex

Meat quality varies among the sexes of meat animals for reasons that relate largely to differences in circulating levels of sex hormones, especially in pigs and occasionally in sheep. Unacceptable meat quality is associated mostly with intact males. Consumers throughout the world express varying tolerance for these quality deficiencies, but intact males are gaining in acceptance because of their superior carcass leanness.

An objectionable onion-like or perspiratory odor is sometimes noted in pork, which represents the most serious sex-associated meat quality problem. This **sex odor** is noted with particular frequency in meat of boars, but sometimes is identifiable in pork of all sex conditions. Offensive odor is detected by some, but not all, consumers and is particularly noticeable during cooking, but may go unnoticed after cooking. Knowledge of the basis for the problem is incomplete, but most evidence suggests that a metabolite of testosterone, 5-androst-16-ene-3-one, is present in the tissues.

TABLE 5.5 Resistance to Shear, and the Tenderness Rating of Selected Muscles of Cooked Beef

Muscle Muscle	Shear,* kg	Tenderness
Rectus femoris	4.3	Slightly tough
Semimembranosus	5.4	Slightly tough
Semitendinosus	5.0	Slightly tough
Biceps femoris	4.1	Medium
Gluteus medius	3.7	Slightly lender
Psoas major	3.2	Very tender
Longissimus	3.8	Slightly tender
Infraspinatus	3.6	Slightly tender
Supraspinatus	4.2	Medium
Triceps brachii	3.9	Medium
Trapezius	6.4	Very tough
Sternocephalicus	7.1	Very tough

*Warner-Bratzler shear, 1.27-cm diameter cylinders. Low values indicate tender meat.
Source: Summarized from the data of Ramsholtom and Strandine (1948).

The odor may come directly from this hormone degradation product or from some other substance produced by the product.

Meat of young bulls and rams does not have undesirable odor or flavor, but it lacks uniformity in tenderness and toughens more quickly with animal age than that of other sexes. Male hormones are responsible for this meat quality problem, but the specific mechanism of action is unknown. Muscles of intact males have increased concentrations of intramuscular collagen, and the collagen may be more mature (greater chemical cross-linking) at uniform chronological ages than that of castrates. Bull and ram meat is also slightly darker than that of steers or wethers of similar ages. The dark color results from higher concentrations of myoglobin as well as a tendency toward stress-induced dark cutting meat.

Diet

The influence of diet on the physical properties of muscle is of minor importance, as long as there are no serious nutritional deficiencies. Any feeding practice in the immediate antemortem period which alters the energy charge of muscles can influence the ultimate physical properties of meat. Holding livestock prior to slaughter provides opportunity for resting and feeding, in addition to improving the ability of animals to withstand later handling. This interval can influence the level of energy stores in muscles. Starchy feeds, and sugar especially, will restore depleted muscle glycogen levels, thus permitting the development of a normal postmortem pH. However, as mentioned previously it is advisable to withhold feed for 24 hours prior to slaughter in order to facilitate the process of evisceration and minimize the chances for microbial contamination of the carcass from the gastrointestinal tract.

Feeding animals higher levels of vitamins results in improved product quality. One, in particular, is vitamin E. Vitamin E is fat-soluble and therefore, is deposited in lipid-rich tissues and cell membranes. Vitamin E is a powerful antioxidant and likely improves the shelf-life and color of fresh meat by inhibiting fat oxidation and stabilizing myoglobin, respectively.

Although not technically a meat quality problem, the development of yellow carcass fat is controlled by diet. When some ruminants are fed forage diets that contain high levels of ß-carotene, this yellow precursor of vitamin A is deposited in the body fat. No scientific basis exists for discrimination against yellow carcass fat. On the other hand, high forage diets produce some undesirable meat flavor compounds. A variety of unusual or "grassy" flavors have been noted, which are caused by a variety of compounds in forages. The problem may be eliminated by feeding grain diets for several weeks before slaughter. For the most obvious reasons, special flavor problems sometimes exist in meat of pastured animals that consume wild onion or wild garlic.

Preslaughter Handling

As outlined in Chapter 4 in some detail, procedures necessary to convert the tissues of living animals into edible food are by nature stressful. Animals are exposed to a myriad of environmental and physiological stimuli. The steps involved in the whole marketing process may include sorting, loading, transportation, weighing, driving, water spraying, and immobilization. The severity of effect depends on the climate, facilities, personnel, and many other factors. Undesirable effects of poor handling procedures include but are not limited to: carcass bruising, dark cutting meat, and PSE meat.

As mentioned, the transportation phase of livestock marketing can be one of the most severe segments of the harvesting process. It is during transit that most death losses and tissue bruising occur. Improperly ventilated trucks, or warm climatic conditions, can result in extreme discomfort for animals. In addition to possible death losses, muscle tissue shrinkage and reduction of dressed carcass weight can result from uncomfortable or prolonged transportation. However, under normal marketing conditions the weight of the muscles is not affected, even though animals usually lose 2 to 5 percent of their liveweight in the marketing process, most often due to loss of gastrointestinal tract contents.

Immobilization Method

In the US, immobilization of meat animals, except under those situations where religious exemptions are involved, is accomplished by accepted methods outlined under meat inspection laws. These include carbon dioxide, electric shock, and captive bolt (or projectile) immobilization.

Although the immobilization process is not free of stress, it reduces stress responses compared to exsanguination without immobilization. However, the overall effectiveness of these procedures is dependent on careful design and operation of equipment and handling facilities used. Most systems are designed to render animals unconscious without stopping action of the heart, so that the bleeding process can be facilitated. Deep stunning systems stop heart action but also minimize reflex struggling, which can produce hemorrhages in muscle tissue.

Properties of muscles are influenced by type and effectiveness of immobilization procedures. The severity of this process is usually expressed in muscles by the degree of glycogen depletion. Figure 5.12 shows glycogen levels in breast muscles of chickens bled under anesthesia, after electrical immobilization, and without immobilization. These differences in glycogen result in differences in ultimate pH and physical properties of muscles.

Immobilization should be followed, as quickly as possible, by rapid bleeding to prevent animals from regaining consciousness, and to release blood pressure. Some immobilization systems, particularly those employing electric shock, raise blood pressure to the point that hemorrhaging can occur in muscles. Unless bleeding is accomplished within a few seconds of immobilization, the meat may exhibit a condition known as **blood splash**, which consists of blood spots (petechial hemorrhages) that cannot be removed. This problem can be minimized by use of proper voltages and correct placement of electrodes.

Postmortem Effects

Temperature

As discussed previously, alterations in carcass temperature declines can affect the rate of chemical reactions occurring in muscle tissue and greatly impact meat quality. Carcass storage temperatures immediately after slaughter can alter dramatically

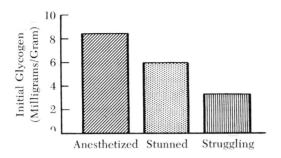

FIGURE 5.12 | Glycogen levels in chicken breast muscles immediately after death: (left) with bleeding under anesthesia, (middle) after electrical stunning, and (right) without stunning. From E. Cosmos, "Ontogeny of Red and White Muscles: The Enzymic Profile and Lipid Distribution of Immature and Mature Muscles of Normal and Dystrophic Chickens," in E. J. Briskey, R. G. Cassens, and B.B. Marsh, eds. *The Physiology and Biochemistry of Muscle as a Food, 2* Madison: The University of Wisconsin Press; © 1970 by the Board of Regents of the University of Wisconsin System, 205.

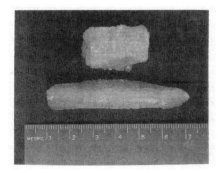

FIGURE 5.13 | Thaw rigor shortening. A freshly excised muscle sample (bottom) is shown in comparison with an identical sample that was frozen pre-rigor and thawed (top). The sample that has undergone thaw rigor (top) is only 42 percent of its original length.

muscle enzyme activities. Temperature differences of 10° C may cause rates of these reactions to change by a factor of three or more. Consequently, it is desirable to reduce muscle temperature after death as quickly as possible to minimize protein denaturation and to inhibit growth of microorganisms. On the other hand, extremely rapid reduction of postmortem muscle temperature can cause undesirable consequences.

Two conditions, known as **thaw rigor** and **cold shortening**, result from low temperature in muscles before the onset of rigor mortis. Thaw rigor is a severe type of rigor mortis that develops when muscle, frozen prerigor, is thawed. Cold shortening, on the other hand, develops when muscle is chilled below 15° C to 16° C before onset of rigor mortis. In thaw rigor, contraction is caused by a cold-induced release of calcium into the sarcoplasm and may cause a physical shortening of 80 percent of original length in unrestrained muscles. A more common degree of thaw rigor shortening is approximately 60 percent of original length. This type of contraction is accompanied by release of large quantities of meat juices and severe toughening. Figure 5.13 illustrates severe physical shortening that occurs in thaw rigor. From this illustration, one can visualize the extent of muscle distortion occurring when meat is frozen and thawed, pre-rigor. Although the degree of shortening of muscles attached to the skeleton is less than in unrestrained muscle, a loss of tenderness and other qualities may take place.

Muscles need not be frozen for some undesirable attributes of thaw rigor to develop. If attained pre-rigor, temperatures above 0° C but below 15° C to 16° C cause a type of contraction known as cold shortening. Although such shortening is less severe than that of thaw rigor, the underlying cause relates to a cold-induced release of calcium from the sarcoplasmic reticulum or a failure of the calcium pump in the sarcoplasmic reticulum to function properly at lower temperatures. Figure 5.14 shows photomicrographs of relaxed and cold-shortened beef muscle. Complete disappearance of the I band is observed frequently in cold-shortened muscle.

Practical implications of prerigor chilling also are important, because many commercial chilling systems remove body heat from the superficial muscles so rapidly that cold shortening is induced. Such shortening is especially severe in unrestrained muscles, but it also occurs in localized areas of muscles attached to the skeleton. The problem may become more serious as efforts are made to produce animals with less fat because the fat serves as insulation to heat transfer from deep within the tissues. Beef and lamb are most susceptible to the condition, but other species may be affected to varying degrees.

Severe shortening and early onset of rigor mortis may be induced by maintaining muscle at relatively high temperatures (up to 50° C). Heat rigor is thus produced, which is the result of a rapid depletion of ATP stores. Consequently, there appears to be an optimum temperature at which muscle should be held during the onset of rigor mortis to minimize shortening, toughening, and other undesirable effects of the rigor process. Although rigor mortis cannot be prevented, carcass chilling systems that maintain temperatures of 15° C to 16° C in the muscle during rigor onset minimize the severity of rigor-associated changes. Figure 5.15 shows the relation between the extent of shortening and the temperature of muscle during rigor onset. Further, in the shortening range of 20 to 50 percent, tenderness decreases as degree of shortening increases.

Accelerated Processing

Accelerated processing is defined as the application of some processing steps, such as cutting, bone removal, or grinding immediately after slaughter. Many related changes in meat quality and product performance result from accelerated processing, most likely because of alterations in tissue pH.

Even though the rate of pH decline in postmortem muscle usually is accelerated by **pre-rigor chopping or grinding**, the extent of pH decline in muscles of some species, such as those with a high proportion of white fibers, is abbreviated when tissue is ground before glycolysis is complete. When the structure of tissue is disrupted extensively, anaerobic glycolysis is abbreviated compared to normal, probably because oxygen enters the tissue from the air and supports continued aerobic glycolysis. Experience shows that the ultimate pH, i.e., pH after glycolysis is complete, of pork and poultry muscle tissue ground within one hour of exsanguination is 0.2 to 0.3 units higher than normal. Certain benefits are realized when meat with elevated pH is used for preparation of processed meat. Such tissue has improved water-binding properties (Chapter 6), which gives better processing yields and improved juiciness.

Sausage makers know that the interval of time between slaughter and meat grinding affects the physical properties of their finished products. In general, meat ground pre-rigor, and mixed with curing ingredients, including salt, has superior water-binding properties and maximum juiciness. The advantage in grinding or processing pre-rigor meat stems from early exposure of undenatured muscle proteins to the action of salt. These proteins become solubilized before denaturing effects of postmortem events can take place. Therefore, they do not undergo the severe transformations to which proteins in intact muscles are subject. Additionally, exposure of pre-rigor meat to salt during grinding inhibits glycolysis and the extent of pH decline even more than grinding alone.

Pre-rigor grinding may protect flavor in certain types of processed meat. In products formulated without salt, lipids of high-pH tissues oxidize more slowly and such products retain a fresh, rancid-free flavor for a longer time. Realization of beneficial effects of pre-rigor grinding on rancidity development requires an ultimate pH of approximately 6.0 or greater. Figure 5.16 shows the relationship between ultimate pH and lipid oxidation (measured as thiobarbituric acid values) in microbe-free ground pork muscle after 12 days of refrigerated storage. Unless the pH remains above 6.0, development of rancid flavor is a serious palatability problem in some unseasoned ground meats.

Fresh meat flavor may be protected by processing some intact cuts immediately after slaughter. Rapid onset and softening of rigor in poultry muscle enables processors to chill, cut, and package poultry parts within two hours or less of exsanguination. Less refrigerated storage space is

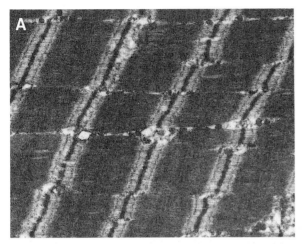

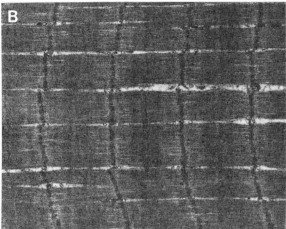

FIGURE 5.14 | Electron micrographs of (A) relaxed and (B) cold-shortened beef muscle. From Pearson and Dutson, Adv. Meat Research 1: 185 (1985). Courtesy of Van Nostrand Reinhold. From *Meat Research* by Pearson/Dutson. Copyright © by Pearson & Dutson. Reprinted by permission of the author.

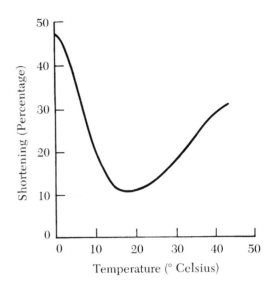

FIGURE 5.15 | The effect of temperature on the degree of shortening in unrestrained bovine muscle. Summarized from the data of Locker and Hagyard, *J. Sci. Food Agric.*, 14, 787 (1963).

required by such rapid processing, and products reach consumers with fresher flavor. Of course the ultimate in flavor protection by accelerated processing is the common practice of cooking lobsters from the live state.

In most kinds of meat, removal of bones and excess fat before chilling reduces energy requirements for chilling and provides for more precise control of chilling rate. When boneless meat is maintained at temperatures that avoid shortening (15° C to 16° C) during rigor development, they do not require the restraint of shortening imposed by skeletal attachments. An alternative method of accelerated processing for intact cuts is to hasten the onset of rigor mortis by carcass electrical stimulation (discussed below) prior to bone removal.

Using accelerated processing enables the industry to minimize time between animal slaughter and meat consumption. These processing systems require less refrigeration than conventional systems, and they provide products to consumers in the freshest form possible. As more knowledge is generated on the control of postmortem changes, greater use of accelerated processing is likely to occur.

Carcass Electrical Stimulation

Electrical stimulation (ES) of freshly slaughtered carcasses has been used successfully to improve tenderness and meat quality in turkey, lamb, beef, and veal. Several mechanisms apparently come into play to produce desirable effects. Even though

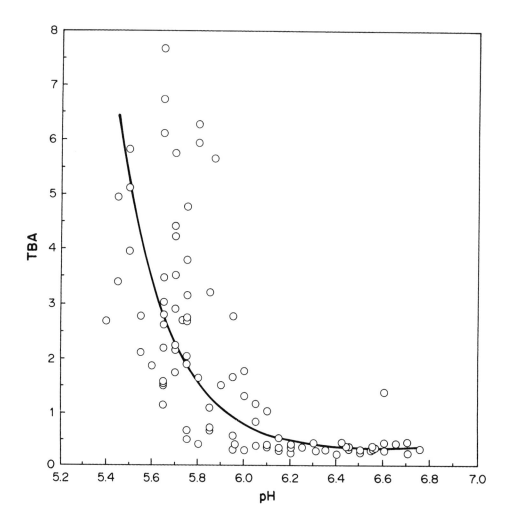

FIGURE 5.16 | Lipid oxidation (TBA values) and ultimate pH of ground pork after 12 days of storage at 2° C. From Yasosky et al., *J. Food Sci.* 49:1510 (1984).

knowledge of underlying bases for the practice is incomplete, widespread industry application has taken place.

Despite the negative attributes of high temperature-low pH combinations described earlier, such conditions early postmortem are associated with improvements in meat tenderness. This does not mean that those conditions cause the improvements, but rather, they enable several other factors to become operative or they accompany changes that are directly responsible for tenderization. Accordingly, one of the most widely recognized effects of carcass ES is acceleration of rate of postmortem pH decline. Figure 5.17 shows the difference in rate of pH decline in ES and unstimulated beef muscles. It should be noted here, however, that ES in pork is generally not accepted primarily because ES hastens postmortem metabolism to a point where it is detrimental to ultimate product quality.

Another effect of ES is hastening of rigor mortis. Violent contractions produced by stimuli utilize great quantities of ATP and deplete energy reserves. In the absence of ATP, muscles quickly develop rigor mortis.

Meat tenderization by ES has been attributed to at least three factors:

- prevention of cold shortening through acceleration of glycolysis and rigor onset before temperatures reach the cold shortening range
- accelerated proteolytic activity through enhanced calcium release; and
- physical disruption of fiber structure through extreme muscle contractions.

The relative importance of these mechanisms depends on chilling conditions. Tenderization by ES is mostly due to prevention of cold shortening (see Figure 5.15). When fat carcasses are chilled or when relatively slow chilling is used, physical disruption of muscle structure is more likely to be predominant. The photomicrographs in Figure 5.18 show normal structure of unstimulated muscle and permanent contracture bands associated with physical damage in ES muscle.

Tenderness improvement occurs in almost all carcasses subjected to ES, but the extent of improvement depends on inherent tenderness of carcasses. Generally, those that are quite tender without ES are not improved appreciably by treatment. On the other hand, aged animals may be tenderized by increments that are similar to those of youthful animals, but the change is usually insufficient to develop acceptable tenderness. Consequently, ES is most appropriate for carcasses of young animals that have not been fed high-energy diets or that lack inherent tenderness.

Further benefits of ES accrue to slaughterers because of rapidly developing quality characteristics in carcasses. Bright red muscle color, muscle firmness, and solidification of marbling develop more quickly in ES than unstimulated carcasses. Incidence of a dark, coarse band formation, found in superficial muscle layers of rapidly chilled carcasses, is greatly reduced by ES. Rapid

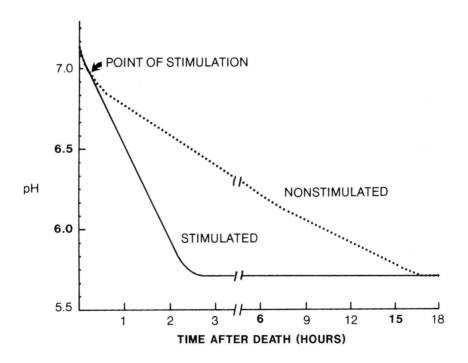

FIGURE 5.17 | Differences in rate of pH decline in electrically stimulated and nonstimulated beef carcasses. From Pearson and Dutson, *Adv. Meat Research 1*: 185 (1985). Courtesy of Van Nostrand Reinhold. From *Meat Research* by Pearson/Dutson. Copyright © by Pearson & Dutson. Reprinted by permission of the author.

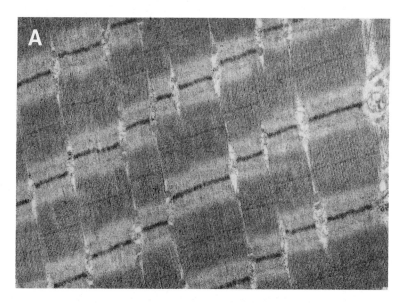

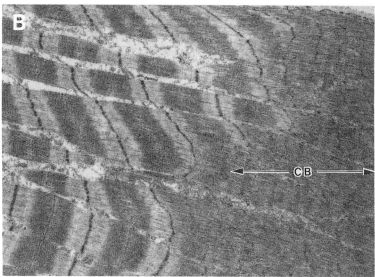

FIGURE 5.18 | Electron micrographs of (A) nonstimulated and (B) electrically stimulated beef muscle. CB = contracture band. From Pearson and Dutson, *Adv. Meat Research 1*: 185 (1985). Courtesy of Van Nostrand Reinhold. From *Meat Research* by Pearson/Dutson. Copyright © by Pearson & Dutson. Reprinted by permission of the author.

development of quality attributes is largely due to hastening of glycolysis and rigor onset by the treatment. Figure 5.19 (see color section) illustrates differences in appearance of ES and unstimulated beef muscles at uniform time (24 hours) postmortem. The effects of ES on meat properties are presented in Table 5.6.

The peak voltage that may be used successfully for carcass ES varies from 30 to 3600 volts. Lower voltages cause muscle contraction by a combination of direct stimulation and propagation of stimuli from the nervous system. High voltages provide direct and uniform muscle stimulation throughout the carcasses. This generalized contraction is shown in Figure 5.20.

Carcass ES is widely practiced and offers several benefits to producers, processors, and consumers of meat. Its application saves livestock feed,

CHAPTER 5 CONVERSION OF MUSCLE TO MEAT: BIOCHEMISTRY OF MEAT QUALITY DEVELOPMENT

TABLE 5.6 Effects of Electrical Stimulatin on Meat properties

Property	Effect
Tenderness	Improved
Color of lean	Improved
Quality grade	Improved
Marbling	More visible
Flavor	Enhanced
Heat ring	Prevented
Aging period	Shortened
Caselife of meat	Extended
ATP depletion	Accelerated
Onset of rigor mortis	Accelerated
"Hot" boning and "hot" processing	Allows these to be performed without adverse effects

Source: Savell (1979).

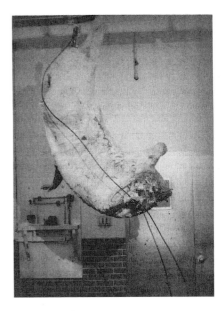

FIGURE 5.20 Beef carcass electrical stimulation.

reduces carcass aging time, enhances raw meat quality development, and improves meat tenderness. The mechanisms of action for these effects are not completely clear, but Figure 5.21 shows many of the known intrinsic changes brought about by this process.

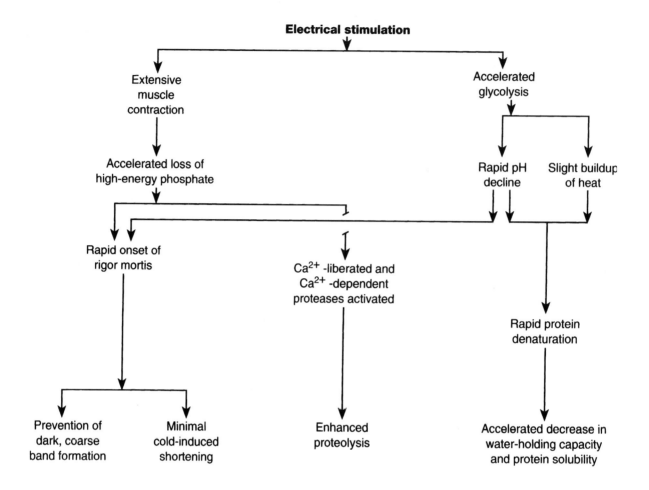

FIGURE 5.21 | Flow diagram of effects of carcass electrical stimulation.

6

PROPERTIES OF FRESH MEAT

OBJECTIVES: *Know those factors that influence a consumer's perception of quality in fresh meat products and understand the physical, structural and chemical properties underlying these factors.*

Key Terms

Water-holding capacity
Shrink
Purge
Bound water
Free water
Immobilized water
Isoelectric point
Net charge effect
Steric effect
Hue

Chroma
Value
Globin
Heme ring
Reduced
Oxidized
Deoxymyoglobin
Metmyoglobin
Oxymyoglobin
Extracellular water

The term "fresh meat" is used in a special context to include products that have undergone the chemical and physical changes which normally occur after the harvesting of animals. To this end, only those meat products minimally processed by procedures such as fabrication into retail cuts, cubing, grinding, marination, or freezing will be considered as fresh meat in this textbook. The major steps of final preparation of fresh meat products prior to eating are left to the cook in the home or the chef in an institutional setting. Examples of such fresh meat products include roasts, steaks, chops, parts, fillets, and ground meats such as hamburger and ground pork patties. The properties of fresh meat dictate its usefulness to the merchandiser, its appeal to the purchaser or consumer, and its adaptability for further processing. Of particular importance to fresh meat quality are: color, texture, firmness, structure, water-holding capacity, and fat quality and quantity characteristics.

MEAT QUALITY

Just as beauty is in the eye of the beholder, meat must please the eye of the shopper before it can satisfy the palate of the consumer. Fresh meat must first look good to the customer who is selecting meat for purchase. Therefore, all characteristics that are apparent from a visual inspection of meat are important to a consumer, specifically, color, textural characteristics, and lean meat yield. Once the consumer buys the meat and decides to eat the product, the palate must be satisfied. At this point, a product must still retain pleasant visual cues but more importantly, the aroma of the entrée must be pleasing and the tenderness, juiciness, and flavor must meet the expectations of the final consumer. Note that expectations and preferences play a major role in this process. Expectations and preferences vary widely among consumers depending on past experiences and cultural background. Recognizing this, processors, retailers, and foodservice operators may elect to place different levels of emphasis on the factors we are about to explore.

WATER-HOLDING CAPACITY

Water-holding capacity, the ability of meat to retain naturally occurring or added moisture during the application of any external force, affects nearly every attribute of fresh meat quality, including color, texture and firmness of raw meat, and juiciness and tenderness of cooked meat. Water-holding capacity is an important characteristic of fresh meat during its further processing into sausages or cured and heat processed products and affects yield and palatability of the finished products. Recall from the previous chapter that water-holding capacity decreases during the transformation of muscle to meat so that meat assumes a normal moist appearance and texture. Forces as gentle as those exerted on a carcass during fabrication and cutting can compel some moisture from the tissue. Processes such as grinding and heating are generally quite harsh and lead to greater losses of moisture.

Water-holding capacity of muscle tissue has a direct effect on shrinkage of meat during storage. When tissues have poor water-holding properties, loss of moisture and, consequently loss of weight during storage (**shrink**) is great. The moisture loss occurs mostly from the cut surfaces of carcasses or cuts. After primal or retail cuts are made, an even greater opportunity for moisture loss exists because of the large muscle surface exposed by cutting. The common practice of placing boneless closely trimmed subprimal cuts in vacuum shrink bags for shipment to retail outlets magnifies the **drip** loss problem because all of the moisture is captured in the bag. The negative pressure produced by the vacuum also contributes to moisture loss from the meat. The accumulated moisture is known as **purge**, and it is quite evident when the bag is opened. High purge or drip losses are economically significant, not only because of loss in weight, but because the lost water contains high quality meat proteins.

Purge losses are not the only negative effect of low water-holding capacity. When purge is high, further loss of natural juices is likely to occur during cooking through evaporation and drip. This may lead to dryness and even the perception of toughness. Depending upon the severity of moisture loss, quality deterioration can be significant.

CHEMICAL BASIS OF WATER-HOLDING CAPACITY

Muscle contains approximately 75 percent water, 20 percent protein, 5 percent fat, 1 percent carbohydrates and 1 percent vitamins and minerals. The majority of moisture held in muscle is retained within and between the myofibrils, and beneath the cell membrane (sarcolemma). Moreover, water is located between muscle cells and between and among muscle bundles. After animals are harvested and muscle is converted to meat, water can migrate more easily within the muscle.

There are no clear lines of division among the locations of water that resides in muscle; such divisions are somewhat empirical. Even so, it is commonly considered that water in meat exists in three forms: **bound water**, **immobilized water**, or **free water**. Recall from chemistry courses that water (H_2O) molecules are not neutrally charged entities but rather they have positively and negatively charged ends, i.e. they are dipolar. Because of this dipolar nature, they associate with electrically charged reactive groups on muscle proteins. Of the total water within muscle, only 4 to 5 percent is located in such a manner and is known as bound water. It is so tightly bound that even during the application of severe mechanical or other physical forces, it remains associated with the proteins. This water is not subject to freezing and many forms of conventional heating. Thus, the amount of bound water changes very little regardless of the conditions.

Additional water molecules are subsequently attracted to the bound water molecules in layers that become successively more weakly bound as the distance from the reactive group on the protein becomes greater. Such water may be termed immobilized water or entrapped water and is largely held entrapped within the spaces between the filamentous structures of the myofibrils. The amount of water immobilized in this manner depends on the amount of physical force exerted on the muscle. Immobilized water is not readily available and is not mobile during the early postmortem period. However, it is extremely subject to the drying process and to freezing. Moreover, this water is affected by rigor mortis development and also by changes that occur during post rigor storage of meat as discussed in the preceding chapter.

The final pool of water held by muscle is retained by weak surface forces and is known as free water. As the name implies, this water is held so loosely that it is readily available whenever a force is applied to meat. Figure 6.1 is a diagram of the principal mechanisms by which muscle proteins retain water. Several factors influence the number and the strength of reactive groups on muscle proteins and subsequently their availability to bind water. The specific conditions are dependent on hydrogen ion production, loss of ATP, onset of rigor mortis, and changes of cell structure associated with the extent of proteolysis.

Net Charge Effect

The continual addition of H^+ to the sarcoplasm that results from ATP hydrolysis and anaerobic glycolysis (see Chapter 5) results in the reduction of pH during the postmortem conversion of muscle to meat. Accumulation of positively charged ions is responsible for reduction in the net charge across all proteins in the muscle cell. This change in overall charge can result in varying amounts of protein denaturation and a subsequent loss in protein solubility. Recall that all proteins have a three-dimensional, folded structure that allows them to perform a specific function. This structure and the collection of charges are dictated largely by the sequence of amino acids in the protein's polypeptide chains and the amino acid side groups that project out from the peptide chain. As a result of these various charges, proteins are capable of folding and binding water, and generally are soluble in an aqueous solution. When a protein has a charge (quite often negative), this allows the structure of the protein to expand because of the repulsive nature of like charges. This allows water

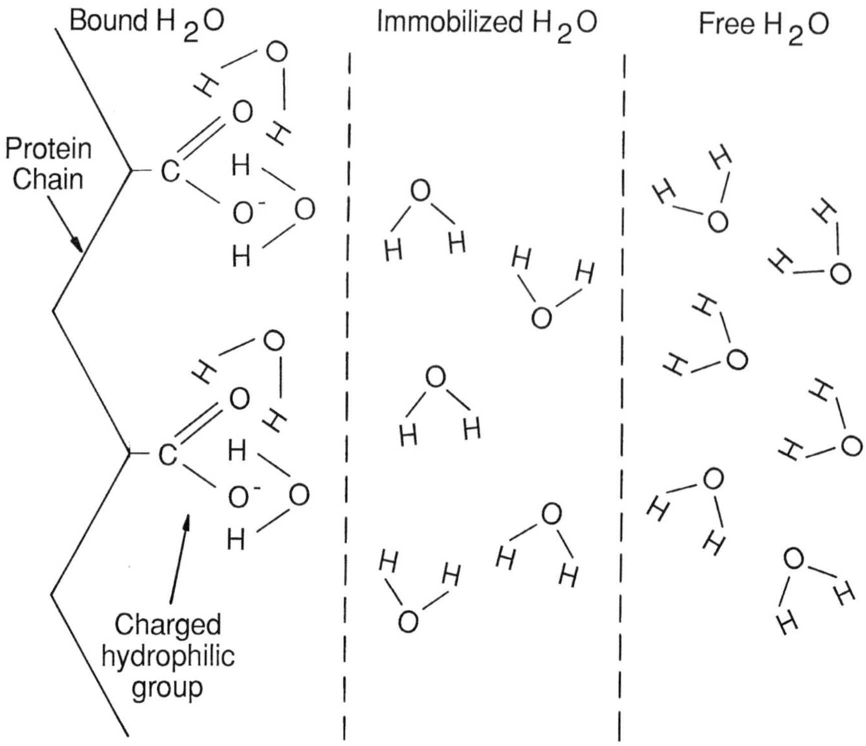

FIGURE 6.1 Charged hydrophilic groups on the muscle proteins attract water, forming a tightly bound layer (left), the molecules of which are oriented by their own polarity and that of the charged group. An immobilized layer (middle) is formed that has a less orderly molecular orientation toward the charged group. The free water molecules (right) are held only by capillary forces, and their orientation is independent of the charged group. Courtesy of W. H. Freeman and Company.

to infiltrate the structure and as a result, makes the protein highly soluble in an aqueous-based system, like the body. As positively charged hydrogen ions flood the system during postmortem energy metabolism, they change the overall nature of the charge by binding negatively charged amino acid residues within a protein. Association of positive ions with a number of negative charges decreases the repulsive action of negative charge, which alters the micro-environment around the protein and its interaction with water. As a result, proteins collapse onto themselves or aggregate with other proteins, which causes them to be less soluble and fall out of solution. This process is a normal event in the transformation of muscle to meat. The net charge of muscle proteins decreases as the pH decreases toward the **isoelectric point** of muscle.

The isoelectric point is the pH at which positive charges equal negative charges for a particular protein, or group of proteins. In the case of muscle, the isoelectric point is 5.0–5.2. To illustrate the complexity of an isoelectric point in a mixed protein tissue like muscle, consider that myosin, the major protein of the myofibril has an isoelectric point of about 5.4, while the iosoelectric point of myoglobin, the major oxygen carrier protein in muscle is greater than 7.0. Regardless, as the pH of muscle approaches its isoelectric point (5.0–5.2) the net charge is nearly zero. As a result, the positive and negative charges attract each other and the entire structure is more compact and less capable of binding water. This influence of pH is called the **net charge effect**. Figure 6.2 shows the relationship between pH and water-binding capacity in meat.

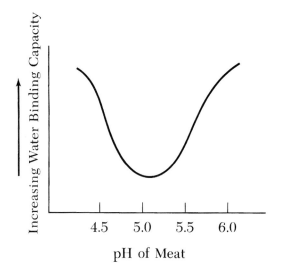
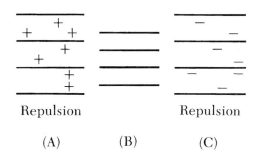

FIGURE 6.2 Effect of pH on the amount of immobilized waste present in meat, due to its impact on the distribution of charged groups on the myofilaments and the amount of space between them. (A) Excess positive charges on the filaments. (B) Balance of positive and negative charges. (C) Excess negatrive charges on the filaments. From Wismer-Pedersen, J., "Chemistry of Animal Tissues-Water," in *The Science of Meat and Meat Products*, 2nd ed. J. F. Price and B. S. Schweigert, eds. W. H. Freeman and Co., San Francisco, 1971.

Steric Effects

In normal meat, only about one-third of the loss of water-holding capacity postmortem is due to the drop in pH, or the net charge effect. **Steric** or **spatial** effects are those physical characteristics that create space within the muscle fiber and facilitate areas for water to occupy. Myofibrils occupy nearly 85% of the volume in a muscle fiber. It is thought that more than 80 percent of the water held in muscle in within the myofibrils where it is held by capillary forces. Any changes to myofibril structure can dramatically affect water-holding capacity. When actin binds to myosin during contraction, the head of myosin rotates, and pulls the actin filament laterally toward the z-line, which tends to squeeze or compress the structure. Additionally, the actomyosin complex formed irreversibly as ATP disappears postmortem creates a tight network among the myofilaments. Certain ions, principally divalent cations such as calcium (Ca2+) or magnesium (Mg2+), have the ability to combine with and neutralize two negatively charged reactive groups on the proteins, which further compresses the structure. In addition, when there is little net charge on the proteins, there are few charged groups available to separate protein chains by repulsive forces. These conditions prevent water binding by bound groups and allow protein chains to pack closely together, preventing those reactive groups that are still available from binding water. This lack of space for water molecules within protein structures is known as a **steric effect** on water binding.

Proteolysis

Just as collapsing the myofibril structure during the conversion of muscle to meat reduces the ability for muscle cells to hold water, opening of the muscle structure post rigor improves water retention through steric effects. Thus, some of the loss of water binding caused by pH decline and rigor mortis is recovered during subsequent storage of meat. As described in Chapter 2, muscle is filamentous in nature, predominately because it is designed to create tension through contraction. The primary organelle responsible for contraction is the myofibril. Yet, the thousand or so myofibrils per cell do not act alone, nor do individual groups of myofibrils act independently of other myofibrils in the cell or in adjacent cells. Rather, each myofibril is tethered to adjacent myofibrils at the Z-line by the intermediate filaments that contain desmin, talin, viniculn. Likewise, myofibrils positioned next to the sarcolemma are attached to the cell membranes and extracellular connective tissues through clusters of proteins that make up a structure known as the costamere. Any process that degrades this elaborate network of proteins postmortem would improve water-holding capacity by allowing the structure to swell or expand to retain water. Postmortem proteolysis indeed increases water-holding capacity during postmortem conditioning of muscles. Myofibrils decrease in volume postmortem due to the onset and completion of rigor mortis. This reduction in size expels water from the myofibrillar component of the cell. However, during postmortem proteolysis, costameric proteins as well as the intermediate filaments holding myofibrils in register across a muscle cell are degraded by calpain proteases. When these proteins are degraded, the lateral organization of the muscle is compromised and the cell is free to expand. This expansion creates additional space for water to occupy within the muscle, even though it has been squeezed from the myofibril during postmortem energy metabolism and rigor mortis develoment.

Exchange of Ions

Proteolytic degradation is not limited to myofibrillar or sarcoplasmic proteins. It also affects some proteins integral to membrane structure. Thus, proteolysis compromises membrane integrity and allows diffusion of ions into areas surrounding muscle proteins. This redistribution of ions across the membrane results in replacement of some divalent ions on protein chains with monovalent ions such as sodium (Na+). As a consequence, for every divalent cation replaced, one reactive group of a protein is freed to bind water, and forces pulling protein chains together are lessened. This exchange of ions in muscle proteins results in improved water-binding capacity. A limited improvement in water-binding capacity also occurs over time due to a slight rise in pH, but this accounts for only a small proportion of the total change.

COLOR

Color is used by consumers more than any other visible meat characteristic as an indicator of the wholesomeness and quality of fresh meat. It has been estimated that 15 percent of all retail beef is discounted due to variations in color. Color, as detected by the eye, is the result of a combination of several factors. Any specific color has three attributes, known as **hue**, **chroma**, and **value**. Hue describes that which one normally thinks of as a color—yellow, green, blue, or red. In reality, it describes the wavelength of light radiation. Chroma (purity or saturation) describes the intensity of a fundamental color with respect to the amount of white light that is mixed with it. The value of a color is an indication of overall light reflectance (brightness) of the color.

The most important contributors to meat color are the pigments that absorb certain wavelengths of light and reflect others. However, other factors influence and modify the manner in which color is visually perceived. Meat color is the total impression as perceived by the human eye, and is influenced by the viewing conditions. There also are marked differences among individuals in their color perceptions. The structure and texture of the muscles viewed also influence the reflection and absorption of light.

Pigments

Pigments in meat consist largely of two proteins: hemoglobin, the pigment of blood, and myoglobin, the pigment of muscles. In well-bled muscle tissue, myoglobin constitutes 80–90 percent of the total pigment. Such pigments as the catalase and cytochrome enzymes also are present, but their contribution to color is minor.

The two major pigments are similar in structure, except that the myoglobin molecule is one-fourth as large as the hemoglobin molecule. Myoglobin consists of a globular protein portion

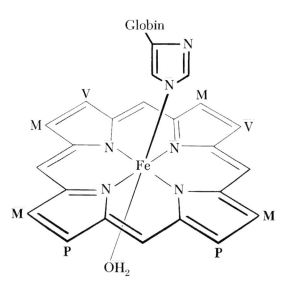

FIGURE 6.3 Schematic representation of the heme complex of myoglobin. The globin and water are not part of the planar heme complex. M, V, and P stand for methyl, vinyl, and propyl radicals attached to the porphyrin ring that surrounds the iron atom. From Bodwell, C. E. and P. E. McClain, "Chemistry of Animal Tissues-Proteins," in *The Science of Meat and Meat Products*, 2nd ed. J. F. Price and B. S. Schweigert, eds. W. H. Freeman and Co., San Francisco, 1971.

(**globin**) and a nonprotein portion called a **heme ring** (Figure 6.3). The heme portion of the pigment is of special interest because the color of meat is partially dependent on the oxidation state of the iron within the **heme ring**.

Species and Other Influences

Myoglobin quantity varies with species, age, sex, muscle, and physical activity. This accounts for much of the variability in meat color. Species differences are apparent when the light color of pork is compared with the bright red color of beef. The pale muscles of veal carcasses are indicative of the fact that the muscles of immature animals have a lower myoglobin content than those of mature individuals. The intact male has muscles that contain more myoglobin than do those of females or castrates at comparable ages. Because of their differences in myoglobin content, the light breast muscles of the chicken contrast strongly with the dark muscles of the leg and thigh. Game animals have darker muscles than domestic animals partially because of myoglobin induction by physical activity. In general, beef and lamb (or mutton) have more myoglobin than pork, veal, fish, or poultry. The following list identifies the most typical color of meat from the various species:

- Beef—Bright cherry red
- Pork—Grayish pink
- Fish—Gray-white to dark red
- Poultry—Gray-white to dull red
- Horse—Dark red
- Veal—Brownish pink
- Lamb and mutton—Light red to brick red

Muscle-to-muscle differences in myoglobin content (and much of the variation among species) are due to the type of muscle fibers present. Those muscles with relatively high proportions (30–40 percent) of red fibers appear dark red in color. However, when they are viewed histologically, these myoglobin-rich fibers are still seen to be mixed with easily distinguishable white fibers. Thus, dark muscle color is often simply a consequence of a relatively high frequency of red fibers (Figure 2.35).

Chemical State

The reaction of pigments with several compounds or ions may result in color changes in meat. However, the ability of pigments to combine with or tie up molecules depends on the proper chemical state of iron in the heme ring. Figure 6.4 depicts two chemical states for myoglobin, based on the valence of iron in the heme ring. When the iron molecule is **oxidized** to the ferric state (+3) it cannot combine with other molecules, including molecular oxygen. Conversely, when the iron is **reduced** (ferrous state, +2) it will readily combine with water (as in uncut meat), or with oxygen (as

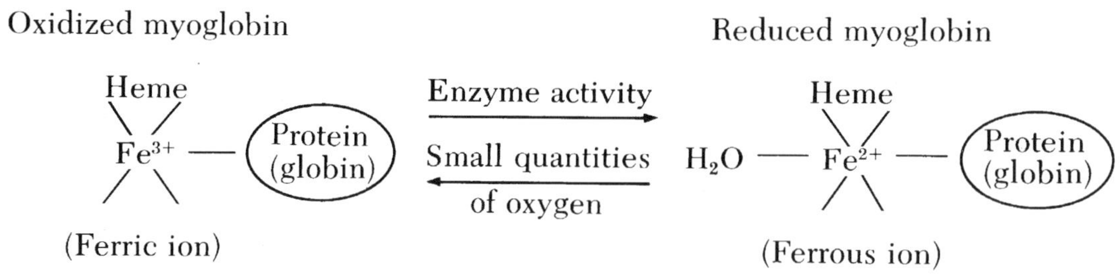

$$\underset{\text{(Ferric ion)}}{\underset{\text{Oxidized myoglobin}}{\text{Heme}\diagup\diagdown\text{Fe}^{3+}\diagdown\diagup-\boxed{\text{Protein (globin)}}}} \underset{\text{of oxygen}}{\overset{\text{Enzyme activity}}{\underset{\text{Small quantities}}{\rightleftharpoons}}} H_2O \; \underset{\text{(Ferrous ion)}}{\underset{\text{Reduced myoglobin}}{\text{Heme}\diagup\diagdown\text{Fe}^{2+}\diagdown\diagup-\boxed{\text{Protein (globin)}}}}$$

FIGURE 6.4 | The oxidation state (valence) of the central iron atom in the heme group of myoglobin under different conditions. Courtesy of W. H. Freeman and Company.

in meat exposed to the air). Thus, the key to maintaining a meat pigment's ability to react with other molecules is to promote reducing conditions in the muscle tissue. This is important because molecular oxygen reacts with the reduced iron of myoglobin, a reaction that provides the desirable red color of fresh meat.

Reducing conditions in meat may occur naturally, as a result of normal enzyme activity (electron transport chain) that takes place continuously. These enzymes use all the oxygen available in the interior of the muscles even long after death. Consequently, the pigment in uncut meat is in the reduced form and has only water with which to react. Such pigment is purple and is called **deoxymyoglobin** (Figure 6.5, see color section).

With cutting, grinding, or exposure to air, pigments in meat undergo color changes due to their exposure to and reaction with oxygen. If only small quantities of oxygen are present, such as in a partial vacuum or a sealed semipermeable package, the iron molecule of the pigment becomes oxidized (Figure 6.4) and changes to a brown color. This oxidized state of the pigment is called **metmyoglobin**, the brownish color in Figure 6.5 (see color section). Formation of this color is a serious problem in merchandising meat because most customers intuitively associate the brown color with a product that is microbiologically compromised or has been stored too long, even though it may be formed in a matter of minutes. It is especially troublesome to processors because the meat remains brown in color indefinitely, or for as long as it is exposed to low oxygen concentration conditions. Only by removing all of the oxygen and developing reducing conditions can it be converted back to deoxymyoglobin.

When freshly cut meat is allowed to come in full contact with the air, the reduced pigments will react with molecular oxygen and form a relatively stable pigment called **oxymyoglobin** (Figure 6.5, see color section). This pigment is responsible for the bright red color in fresh meat. Oxymyoglobin will form within 30–45 minutes after exposure to the air. This bright red color development is known as bloom. In this reaction, deoxymyoglobin (purple) is oxygenated (atmospheric or molecular oxygen is added). Metmyoglobin (brown) is the oxidized form of the pigment (chemical state of iron ion is changed from ferrous to ferric). Under atmospheric conditions, oxymyoglobin (oxygenated pigment) is stable and not easily oxidized to metmyoglobin.

The three common forms of myoglobin in fresh meat are easily observed in fresh ground beef when it is removed from the packaging material and physically separated into two portions. The outer surface will be cherry-red (oxymyoglobin) because it has become oxygenated, while the innermost aspects of the product will be dark purple, primarily consisting of deoxymyoglobin, because no oxygen is present. Yet, at the interface, just below the surface, the ground meat will appear brown (metmyoglobin) because some oxygen is available to the pigment but not at levels normally present in the atmosphere.

Oxymyoglobin is formed spontaneously when meat is exposed to air, but its stability depends on a continuing supply of oxygen since the enzymes involved in oxidative metabolism rapidly use the available oxygen. As gases in the air diffuse inward, an oxygen gradient, and consequently, a color gradient as illustrated in Figure 6.5 (see color section), is established from the surface of the muscle inward. The bright red color on the surface depends on the availability of oxygen in the superficial layers of the tissues.

Muscles differ in their rates of enzyme activity which, in turn, regulates the amount of oxygen available in the outermost layers of tissue. As the pH and temperature of the tissues increase, enzymes become more active and the oxygen content is reduced. Consequently, maintaining the temperature of meat near the freezing point minimizes the rate of enzyme activity and oxygen utilization and helps maintain a bright red color for the maximum possible time.

Unusual color development may occur in meat in several ways, some of which are unrelated to normal chemical reactions of the pigments. The PSE and dark cutting conditions in meat (Chapter 5) are partially the result of unusual degrees of water binding in the muscles, and consequent alteration of light reflection. The paleness of PSE meat is caused largely by a high proportion of free water in the tissues, which is located between the muscle cells rather than within them. Tissues containing a great amount of **extracellular water** have many reflecting surfaces that reflect light, but have only a limited light absorption capability. Color intensity therefore is greatly reduced. In dark cutting meat, high water-binding capacity maintains an unusually large proportion of water as intracellular water. Because of this, white light reflection is minimized and color absorption is enhanced. Dark cutting tissue also has a high rate of oxygen-using enzyme activity, due to its high pH. This reduces the proportion of the pigment in the red oxygenated (oxymyoglobin) state.

STRUCTURE, FIRMNESS, AND TEXTURE

Some of the physical properties of fresh meat, such as structure, firmness, and texture, are difficult to

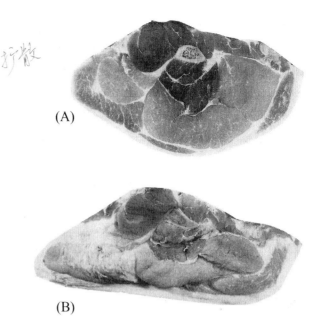

FIGURE 6.6 | An example of the varying firmness of pork muscle; butt end views of two fresh hams. (A) A firm, rigid structure. (B) A soft structure with pronounced muscle separation. From Briskey, E. J. and R. G. Kauffman, "Quality Characteristics of Muscle as a Food" in *The Science of Meat and Meat Products*, 2nd ed. J. F. Price and B. S. Schweigert, eds. W. H. Freeman and Co., San Francisco, 1971.

measure objectively. These factors are usually evaluated by consumers with their visual, tactile, and gustatory senses. However, these properties are no less important than many other more easily measured properties of meat. Figure 6.6 illustrates the visual contrast between two meat cuts that differ in structure, firmness, and texture (see also Figure 14.5 in the color section). The extent to which these cuts conform to the surface on which they rest, the degree of separation between their muscles, and the prominence of muscle bundles make a composite impression that can be either attractive or unattractive. Many factors within muscles, such as rigor state and associated water-holding properties, intramuscular fat content, connective tissue content, and bundle size contribute to these physical properties. Several of these factors have been referred to in discussions of water-holding capacity, color, and postmortem changes.

Rigor State

If, during the course of carcass chilling, muscle firmness is compared to that of freshly slaughtered animals, there is an evident progressive development of rigidity. The change is said to result from carcasses setting-up during chilling. This increased firmness develops from loss of extensibility accompanying rigor mortis, and solidification of fat within and surrounding the muscles. In some muscles the sarcomeres are very short (.6–1.8 µm), indicating that the filaments have a high degree of overlap and the microstructure is very tight and dense.

During the storage and processing of fresh meat, the muscles retain a firm, somewhat moist condition unless they have undergone unusual rates of postmortem change, as described in Chapter 5. Yet, certain subtle changes occur that result in an improved palatability, especially in tenderness. These changes are partially responsible for improvements in palatability associated with meat aging.

Water-Holding Capacity

The degree of water-holding capacity associated with each rigor stage, or with the rate of postmortem change, is observable because of large-scale effects on firmness, structure, and texture. Those muscles with extremely high proportions of bound water are firm, have tight structure, and have a dry or sticky texture. Conversely, tissues with poor water-binding ability are soft, have a loose structure, and have a wet or grainy texture. The prevalence of intracellular water in the former case and extracellular water in the latter case explains these differences associated with water-holding capacity.

Intramuscular Fat

Intramuscular fat (marbling) contributes to firmness of refrigerated meat. Solidification of fat occurs during chilling and helps retail cuts, such as steaks and chops, retain a uniform thickness and characteristic shape during handling and storage. Marbling is also a visual factor, and as such it may be pleasing to some consumers who associate it with flavor and juiciness, while others may object to any visual fat in the meat they buy.

Connective Tissue

The size of individual muscle bundles (fasciculi) as well as the amount of connective tissue in muscle affects the texture of meat. Large bundles of muscle fibers and large amounts of perimysial connective tissue surrounding primary and secondary bundles are associated with coarse-textured meat. Muscles of the limbs that are extensively involved in locomotion have relatively high amounts of connective tissue and tend to be coarse, whereas the much less used muscles of the back, such as the longissimus, and particularly the psoas muscle, have less abundant connective tissue and are finer textured. Coarse muscles are less tender than fine-textured ones, unless special methods of cookery are used to break down connective tissue collagen and thereby tenderize the meat.

The perimysium surrounding primary and secondary muscle bundles is composed primarily of collagen. Collagen content of the perimysium ranges from 54 to 98 percent among the major muscles of beef. The collagen content of the epimysium surrounding whole muscles ranges from 13 to 24 percent in beef, whereas that in the endomysium surrounding individual muscle fibers ranges from 24 to 42 percent.

Connective tissues play a major role in the tenderness/toughness of meat. However, the overall tenderness of a meat product results from a complex combination of factors any one of which can detract seriously from palatability of a given product. Please see the discussion beginning on page 261 in Chapter 11, where factors that are discussed throughout the textbook are summarized and reviewed.

7

MEAT MERCHANDISING

OBJECTIVES: *Explore the retail and food service merchandising of meat. Explore the distribution system and types of retail stores for merchandising fresh and processed meat for home consumption. Describe the fabrication, pricing, and labeling of fresh meat cuts for retail sale and the packaging and display systems for merchandising fresh, frozen, and processed meat. Describe the use of meat in food service establishments, standards and specifications for institutional meat purchases, and cookery of meat for food service.*

Key Terms

Meat merchandising
Boxed meat
Branded
Self-service
Service counters
Deli departments
Ribbed
Primal
Sub-primal
Parts
Water vapor transmission rate
Weep
Purge
Oxygen transmission rate
Laminates
Modified atmosphere packaging
Cutting tests
Uniform retail meat identity standards
Universal product code
Color balance or color rendition
Forming film
Non-forming film
Institutional meat purchase specifications
Portion cuts
Peeled/denuded
Peeled/denuded, surface membrane removed

In a broad sense, **meat merchandising** occurs from the time livestock is offered for sale until meat is placed before consumers. Livestock producers merchandise living animals, slaughterers merchandise carcasses and wholesale, primal, and sub-primal cuts, meat processors merchandise processed meats, food retailers merchandise meats for home consumption, and food service outlets merchandise cooked meat as meal components. Most meat reaches the consumer through retail outlets or food service establishments.

Increasing volumes of fresh and processed meat are sold through mail order catalogs and through the Internet. For some companies, internet and catalog sales of specialty sausages, fresh meats or game meats have become primary methods of merchandising. These types of sales generally increase before and during holiday seasons in the form of a variety of gift boxes. Merchandising through the internet and catalogs allows many small or regional companies to access national and international markets. It is outside the scope of this text to discuss in detail all of the methods for meat merchandising or the economics of meat marketing. Hereafter, discussion of meat merchandising will be restricted to the technology involved in presentation of meat to consumers in retail stores and food service units.

RETAIL MEAT DISTRIBUTION SYSTEMS

Slaughter Plants To Retail Stores

Large volumes of fresh (unprocessed) meat move directly from slaughter plants to retail stores. Wholesale cuts and some carcasses are shipped in refrigerated vans to retail stores, which maintain well-equipped meat cutting departments. When untrimmed cuts or intact carcasses are received, they are fabricated to retail cuts and various amounts of excess fat and bone. Most carcasses are reduced to wholesale, primal, or subprimal cuts by the slaughterer, who removes some of the excess fat and bone; the meat is said to be "saw ready," requiring less retail labor and refrigerated storage space. Such trimmed primal or subprimal cuts are usually vacuum packaged and placed in boxes for shipment, and are referred to as **boxed meat**. Examples of boxed meat are beef subprimal cuts derived from rounds, loins, ribs, and chucks, pork loins and spareribs, and lamb primal cuts (legs, loins, racks, and shoulders) and subprimal cuts. (See later discussion of primal and subprimal cuts in this chapter.) Whole birds are individually packaged, boxed, and shipped in refrigerated or frozen form. However, few whole birds are sold at retail compared to the amount of poultry sold as parts, i.e., boneless or bone-in breasts, thighs, or drumsticks. Imported meats, such as lamb legs, are shipped in frozen form.

Boxed cuts of fresh pork represent a traditional form for transferring highly perishable products to retail sites. Traditionally, economical packaging materials, such as waxed paper, have been used for shipment of fresh pork products. However, vacuum packaging of fresh pork products for shipment to the retailer is becoming a common current practice. Storage life of fresh pork is limited because of rapid deterioration of color and fresh flavor. Similar restrictions exist for poultry because of rapid dehydration and loss of soluble protein during storage. Consequently, most fresh poultry parts are prepackaged in vacuumized retail packages and frozen by slaughterers. Boxed beef subprimal cuts are packaged in vacuumized bags because the products may remain in packages for relatively long periods without serious deterioration. Since beef usually benefits from some aging before consumption, the time in protective bags is sometimes longer than that required for transportation only.

Fresh meat is also converted to retail cuts and packaged for retail sale at the slaughter plant or at a processing plant dedicated to cutting and packaging. The amount of fresh meat handled in this way is increasing, particularly for beef, pork, and lamb. A large amount of poultry, particularly chicken, is pre-packaged for retail sale at the slaughter plant.

Processing Plants To Retail Stores

Although animal slaughter and further carcass processing occur at the same location in many instances, these two industrial areas tend to be separated managerially if not physically. Processing plants add value to meat raw materials by manufacturing smoked meat, sausage, prepackaged meat, precooked meat, and other products. Historically, they have produced smoked meat and sausage, but with increasing consumer demand for highly processed convenient forms of fresh meat, product lines have expanded.

Most smoked meats and sausages are placed in consumer-sized packages by processors. Products are sliced or portioned and packaged in quantities that are consistent with product storage life and consumer requirements. Items are **branded** (identified with the processor's brand name) and, in some cases, widely advertised. They are pre-weighed and usually pre-priced, with little handling required by retailers. In-store promotional materials and consumer information frequently are supplied by processors.

While most processed meat (cured, smoked, sausages) are packaged for retail sale at the processing plant, this has not been the case for fresh meats. The concept of further processing of fresh meat originated largely in the poultry industry. Prepackaged branded products such as chicken and turkey parts or boneless pieces are presented in a variety of package and cut sizes. Packaging systems that inhibit microbial growth, moisture loss, and oxidative changes (vacuum packaging) are required for such products. Such packaging, when applied to red meat cuts, has only recently received consumer acceptance because of its effects on meat color (to be discussed later in this chapter). Nevertheless, centralized packaging of red meats has been used with conventional packaging systems (non-vacuumized) by many retailers. Centralized packaging was originally established by retailers operating multiple stores in a limited geographical region. Such operations could provide frequent deliveries of freshly prepared cuts. In recent years, centralized or "case-ready" packaging has been increasingly provided by meat packers who build specialized cutting and packaging plants within easy driving distance of major population centers. Deliveries from these plants go to many different retail chains. Centralized/case ready packaging affords many efficiencies of labor, equipment, packaging material, and by-product utilization. It requires careful inventory control and prompt movement of products through the systems. Centralized packaging of fresh meat is increasing, however, because of the efficiencies gained.

Precooked fresh meats have entered retail markets because of consumer desires for quick meal preparation. Products are packaged in flexible film, which serves as the display package. In most cases, the film also will withstand heat so that the product can be heated in the package before serving. Turkey breasts, boneless pork loins, spareribs, and beef roasts are examples of precooked unfrozen items. Many of these items also are merchandised through deli departments along with freshly sliced sausages. Frozen precooked products appear in an endless variety as well.

Precooking frequently is accomplished in retail stores in the deli department. Barbecued items such as chicken, beef ribs, and spareribs are particularly popular and are sold at cooked meat temperatures for immediate consumption in the home.

Distribution Centers to Retail Stores

As discussed earlier, some centralized processing facilities also are distribution centers for major retailers. In some instances, however, the distribution function is performed with or without additional processing by meat wholesalers. These firms service independent retail stores as well as food service units.

RETAIL STORE TYPES

Supermarkets

Far more meat is sold in supermarkets and warehouse stores in the US than in any other type of retail market. Large amounts of refrigerated display space are devoted to meat, giving shoppers a wide choice of meat products. Experience has shown that the opportunity to choose among displayed packages is important to shoppers. Therefore,

self-service is the prevailing type of selling employed in supermarkets and warehouse stores. This allows shoppers to carefully observe the perceived quality of meat as well as unit price before purchase is actually made. On the other hand, some shoppers prefer the personal service available at **service counters** where a clerk is always available to provide assistance. Consequently, most supermarkets and many warehouse stores provide a station at which such service is offered. Special cuts as well as consumer information are provided in this way, but supermarket sales volume at service counters is low compared to self-service counters.

Deli departments are usually prominent in supermarkets and many warehouse stores; the deli may be closely integrated with meat service counters or may be stand-alone to provide better separation between raw and ready-to-eat meats. Delis provide meat in many forms that may include pre-cooked items, freshly prepared meat entrees, freshly sliced processed meats, or even entire meals.

Butcher Shops

Meat merchandising in many parts of the world is largely through butcher shops. Unwrapped, and sometimes uncut, meat is usually displayed in glass-fronted display cases. Selections are made by shoppers after which sales personnel cut, weigh, package, and price the items. In this process, shoppers frequently receive information on meat selection, cookery, and serving methods.

Butcher shops in the US sometimes include small-scale slaughter facilities, which provide fresh meat supplies. In Europe, butcher shops usually include small processing plants, which produce most of the processed meats in those countries. Because of this integration of processing and merchandising, butcher shops or similar units sometimes are perceived by consumers to provide unusual freshness and uniqueness in meat products. Consumers with home freezers often purchase carcasses or portions of carcasses through these types of direct marketing operations.

Convenience Stores

Convenience stores represent important outlets for meat merchandising because of their neighborhood locations. Even though meat selections are usually limited to vacuum packaged processed or frozen meats, the convenience associated with these stores results in significant meat sales. Deli counters and food carry-out operations frequently are located in these types of stores.

Specialty Shops, Catalog and Internet

Food specialty shops offer attractively packaged items with unique palatability characteristics. Dry sausages, country hams, jerky, and other non-refrigerated meat products often are merchandised through these outlets. Sausages, jerky, and cured and smoked meats manufactured from buffalo, venison, elk, and game birds are often merchandised through specialty shops or catalog sales. Also, special steaks and roasts are often available from these outlets. These types of products have become widely available on the internet as electronic commerce has grown and the availability of reliable expedited shipping has increased.

FRESH MEAT DISPLAYS

Fresh meat displays perform the primary selling function in merchandising fresh meat. If products are themselves appealing and they are presented in a manner to convey an image of wholesomeness, nutritive value, and eating satisfaction, a major part of the persuasion of potential buyers will have been accomplished. Packaging systems and display conditions play major roles in these presentations.

Meat Cutting and Identification

Standardized cutting and nomenclature for retail cuts are prerequisites for good merchandising and proper use of meat. Because various cuts differ in composition and tenderness, lean cuts are separated

TABLE 7.1	Anatomical and Common Terms Used for Bones from the Beef, Pork, and Lamb Carcass*	
Bone Number (see Figure 7.1)	**Anatomical Term**	**Common Term**
1	Cervical vertebrae	Neck bones
1a	First cervical vertebrae	Atlas
2	Thoracic vertebrae	Back bone
2a	Spinous process of vertebrae	Feather bone
2b	Cartilage of spinous process	Button
3	Lumbar vertebrae	Back bone
3a	Transverse process of vertebrae	Finger bone
4	Sacral vertebrae	Back bone
5	Coccygeal vertebrae	Tail bone
6	Scapula	Blade bone
7	Humerus	Arm bone
8	Radius	Foreshank bone
9	Ulna	Foreshank bone
9a	Olecranon process	Elbow bone
10	Sternum	Breast bone
11	Costae	Ribs
12	Costal cartilages	Rib cartilages
13	Pubis (pubic symphysis)	Aitch bone
13a	Ilium	Part of pelvic bone (hip bone)
13b	Ischium	Part of pelvic bone (pin bone)
14	Femur	Leg (round) bone
15	Patella	Knee cap
16	Tibia	Hind shank bone
17	Tarsal bones	Hock bones
18	Metatarsal	Hind shank bone
19	Phalangeal bone(s)	Hind shank bone (lamb) foot bones (pork)
20	Fibula	Hind shank bone
21	Carpal bones	Shank bones (lamb), front foot bones (pork)
22	Metacarpal bone(s)	Shank bones (lamb), front foot bones (pork)

*Bones of veal and mutton carcasses are the same as those for beef and lamb, respectively.

146 **CHAPTER 7** MEAT MERCHANDISING

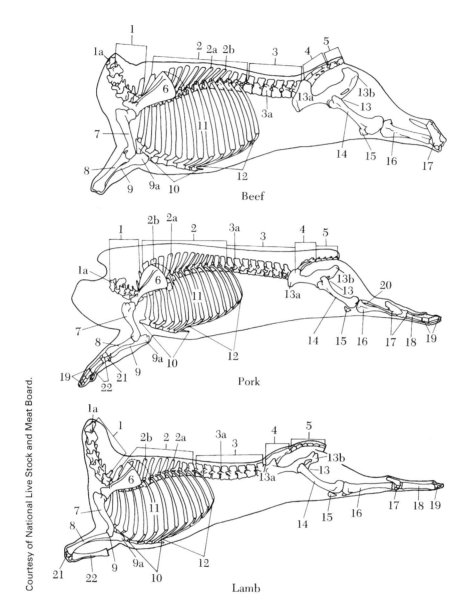

FIGURE 7.1 | Beef, pork, and lamb skeletal diagrams. The skeletal anatomies of veal and mutton are the same as those of beef and lamb, respectively. See Table 7.1 for anatomical and common terms used for bones.

from fat cuts, more tender from less tender cuts, and thick from thin cuts. In all domestic mammalian species, the most tender cuts come from the loin and rib, which contain muscles used for support. By comparison, cuts from the leg, containing muscles used for locomotion, usually are less tender. The belly of the pork carcass is a fat cut, and is separated from lean cuts—ham, loin, picnic shoulder, and Boston-style shoulder. The breast, thigh, and drumstick of poultry are thick cuts as compared to the thin back and wing.

Skeletal diagrams of beef, pork, and lamb carcasses are presented in Figure 7.1. The anatomical and common names for the major bones are presented in Table 7.1. The bone names are the same for the three species, but the styles of dressing result in more bones being removed from the fore and hind legs of beef carcasses than from lamb and pork carcasses. Also, the fore and hind feet are removed from lamb, but not from pork, during slaughter. There are some differences among the three species in number of bones. For example, beef and lamb carcasses each have 13 thoracic vertebrae (sometimes 14 in lamb), but their number is variable in pork, usually 13 to 15. Lamb and pork have seven lumbar vertebrae, and beef have six.

The names of retail cuts frequently are related to and named for bone structures. These include rib roast, rib chops, rib steak, blade roasts, arm roasts, arm chops, short ribs, neck bones, spareribs, riblets, T-bone, wedge bone sirloin, flat bone sirloin, and pin bone sirloin.

A skeletal diagram of the chicken is presented in Figure 7.2. Many chicken bones have the same names as those of beef, pork, and lamb. However, in contrast to the skeletons of mammals, the vertebral trunk of chickens and other birds has several fused vertebrae. The last four thoracic, three lumbar, two sacral, and five coccygeal vertebrae are fused together to form a rigid bone called the lumbosacral. This bone gives rigidity to the body in flight, and enables birds to walk with two legs while maintaining a horizontal position. The strength of the flexible back of mammals is produced by the combination of an articulated backbone with a complex system of attached muscles. In birds, the backbone is rigid and the muscular system connected to it is extremely reduced in size compared to that of pigs, lambs, and cattle. The sternum (keel bone) of the chicken is its largest bone, and is not segmented as in mammals. The two clavicular bones are fused with each other to form the forked clavicle, commonly known as the wishbone. Chickens have 13 cervical and seven thoracic vertebrae, of which four of the latter are fused together.

Certain locations on the skeleton, as illustrated in Figures 7.2 and 7.3, are used as guides in making wholesale cuts. For examples, the chuck of beef and shoulder of lamb and veal are removed from the wholesale rib between the fifth and sixth ribs. The Boston-style shoulder and picnic shoulder of pork are separated from the loin and belly between the second and third ribs. The quarters of chickens

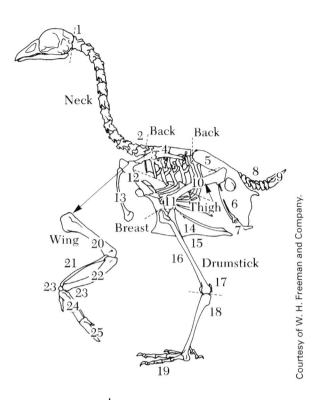

FIGURE 7.2 | Skeleton of the chicken with parts and bones identified. The parts are named on the illustration, and dashed lines indicate locations where the parts are cut. Bones are identified with numbers, as follows: (1) first cervical vertebra (atlas), (2) last cervical vertebra, (3) scapula, (4) fused thoracic vertebrae, (5) ilium, (6) ischium, (7) pubis, (8) coccygeal vertebrae, (9) pygostyle, (10) femur, (11) patella, (12) coracoids, (13) clavicle, (14) sternum, (15) fibula, (16) tibia, (17) sesamoid, (18) metatarsus, (19) bones of toes (phalanges), (20) humerus, (21) radius, (22) ulna, (23) carpal bones, (24) metacarpus, and (25) finger bones (phalanges).

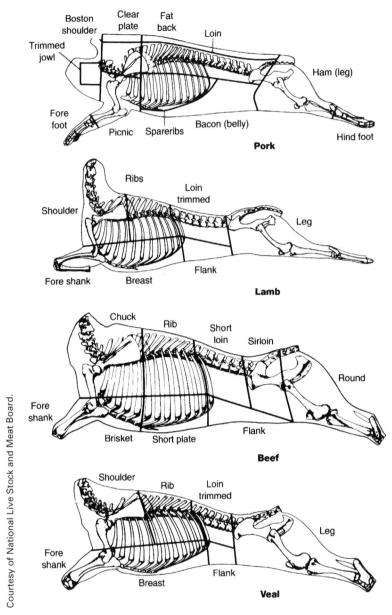

FIGURE 7.3 | Diagrams of beef, veal, pork, and lamb carcasses illustrating wholesale cuts in relation to the skeleton.

are separated between the front and hind portions of the back.

Because beef carcasses are so large, it is customary to split them into sides, and the sides are **ribbed** by cutting them into fore- and hindquarters between the twelfth and thirteenth ribs. Much of the beef destined for retail use is processed by slaughterers into primal or subprimal cuts. **Primal** cuts (round, loin, rib, chuck, flank, plate, and brisket) are major wholesale cuts and are frequently too large for easily handling. **Subprimal** cuts are prepared from primal cuts by removal of fat and bone and sometimes by dividing the primal into multiple pieces. Veal and lamb carcasses are small enough to be handled without splitting into sides but these carcasses are commonly cut into fore- and hindsaddles, or even into wholesale cuts to facilitate merchandising. Pork carcasses are split into sides

(a) Sternum
(b) Rib
(c) Costal cartilages

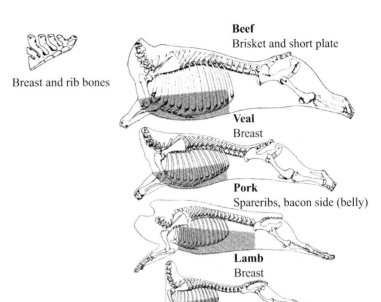

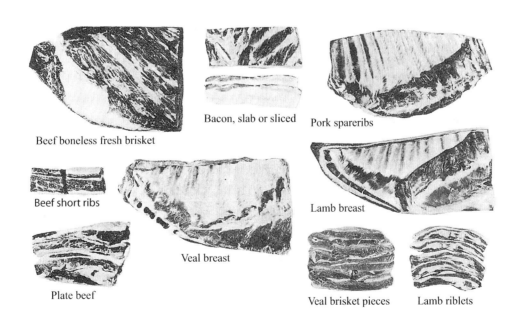

FIGURE 7.4 | Retail cuts from the breast, brisket, spare ribs, bacon side, and short plate.

TABLE 7.2 Approximate Yields of Wholesale Cuts, Parts, or Edible Portions from Beef,* Veal,* Lamb,* Pork,* Fish, and Poultry Carcasses and Whole Seafood

Wholesale cut, part, or edible portion	Percent	Wholesale cut	Percent
Beef			
Hindquarter (48 percent)		Forequarter (52 percent)	
Round	23	Chuck	26
Loin	17	Rib	9
Flank	5	Shank	4
Kidney, kidney fat, and hanging tender	3	Brisket	5
		Short Plate	8
Veal			
Shoulder	28.0		
Rib	7.3		
Loin	7.7		
Leg	34.0		
Shank	3.8		
Breast and flank	13.4		
Kidney and kidney fat	5.8		
Lamb			
Hindsaddle (50 percent)		Foresaddle (50 percent)	
Leg	39	Shoulder	26
Loin	7	Rib	9
Flank	2	Shank	5
Kidney and kidney fat	2	Breast	10
Pork**			
Ham	21		
Loin	18		
Boston shoulder	6.6		
Picnic shoulder	8.8		
Belly	17.3		
Spareribs	3.8		
Jowl	3.0		
Feet, tail, and neck bones	6.0		
Fat back, clear plate, and other fat trim	11.2		
Lean trim	4.3		

Dressed Fish***	
Edible portion	67
Blue Crab	
Edible portion	10
Oysters	
Edible portion	12
Shrimp	
Edible portion	48
Lobster (spiny)	
Edible portion	40
Scallops	
Edible portion	20
Chickens**** (broilers), 56 days	
Breast	27.5
Drumstick	15.9
Thigh	17.4
Wing	11.9
Back	18.2
Neck	3.7
Turkeys*****, 84–112 days	
Breast	36.1
Drumstick	13.2
Thigh	15.7
Wing	13.0
Back	13.4
Neck	7.8

*No allowance for cutting shrinkage.
**Packer style, dressed without leaf fat.
***Head, tail, fins, scales, and entrails removed.
****From Hayse and Marion, Poul. Sci. 52:718 (1973).
*****From Leeson and Summers, Poul. Sci. 59:1237 (1980).

and are small enough to be transported easily as such, but generally they are reduced to wholesale cuts because many pork products are further processed at slaughter plants. Whole turkeys and broilers are transported to retail stores but many poultry halves, quarters, and **parts** are prepared by processors before shipment. Whole fish are often transported to retail outlets, but shellfish usually undergo dressing for separation of edible portions. The approximate yields of wholesale cuts, parts, or portions from beef, veal, lamb, pork, poultry, and seafood are given in Table 7.2.

In some instances, wholesale cuts are made in a different manner than that illustrated in Figure 7.3. For example, the beef chuck may be cut to contain the arm and, in this case, is called an arm chuck. Many of these special cuts are boneless or partially boneless and are processed into special roasts, steaks, or chops for food service use.

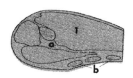 Arm bone

(a) Humerus
(b) Ribs
(1) Triceps brachii and other muscles (beef boneless shoulder steaks and pot-roasts from this section)

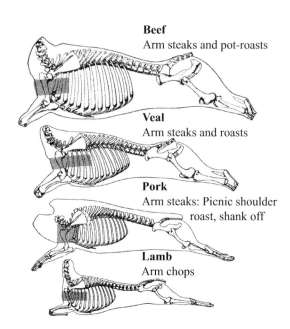

Beef
Arm steaks and pot-roasts

Veal
Arm steaks and roasts

Pork
Arm steaks: Picnic shoulder roast, shank off

Lamb
Arm chops

Beef boneless shoulder steak

Veal arm steak

Beef chuck short ribs

Beef English (Boston) cut

Beef arm steak (short cut)

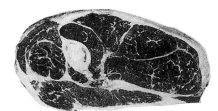
Beef arm steak (full cut)

Pork arm steak

Lamb arm chop

FIGURE 7.5 | Retail cuts from the shoulder (chuck) arm.

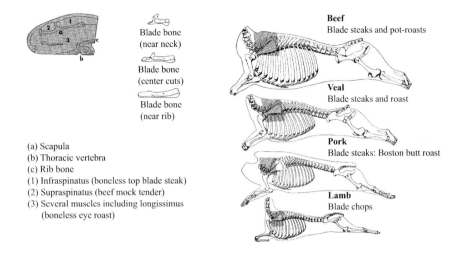

(a) Scapula
(b) Thoracic vertebra
(c) Rib bone
(1) Infraspinatus (boneless top blade steak)
(2) Supraspinatus (beef mock tender)
(3) Several muscles including longissimus (boneless eye roast)

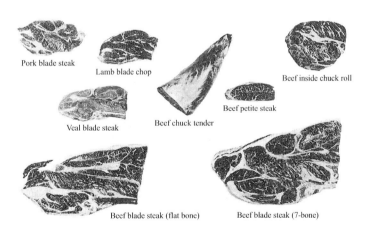

FIGURE 7.6 | Retail cuts from the shoulder blade.

Considerable quantities of fat and bone may be removed from cuts during preparation as preferred by consumers. Historical trends have resulted in more boneless cuts and less external fat on cuts. Retailers seldom leave more than 1.25 cm of fat on any surface and most leave no more than 0.6 cm. Removal of essentially all external fat is quite common in the preparation of retail cuts. Bone removal adds convenience to the use of meat cuts but renders them more difficult to identify.

Whether a retail cut is called a roast or steak is based mainly on the thickness of the cut and the recommended cookery method. Roasts are generally thicker than steaks and chops. There are essentially seven basic retail cuts of meat: the blade, arm, rib, loin, sirloin, leg, and breast cuts (Figures 7.4 through 7.10). The retail cuts from these seven basic areas are common to beef, veal, pork, and lamb. Among these four species, breast cuts vary most, due to retail cutting methods. For example,

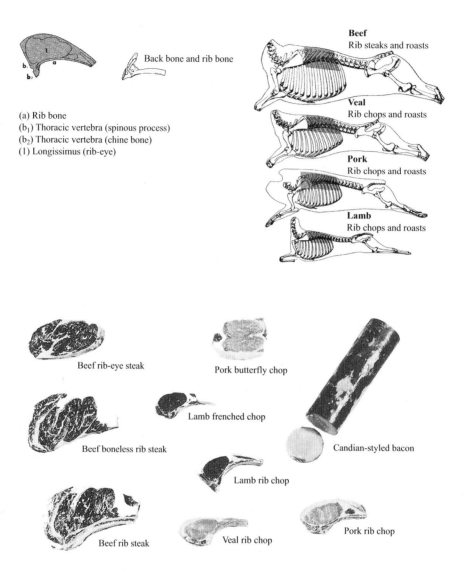

FIGURE 7.7 | Retail cuts from the rib.

in lamb and veal the breast is cut as one piece, or converted into a number of retail cuts. In beef, the breast is cut into two pieces, the brisket and short plate; in turn, these are made into smaller retail cuts. The flanks of beef, veal, and lamb usually are processed into ground meat. The breast and flank of pork are processed as one piece (belly) into bacon, after the rib bones (spareribs) are removed.

When retail cuts are processed as boneless cuts, the recognition of muscle structure becomes more important as a means of identification. Photographs of the most commonly made boneless retail cuts are presented in Figures 7.4 through 7.10, along with the bone-in cuts from the seven basic areas.

Retail cuts from the breast area are presented in Figure 7.4 and may include the sternum, ribs, and rib cartilages. The most obvious characteristics in the diagrams and photographs are:

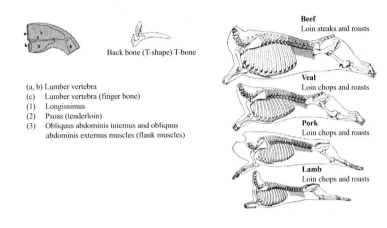

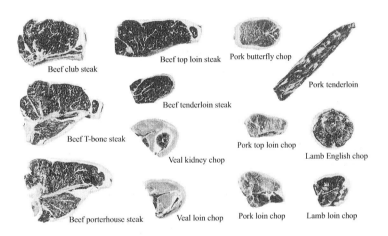

FIGURE 7.8   Retail cuts from the loin (short loin).

- the similarity in contour of the beef boneless brisket, veal breast, lamb breast, and pork spareribs,
- the similarity of bones in pork spareribs and veal and lamb breasts,
- the alternating layers of fat and lean in sliced bacon, beef short ribs, and lamb riblets, and
- the presence of the diaphragm muscle in lamb breast and pork spareribs and the absence of this muscle in veal breast.

Shoulder arm cuts are illustrated in Figure 7.5. These cuts contain the humerus and cross-sections of the rib bones. Obvious features of these cuts in the diagrams and photographs are:

- the similarity in bone and muscle structure of beef arm steak, veal, and pork steaks, and lamb arm chop,
- the shape of the bone-in cuts, and the muscle structure and differences in size of the respective cuts, and

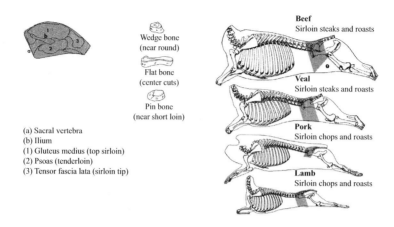

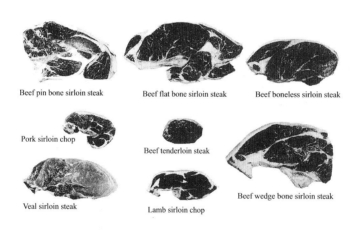

FIGURE 7.9 | Retail cuts from the sirloin.

- the absence of ribs in beef and pork arm steaks, and the presence of ribs in the beef chuck cross rib pot-roast, and beef chuck short ribs.

Most of the shoulder blade cuts can be identified by the blade bone (Figure 7.6). The scapula may be flat, or it may have a "spine" resembling the figure seven, depending on whether the cut comes from the center or the posterior portion of the bone. Beef, veal, and lamb blade cuts may also contain ribs. The pork shoulder blade steak contains no ribs because they are removed with the neck bones. When the beef blade chuck is boned, several cuts can be made. The three most common ones are pictured. Each has a characteristic muscle structure, which aids in its identification. The whole beef mock tender has a shape similar to that of the whole beef tenderloin.

The identifying bone characteristics of retail cuts from the rib area are the rib and thoracic vertebra consisting of the chine bone and feather bone (Figure 7.7). When all of the bones are removed,

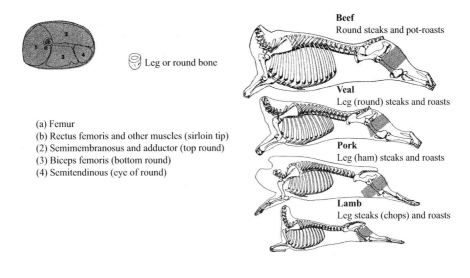

(a) Femur
(b) Rectus femoris and other muscles (sirloin tip)
(2) Semimembranosus and adductor (top round)
(3) Biceps femoris (bottom round)
(4) Semitendinous (eye of round)

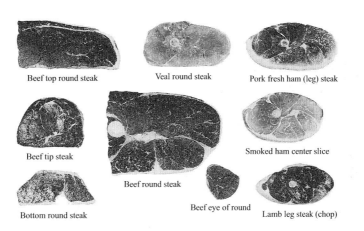

FIGURE 7.10 | Retail cuts from the leg, round, or ham.

cuts from the rib are easy to recognize by the appearance of the ribeye muscle, which extends along the backbone in the rib and loin.

Bone-in cuts from the short loin contain only the lumbar vertebrae, except for the beef top loin steaks from the anterior position, which contain one rib (Figure 7.8). The lumbar vertebrae, which are cut in half when carcasses are split into sides (halves), consists of the spinous process, chine bone, and finger bone, and forms a characteristic T-shape.

Two muscles in particular, the longissimus (loin eye or ribeye) and the psoas (tenderloin), assist in identifying either bone-in or boneless cuts from the short loin. The size of the tenderloin muscle in the beef porterhouse steak is the distinguishing characteristic between this steak and the beef T-bone steak. The center of the tenderloin muscle must be 3.2 cm in diameter, or more in porterhouse steaks, and 1.5 to 3.2 cm in diameter in T-bone steaks. The tenderloin is a tapering muscle that extends the full length of the loin (short loin

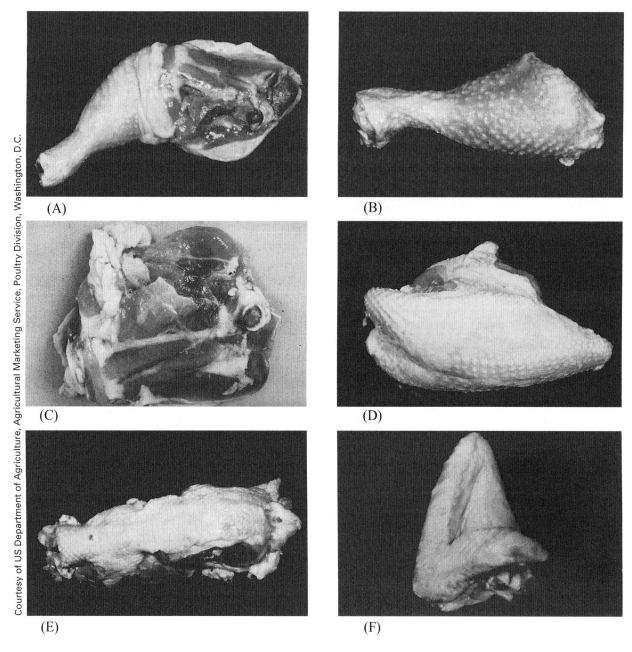

FIGURE 7.11 | Chicken broiler parts: (A) leg, (B) drumstick, (C) thigh, (D) breast with ribs, (E) back, and (F) wing.

and sirloin). The diameter of this muscle is greatest in the sirloin.

The shapes of bones in cuts from the sirloin area vary, as illustrated in Figure 7.9. Skeletal diagrams reveal that cuts may contain lumbar or sacral vertebrae and a portion of the ilium. The shape of the ilium determines whether the steak is a wedge, round, flat, or pin bone sirloin. Veal steaks and lamb and pork chops from the sirloin area are identified only as sirloin steaks or sirloin chops. The beef sirloin, when processed into boneless cuts, yields the boneless sirloin (top sirloin)

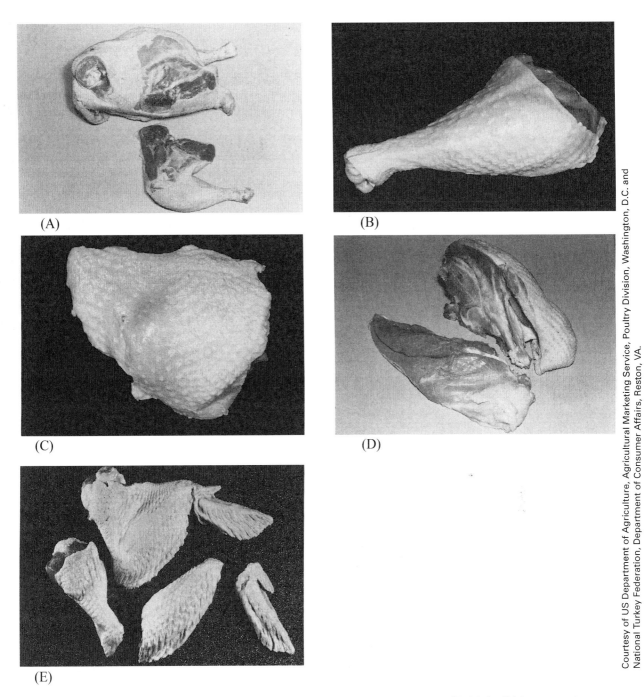

FIGURE 7.12 | Turkey parts: (A) leg and site of removal, (B) drumstick, (C) thigh, (D) intact and boneless breast, and (E) intact and disjointed wing.

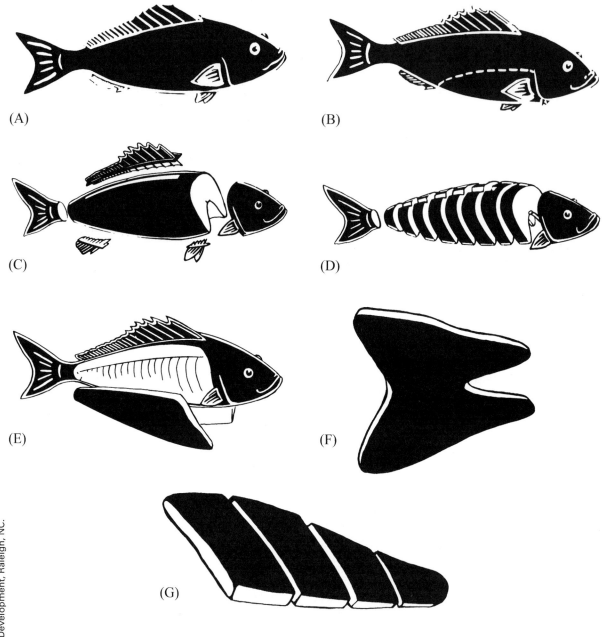

FIGURE 7.13 | Market forms of fish: (A) whole or round, as from the water; (B) drawn, with entrails removed; (C) dressed, scaled and eviscerated; usually head, tail, and fins removed; (D) steaks, cross-section slices of dressed fish; (E) fillets, sides of dressed fish cut lengthwise away from the backbone; (F) butterfly fillets, two sides or fillets held together by the uncut belly skin; (G) portions or sticks, cut from frozen fish blocks, sometimes coated with batter and breading.

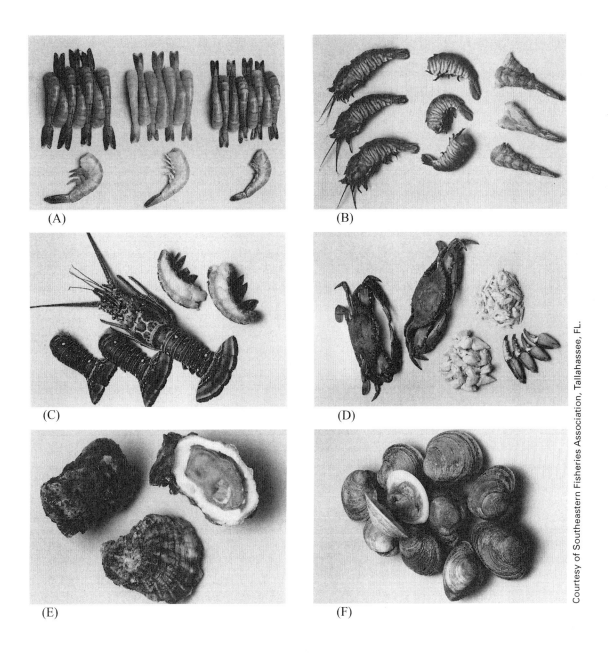

FIGURE 7.14 | Shellfish and edible portions: (A) headless shrimp, left to right, white, pink and brown; (B) rock shrimp, left to right, whole, headless, and peeled and deveined; (C) spiny lobster; (D) blue crab, left to right, raw crab, cooked crab, backfin meat, claw meat, and claws; (E) oysters; and (F) hard clams.

and tenderloin. These cuts can be identified by muscle structure, as illustrated by photographs in the lower portion of Figure 7.9.

Only a cross-section of the femur is present in cuts from the center portion of the leg, round, or ham (Figure 7.10). This bone is round and could be confused only with the humerus from the arm cuts. However, muscle structure of the round is very different from that of arm cuts. The beef round often is divided into top round, bottom round, and eye of round steaks. Their individual muscle structures are their identifying characteristics. The round tip or sirloin tip is a separate cut in beef, but it is a component of leg or round steaks in lamb, veal, and pork.

Poultry carcass areas are traditionally referred to as parts rather than wholesale or retail cuts. Figures 7.11 and 7.12 show typical parts of chicken broilers and turkeys respectively. Breasts and wings are composed of mostly white muscles, whereas drumsticks and thighs consist mainly of dark muscles. Backs and necks are typically subjected to mechanical deboning to provide raw materials for processed meat and only occasionally are available in retail stores. Parts presented in retail stores usually have skin intact, but breasts are sometimes sold with skin removed.

Figure 7.13 shows the variety of forms in which fresh fish are available in retail markets. Whole or round fish are as they come from the water; drawn fish have been eviscerated; dressed fish have viscera, head, tail, and fins removed; steaks are cross-sectional slices of large fish; fillets are the sides of fish cut away from the backbone; butterfly fillets are the two single fillets held together by the belly skin; and portions or sticks are cut from frozen blocks and may or may not be coated with batter or breading.

Many kinds of shellfish are merchandised, some of which are pictured in Figure 7.14. The shells of crabs, oysters, clams, and scallops usually are removed before presentation to consumers. Shrimp may be merchandised with only heads removed or with heads and shells removed.

Packaging Functions and Materials

Fresh meat packages must provide protection from product contamination and deterioration, product visibility, and label information. Packaging materials are available that differ in gas barrier properties, sealability resistance to puncture, opacity, printability, shrinkage properties, resistance to oil, and other properties. Greatest use is made of the copolymer films, which contain up to three or more layers and which may be engineered to have almost any combination of the above listed properties.

Tray and Overwrap

Most fresh meat sold in self-service stores is packaged in rigid plastic trays with a film such as polyvinyl chloride as an overwrap. Trays provide strength to the package, a function that is lost if the cut is too large for the tray. Likewise, cuts that are too small for the tray are overshadowed by their package. Properties of overwraps are especially important because they regulate the gaseous environment in the package. To minimize moisture loss, the overwrap should have a low **water vapor transmission rate**. However, retail meats may lose some of their moisture even after packaging in moisture-proof wrapping. Free water and associated soluble proteins may exude from cut surfaces and accumulate around the meat causing a wet, unattractive retail package. This production of visible meat juice is known as **weep** or **purge** and represents losses in palatability and nutritive value. The problem is minimal in meats having high water-binding capacity and in packages that fit tightly around the meat. However, purge can be a serious problem in vacuum packaged pork with low water holding capacity.

Fresh meat overwraps are designed to provide an abundant amount of oxygen at meat surfaces. The requirement for a high **oxygen transmission rate** is based on the need for abundant oxygen to form oxymyoglobin, the bright red pigment. When very low levels of oxygen are present, the brown pigment, metmyoglobin, is formed (Figure 6.5). Because of high oxygen transmission rate through materials used for fresh meat packages, growth of psychrophilic aerobic bacteria is sustained. Consequently, the existent bacterial load when meat is packaged largely determines the color shelf life. These organisms compete for available oxygen and thereby shorten the time that oxymyoglobin persists. With good sanitation and maintenance of cold temperatures, bright color in retail meat packages may be maintained 5 to 7 days. But, while color shelf life may be as long as 7 days, such meat should not remain in display cases longer than 3 to 4 days so that consumers can keep the product for 2 or more days without discoloration. Another problem associated with packaging films that have a high oxygen transmission rate is the enhanced opportunity for oxidative rancidity to develop. Consequently, retail meat wrapped with trays and overwraps is not suited for frozen storage.

In addition to the above requirements for fresh meat wraps, films used should be strong, have good stretch and heat-sealing properties, and retain the seal under normal storage and handling conditions.

Vacuum Packaging

Most of the problems encountered with tray and overwrap packaging of fresh meat may be overcome with vacuum packaging. However, until recently, consumer objections to the color of meat in such packages have precluded widespread use of vacuum packaging. It is likely that informed consumers will accept greater quantities of meat packaged in vacuum or controlled atmospheres and that these new methods will eventually replace existing methods of packaging.

Composite polymer films known as **laminates**, which have low water vapor and oxygen transmission rates, are used for vacuum packaging. Films that are good barriers to water vapor include polyethylene and oriented polypropylene. Those that are good barriers to oxygen include polyvinylidene chloride and ethylene vinyl alcohol. Combinations of such films are formed into enclosures that, after air evacuation and sealing, conform tightly to products and provide effective barriers to oxygen and moisture.

The atmosphere in vacuumized packages of fresh meat changes in the hours immediately after packaging. Low levels of oxygen entrapped in superficial tissue layers result in some metmyoglobin formation. However, essentially all pigment forms are gradually transformed to the purple reduced myoglobin, as respiratory enzymes in muscle and microorganisms use the remaining oxygen. This respiratory activity results in carbon dioxide production and some pH lowering from carbonic acid formation. Facultative anaerobic bacteria, though slowed by refrigeration, may thrive and produce lactic acid as well. Low pH and absence of oxygen inhibit further bacterial growth and significantly extend the shelf life of vacuum packaged meat up to 3 weeks. However, meat that is heavily contaminated with microorganisms may undergo pigment decomposition, discoloration, and off odors from growth of anaerobic bacteria.

The purple color of vacuum packaged meat has been a significant merchandising problem for retailers in the past. Some consumers associate the darker color with a lack of freshness. Yet, as discussed above, the microbiological quality is greater than that of meat packaged in oxygen permeable wraps. Additionally, when vacuumized meat packages are opened and exposed to air, the oxymyoglobin "bloom" will appear.

Modified Atmosphere Packaging

Because a major portion of the protective effect of vacuum packaging relates to carbon dioxide formation and the development of bright red color depends on the presence of oxygen, efforts have been made to control the gaseous atmosphere in meat packages to achieve both objectives. Accordingly, gas mixtures of 80 percent oxygen and 20 percent carbon dioxide have been used for flushing packages prior to sealing. Even though carbon dioxide helps control microbial growth, oxygen reduces the beneficial effects. Consequently, shelf life of meats packaged in this way exceeds that in oxygen permeable packages by only a slight degree. Nitrogen gas flushing has been used but apparently offers no advantages over vacuum packaging.

Carbon monoxide has been approved in the US as a component of the head space gases of a **modified atmosphere package (MAP)** for fresh meat. A small amount of CO (0.4%) in the gas mix stabilizes the meat pigment and helps prevent color fading. This packaging system commonly utilizes nitrogen (~80%), carbon dioxide (~20%) and carbon monoxide (<0.4%) in the package headspace. The resulting package maintains acceptable color much longer than a high oxygen MAP package under similar conditions. When using carbon monoxide in a MAP package it is important to pay close attention to the sell-by date on the package since the color may still be attractive even after the sell-by date.

Weighing and Pricing

Accurate weights and competitive prices on fresh retail meats are essential to maintain consumer confidence and desired volume of business. The perishability of meat dictates that prices stimulate quick sales.

Automated Systems

After retail meat is cut and packaged, computerized scales determine product weight from multiple scaling on moving conveyors. Package material tare weight is subtracted to determine net product weight. Package price is determined from preprogrammed unit prices and a label is printed that shows, among other things, product weight, unit price, and package value. Such systems usually are interfaced with computers that maintain inventory control and sales records.

Pricing Policy

Because of the wide range of attributes in retail meat cuts, discovery of prices for all products that accurately reflect supply and demand requires complete knowledge of products and market conditions. It is beyond the scope of this text to describe the multitude of factors that impact the price discovery process. However, because of constantly changing market conditions, retailers find it essential to conduct frequent tests of retail yield and value for the purpose of maintaining operating margins. Such tests are called **cutting tests**.

Cutting tests may be performed on carcasses or wholesale cuts. They represent pricing experiments in which the actual yield of each retail product is priced using adjusted prices based on new market conditions. Factors used to reassign prices include wholesale cost, desirability of cuts, competitor prices, and other factors. Total retail value of the carcass or wholesale cut is calculated and compared to wholesale cost. The difference is the margin, and it represents the money available to pay all costs associated with meat retailing plus profit. Required margins are usually 25 to 40 percent of wholesale cost. Figure 7.15 is a sample cutting record form used to perform cutting tests. Each retail cut from a given wholesale cut is prepared, weighed, and its current dollar return calculated, based on current retail price. A dollar adjustment is calculated by comparing the current total return with the desired total return. Prices of each individual cut are adjusted accordingly based on their relative popularity. These adjustments may be indexed by the meat manager to reflect knowledge of the tolerance of cuts to price change. It is important that carcasses or cuts used in cutting tests be typical of those to be merchandised. When supplies of uniformly yielding carcasses or cuts are available, retailers may use standard retail cut yields for each new combination of prices. As additional cutting tests are performed, the results are added to a computer data bank. Cutting tests are used frequently to test alternate methods of cutting, evaluate the skill of employees, or assess the usefulness of different kinds of cutting equipment.

Labeling

Retail meat labels vary from simple to complex with respect to the amount of information contained. As consumers have requested more information on composition of meat they purchase, labels have become more informative. Depending on location, some information items are required by law and others are optional with the retailer. Labeling regulations and enforcement are inconsistent for packages labeled at the retail store versus those labeled in a USDA inspected facility. USDA mandated labeling, also described in Chapter 13 for processed meats, must include the following items: product name, company name and address, ingredient statement (for multiple ingredient items), net weight statement, safe handling instructions, nutrition facts panel and a USDA inspection legend showing the establishment number where the product was produced. Most of these items along with others not required by USDA, are commonly included in retail product labels as discussed below.

Name of Cut

A variety of names are used for fresh meat cuts throughout the US. Many of these names have regional significance and are well recognized only in restricted geographical areas. To standardize names of retail cuts, the industry has adopted a

RETAIL MEAT PRICING

Cutting Test

Wholesale cut _____

Wt. _____   Purchasing price _____ /lb.   Total cost _____

% Margin desired _____

Total retail dollar return desired _____

Retail cut	Wt.	Current Selling Price	Current Dollar Return	Index	New Selling Price	New Dollar Return
_____	____	_____	_____	____	_____	_____
_____	____	_____	_____	____	_____	_____
_____	____	_____	_____	____	_____	_____
_____	____	_____	_____	____	_____	_____
_____	____	_____	_____	____	_____	_____
_____	____	_____	_____	____	_____	_____
_____	____	_____	_____	____	_____	_____
_____	____	_____	_____	____	_____	_____
_____	____	_____	_____	____	_____	_____
_____	____	_____	_____	____	_____	_____

Total Retail Wt. _____

By-Products

Fat _____ ____ _____ _____ ____ _____ _____

Bone _____ ____ _____ _____ ____ _____ _____

Cutting Shrinkage ____ _____ _____ ____ _____ _____

TOTALS ____   Current Dollar Return   New Dollar Return

FIGURE 7.15 | Retail cutting test record form.

---

common list of names called the **Uniform Retail Meat Identity Standards**. Meat cut names consist of three parts under this system, which are, in order, meat species, wholesale cut, and retail cut. If desired by the retailer, a so-called fanciful name may be added to gain additional consumer recognition.

Weight

Retail meat weights must equal the specified weight on the label, but for profitable operations they should not exceed that weight. Any juice accumulation in the package from weep of cuts technically may be regarded as product weight, but good retail practice would assure consumers that intact products weigh the amount indicated on the label. Package weight also is important to consumers who purchase specific quantities for specific needs.

Price per Unit of Weight

As discussed earlier, the price per unit weight of product must result in optimum rates of product movement. Generally speaking, within a grade or type of meat, the highest prices will appear on cuts that have maximum tenderness, are boneless, have minimal excess fat, and offer great convenience to the consumer. In addition, some consumers are particularly price conscious and make purchasing decisions based on this factor.

Package Cost

One of the most important items of information on meat labels is the unit cost of the package of meat. Many purchases are made on the basis of previously budgeted expenditures for meat for a specific meal or time span. Consequently, a variety of package sizes and unit costs within similar products is essential. Studies have shown that many consumers make purchases based largely on number of servings and cost of package.

Date

The date on which meat is packaged should appear on labels for the benefit of both consumers and retailers. Meat that remains in display cases for more than a few days may be overpriced or unattractive. The date is an index of freshness on which many consumers rely.

Cooking Assistance

Brief cooking instructions are helpful to many meat cooks. Labels or printed materials accompanying products may suggest appropriate cooking methods or provide recipes. Some whole turkeys and fresh meat roasts are packaged with a temperature sensitive device that indicates an acceptable end-point cooking temperature.

Grade or Brand

Some kinds of meat, particularly beef, lamb, and whole turkeys, are promoted and merchandised on the basis of US Department of Agriculture grade. In certain geographical regions and among high-income groups, the grade is an important label component. On the other hand, some consumers rely on brand names of products as applied by the slaughterer or processor. The criteria for these designations are frequently the same as those used in US government grading but are less rigid in requirements. Most fresh poultry and some fresh products of all species carry brand names. Experience has shown that branded products often engender consumer loyalty that does not develop when meat is sold merely as a commodity.

Nutrients

Nutritional labeling of fresh meat is a relatively recent innovation in merchandising. Nutrition information is supplied by many retailers, usually in the form of point of purchase information. Consumers are particularly concerned about fat content of meat but they also prefer to have available such information as protein, vitamin, and mineral content.

Name of Retailer

The name of the retail company from which the meat was purchased allows products to be traced to their retail origin. This protects consumers in the case of unsatisfactory products and advertises for retailers when products serve consumers well.

Universal Product Code

In addition to the preceding information, almost all retail meat packages also contain the **universal product code**. Use of universal product codes gives retailers the option of not printing labels that show the unit price and package cost. However, most retailers include the code along with the other information described. The universal product code improves efficiency of handling retail packages. Packages are scanned at the checkout counter and the item and its cost is printed on the cash register tape, which the buyer receives. Use of universal product codes, scanning systems, and computer acquisition and processing of the data provides immediate information on sales and inventory. Scanning systems also can provide great flexibility to retailers to adjust prices without re-labeling products.

Safe Food Handling Instructions

Increased incidence of food poisoning from fresh meats accidentally contaminated with food pathogens such as Salmonella and E. coli O157:H7 in recent years has heightened concern about safety of meat products. Meat handling practices to minimize microbial contamination and microbial growth and for safe storage and handling are discussed in Chapters 9 and 10. To help insure that incidents of food poisoning are minimized as a result of improper handling of fresh meat, most retail packages contain a label with instructions for proper handling and storage after the consumer has purchased the product.

Display Systems

Refrigerated display space is very expensive to maintain. Consequently, meat items compete with each other and with other refrigerated foods for this space. Their attractiveness may depend on the appearance of other items in the same display area. In addition, displays must be organized so that shoppers will locate desired items readily.

Most retailers place fresh meat cuts of one species in a common display area since most shoppers, to some degree, are species oriented when buying meat. Beyond this, they may arrange meat by cooking method, a practice that tends to segregate cuts by price as well. Still other display systems arrange meat by increasing color intensity so pale meats will not contrast sharply with bright red products.

Display Lighting

The appearance of meat depends on the type of light under which it is displayed. The type of light illuminating the meat case affects the light which is reflected to the eye. Certain wavelengths of light, especially in the ultraviolet range, contribute to pigment oxidation and color fading. Additionally, heat produced by light may warm the product and accelerate the fading process.

Most meat display case lights are of the fluorescent rather than incandescent type because the former produce about one-fifth as much heat as the latter. The specific type of fluorescent lighting is important because of differences in **color balance** or **rendition**, i.e., the degree to which spectral energy of the light coincides with spectral reflectance of meat. Warm white to neutral white (2800–3500K) fluorescent illumination with an intensity of 150–200 foot candles is reasonably effective for rendering color of red meat. Intensity in the red light range is essential but cannot be excessive or meat will appear misleadingly red. Sources that provide good color balance have about 20 to 35 percent of their emission in the longer wave length red range.

Display Temperature

Fresh meat deteriorates at rates that are directly proportional to temperature in the unfrozen range. Low display temperatures suppress enzyme activity, color change, oxidation, dehydration, and weep. Because the freezing point of meat is about –2 °C, display temperatures approaching that temperature are most advantageous. However, retail display cases vary widely in temperature within and between cases and temperatures up to 6 °C are not uncommon.

FROZEN MEAT DISPLAYS

Many problems associated with fresh meat merchandising may be eliminated by distributing it in frozen form. Because of tradition, added cost, product appearance, and other factors, frozen meat represents a small segment of retail merchandising. However, examples of successful merchandising of frozen meat exist including several fish and poultry products. These products are prepared by processors and transported to retailers in frozen form.

Packaging Requirements

Packaging materials used for frozen meats have barrier properties similar to those used in fresh meat vacuum packaging because prevention of moisture loss and oxidation is essential to avoid freezer burn (Chapters 9 and 10). Skin-tight wraps are required to avoid accumulation of ice crystals inside packages. Because of low temperatures and frozen product texture, packages must be very pliable, durable, and puncture resistant. Such packages are more expensive than fresh meat wraps and add to the cost of frozen meats.

The frozen state of meat and the oxygen-free conditions result in meat color that is darker and less appealing than that of fresh meat. This color is partially converted to bright red oxymyoglobin when the cuts are thawed but some brown metmyoglobin may persist, especially if slow freezing rates were utilized.

Temperature

The most critical phase of handling frozen meat is the storage phase, making proper display temperatures essential for successful frozen meat merchandising (Chapter 10). Temperatures of –18° C or lower are recommended. Of equal importance is the need to maintain a stable display temperature to avoid formation of large ice crystals and excessive drip loss on thawing.

PROCESSED MEAT DISPLAYS

Most processed meats are sliced and packaged for consumers at the processor level with the only retail function being that of display and associated services. These products include sliced bacon, cooked sausages, sliced luncheon meats and ham, and marinated products. Packaging of these products is described in the following section. However, many retail stores also have a deli service counter where a variety of cured meats, sausages, precooked boneless roasts, cooked turkey breasts and other processed meats are displayed in enclosed, refrigerated cases. These products are sliced to order, weighed, packaged, and priced for customers. Cured products in these displays are subject to color fading because the cut surfaces are exposed to light and air. Thus, these products usually are not sliced until the consumer places an order. The quantity of processed products sold from deli cases is small compared to that which is sliced and packaged in consumer-sized packages by the processor.

Packaging Requirements

Retention of color in cured meats depends on the absence of oxygen or exclusion of light. As discussed in Chapter 8, cured pigment fading occurs under strong fluorescent light if oxygen is present. Therefore, packaging requirements for cured meat necessitate the use of materials that are oxygen or light impermeable. Most cured products are vacuum packaged, but some are packaged in non-vacuumized opaque materials such as cardboard with only a limited portion of the product visible through a "window." Still others are placed in packages that are both vacuumized and opaque. As in fresh meat, the packaging materials for cured meat products must be moisture proof.

Most processed meat is packaged in systems that require two types of laminated film. One film is a **forming film** that when heated forms a cavity into which meat products are placed. A **nonforming film** is placed over the cavity and product, vacuum is drawn, nitrogen gas is infused (in some cases), and the package is sealed. With good sanitation and proper refrigeration, meat so packaged remains wholesome for 60 days or longer, but should not remain in meat displays for longer than 2 to 3 weeks. Precooked fresh meats and marinated meats are generally in vacuumized packages that may be either clear or opaque. They have similar storage times to other processed meats. In addition, these packaging methods lend themselves to frozen storage of the products after the consumer purchases them. Other materials are available for packaging processed meat, including glass and metal containers and aluminum foil.

The manner in which processed meats are consumed influences the size of packages used. These meats are often used for snacks or lunches when relatively small quantities are consumed. When packages are opened, deterioration rate advances and spoilage may occur. Consequently, small packages are consistent with eating habits and needs of many consumers. Accordingly, packaging costs per unit weight of meat are higher in processed meats than fresh meats.

Most processed meat packages are printed with the universal product code by processors. This affords retailers great flexibility in assigning prices and enables them to precisely regulate flow of products through outlets. Since processed meat packages are typically made under USDA inspection they must include all of required label information including the inspection legend with the establishment number.

PROCESSING AND DISTRIBUTION FOR FOOD SERVICE

In many countries, up to one-third of all meat consumed is prepared by the food service industry. Restaurants, hotels, and institutional food service units play important roles in maintaining quality of a large portion of our meat supply as it reaches consumers. Eating outside the home enhances enjoyment for a multitude of reasons, and the height of satisfaction usually depends on the quality of meat consumed.

TYPES OF FOOD SERVICE

Restaurants

Restaurant food service includes traditional dining rooms, snack bars, cafeterias, street vendors, and fast food outlets. They vary in menu choices, speed of service, cost, and other attributes, but their success depends on the degree to which consumer satisfaction is achieved. With few exceptions, that success is greatly influenced by quality and cost of meat served. In development of menus, restaurants emphasize meat items that can be produced throughout the day with minimal numbers of staff in the kitchen. Restaurants also are constantly changing the focus of the plate served. In many cases, the traditional meat item, which made up the center of the plate, now serves as a support item providing texture, additional value, or a flavor compliment.

Hotels

Food service in hotels usually is a component of total available hospitality and amenities. Facilities usually include capabilities for meal service throughout the day. Consequently, a wide range of meat items are required, menus are less specialized than those found in many restaurants and menus change less often. A large proportion of meals is those served at banquets. Because of challenges with adequate staffing, unusual time demands from guests, and potential large fixed costs, hotels work hard to establish menu items that can be prepared in advance and held for long periods of time. Slow cooking of prime rib roasts over a 24-hour period and advance preparation of sauces are examples. Items that provide simple and attractive plate presentation also are of importance.

Other Providers

Once called institutional dining facilities or providers, these businesses are changing more rapidly than other segments of the food service industry. These businesses include contract food companies serving large business buildings or individual corporations, hospitals or other health care providers, schools and universities, airlines, contract commissary providers that provide food items for many quick service operators, and other mass feeders. These operators often provide food in non-traditional settings, often have high numbers of daily/repeating customers, and have high demands for nutritional offerings. Menu planning requires dietetic and variety considerations beyond other food service units. There is a high demand for new and unique products. Purveyors of meat products to these operators must be sensitive to wide variances in labor skill, available production periods, storage capacities, holding times for products, and expenditure profiles of the final customer.

MEAT ACQUISITION FOR FOOD SERVICE

The first step in acquisition of meat for food service is menu development. Wise choices of menu items followed by careful preparation and serving build good reputations for eating establishments. An optimum number of menu items gives variety but allows thorough development of methods of preparation for each item. Meat is the focus for nearly every entree and is the food around which most meals are planned.

Institutional Meat Purchase Specifications

Specifications for many fresh and processed meat items normally used by food service units have been developed by the US Federal Meat Grading Service to assist purchasers and purveyors of food service meats. Meat items have been standardized, described, and identified to facilitate communication throughout the industry. These specifications are called the **institutional meat purchase specifications (IMPS)** and are available from the Superintendent of Documents, US Government Printing Office, Washington, D.C. 20402. They are numbered by species and type of cut as shown in Table 7.3.

The North American Meat Processors Association has published The Meat Buyers Guide, an illustrated reference to these standardized products for beef, veal, lamb, and pork, and the Poultry Buyers Guide, a similar reference for products from chicken, turkey, duck, goose, and game birds. Figures 7.16 and 7.17 (see color section) show selected examples of beef products for which IMPS standards have been established. Figures 7.18 and 7.19 (see color section) show selected examples of poultry products for institutional meat buyers. Meat cut standardization is based on anatomical guidelines for removal from the carcass and subsequent trimming of fat, bone and superficial muscles. Various characteristics may be specified beyond standardized procedures.

Quality Grade

Although quality grade information is not used widely to merchandise food service meat, it is important in some species to assure eating and cooking qualities. Most significant in this regard are beef quality grades. Those of lamb, veal, and poultry are of secondary importance. Fresh pork is not sold with quality grade designations, but some attention to meat color and structure is important.

Intact beef cuts used for food service should come from carcasses in youthful maturity groups unless some method of tenderization is used. Youthful US grades normally used are Prime, Choice, and Select but tenderized steaks are prepared from Standard, Utility, and Commercial grade carcasses (Chapter 14). Prime grade beef is used infrequently in food service because of fat content; however, dining rooms specializing in beef often serve Prime beef for those customers desiring the ultimate in beef flavor. Choice or Select grade beef is most often used in food service because these grades are acceptable to most consumers in both palatability and leanness. Food service operators usually specify Choice grades for lamb and veal and grade A for poultry.

Fat and Skin Limitations

Fat is trimmed from meat cuts by purveyors based on buyer specifications. Average fat covering with tolerance of a maximum thickness at any point usually is specified for primal and subprimal cuts and roasts of beef, pork and lamb. For example, if maximum average fat thickness for beef roasts is specified to be 0.6 cm, the maximum fat thickness at any one point may be 1.3 cm. Fat trim specifications on steaks, chops, cutlets, and fillets usually are for a maximum (at any one point) thickness of fat on the edges of the cut. If not specified, fat thickness may not exceed 0.6 cm. Alternatively, purchasers may specify that primal and subprimal cuts, roasts, and many **portion cuts** (steaks, chops, cutlets, and fillets) be practically free of surface fat, that surface fat and most seam fat is removed (**peeled/denuded**), or that surface fat, most seam fat and the muscle membrane (epimysium) is removed (peeled/denuded, **surface membrane removed**). Specifications for poultry may include whether abdominal fat is included in cuts where the backbone is present and whether cuts are skinless or skin-on and the amount of skin. Fat content of ground meat or meat patties is specified as average percent of raw weight with accompanying tolerance range. Considerations of fat or skin level to be specified include cost (labor and weight losses), functions of fat during cooking (basting action and protection from dehydration), and nutritive effects (caloric content).

Refrigeration State

Food service meat is available as fresh or frozen cuts. Because properly frozen meat is essentially equal in quality to unfrozen meat, specifications

TABLE 7.3 Institutional Meat Purchase Specification Identifying Numbers for Various Species

	Species primals/subprimals	Portion cuts or parts
Fresh beef	100	1100
Fresh lamb and mutton	200	1200
Fresh veal and calf	300	1300
Fresh pork	400	1400
Cured, cured and smoked, and fully cooked pork products	500	
Cured, dried, cooked, and smoked beef products	600	
Edible by-products		
Beef	1700	
Veal	3000	
Sausage Products	800	
Chicken	P1000	
Turkey	P2000	
Duck	P3000	
Goose	P4000	
Game Birds	P5000–P7500	

are based largely on circumstances and convenience associated with its utilization. Fresh meat must be purchased more frequently and in smaller quantities than frozen meat because of its limited shelf life. With this practice is risk of depleted supplies if rate of use is greater than anticipated. Freezer facilities must be available to use frozen meat and, unless cooked from the frozen state, time must be allowed for thawing. Cooking steaks and chops from the frozen state requires greater time than cooking similar unfrozen meat and often is not compatible with prompt food service.

Cooking from the frozen state is feasible for patties, nuggets, and similar products in which cooking time is relatively short. Inventories are easier to control and microbiological quality is easier to maintain in frozen meat than fresh meat.

Size of Cut

Uniformity of cut size is especially important in food service meats because of the need for predictable cooking times, serving yields, serving size, and meat cost. Specifications for roasts normally prescribe weight ranges within which they are acceptable. Size of primal cuts, subprimal cuts, and roasts is governed largely by size of carcasses from which they are prepared. Consequently, purveyors use carcasses of prescribed weights to prepare roasts of desired weight ranges.

Size of portion cut bone-in meats (steaks and chops) is controlled by cutting procedures. This allows purveyors to meet more uniform portion specifications for weight and thickness. Because of variation in muscle size, purchasers may specify weight or thickness of cuts but not both. For example, when 227-gm portions are specified, those weighing 213 gm to 241 gm are acceptable. Likewise, when 3.17-cm thick portions are specified, those measuring 2.54 cm to 3.81 cm are acceptable. Standardized weight results in variation in thickness. Such variation may be beneficial in preparation of beef steaks wherein variable doneness is often required and thus, cooking times and serving sizes may be standardized. Standardized thickness results in variation in weight. Such variation results in different serving sizes that may be advantageous in satisfying different appetites in some buffet or banquet situations. Of course, with either system, specific tolerances are accepted by both purchaser and purveyor as terms of sale.

Boneless primal and subprimal cuts may be frozen, tempered and then formed with pressure to uniform cross-sectional dimensions before they are cut into steaks or chops (See Chapter 8 for a discussion of meat forming technology). This allows production of steaks and roasts with more specific dimension and weight tolerances than for bone-in cuts.

Size of chicken broiler parts and quarters is controlled by weight alone. Very low tolerances are prescribed and most products are available in a variety of weights.

Meat patty size is controlled by both thickness and weight. Specifications usually indicate number of patties per unit of weight and tolerances are given as number of patties over or under the specified number in a 4.54-kg unit.

Aged Beef

Purchasers may specify aged beef as accomplished by dry aging or aging in plastic bags. Desirable tenderization and flavor development occurs in aged products but these advantages must be balanced with costs of shrinkage from moisture loss and trimming of contaminated surfaces.

Processed Meat

Smoked Meat and Sausage

Institutional Meat Purchase Specifications exist for some smoked meats and sausages. Certain standardized products are described, but the infinite variety of these products makes complete standardization difficult. In general, products in these categories are identical with those sold in retail stores, the processing of which is described in Chapter 8. However, specialized packaging is applied to products destined for food service use. This reduces packaging cost because of larger package size and elimination of product display requirements.

Cured and smoked meats are popular in food service units. High proportions of pork presented are in cured and smoked forms. Hams, bacon, Canadian-style bacon, smoked pork chops, precooked bacon, corned beef, and smoked turkey are found frequently on food service menus. Extended shelf life of these products as compared to fresh meat minimizes problems associated with inventory maintenance.

Breakfast sausages are among the most used sausage products for food service. Yet, these non-cured, seasoned products have limited stability of fresh flavor and must be received frequently in small quantities. Some batter-type and chopped sausage products are useful for specialized menu items, but their convenience and ease of use in home meal service limits their demand when consumers dine out.

Precooked and Restructured Meat

Precooked frozen entrees offer significant savings in labor for food service establishments. Products are available that require only brief reheating before serving. Some meat stews, pies, rolls, and roasts are available but greatest acceptance has been experienced with battered and breaded fish and chicken fillets or nuggets, sliced restructured roast beef, and toppings for pizza and tacos. Limited success also has been experienced with precooked meat patties sold in vending machines.

Many precooked items also are restructured. Chunked and formed precooked beef rolls are reheated and served as sliced roast beef. Formed steaks and nugget-like products often are received in precooked form, then reheated before serving. In any form, however, success of precooked products largely depends on their freedom from warmed-over flavor. White meats such as poultry breast and fish are less vulnerable to this problem, probably because of their low content of iron-containing heme compounds, which are catalysts to lipid oxidation, or to their low content of phospholipids, which are readily oxidized.

Meat Orders

Obviously, meat quantities to be ordered must be based on projections of business volume or number and type of consumers expected. This information must be combined with knowledge of cooked meat yields and serving sizes anticipated. Typical cooked meat servings weigh approximately 100 gm. Amount of raw meat required to provide such a serving depends on fat content, water-binding capacity, degree of doneness, and other factors that affect cooking shrinkage. Cooking losses usually range from one-fourth to one-third of raw meat weight.

Portioned meat sizes should reflect both meat cost and menu prices. Some variety of portion sizes enables consumers to choose appropriate amounts for consumption. Menu statements regarding portion size should indicate whether a raw or cooked basis is used. Roast sizes should be based on quantities needed over the limited time during which fresh flavor and juiciness can be maintained after cooking (usually not more than 2 to 4 hours). Small roasts provide cooked products wherein uniform doneness is desired, and large roasts usually serve best when variable degrees of doneness are desired. Cooking losses and crusted areas tend to be in greater proportions in small than large roasts because of their greater surface area per unit of mass.

Delivery schedules should complement serving schedules, meat quantities required, storage space available, and product shelf life. Frequent deliveries may assure product freshness, but less frequent deliveries reduce delivery costs that are ultimately paid by consumers. Ideally, optimum delivery schedules maintain low inventories at reasonable cost without exhausting supplies.

Receiving and Storing Meat

When meat is received, it should be examined and weighed to determine its overall condition and degree to which it meets purchase specifications. Samples should meet specified tolerances for fat trim, weight, and thickness. Grade stamps should be clearly evident. Products should be accompanied by data on count, total weight, time shipped, and other information.

The most urgent consideration upon receipt of meat shipments is that of product temperature. Deterioration occurs rapidly if refrigerated vans are operated at warm (>5 °C) temperatures, if products remain out of refrigeration for extended time, or if storage areas are improperly cooled. If meats are properly chilled on receipt, they may remain in their packaging materials; otherwise, individually wrapped products should be removed from shipping cartons to remove packaging insulation and allow cool air to reach the products.

Meat products, especially large cuts without packaging, should be stored in separate areas from aromatic vegetables, fruits, and other foods, which might impart off flavors to the meat. Unless it is vacuum packaged, cured meat should be separated from fresh meat to avoid nitrite absorption and cured color development in uncured products. Refrigerators should be free of microbial contamination in general, and yeast and molds in particular, since they are spread easily through the air. Some air circulation is desirable to maintain uniform temperatures. High humidity minimizes product dehydration from exposed surfaces. Inventory control during storage should prioritize products for use so those in storage for the longest time are used first.

In-Unit Processing

It may be advantageous to perform limited meat processing in food service units if skilled labor, equipment, and refrigerated working areas are available. Primal or subprimal cuts may be processed further into oven-ready cuts, steaks, chops, or ground meat using labor available between peaks of food service activities. Careful economic evaluation of such processing is required, however, to assure proper use of all products and by-products.

In-unit processing may serve as a focal point for consumers when activities are performed in visible areas and products are displayed. Additionally, a variety of cut selections may be made available for increased consumer choices.

8

PRINCIPLES OF MEAT PROCESSING

OBJECTIVES: *Investigate processes and ingredients used in manufacturing a variety of meat products including smoked, cured, cooked, seasoned, dried, fermented, comminuted or whole muscle products and consider functional constraints of meat ingredients in various manufacturing operations.*

Smoked meats
Enhanced
Deep basted
Comminuted
Sausage
Meat curing
Salt
Nitrite
Reductants
Phosphates
Seasonings
Bone sour
Curing brine
Pickle
Immersion curing
Artery pumping
Stitch pumping
Multiple stitch injection
Dry cure
Tumbling

Massaging
Salt-soluble proteins
Maceration
Nitric oxide myoglobin
Nitrosylhemochromogen
Nitric oxide
Residual nitrite
Cure accelerators
Oxymyoglobin
Metmyoglobin
Autoreduction
Naturally cured
Light fading
Hemichrome
Nitrite burn
Nitrosamines
Comminution
Meat batter
Meat grinder
Bowl chopper
Emulsion mill
Flaking machine

Blending
Preblending
Swelling
Emulsification
Emulsion
Dispersed phase
Continuous phase
Emulsifying agent
Hydrophilic
Hydrophobic
Fat pockets
Fat caps
Variety meats
Mechanically separated (species) meat
Advanced meat recovery systems
Surimi
Bind
Filler meats
Emulsifying capacity
Moisture:protein ratio

Key Terms

Added water	Irradiation
Protein fat-free	Sweeteners
Natural juices	Dextrose equivalent
Water added	Flavorings
Ham and water product with X percent added ingredients	Food allergies
	Alkaline phosphate
Extenders	Molds
Binders	Casings
Fillers	Stuffing
Spices	Coextrusion
Seasoning	Desinewing
Aromatic seeds	Restructured meats
Herbs	Liquid smoke
Microground	3, 4-benzopyrene
Ethylene oxide	Case hardening

Key Terms (continued)

Processed meat and poultry products are defined as those muscle-derived products in which properties of fresh meat have been modified using one or more procedures, such as grinding or chopping, addition of seasonings, alteration of color, or heat treatment. These modifications contribute to preservation, convenience, appearance, palatability, variety, and safety, giving consumers a wide choice of meat products. Typical processed meat products include items such as cured ham, bacon, corned beef, restructured steaks, restructured roasts, brine-enhanced, marinated or precooked cuts, and a variety of luncheon meats and sausages. Most of these products are subjected to a combination of several basic processing steps before reaching their final form. Because there are literally hundreds of different processed meat products, each with its own characteristics, it is impractical to discuss completely all of the procedures followed in their manufacture. However, most products undergo basic processing steps in common, and it is our purpose to discuss these basic procedures in this chapter.

TYPES OF PROCESSED MEAT PRODUCTS

Whole Muscle

Many processed products are prepared from large pieces of meat or even whole, intact cuts (bone removed in some cases), examples of which are hams of all types, bacons, Canadian-style bacon, and smoked turkey. In the meat industry, many of these products are commonly referred to as **smoked meats**. These products usually are cured, seasoned, heat processed, and smoked, and often they are molded, shaped, or formed.

In addition to smoked meats, several types of marinated whole muscle products are produced including boneless pork loins, beef cuts, chicken and turkey breasts and whole birds among others. These products are injection marinated with solutions containing salt, phosphate, and flavoring or tenderizing ingredients but no nitrite (cure). In the retail trade they are often referred to as **enhanced** or **deep basted** products and may be marketed either fresh, precooked, or from the hot food deli.

Sausage

Comminuted products are those made from raw meat materials that have been reduced into small meat pieces, chunks, chips, or flakes by grinding, chopping or flaking. Some comminuted products may be classed as sausages. **Sausages** are comminuted seasoned meat products that also may be cured, smoked, shaped, and heat processed. The degree of comminution varies widely. Some sausages are very coarsely comminuted with individual particles easily visible; examples are salami, pork sausage, smoked dinner sausage and summer sausage. In other sausages, the meat particles are so small that the sausage mix is a viscous mass with many characteristics of an emulsion, although in a strict sense, those products are not true emulsions. These sausages are referred to as batter-type sausages; examples are frankfurters and bologna.

Sausages are commonly classified based on the type of meat ingredients and processing methods used in their manufacture. Some products may be made from meat of only one species: examples are beef summer sausage, turkey frankfurter or pork sausage. However, it is very common to use two or three types of red meat and poultry ingredients in many sausage formulations. Most sausages may be classified into one of the following five classes according to processing methods (examples of each class are given):

- **Fresh**—Fresh pork sausage
- **Cooked, Smoked**—frankfurter, bologna, knackwurst, mortadella, berliner
- **Cooked, Not Smoked**—liver sausage, braunschweiger, beer salami
- **Dry, Semidry, or Fermented**—summer sausage, cervelat, dry salami, cappicola, pepperoni, and Lebanon bologna
- **Cooked Meat Specialties**—luncheon meats and loaves, sandwich spreads, jellied products.

Naming of sausage products is not consistent with the product's manufacturing classification. Names of many European derived sausages can be traced back to a city or region in Western Europe where the sausage originated. However, today, the properties of such products may vary considerably among manufacturers with differences in degree of comminution, smoking, cooking, and casing size.

Non-Sausage Processed Meats

There also are many comminuted or sectioned products that are not classed as sausages. Hamburger, ground beef, formed portioned steaks, and chicken nuggets are probably the most common. Many commercial products begin as ground meat, chunks, pieces, flakes, slices or fillets that are then formed into roasts, steaks, sticks, patties, or nuggets. Some of these products are formed by machines using sufficient pressure to force ground meat tightly together. These often are marketed under the general titles of "burgers" or "steaks" and may even be breaded and precooked. Other products, sometimes called "restructured" meats, are produced by using solubilized and gelled proteins or other non-meat ingredients to bind the meat pieces together.

All canned meat products are classified as processed meat, from canned hams requiring refrigeration because their heat treatment is sufficient only to cook and pasteurize the product, to fully sterilized meat products that need not be refrigerated until the container is opened. Precooked meats, such as boneless cuts or frozen entrees, also must be included as processed products. Thus, it is apparent that processed meat products take many shapes, sizes, and varieties. Their number and variety is limited only by the processor's imagination.

HISTORY OF MEAT PROCESSING

Meat processing probably began soon after people became hunters. The first type of processed product may have been sun-dried meat, and only later was meat dried over a slow-burning wood fire to produce a dried, smoked meat similar to jerky. Salting and smoking of meat was an ancient practice even in the time of Homer, 850 BC. Early processed meat products were prepared mainly for one purpose, preservation for use at some future time. People learned very early that dried or heavily salted meat would not spoil as easily as the fresh product. Meat processing probably evolved from such practices, coupled with the necessity for storing meat for future use. Once advances in preservation technology occurred, especially in refrigeration and packaging, meat processors no longer were restricted to preservation by high salt concentration or drying. They were free to explore development of products with lower salt and higher moisture concentrations, and with new seasonings and combinations of meat ingredients, thereby creating many new processed products. Present day manufacture of processed meat products is driven largely by consumer demands for safety, convenience, unique flavors, distinctive product forms and imaginative packaging. The extended shelf life of many processed meat products contributes to their appeal allowing them to be widely distributed and utilized in many different ways.

Many of today's products were known to the ancient Egyptians and Romans. Roman butchers prepared cracklings, bacon, tenderloins, oxtails, pigs' feet, salt pork, meatballs, and sausage of many varieties. Writings dating from the reign of Augustus (63 BC–14 AD) contain directions for preservation of meat with honey (using no salt), and for preservation of cooked meats in a brine solution containing water, mustard, vinegar, salt, and honey. Recipes also are given for liver sausage, pork sausage, and for a round sausage of chopped pork, bacon, garlic, onions, and pepper that was stuffed in a casing and smoked until the meat was pink. These meat product references from earliest recorded history indicate that meat processing played an important role in development of civilized cultures.

BASIC PROCESSING PROCEDURES

Curing

Modern **meat curing** involves the application of salt, nitrite, seasonings, and other ingredients to meat to develop unique color, flavor, and texture properties and provide improved product safety and shelf-life extension.

History and Present Application

Cured meat products were originally prepared by addition of salt at concentrations that were high enough to preserve meat. Salt inhibits spoilage largely by reducing the water activity thus limiting microbial growth. Meat preserved by salting has an unattractive brownish-gray color due to oxidation of myoglobin (the pigment in muscle). The use of nitrate to fix the pink color of cured meat probably evolved more by accident than by design. Potassium nitrate (saltpeter) was most likely present as an impurity in the salt (especially sea salt) used as a preservative, and early meat processors eventually learned that, when it was present, a more pleasing bright reddish-pink color resulted. Today, color development during curing is equally as important as flavor and texture changes. The most notable difference between cured meats today that are stored under refrigeration and those of the past is the lower salt content and lower flavor intensity of the modern product.

Meat Curing Ingredients

Two main ingredients used in order to cure meat are **salt** and **nitrite**. However, other substances are added to accelerate curing, stabilize color, modify flavor and texture, and reduce shrinkage during processing. Ingredient use in cured meat products is regulated in the US indirectly by the Food and Drug Administration of the US Department of

Health and Human Services and directly by the Food Safety and Inspection Service of the US Department of Agriculture.

Salt (sodium chloride) is included in essentially all meat-curing formulas. Because it is not generally used at concentrations high enough to effect complete preservation, its main functions are to solubilize proteins and develop flavor. However, even at concentrations as low as 1.5 percent of finished product, salt may provide some preservative action leading to extended shelf.

Nitrite, either as a potassium or sodium salt, is used to preserve desirable meaty flavor, prevent warmed-over flavor, fix a bright reddish pink color and inhibit microbial growth, particularly outgrowth of *C. botulinum* spores. Sodium or potassium nitrates were the first compounds used for this purpose. However, in the early 1900's it was discovered that nitrate is reduced to nitrite by microorganisms in the meat and that direct nitrite addition results in the desired flavor, color fixation, and protective effects.

Nitrate, a source of nitrite when chemically reduced, currently has very limited use in meat curing and is generally restricted to slow cured products such as dry cured ham, fermented sausage or for use in natural curing. Natural sources of nitrate (ex. celery juice powder, sea salt, turbinado sugar, etc.) are used in the natural curing process. A lactic acid starter culture is used to convert the nitrate to nitrite. Nitrate reductase enzyme produced by the starter culture catalyzes the reduction of nitrate to nitrite.

Several **reductants**, (compounds capable of donating electrons), are incorporated in meat-curing mixtures to accelerate color development. Nitrite must be reduced to nitric oxide and ferric iron of muscle pigments must be reduced to ferrous iron before proper color development can occur. The most commonly used reductant is sodium erythorbate. Sodium erythorbate (isoascorbate) is an isomer of sodium ascorbate (sodium salt of ascorbic acid—Vitamin C). Also, enzymatic reducing activity is present naturally in muscle in the form of reduced nucleotides, amino acids, and metabolic cofactors. Other factors that contribute to reduction reactions are discussed in a subsequent section.

Alkaline or neutral **phosphates** are often incorporated into curing mixtures for several reasons. Phosphates increase the water-binding capacity of meat and reduce shrinkage of meat products during subsequent heat-processing. They also retard development of oxidative rancidity by sequestering pro-oxidant metal cations including iron, copper, and magnesium found in meat or present in water or salt added during processing. Phosphate products, especially those containing diphosphates (pyrophosphates) also can improve texture. Finally, phosphates may increase the rate of color development, particularly if neutral pH-adjusted phosphate blends are used as compared to more alkaline (pH > 8.0) phosphate. While phosphates have only a slight effect on the rate of color development, they may be important for small diameter products that are cooked very quickly during processing.

Seasonings, including spices, herbs, vegetables, and sweeteners, often are incorporated in meat along with curing ingredients. They do not enter into the curing reaction, but do impart unique flavors. See the "Formulation" section later in this chapter for more detailed discussion of ingredients.

Meat Curing Methods

Several techniques are used to incorporate curing mixtures into meat products—injection, direct addition, immersion and dry curing. Irrespective of the method used, the important requirement is to distribute cure ingredients evenly throughout the entire product. Inadequate or uneven cure distribution results in poor color development and the possibility of microbial and oxidative spoilage in areas that are not penetrated by the curing mixture. **Bone sour** in hams, and gray areas in the interior of cured products are examples of problems resulting from improper distribution of the curing mixture.

Cure ingredients are incorporated into sausage products during the mixing or comminution processes. The ingredients are added in dry form, or as a concentrated solution, and are uniformly distributed through the product during the blending and comminution steps.

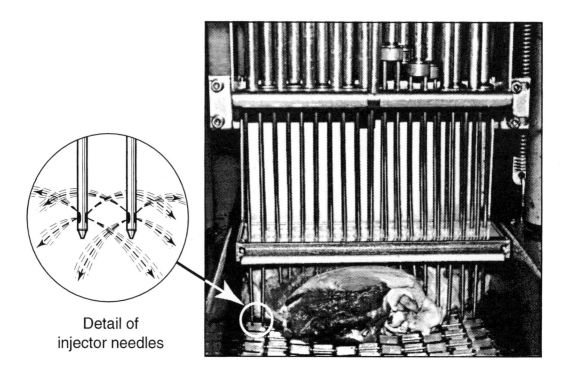

FIGURE 8.1 Multiple needle injection of curing solution into a bone-in ham. Inset shows injection needles with four holes in a single plane.

In the preparation of smoked meat products, curing agents may be incorporated by several methods. The cuts are sometimes immersed in a **curing brine** or **pickle**, a solution composed of curing ingredients dissolved in water. This is a slow method of curing, since extended periods are required for the curing ingredients to diffuse through the entire product. In addition, microbial growth and spoilage can occur during curing, even though the product is refrigerated. At present, this so called **immersion curing** is used mainly for specialty products such as neck bones, tails, pigs' feet, and salt pork or other thin cuts and in very small plants for curing of pork bacon.

The curing procedure may be shortened greatly by injection of brine directly into meat cuts using one of several methods (artery, stitch, multiple stitch), all of which achieve a more rapid and uniform distribution of brine throughout the tissues. In hams, a brine solution can be pumped directly into the vascular system, (**artery pumping**). Successful use of artery pumping requires care, so that blood vessels are preserved during carcass cutting and not ruptured by excessive pumping pressure. A very popular method of pumping that can be used on any type of meat cut is called **stitch pumping**. In stitch pumping, brine is injected through a hollow needle into various parts of the cut, especially the thickest part, and near joints. **Multiple stitch injection** is a widely used variation of stitch pumping. With this technique, brine is injected simultaneously and automatically through a series of hollow needles (Figure 8.1).

An older method for cure introduction is the **dry cure**, in which curing agents are rubbed in dry form over the meat surfaces. The cuts are then placed in refrigeration on shelves or in containers and allowed to cure. For large cuts such as hams, cure ingredients must be applied more than once during the curing period and the cuts must be overhauled, turned over and restacked. Since dry curing is slow (cure penetration is about 2.5 cm/week) and requires a large amount of hand labor, it is now used only on specialty items such as country-cured

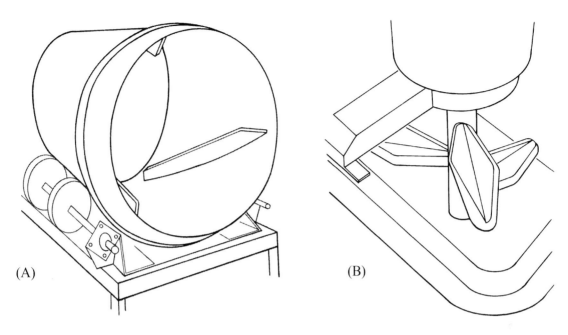

FIGURE 8.2 | Basic design of (A) tumbling and (B) massaging machines.

hams and bacons, and European-type dry cured hams such as Italian Prosciutto and Parma hams or German Westphalian and Black Forest hams. Nevertheless, worldwide production of dry-cured products represents an important segment of the processed meat industry. These traditionally-crafted products possess unique flavor and texture qualities that apparently cannot be duplicated using modern automated, high-speed meat curing methods.

Mechanical Methods to Improve Cure Distribution

Diffusion and binding of pickle within cure-injected meat cuts are facilitated by mechanical action. Processes known as **tumbling** or **massaging** subject products to agitation, which helps disrupt tissue structure and hasten distribution of cure ingredients. Bringing salt into contact with **salt-soluble proteins** (myosin, actin and others) results in greater water binding and solubilization of proteins. Tumbled or massaged products also develop a protein exudate that binds meat chunks together when later they are formed and cooked. Vigorous mechanical treatment such as tumbling may lead to foam production as air is entrapped in the protein exudate. Application of vacuum during tumbling eliminates foaming and associated color and binding problems. Figure 8.2 illustrates the design of tumbling and massaging equipment used frequently by processors.

Tumbling or massaging improves smoked meat yields because of improved water-binding capacity. These mechanical treatments improve tenderness due to increased hydration of meat proteins and physical disruption of muscle fibers. The loss of normal muscle fiber structure and the destruction of cellular barriers facilitate thorough and rapid distribution of cure ingredients. Thus, the tumbling and massaging processes improve uniformity of color development and tenderness.

Tenderization and cure distribution in boneless smoked meat products may also be improved by **maceration** before or after brine injection. Maceration involves cutting of the meat surface by rotating knives. Cuts are made about 1 cm apart and may penetrate nearly through the muscle. For muscles with a fat cover on one side, the macerator may be equipped with one cutting roller and one roller with no blades so only the lean side is macerated. In many meat curing operations muscle maceration and vacuum tumbling or massaging are combined to allow effective curing of boneless muscles without brine injection. In these processes the curing solution is metered into the massager and quickly absorbed by the macerated muscle during massaging.

Chemistry of Cured Meat Color

Since one purpose of curing meat is to develop an attractive, stable color, the chemistry of color fixation is important and is discussed in some detail. The basic reaction occurring during color development is represented by the following equation:

Myoglobin + Nitic Oxide ↔ Nitric Oxide Myoglobin +heat → Dinitrosyl Hemochromogen

Nitric oxide myoglobin has an attractive, bright red color, and is the pigment present in the interior of cured products before heat processing. The color hue is changed and stabilized upon heat denaturation of the protein portion of myoglobin. The resulting pigment is **dinitrosylhemochromogen**, which is responsible for the bright pink color characteristic of cured meat. See Chapter 6 for more information about meat color.

Nitric oxide is produced by the reduction of nitrite. If nitrate is used in curing, it must first be converted to nitrite, which is, in turn, reduced to nitric oxide. Nitrate-reducing bacteria are responsible for the conversion of nitrate to nitrite. If nitrite is added directly to the cure, the necessity for nitrate reduction is eliminated and color development is much more rapid. Nitrite is added to nearly all commercial curing mixtures, while nitrate is being used less and less, usually for products that traditionally undergo a long curing process, such a dry cured hams. Regardless of source, US meat inspection regulations limit the amount of nitrite

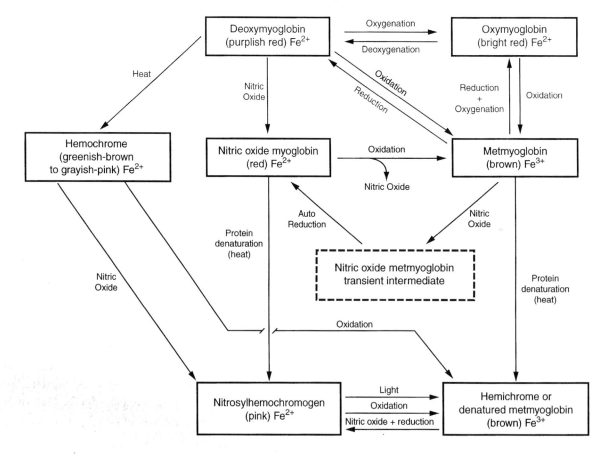

FIGURE 8.3 | Chemical reaction pathways for myoglobin in meat products.

permitted (based on meat and meat by-product in the formulation) to 200 parts per million (ppm) for most cured meat products, 120 ppm for bacon and 156 ppm for comminuted meats. **Residual nitrite** concentrations in finished products are much lower than those initially added since nitrite is depleted as curing reactions progress.

There are several mechanisms by which nitrite can be converted to nitric oxide. In water solution, at the pH of meat (5.5 to 6.0), less than 1 percent of the nitrite is present as nitrous acid (HNO_2) while most is thought to be present as nitrous acid anhydride, N_2O_3, (dinitrogen trioxide). Dinitrogen trioxide is reduced in muscle by endogenous or exogenous reductants (HRd) to produce nitric oxide (NO). Nitric oxide is the reactive species which ultimately bonds with myoglobin to produce the cured meat pigment. The following equations represent the production of NO from nitrite in meat:

$$N_2O_3 + HRd \leftrightarrow RdNO + HNO_2$$
$$RdNO \leftrightarrow Rd + NO$$

Many reducing compounds are present and active in fresh meat, such as cysteine, cytochromes, quinones and NADH (reduced form of nicotinamide adenine dinucleotide [NAD+]) among others. These endogenous compounds may provide the reducing potential for conversion of nitrite to nitric oxide, however, their effectiveness decreases during storage of meat. A more significant source of endogenous nitrite reduction is probably myoglobin itself. Upon nitrite addition to meat the myoglobin is quickly oxidized to metmyoglobin while nitrite is reduced to nitric oxide.

In order to assure an adequate supply of reducing potential for curing reactions, exogenous reducing agents are commonly added. Nitric oxide formation may be greatly accelerated by addition of reducing agents to the curing mixture. The sodium salts of ascorbic acid (sodium ascorbate) or erythorbic acid (sodium erythorbate) are the most widely used reducing agents that function as **cure accelerators**. Some chemical acidulants such as acid pyrophosphate also may be added to accelerate reduction and stabilize color slightly.

Figure 8.3 illustrates some reactions that heme pigments undergo during color development in cured meat. Since nitrite is a very efficient oxidizing agent for myoglobin, the initial reaction is probably the conversion of myoglobin and **oxymyoglobin** to **metmyoglobin**. Such a conversion is easily observed as a distinctive brown color at the surface of freshly cured meat products. The exact pathway of subsequent reactions is not clear. Nitric oxide appears to react with metmyoglobin and contributes to a rapid **autoreduction** of metmyoglobin to form nitric oxide myoglobin. This reduction may be accomplished either naturally in meat or by reductants included in the cure. Natural reduction is a slow process, and plays a significant role only in lengthy curing processes. Reducing activity of ascorbic/erythorbic acid salts (ascorbates/erythorbates) accelerates the reduction process. Sulfhydryl groups released during heat processing of cured meat also are very strong reducing compounds and can contribute significantly to the reduction. When exogenous reductants are used, the time for cured meat color development can be shortened to several minutes instead of several hours or days, as would be required if natural reducing activity were solely responsible for nitric oxide myoglobin production. Use of reductants in meat curing forms the basis for modern, continuous-process frankfurter production, and accelerated ham and bacon processes.

The final reaction shown in Figure 8.3, formation of nitrosylhemochromogen, involves denaturation of the protein portion of myoglobin and separation of the mono-nitrosated heme structure from its covalent attachment to the protein. In the process, a second nitric oxide is added on the opposite side of the iron at the site where the protein was attached. The free heme structure, with two nitric oxides attached, likely remains within the denatured globin moiety. But this is not necessary for achieving the typical pink cured meat color. The pink of the nitrosylhemochromogen contrasts to the more reddish nitric oxide myoglobin. The denaturation is usually caused by heat in the cooking process but may also be caused by acid accumulation in fermented products.

Naturally Cured Meat Products

Consumer interest in ham, bacon and cured sausage products made without nitrite has led to production of a variety of uncured versions of traditionally cured meat products. In this effort manufacturers generally take one of two approaches. One approach is to simply remove nitrite from the formulation and manufacture a product lacking the cured color and flavor for the corresponding cured product. Another more widely practiced approach is to remove nitrite from the formulation but replace it with food ingredients in which nitrate occurs naturally. This so-called Natural Curing process utilizes an ingredient which supplies nitrate and a nitrate reducing starter culture to convert the nitrate to nitrite. The ingredient supplying nitrate may be any of a number of items including celery juice powder, sea salt, and turbinado sugar among others. The starter culture is usually *Staphylococcus carnosus* or another culture which produces nitrate reductase, an enzyme which catalyzes conversion of nitrate to nitrite.

During manufacture of the naturally cured product some incubation time must be provided at a suitable temperature to allow for the conversion of nitrate to nitrite. Lengthy curing processes such as ham manufacture may not require any additional time if nitrite production from nitrate is adequate during the normal thermal processing schedule. However, short processes such as frankfurter manufacture may need to be extended by up to one hour to allow for adequate nitrite production and cured color development. In order to shorten the processing time some suppliers offer celery juice powder in which nitrate is already reduced to nitrite.

Naturally cured meat products may exhibit cured meat color and flavor very much like that of their traditionally cured counterparts. However, color stability is often reduced with the result that naturally cured meats have a shorter display life. When naturally cured products are made using celery juice concentrate as the nitrate source, a prominent celery flavor may be imparted to the product. In lightly seasoned or unseasoned products such as cooked ham this may be quite noticeable but in seasoned sausages the celery flavor is not obvious.

In the US, product types which are usually cured must be labeled as "Uncured" and "No Nitrate or Nitrite Added" if they are made without the addition of pure nitrate or nitrite. For those "Uncured" items made using a natural nitrate or nitrite source such as celery juice powder, that ingredient must be featured by listing "No Nitrate or Nitrate added except for that naturally present in celery juice powder". The term **Naturally Cured** currently is not accepted for labeling of these products.

Flavor Stability

An important quality benefit of curing meat with nitrite is flavor stability, which results from antioxidant functions of nitrite. First, the formation of nitrosylhemochromogen and the associated immobilization of iron in the complex slows the initiation of lipid oxidation. Reactions between iron porphyrins in meat and nitric oxide prevent iron catalyzed oxidation of unsaturated fatty acids and production of volatiles responsible for undesirable warmed-over flavor in uncured meat. The second antioxidant function of nitrite results from the formation of nitric oxide that acts as a free radical acceptor. Nitric oxide effectively stops lipid oxidation already in progress by terminating the free radical chain reaction. The resulting stabilization of desirable flavor increases consumer demand for cured meat as a precooked, table-ready food. Nitrite is the only approved additive available that complexes iron, stabilizes flavor, and produces desirable color.

Stability of Cured Meat Pigment

Nitrosylhemochromogen is a heat-stable pigment and does not undergo further color change upon additional cooking of cured products. However, cured meat pigments undergo other reactions that result in color changes. Often these changes include a loss of color. Nitrosylhemochromogen is very susceptible to oxidation when oxygen is present. This effect can become important if cured meat cuts are displayed under strong fluorescent illumination while they also are exposed to air. Under these conditions the surface color of cured meats will fade in as little as one hour, whereas under the same conditions, fresh meat will hold its color for three days or longer. **Light fading** is a two-step process that includes (1) dissociation of nitric oxide from heme groups catalyzed by light, followed by (2) oxidation of nitric oxide and heme groups by oxygen. A brownish-gray color develops on exposed meat surfaces during light fading because faded pigment (sometimes called **hemichrome**) has its heme group in the ferric (Fe^{3+}) state.

The most effective way to prevent light fading is to exclude oxygen from contact with cured meat surfaces. This can be done by vacuum packaging meat in oxygen impermeable films, in which case, clear films and illuminated display methods can be used without significant fading. In the absence of oxygen, nitric oxide split from heme groups by light is not oxidized, and can recombine with the heme. A less frequently used method to prevent light fading is to package cured meat in opaque materials that shield the product from display lights.

The presence of rancid fat in cured meat products results in color instability. Unsaturated fatty acids, like cured meat pigments, oxidize in the presence of oxygen and will accelerate the oxidation of cured meat pigment. Cured meat pigment also is susceptible to bacterial degradation of the heme ring producing biliverdin, a distinctly green pigment. This greening usually results from poor sanitation or improper storage conditions where products are contaminated and held under conditions favorable for microbial growth. Under aerobic conditions, the bacteria responsible for the greening effect produce hydrogen peroxide, which directly oxidizes cured meat pigment.

Excessive amounts of nitrite in cures also may cause a greening of cured meat pigment, called **nitrite burn**. Since nitrite is highly reactive in an acid environment, nitrite burn is especially troublesome in fermented sausages and pickled pigs' feet because of their high acidity. Greening due to nitrite burn results from oxidation of cured meat pigments, but exact details of the reaction are unknown.

Heat Stable Pink Color

Heat-stable pink color has been identified as a problem in a variety of fresh meat products without added nitrite. The cause is sometimes nitrite, which is inadvertently added to the product during manufacture. Use of the same injection equipment for cured and uncured products increases the risk of nitrite contamination. However, other causes have been identified or suggested. Heat-stable pink color may occur at the surface of uncured meat upon cooking in gas ovens, over charcoal, or in heavy smoke. Exposure of the cooked meat to carbon monoxide, produced in the gas oven or by incomplete combustion of charcoal, causes formation of globin carbon monoxide hemochrome, a red or pink color similar to that of cured meat. A persistent internal red color in cooked hamburger patties is associated with high pH raw meat, such as bull meat. Globin hemochromes have been identified as the pink pigments of turkey rolls and canned tuna. Meat loaves or meatballs often exhibit heat-stable pink color possibly due to nitrate in fresh onion, or celery included in the formulation. The occurrence is especially common when an extended holding time of the raw product allows time for microbial reduction of the nitrate.

Public Health Aspects of Nitrite Usage

Nitrite is toxic if consumed in excessive amounts. A single dose of nitrite in excess of 15 to 20 mg/kg of body weight may be lethal. However, the maximum level of nitrite permitted in cured meat products is twenty to forty times below this lethal dose. Also, residual nitrite is further depleted in the normal course of cooking and smoking meat products. Thus, there is essentially no danger of nitrite toxicity when regulatory limits are observed.

Under certain conditions (high temperature or low pH), a class of carcinogenic compounds known as **nitrosamines** can be formed in food products by reactions between nitric oxide and secondary or tertiary amines. Some classes of nitrosamines are known to be carcinogenic when ingested by animals, but their carcinogenicity is quite variable among species and organs. Secondary and tertiary amines are always present in protein foods as amino acid side chains of proline, hydroxyproline, histidine, arginine, and tryptophan. Therefore, the potential for nitrosamine formation exists in cured meats. Even though a large majority of tested cured meat products have been negative for nitrosamines, some have been found to contain these carcinogens at levels of approximately 50 parts per billion (ppb) or less. These compounds are formed more rapidly under conditions of high heat such as those used to fry bacon. Consequently, concentrations of nitrite permitted in bacon in the United States are lower (120 ppm ingoing nitrite) than in other cured products (156 or 200 ppm ingoing nitrite). In addition, for many years, beginning in 1978, FSIS monitored nitrosamine production in sliced bacon subjected to a controlled frying procedure. In the early years of bacon nitrosamine testing many companies modified their manufacturing procedures in order to reduce residual nitrite and the associated potential for nitrosamine production. Eventually, detectable nitrosamine production in bacon samples dropped to near zero. The bacon nitrosamine monitoring process was deemed obsolete and officially terminated in 2009.

Risk involved with nitrite use must be evaluated with a view toward benefits derived from its use. The characteristic color that is produced when nitrite is added to meat already has been discussed. Nitrite also limits lipid oxidation and prevents rancid flavor development. Of greatest importance is nitrite's inhibition of *Clostridium botulinum* spore outgrowth and prevention of toxin production in cured meat products that are heat pasteurized during processing and require post processing refrigeration. Wieners, bologna, and canned hams are examples of such products and, in the US, a residual level of 200 ppm of nitrite has been established as safe. It is generally recognized that 40 to 80 ppm of residual nitrite is the minimum level required to inhibit outgrowth of *Clostridium botulinum* spores in meat products. When bacon is manufactured with 40–80 ppm nitrite, it is required to have a second protective mechanism. This protection is provided by introduction of lactic acid-forming bacterial cultures into products because, upon temperature abuse, the organisms will grow and produce lactic acid, providing low pH protection against outgrowth of botulism spores. Without nitrite, the hazard of botulism food poisoning is increased in brine-injected cooked meat products, especially those that are vacuum packaged. Products such as roast beef, pork, or turkey, manufactured without nitrite, and sold in vacuum packages must be carefully refrigerated or frozen to assure safety.

To further assess risk, nitrite intake from consumption of cured meat should be compared to that from other sources. Particularly significant is the intake of nitrates in drinking water and green vegetables because dietary nitrate can be converted to nitrite by bacteria in saliva. Researchers have estimated that less than 5 percent of total nitrite exposure of the average consumer originates from cured meat.

Comminution, Blending, and Emulsification

Comminution

The process by which particle size is reduced for incorporation of meat raw materials into finished products is called **comminution**. The degree of comminution (or particle size reduction) differs greatly between various processed products and is often a unique characteristic of a particular product. Some items are coarsely comminuted, while others are so finely divided that they form a **meat batter**. Two main advantages are gained from all comminution processes: an improved uniformity of product due to more uniform particle size and rapid distribution of ingredients, and an increase in tenderness.

Equipment commonly used for comminution includes the **meat grinder**, **bowl chopper**, **emulsion mill**, and **flaking machine**. Figure 8.4 illustrates the cutting or shearing action and product flow of these types of comminution equipment. Grinders are usually employed for the first step in comminution of sausage and some restructured products. For ground sausages and fresh ground meats, grinding is often the only form of

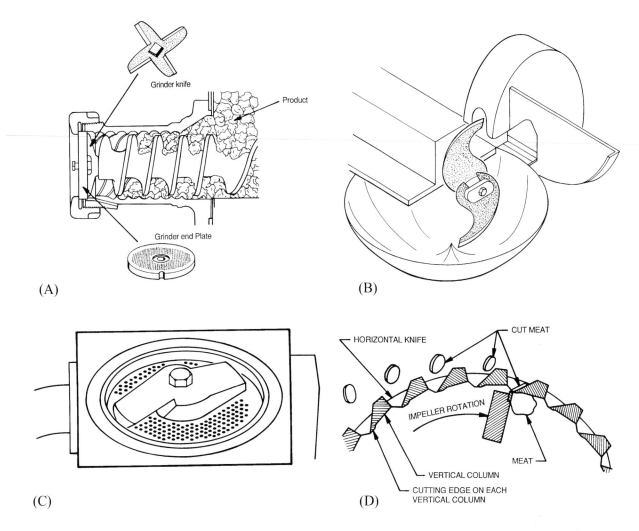

FIGURE 8.4 | Designs of various comminuting devices. (A) grinder, (B) bowl chopper, (C) emulsion mill, and (D) flaking machine.

comminution employed. In the past, the bowl chopper was used to form meat batters, but it is now usually used to reduce particle size of meat and fat, and for mixing ingredients in preparation for final comminution in an emulsion mill. Compared to the bowl chopper, emulsion mills operate at much higher speeds, form batters in much less time, and produce batters that have smaller fat particle sizes. Because emulsion mills operate at high speeds, meat materials are subjected to considerable friction and relatively high temperatures. Great care must be taken during this process because excessively high temperatures can prematurely denature soluble proteins leading to reduced stability of meat batters.

Blending

As a separate processing step, **blending** refers to additional mixing for the purpose of solubilizing and swelling meat proteins before further processing. A separate blending step ensures a more uniform distribution of ingredients, especially of the cure and seasoning, than could be achieved by grinding alone. Coarsely ground sausages are blended prior to being stuffed into casings. Large batch blending of meat, seasonings, and other ingredients is a common procedure prior to comminution of meat batters.

When raw materials are ground and mixed with salt, phosphate and water several hours before batter production, the process is known as **pre-blending**. This allows additional time for protein solubilization and swelling to occur at elevated ionic strength. In addition, this time allows for sampling and analysis of protein, moisture, and fat content of raw materials. Preblends of differing fat content can then be combined using knowledge of their composition to precisely control the composition of finished products.

Protein Matrix Formation and Emulsification

When muscle, fat, water, and salt are mixed together and subjected to high-speed cutting and shearing action, a batter is formed that has some of the characteristics of a fat-in-water emulsion. The microstructure of this type of product is of interest because the texture and stability of these meat products are closely related to physical relationships among ingredients. Formation of a typical meat batter consists of two closely related transformations:

- **swelling** of proteins (hydration and solubilization) and formation of a viscous matrix which ultimately forms a heat-set gel upon cooking, and
- **emulsification** of dispersed fat droplets and air by solubilized proteins.

The relative importance of each of these transformations to the quality and stability of batter-type sausage is unclear.

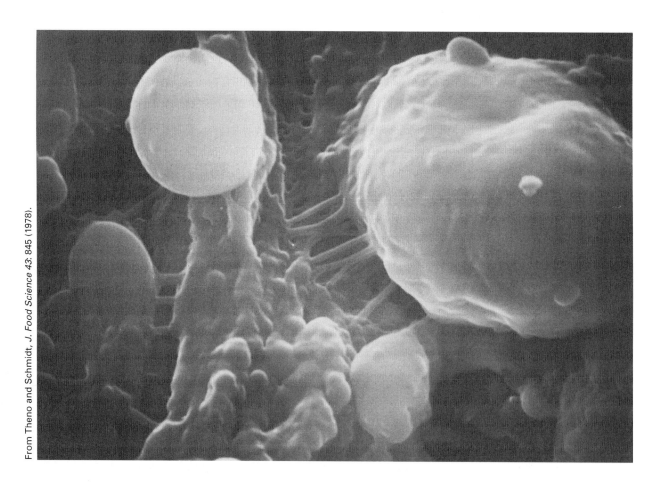

FIGURE 8.5 | Scanning electron micrograph of batter-type sausage showing gelled matrix and emulsified phases.

Disruption of the fibrous structure of muscle increases exposure of the proteins to extracellular and added water. The myofibrillar proteins (primarily myosin and actin or actomyosin) exist as gel networks capable of absorbing this water if ionic strength and other conditions are favorable. Water uptake by myofibrillar proteins in the presence of added salt results in swelling of proteins, which produces a viscous matrix. Of course some of the proteins remain reasonably intact (unswollen) in muscle fiber or connective tissue segments, whereas others are solubilized in meat batters and, as discussed later in this section, serve as emulsifying agents.

Formation of a matrix in meat batters stabilizes the structure in finished products by immobilizing free water and limiting moisture losses during heat processing. The matrix also helps stabilize the particles of fat formed during comminution against coalescence when they melt during heating. Matrix immobilization of free water results from water binding by proteins as described in Chapter 6, whereas immobilization of fat is a mechanical entrapment by the viscous matrix. Figure 8.5 is an electron micrograph of a batter-type sausage showing the viscous structure of the protein matrix.

An **emulsion** is defined as a mixture of two immiscible liquids, one of which is dispersed in the form of small droplets or globules in the other liquid. Liquid that forms the small droplets is called the **dispersed phase**, whereas the liquid in which the droplets are dispersed is called the **continuous phase**. The size of the dispersed-phase droplets ranges from 0.1 to 5.0 micrometers (um) in diameter.

Two examples of typical emulsions are mayonnaise and homogenized milk; in both, fat droplets are dispersed in an aqueous medium. Meat emulsions are two-phase systems, with the dispersed phase consisting of either solid or liquid fat particles, and the continuous phase being water-containing salts and dissolved, gelled, and suspended proteins. Thus, they can be classified as oil-in-water emulsions, as illustrated in Figure 8.6. Many fat particles in commercial meat emulsions are larger than 50 um in diameter, and thus do not conform to one requirement of a classical emulsion. Also, like mayonnaise, meat emulsions typically include small, entrapped air bubbles that are coated with soluble protein.

Emulsions are generally unstable unless another component, known as an emulsifying or stabilizing agent, is present. When fat is in contact with water, there is a high interfacial tension between the two phases. An **emulsifying agent** functions to reduce this interfacial tension, thus permitting formation of an emulsion with less energy input as well as increasing its overall stability. A distinguishing characteristic of emulsifying agents is that their molecules have an affinity for both water and fat. The **hydrophilic** (water-loving) portions of such molecules have affinity for water, while their **hydrophobic** (water-hating) portions have affinity for fat. These affinities are best satisfied when the hydrophobic and hydrophilic portions of the emulsifying agent can align themselves between both the lipid and aqueous phases. If enough of the emulsifying agent is present, it will

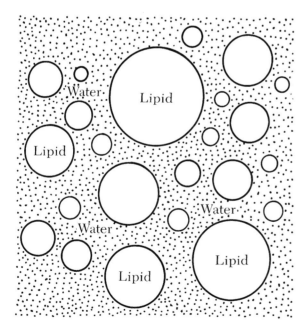

FIGURE 8.6 | Schematic drawing of an oil-in-water emulsion showing the dispersion of lipid droplets (dispersed phase) in water (continuous phase).

form a continuous layer between the two phases (Figure 8.7), thereby helping to stabilize the emulsion by separating the two phases.

Components of a meat batter are shown diagrammatically in Figure 8.8 and microscopically in Figure 8.5. A matrix of muscle and connective tissue fibers, segments of fibers, and gelled proteins is suspended in an aqueous medium containing soluble proteins and other soluble muscle constituents. Spherical fat particles coated with soluble protein are dispersed in the matrix. In sausage emulsions, soluble proteins dissolved in the aqueous phase act as emulsifying agents by coating all surfaces of the dispersed fat particles. The soluble proteins may be either sarcoplasmic or, in the presence of salt, myofibrillar proteins. However, myofibrillar proteins are much more efficient emulsifying agents and thus are almost completely

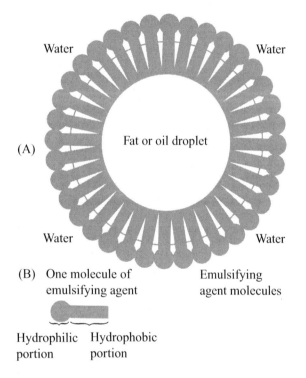

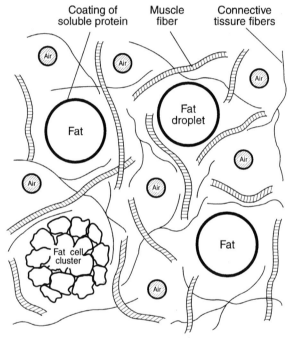

FIGURE 8.7 — A) The emulsifying agent is located at the interface between the lipid and water phases. These molecules are oriented so that their hydrophilic portions are in contact with the water phase while their hydrophobic portions are in contact with the lipid phase. The emulsifying agent forms a mono-molecular layer surrounding the lipid droplet. (B) A molecule of emulsifying agent with its hydrophobic and hydrophilic portions indicated.

FIGURE 8.8 — A diagrammatic illustration of raw meat batter. Fat droplets and microscopic air vacuoles are dispersed in an aqueous matrix that contains soluble proteins and suspended cellular components including muscle fiber fragments and connective tissue structures. Note that the fat droplets and air vacuoles are coated with a thick layer of soluble protein (emulsifying agent) which has been released into the aqueous matrix from the muscle fibers.

responsible for emulsion stability. The myofibrillar proteins, actin, myosin, and actomyosin, are insoluble in water and dilute salt solutions, but are soluble in more concentrated salt solutions. One major function of salt in sausage batters is to help solubilize these proteins into the aqueous phase so that they become available for coating fat particles.

Factors Affecting Meat Batter Formation and Stability

The extent to which fat is incorporated into a stable batter is influenced by several factors, including: temperature during matrix formation and emulsification, fat particle size, pH, amount and type of soluble protein, and batter viscosity. During comminution, the batter temperature increases due to friction in the chopper and emulsion mill. This leads to partial melting of fat and preliminary denaturation or unfolding of protein occurs, which promotes adsorption of proteins to the dispersed fat particles. Warming is beneficial in other ways, because it helps release soluble protein, accelerate cured color development, and improve flow characteristics. However, if the temperature becomes too high during comminution, emulsion breakdown may occur during subsequent heat processing. The maximum permissible temperature depends on the type of equipment used and melting point of fat. Final batter temperatures of 15 to 16° C for poultry, 20° to 21° C for pork and 26° to 27° C for beef may be reached, without dramatically affecting emulsion stability. Lower temperatures may be needed when batters are formed using bowl choppers. There are several possible explanations for the deleterious effects of excessive temperature in batter preparation. These include protein denaturation, decreased batter viscosity due to disruption of the protein matrix, and melting of fat particles. Because soluble proteins serve as emulsifying agents, denaturation during comminution can cause breakdown during heat processing. As batter viscosity decreases, a reduction in emulsion stability can occur because dispersed fat particles, being less dense than the continuous aqueous phase, tend to float to the surface. Fat particle migration is opposed by the viscous nature of the matrix phase and, as the continuous phase becomes less viscous, the tendency toward separation increases. Because melted fat separates easily into smaller particles, this can affect emulsion stability, as is discussed later in this section.

Controlling the temperature of meat batters is essential to assure desirable texture and consistency of the finished product. Temperature increases can be controlled or reduced by adding ice, rather than water, to the meat ingredients during chopping. Ice is superior to water because of the additional latent heat of fusion it must absorb in order to melt. Approximately 80 calories are needed to convert one gram of ice at 0° C to water at 0° C. Above 0° C, only about one calorie of heat per gram of water is needed to raise its temperature one degree Celsius. Thus, the amount of heat used in melting a gram of ice is sufficient to raise the temperature of one gram of water 80° C. Other effective means for reducing temperature during comminution include addition of carbon dioxide "snow," or use of partially frozen meat ingredients. If the batter is too cold due to use of partially frozen raw materials it may be necessary to use warm water in the formulation. In addition, it is possible to achieve the desired temperature rise by extending the chopping time, but this may lead to unintended textural changes or emulsion break.

During batter production, fat in meat ingredients must be subdivided into smaller and smaller particles until an emulsion is formed. However, as fat particle size decreases, there is a proportional increase in total surface area of fat particles. For example, if a sphere of fat with a diameter of 50 μm is chopped into spheres with a diameter of 10 μm, 125 fat particles will result. However, their total surface area will have increased from 7850 sq. μm to 39,250 sq. μm. This five-fold increase in surface area requires five times as much soluble protein to completely coat the surfaces of the smaller particles. Thus, over-chopping batters can increase the fat particle surface area to a point where the soluble protein present cannot adequately stabilize the emulsion.

In preparation of meat batters, lean meat ingredients are chopped with salt to facilitate protein solubilization and swelling prior to addition of ingredients having high fat content. Batter stability

FIGURE 8.9 | Emulsion break in frankfurters. The product on the left was produced from a stable batter, whereas the others are examples of fat caps and fat pockets that develop during heat processing.

increases as the amount of protein available for matrix formation and fat emulsification increases. The amount of protein extracted from tissues is affected by several factors. Protein extraction and batter stability increase with pH. Prerigor meat is superior to postrigor meat, because more salt-soluble protein can be extracted and the pH is high.

Meat batter stability is affected by the source of salt-soluble proteins: the species, muscle location, animal age, and other factors. Proteins of bull meat are well known for their excellent binding qualities. On the other hand, salt-soluble proteins of poultry light muscles are superior to those of dark muscles in forming stable meat batters. These differences may be due to different myosin isoforms known to exist in various muscles.

Meat batter breakdown is actually a re-aggregation (coalescence) of the finely dispersed fat particles into larger, easily visible fat particles. Of course, this breakdown is undesirable and must be avoided if high-quality products are to be prepared. If breakdown occurs, it usually occurs during heat processing but may not be observed until the products cool. During heating, dispersed fat particles melt, and in the liquid state, they exhibit a greater tendency toward coalescence. Particles that are completely coated with soluble protein (emulsifying agent) and well dispersed by the viscous matrix are unlikely to coalesce. The enlarged fat particles can appear as **fat pockets** on the surface or interior of products (Figure 8.9). Large fat particles also can coalesce at the ends of sausages into what are known as **fat caps**. However, less dramatic fat separation, such as greasy surfaces on finished products, is more common and causes significant problems in sausage manufacture.

Batter Preparation in Low Fat Products

Production of low fat cooked sausage products commonly involves the incorporation of additional added water to replace fat. This practice dramatically changes the conditions in the product. Lower brine concentration leads to reduced protein solubility and lower water binding in the batter. The lower fat content requires less protein emulsifier for stabilization in the matrix. The increased protein content of the matrix often leads to a dense rubbery product texture. In manufacturing such products there is less focus on emulsification of fat and more emphasis on water binding. Batter temperatures are generally kept lower to preserve as much native protein as possible. Entrapped air in such products is particularly important as a way to reduce rubbery texture. Along with changes in manufacturing procedures, various non-meat ingredients are commonly used in low fat products to improve water binding and texture properties. These will be discussed later in the section entitled, "Binders and Extenders."

Air in Meat Batters

Chopping of meat batters in the presence of air leads to incorporation of significant volumes of gases into the product. Entrapped air is emulsified by soluble proteins in much the same fashion as fat. Coating or emulsification of small air bubbles may be accomplished by the same proteins that might otherwise be used to coat fat droplets. Thus, excessive air in a meat batter may lead to destabilization of the emulsion. However, some entrapped air is needed to achieve desirable texture and elasticity of the finished product. This is especially important in low fat products and hot dogs that are expected to plump up upon heating. Control of air incorporation is achieved by application of vacuum during blending, chopping, or emulsification. Using a combination of vacuum and non-vacuum processing steps allows the manufacturer to maintain a stable emulsion while incorporating the right amount of air for desirable product texture.

Formulation

Many different ingredients are incorporated into processed meat products including meat, curing mixtures, seasonings, binders, fillers, and water. In preparation of a particular product, the manufacturer is not restricted to a set recipe. The specific ingredients, and their respective amounts, can be varied to produce a desired product, as a diet can be balanced by varying foods and their relative proportions. In the process of formulation, meat processors select the amount and type of ingredients used. The first goal of formulation is to produce products of uniform appearance, composition, taste, and physical properties from batch to batch or day to day. Successful formulation depends on availability of accurate information on composition, color, and chemical and physical properties of potential raw materials. Spices vary in purity and strength. The amount of binders, water, and fat that may be included is restricted by meat inspection regulations. The second goal of formulation is to produce products that meet preset quality standards at the least cost in raw materials. Because of fluctuations in the cost of various meat ingredients, it is often economically desirable to partially or completely substitute one ingredient for another. The process of formulation must determine the extent to which substitutions can be made, and when, in economic terms, to do so. Linear programming procedures for least-cost formulation are widely used in the meat industry to help accomplish these goals.

A quality control function in least-cost formulation programs ensures that the formula used will produce a product that meets quality standards. Raw materials costing the least will not always produce acceptable finished products. Consequently, computer programs must contain complete information on performance attributes of raw materials such as water-binding capacity, color, fat content, bind properties (bind units) or collagen content. They also must include specified limits on quantities of each potential ingredient that may be included without undesirable effects. Accordingly, quality control personnel must have sufficient knowledge about the way potential raw materials interact and their effects on quality and yield of finished products.

Meat Ingredients

A basic requirement for producing uniform processed meat products is proper selection and preparation of meat ingredients. Animal tissues vary widely in moisture, protein, and fat content, fat quality (degree of saturation), pigmentation, and ability to bind water and fat. Thus, the processor must know the properties and composition of the various available meat tissues in order to arrive at the correct meat formulation. In addition to skeletal muscle meat, by-product meats (variety meats) of non-skeletal muscle origin are potential sausage ingredients and are often used in cooked sausages. Not all sausages contain **variety meats**, but those that do sometimes have an exceptional nutritional value. For example, liver in Braunschweiger contributes iron and B vitamins. Meat inspection regulations require sausages containing these meat ingredients to be clearly labeled, for example, "Frankfurters with Variety Meats." In all cases, the ingredient statement must clearly list the presence of all ingredients.

Skeletal muscle to be comminuted is prepared mostly from carcasses or portions of carcasses that have limited use as fresh meat cuts. Carcasses of older animals do not possess sufficient tenderness to be desirable in non-comminuted forms because of extensive cross-linking of intramuscular collagen. The thin cuts and chucks of USDA Select or Choice beef carcasses often exhibit considerable seam fat and lack tenderness because of abundant amounts of connective tissue. Meat from these carcasses and cuts is well suited for use in comminuted products.

The time postmortem is an important consideration in selection of carcasses as sources of meat ingredients for processing. The advantages of prerigor salted meat were discussed in Chapter 5. When postrigor tissues are used, they usually are prepared for processing as soon after complete chilling as possible to minimize moisture loss and microbial growth.

Bone removal is the first step in preparation of skeletal muscle for processing. Much of this work is accomplished manually in labor-intensive operations. Mechanical deboning equipment is available, however, which drastically reduces boning labor costs and in some instances retrieves meat that cannot be removed from bones by hand deboning. Two general categories of deboning equipment are used in the meat and poultry industries. A functional classification of deboners is based on whether the bones are crushed or remain intact. **Mechanically Separated (species) Meat** is produced by machines which grind or crush bones and subsequently separate bone, cartilage, ligament, and tendon from soft tissue (muscle and fat) by forcing the tissues through a sieve to produce a meat paste (Figure 8.10). This method is particularly suited for deboning poultry carcasses as well as fish because the pliable bones do not often shatter or produce small fragments that might pass through the sieve. The boneless meat produced with this equipment must be specifically identified in the ingredient statement. Because of concerns about spread of bovine spongiform encephalopathy, mechanically separated beef is not considered suitable for human consumption. Mechanically separated pork is limited to 20 percent of the meat in hotdog and bologna products while mechanically separated poultry use is not limited. **Advanced Meat Recovery Systems** (AMR) equipment (Figure 8.10-C) also uses high pressure to force soft tissue through small holes

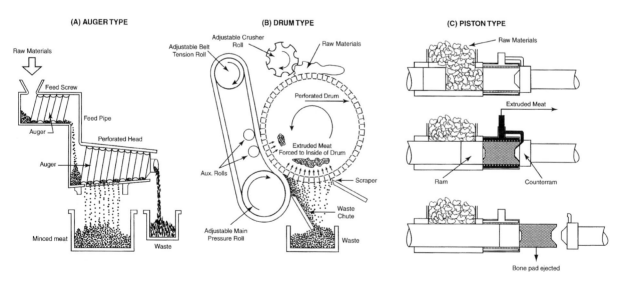

FIGURE 8.10 | Equipment for deboning meat. (A) Auger type, (B) Belt and drum, and (C) Piston Type (Advanced Meat Recovery). Boneless meat from equipment A and B must be labeled as Mechanically Separated. Meat from equipment C requires no special labeling.

or slots. Whole bones, especially blade bones or vertebra with adhering meat, are squeezed in a perforated stainless steel cylinder. The bones in the AMR system remain relatively intact through this process. The boneless meat from AMR equipment looks like fibrous ground meat. Since no bone particles are normally present it requires no special labeling or use restrictions. However, beef vertebra and skulls from mature animals (over 30 months of age) are excluded from the AMR process and all spinal cord must be removed from the vertebra of younger animals. Due to these restrictions it is uncommon to include any beef vertebra in the AMR process. Nevertheless, boneless meat from the AMR process is regularly tested for presence of spinal cord and related nerve tissue.

Another automated deboning system is intended for use with cuts such as pork shoulder or chicken thigh which contain a few large bones. Top and bottom dies are designed to match the shape of the bones in the cut. The cut is positioned on the bottom die while the top die forcefully presses the bone out of the surrounding muscle. This system works best when cuts are of uniform size. The boneless muscle has properties similar to hand deboned meat. Additional lean may be recovered by subsequently passing these bones through an AMR or other deboner.

Mechanical deboning produces meat raw materials economically and represents a means by which large quantities of meat may be salvaged that might otherwise be lost for human consumption. However, these methods result in raw materials with properties that dictate their use in selected manufacturing procedures. The mechanical action to which the meat is subjected may result in partial emulsification of products. Consequently, less comminution is required in later processing. When bones are crushed or ground during deboning, heme iron from bone marrow and oxygen from air are incorporated into the product, requiring special protection against development of oxidative rancidity. To minimize this problem the mechanically separated meat should be used quickly or antioxidants may be added. Freezing is not a preferred option since frozen, mechanically separated meat has a very short frozen shelf life. Some bone particles up to the size of table salt crystals may be present and produce a grainy texture in the finished products. The nutritional value of mechanically separated meat is slightly superior to conventional raw materials because of its abundant amounts of heme iron and elevated calcium content. Nevertheless, bone particles and calcium are restricted because they are not within the definition of meat.

Mechanical separation is used to debone fish for production of minced muscle and a product known as **surimi**. Different species of fish may be used but a cold-water species such as the Alaskan Pollock is especially suited for this purpose. Surimi is produced by mechanically separating minced fish muscle from bones and skin, then washing the muscle with water or dilute salt water to remove blood, fat, and water-soluble components. After removal of excess water, the material is mixed with small amounts of sugars and phosphates to protect the tissue proteins, particularly myosin, from denaturation during frozen storage. Because of its high concentration of salt soluble proteins, surimi is an excellent raw material for products requiring strong binding properties. It undergoes heat gelation at lower temperature (to be discussed later in this chapter) than other meat proteins. Because of its bland flavor, it may be combined with other meats as a binding agent without contributing species flavor. Surimi is commonly texturized by extrusion and heating processes to produce simulated shellfish of many types.

Functional Properties of Meat Ingredients

Functional properties of fresh meat are those characteristics which contribute to the palatability and performance of the final processed meat product. Muscle pigment (myoglobin) content influences finished product color. Myofibrillar protein contributes to stability of the end product because of its ability to bind water or emulsify fat. Heat-set or cold-set gelation of proteins may be required for meat particle adhesion and establishment of the final product texture. Thus, variation in functional properties, beyond simple differences in proximate composition, help determine the relative values of raw ingredients used in meat processing.

Following is a discussion of a variety of terms and issues that arise as one attempts to incorporate meat functional property considerations into product formulation and manufacturing decisions.

As used in the meat industry the terms "bind" or "binding" take on various meanings depending upon the setting. **Bind** as used in relation to sausage and comminuted products is not well defined but, in general, refers to the ability of meat to hold fat and added water during processing and avoid emulsion break. Protein content, ease of solubilization, and gelation properties all contribute to the binding ability of a particular raw material. Some meat ingredients have very high binding capacity, and others are inferior. For purposes of least cost formulation calculations, numerical bind values are often assigned to ingredient types based on empirical measurement of their performance in emulsified sausage systems. Table 8.1 gives a general classification of some meat and poultry ingredients based on their bind values. Beef skeletal muscle is an example of meat with high binding capacity; head and cheek meat, on the other hand, are intermediate in binding ability. Meat with high fat content, high amount of connective tissue, or large proportions of smooth muscle, is low in binding ability, such as regular (50 percent fat) pork trimmings, beef briskets, and tongue trimmings.

Collagen, the connective tissue protein, can contribute to fat binding during sausage mixing or batter formation. However, during heat processing, collagen shrinks generating localized distortion and causing liquid (fat and water) movement within the meat batter. This may lead to emulsion breakdown if fat droplets are not adequately coated with emulsifier. As heating continues hydration of collagen increases and individual collagen molecules or small aggregates of molecules are released from connective tissue structures. The resulting collagen (gelatin) solution is thought to contribute very little to fat emulsification but is an important contributor to water binding. Gelation of the collagen solution upon cooling can lead to considerable changes in the elastic and textural properties of the product. In spreadable pates or liver sausage, the collagen gel is the basis for typical product texture. Tripe, lips, stomachs, skin, and snouts, commonly referred to as **filler meats**, may be included in sausage, but the amount must be limited since these materials are rich in collagen and have very poor binding properties. The binding ability of mechanically deboned poultry meat may be limited if the skin is left on because of the high proportion of collagen and fat in skin.

Proteins originating from meat ingredients are responsible for both water binding and fat stabilization. Variation in the ability to emulsify and entrap fat in meat batters is due to the amount and type of salt-soluble protein available. The total capacity of meat ingredients to stabilize fat is sometimes called

TABLE 8.1	Binding Classification of Meat Raw Materials		
High binding meats	**Intermediate binding meats**	**Low binding meats**	**Filter meats**
Bull meat	Beef cheek and head meat	Hearts	Ox lips
Cow meat	Pork cheek and head meat	Beef hanging tender	Tripe
Beef chunks	Beef flanks, plates, navels	Weasand meat (esophagus muscle)	Pork
Boneless pork	Beef shanks		Stomachs
Shoulders	Regular pork trimmings (50 percent fat)	Giblets	Skin
Lean pork		Tongue trimmings	Snouts
Trimmings (80 percent lean)	Pork jowls	Deboned poultry backs and frames	Lips
	Beef briskets		Livers
Poultry meat (without skin)			

emulsifying capacity. Such capacity is higher in some ingredients than others because of differences in collagen and myofibrillar protein content. Emulsification values have been determined for many of the common meat ingredients, based largely on the amount of salt-soluble proteins available per unit of meat. Tables of these values are sometimes used for linear programming of sausage emulsion formulations, if the total amount of protein in the meat batch is known. However, knowledge of binding capacity, if expressed numerically, may reduce or eliminate the need for emulsification values because water binding is associated with, and responsible for, fat stability. Another valuable input for linear programming is collagen content of ingredients because of collagen's limited ability to stabilize fat.

Fresh meat proteins are always hydrated to some extent. Thus, moisture and protein tend to share the same space within the meat product and exist in a reasonably constant ratio. For this reason **moisture:protein (M:P) ratio** is often used in describing meat ingredients. A typical M:P ratio for beef skeletal muscle is about 3.5:1. The M:P ratio may be utilized to simplify prediction of proximate composition of a meat ingredient. If one measures the fat content of boneless beef and assumes that the ash content is 1 percent then the remainder is moisture and protein in a ratio of 3.5:1. This shorthand prediction of composition can be quite useful in a manufacturing environment. However, it is important to keep in mind that there may be some seasonal variation in M:P ratio of raw meats. For example, the M:P ratio of pork muscle tends to increase somewhat in the Fall.

Fat is an important constituent of processed meat products and makes a large contribution to their palatability. In sausages, fat contributes to tenderness and juiciness. Fat content of meat ingredients varies widely and depends on type of cut or trimmings as well as carcass grade. Cost of meat ingredients is generally inversely correlated with fat content. Products such as cooked sausage and fresh sausage have regulatory limits for fat content. Thus, knowledge of fat content in meat ingredients is important for management of palatability, cost, and regulatory status.

Additionally, fat contributes to structural stability of batter-type products. Fat exists in such products as finely comminuted particles (1 to 20 μm diameter) and as intact fat cells (50 to 100 μm diameter). Emulsified zones in the protein matrix are made up of finely comminuted fat particles with soluble protein shells as described earlier. The existence of emulsified zones seems to improve the water-binding properties of salt-soluble proteins because of affinity of these proteins for fat-water interfaces. Thus, an indirect contribution toward stable structure is made by finely comminuted fat particles. Consequently, properly designed comminution systems are essential to achieve the optimum degree of particle size reduction.

Moisture

Moisture accounts for 45 to 80 percent of the finished weight of processed meat products, more than any other single component. Most of the moisture originates from the lean meat ingredients. However, processors add water to many products as part of the formulation. There are several reasons for adding water. Many products would be dry and less palatable if only the moisture contained in the meat ingredients were present in the final product; additional water improves tenderness and juiciness. Water may be used as a partial substitute for fat in low fat products. As previously mentioned, moisture, added as ice, also helps to keep product temperatures down during comminution. Water serves as the carrier for distributing the curing ingredients into non-comminuted smoked meat items such as ham and bacon. The moisture added during pumping, on immersion of cuts in pickle, or to sausage formulations, also serves to replace moisture lost during subsequent processing operations, particularly in heat processing. Thus, by adding water, the yield of finished product can be improved. The amount of moisture added during processing depends on the composition and properties of raw materials, losses anticipated during all stages of processing, and moisture level desired in the final product.

Processed meat products must comply with meat inspection regulations with respect to moisture and fat contents. In the US, the normal moisture/protein ratio for comminuted products is considered to be 4:1. **Added Water** is the amount of moisture in excess of this ratio. In other words, if a product contains 11 percent meat protein and 60 percent water, then the Added Water would be 16 percent (60-(4×11)). For cooked sausage products, manufactured under red meat inspection regulations, the combination of fat percent and Added Water percent must not exceed 40 percent. Thus, the product containing 11 percent protein and 60 percent water (16 percent Added Water) would be limited to 24 percent fat. Nevertheless, for a product with less than 10 percent Added Water, the maximum fat content of cooked sausage must not exceed 30 percent. In fresh sausage that is not heat processed, water added to facilitate processing may constitute a maximum of 3 percent of the formulation.

Regulations for added water in many smoked meats, such as cured or cured and cooked hams, pork loins, and pork shoulder cuts, are based on different criteria than those used to regulate added water in comminuted products. The standard is based on the amount of fat-free protein in the finished product and requires a minimum protein content without directly considering water content. Minimum **protein fat-free** (PFF) values have been established for cured pork products, and analyses of protein and fat performed in regulatory laboratories determine if products are in compliance with these minimum values. The PFF value of tested products is determined by the following formula:

PFF value = % protein/(100- % Fat) x 100%

To qualify for specific product labels, the PFF value must equal or surpass certain predetermined standards. For example, to qualify for the label "Ham" without any disclaimer, the PFF value must be at least 20.5 percent. For labels of "Ham with **Natural Juices**" or "Ham with **Water Added**" the PFF values must be at least 18.5 or 17.0 percent, respectively. Ham products with PFF values less that 17.0 percent are labeled "**Ham and Water Product with X percent Added Ingredients**" where X is the added ingredient content expressed as percent of the finished product weight. Other minimum PFF values have been developed for other cured pork products. Processors must carefully regulate their pickle pumping operations to assure that water added during pumping is sufficient to replace that which will be lost during the remainder of the processing operation and to provide the correct amount of added water for particular labels. Water-Added products generally are more tender and juicy than those without added water and have gained consumer acceptance.

Other Non-Meat Ingredients

A variety of non-meat products are incorporated into sausage and loaf items as well as cured meat products. These materials are commonly referred to as **extenders**, **binders**, or **fillers**. Nevertheless, they perform a variety of functions when included in product formulations. Following are some of their functions:

- improve meat batter stability
- improve water binding capacity
- enhance texture or flavor
- reduce shrinkage during cooking
- improve slicing characteristics
- reduce formulation costs.

Extenders commonly used in sausage formulations are characterized by high protein content, and usually are either dried milk or soybean products. Soy products include soy flour, soy grits, soy protein concentrate, and isolated soy protein. Grits and flour contain 40 to 60 percent protein and are identical, except that the flour is more finely ground than the grits. Both products impart a distinctive flavor to meat items, which has limited their application. On the other hand, soy protein concentrate (70 percent protein) and isolated soy protein (90 percent protein) have bland flavors. Isolated soy protein is also dispersible in water and possesses good water- and fat-binding capabilities.

Textured vegetable protein is generally produced by simultaneously cooking and extruding a mixture of soy flour and other ingredients, such as flavoring and coloring, into particles. Textured vegetable proteins are used as extenders for ground beef, and as binders and extenders in meat patties. A typical formulation is a mixture of three parts water and one part textured protein added to ground beef, at a rate of 20-25 percent extender mixture and 75–80 percent ground beef. Use of vegetable protein in this manner significantly reduces formulation costs. Extruded protein products have a meat-like texture, hydrate rapidly, and have affinity for juice retention. Meat patties containing textured protein shrink less during cooking than do all-meat patties since soy particles interrupt the meat protein gel and increase retention of fluids. Textured vegetable proteins simulating beef, ham, chicken, and bacon are found in such items as salad dressings, pizza toppings, seasoning mixes, casseroles, and many types of prepared foods and snacks.

Nonfat dried milk solids (NFDM), calcium-reduced nonfat dried milk, dried whey, and reduced whey are products derived from milk and are used as extenders. NFDM contains approximately 35 percent protein, of which 80 percent is casein and the remainder is largely β-lactoglobulin and lactalbumin. NFDM has limited ability to emulsify fat because the casein is combined with large amounts of calcium, making it poorly soluble in water. You will recall that proteins must be solubilized in order to function as emulsifying agents. If sodium replaces most of the calcium, the water solubility and emulsification capacity of the casein is improved. This sodium replacement product is called calcium-reduced NFDM. Dried whey, from which a part of the lactose has been removed, has a higher emulsifying capacity than NFDM, because casein is not present and β-lactoglobulin and lactalbumin are readily soluble proteins.

Other extenders used in sausages and nonspecific loaves include:

- cereal flours obtained from wheat, rye, barley, corn, or rice
- starch extracted from these flours, or from potatoes; and
- corn syrup or corn syrup solids.

The flours are high in starch content but relatively low in protein. Therefore, they are able to bind large amounts of water. Corn syrup and corn syrup solids are derived from corn starch by partial degradation of starch into glucose, maltose, and dextrin units. They are less sweet than sucrose and are added for flavor and textural properties.

Binders used in cured pork products are sodium caseinate, carrageenan, isolated soy protein, and modified food starch. These ingredients may be used to reduce fluid purge from hams and other cured pork products which, by virtue of added ingredients, fit into "Natural Juices", "Water Added", or "X percent Added Ingredients" label categories. Use of these binders is limited to 2 percent (1.5 percent for carrageenan) of the finished product.

The amount of extenders permissible in sausages and loaves is specified by meat inspection regulations. Soy products (with the exception of isolated soy protein), cereal flours and starches, NFDM or calcium-reduced NFDM, and dried or reduced whey products may be added to cooked sausages, either singly or in combination, to a maximum of 3.5 percent by weight of the finished product. Isolated soy protein may account for no more than 2 percent of the sausage. Sausages (except reduced fat products) containing more extenders than these limits must be labeled as "imitation." Nonspecific loaves that do not include the word meat in the product name are not limited with respect to extender content; examples are pickle and pimento loaf, and macaroni and cheese loaf. Reduced fat sausage including fat replacer ingredients may include non-meat ingredients as roughly 1:1 replacement for fat. The specific ingredients which may be used for fat replacement in these so called "Modified Standard Products" are not listed by USDA. The manufacturer has the option of using a variety of ingredients including but not limited to those listed by USDA as binders or extenders. Such ingredients appear on the product label along with a disclaimer indicating that this ingredient is "Not in regular product" or "In excess of the amount permitted in regular product".

Seasonings and Flavorings
Seasoning is a general term applied to any ingredient added to improve or modify flavor in processed meat products. Thus, the obvious reason for incorporation of seasonings into processed meat products is to create distinctive flavors. The use of

TABLE 8.2 Types and Origins of Spices Commonly Used in Processed Meats

Spice	Part of plant	Region grown	Use
All Spice	Dried, nearly ripe fruit of *Pimpinella officinalis*	Jamaica, Cuba, Haiti, Republic of Trinidad, and Tobago	Bologna, pickled pigs' feet, heat cheese
Anise (seed)	Dried ripe fruit of *Pimpinella anisum*	Russia, Germany, Scandinavia, Czechoslovakia, Netherlands, Spain	Dry sausage, mortadella, and pepperoni
Bay Leaves (Laurel Leaves)	Dried leaf of *Laurus nobilis*	Mediterranean region, Greece, Italy, Great Britain	Pickled pigs' feet, lamb, pork tongue
Cardamom	Dried ripe seeds of *Elettaria cardamomum*	Malabar coast of India, Sri Lanka (Ceylon), Guatamala	Frankfurters, liver sausage, head cheese
Cassia	Dried bark of *Cinnamomum cassia, C. Loureirii, C. burmanni*	People's Republic of China, India, Indochina	Bologna, blood sausage
Celery Seed	Dried ripe fruit of *Apium graveolens*	Southern Europe, India	Pork sausage
Cinnamon	Dried bark of *Cinnamomum zeylanicum* and *C. loureirii*	Sri Lanka (Ceylon), Sumatra, Java, Vietnam, Borneo, Malabar coast of India	Bologna, head sausage
Clove	Dried flower buds of *Eugenia caryophyllata*	Brazil, Sri Lanka (Ceylon), Tanzania (Zanaibar), Malagasy Republic (Madagascar)	Bologna, head cheese, liver sausage
Coriander (seed)	Dried ripe fruit of *Coriandrum sativum*	England, Germany, Czechoslovakia, Hungary, Russia, Morocco, Malta, India	Frankfurters, bologna, Polish sausage, luncheon specialties
Cumin	Dried ripe fruit of *Cuminum cyminum*	Southern Europe, India, Mediterranean areas of North Africa, Saudi Arabia, India, and People's Rupublic of China	Curry Powder
Garlic	Fresh bulb of *Allium sativum*	Sicily, Italy, Southern France, Mexico, South America, India, United States	Polish sausage, many types of smoked sausage
Ginger	Dried rhizome of *Zingiber officinale*	Jamaica, West Africa, West Indies	Pork sausage, frankfurters, corned beef
Mace	Dried waxy covering (aril) that partly encoloses the seed of the nutmeg *Myristica fragrans*	Southern Asia, the Malay Archipelago, (Indonesia and East Malaysia	Veal sausage, liver sausage, frankfurters, bologna

Marjoram	Dried leaf (with or without flowering tops) of *Marjorana hortensis* or *Origanum vulgare*	Northern Africa, Greece, and other Mediterranean countries	Liver sausage, Polish sausage, head cheese
Mustard (black, white, yellow, brown, red)	Dried ripe seed of *Brassica nigra B. juncea, B alba*	People's Republic of China, Japan, India, Italy, Russia, The Netherlands, England, United States	Good in all sausage
Nutmeg	The dried and ground ripe seed of *Myristica fragrans*	Southern Asia, Malay, Archipelago (Indonesia and East Malaysia)	Veal sausage, bologna, frankfurters, liver sausage, head cheese
Onion	Fresh bulb of *Allium cepa*	United States, worldwide	Liver sausage, head cheese, baked loaf
Paprika	Dried, ripe fruid of *Capsicum annuum*	Hungary, Spain, United States, Ethiopia	Frankfurters, Mexican sausage, dry sausage
Pepper (black)	Dried unripe fruit of *Piper nigrum*	The Republic of Singapore, Lampung, Sumatra, Thailand, India, Phillipines, Indonesia, Tanzania (Zanzibar)	Bologna, Polish sausage, head cheese
Pepper (cayenne or red)	Dried ripe fruit of *Capsicum frutescens*	Tanzania (Zanzibar), Japan, Mexico, United States	Frankfurters, bologna, veal sausage, smoked country sausage
Pepper (white)	Dried, ripe decorticated fruit of *Piper nigrum*	The Republic of Singapore, Thailand	Good in all sausage
Pimento (pimiento)	Ripe, undried fruit of *Capsicum annuum*	Spain, United States	Loaves
Sage	Dried leaves of *Salvia officinalis*	Dalmatian coast, United States	Pork sausage, baked loaf
Savory	Dried leaves of either *Saturaja hortensis* or *S. montana*	Mediterranean countries, United States	Good in all sausages
Thyme	Dried leaves and flowering tops of *Thymus vulgaris*	Mediterranean coast	Good in all sausages

seasonings allows processors to create new products and provide variety in existing products. A great amount of artistry is required for imaginative and successful use of seasonings, and meat processors guard the secrecy of their seasoning formulations very closely. In addition to flavor, seasonings contribute to preservation of meat. Preservative action of salt, especially in high concentrations, has been mentioned previously. Certain spices possess antioxidant properties and thereby reduce the rate of oxidative rancidity development. On the other hand, spices may carry excessively high bacterial loads that shorten the shelf life of products. Because of this, spice companies supply many sterilized spices to the meat industry. Table 8.2 lists many common spices used in meat products, their origin, and some typical examples of their use in processed meat.

Salt and pepper form the basis for sausage seasoning formulas. All other seasoning ingredients are supplementary but are very necessary to obtain distinctive flavors associated with various products. These seasonings include spices, herbs, vegetables, sweeteners, and other ingredients, such as monosodium glutamate, that contribute to flavor enhancement.

Spices are aromatic substances of plant origin. The parts of plants used to obtain spices vary. For example, the fruit is used for pepper, allspice, and nutmeg; the flower bud for clove; the aril (fleshy seed covering) for mace; the rhizome for ginger; and the bark for cinnamon and cassia. **Aromatic seeds such as cardamom, dill, mustard, and coriander also are classed as spices. Herbs** are dried leaves of plants, and those used in sausages include sage, savory, thyme, and marjoram. Seasonings originating from vegetable bulbs are onion and garlic. Because all the above seasonings are natural products, they are quite variable in flavor, strength, and quality due to variation in weather during growth, local soil fertility, and storage conditions. Thus, a large amount of empirical experience is necessary for proper selection of natural seasonings for use in processed meat.

Natural seasonings may be used whole; for example, whole pepper corns are included in certain dry sausages and peppered loaves. But usually seasonings are used in a processed form, either ground or as essential oils and oleoresins. Ground spices are available to processors in several degrees of fineness. The most finely ground (**microground**) spices are sometimes invisible, and do not add to or detract from product appearance. Extractives, the essential oils and oleoresins, are obtained from natural spices by distillation and solvent extraction, respectively. Extractives have some advantages over ground spices because they are free of microbial contamination and are invisible in finished meat products. Extractives can be blended in the same manner as ground spices— usually combined with a carrier, such as salt or dextrose, for incorporation into processed meat.

Most spices and many herbs are grown in warm climates, picked by hand and dried in open air with minimal heat. Thus, high microbial loads are common on natural spices and herbs. Anti-microbial treatments are commonly (but not always) applied to these items before they are sold for use in the food industry. Since high temperature drives off much of the spice flavor, low temperature sterilization treatments are used. **Ethylene oxide** gas has been the most common sterilization procedure for spices. However, **irradiation** is now widely used and continues to gain acceptance by industry and consumers.

Sweeteners used in meat products are sucrose, dextrose, corn syrup or corn syrup solids, and lactose. Lactose has little sweetening ability and may contribute bitterness in certain products but is present in sausages only when nonfat dried milk is included in the formulation. NFDM may contain approximately 50 percent lactose, but some lactose-reduced milk products are available. Corn syrup and corn syrup solids are only 40 percent as sweet as sucrose, and they therefore are often classified as extenders. They are sometimes identified with their **dextrose equivalent** (DE), which is a measure of reducing sugar content of products. Sucrose and dextrose are used primarily for their sweetening ability, which contributes directly to flavor and counteracts the astringency of salt. In addition, they participate in surface browning reactions during meat cooking and are readily available for fermentation by lactic acid-producing organisms used in preparing some dry sausages. Fermentation products are responsible for the characteristic flavors of these products.

Flavorings are commonly manufactured by partial hydrolysis of soy, milk, or even blood proteins. The resulting peptides may impart distinctive, intense flavor notes. In addition to peptides, the hydrolysis process may produce a significant amount of free amino acids especially glutamic acid. Glutamic acid is easily converted to monosodium glutamate, an effective flavor potentiator. By controlling the hydrolysis process a wide variety of flavor profiles may be created including beef, pork, or poultry flavors among many others. Flavorings produced in this way are commonly used in meat products to intensify meaty flavor or to add a unique flavor to a product. The ingredient statement of such products shows the flavoring as, "Partially Hydrolyzed Soy Protein" or "Soy Protein Flavoring".

Food Allergens

Food allergies are of considerable concern in the US. FDA has identified the eight most common causes of food allergies (milk, eggs, peanuts, tree nuts, fin fish, shell fish, soy, and wheat) and requires food manufacturers to identify these on the product label. Several of these food allergens are commonly used as ingredients in meat products. In addition to appropriate product labeling, meat processors using allergen ingredients should pay close attention to production sequencing and opportunities for product cross contamination.

Phosphates

Many forms of phosphates are approved for use in the US in pumping pickle for hams, bacons, picnics, and similar products and for direct addition to cooked sausage and other prepared meat and poultry products. Their levels may not exceed 0.5 percent in finished products.

Several beneficial effects of phosphates in meat curing can be cited. Phosphates increase the water-holding capacity of meat. When phosphates are used, there is less moisture loss during cooking. For example, there is less purge, or water and gelatin released from meat during cooking in processed canned hams and less loss in fully cooked hams (hams reaching an internal temperature of 68° C [155° F] or greater). As a consequence of the improved water-binding capacity there is some increase in tenderness and juiciness of the cured product when phosphates are used. The basis for improvement of water-binding capacity in phosphate-treated meat includes changes in ionic strength, pH, and cleavage of some actomyosin cross bridges. Phosphate's effect on ionic strength is somewhat limited especially for polyphosphates with large molecular weight. The buffering effect of alkaline phosphates is likely the most important factor with increased pH affecting protein swelling or solubilization. Phosphate also has been shown to cleave certain actomyosin cross bridges which have not fully converted to rigor cross bridges.

When using an **alkaline phosphate** (pH > 7) an increase in pH and the associated increase in net negative charge on meat proteins results in more charged groups available for water binding. The increase in net negative charge also leads to increased repulsion among meat proteins. As a result the tight network of protein produced by formation of actomyosin crossbridges during rigor mortis is partially dissociated by phosphates. Some researchers have concluded that diphosphates (pyrophosphates) are capable of cleaving the bonds between myofilaments, allowing more space for water to enter the network and to bind to charged groups on proteins.

Phosphates influence cured meat color and flavor in several ways. Cured meat color is more acceptable, uniform, and stable because phosphate reduces pigment oxidation and less light is reflected by muscles with highly hydrated proteins. Phosphates offer protection against browning during storage, and act synergistically with ascorbates to protect against rancidity in cured meat: the protection offered by both compounds is greater than the sum of the separate protection offered by each alone. The antioxidant activity is due to inhibition of lipid oxidation by high pH conditions and sequestration of metal catalysts by phosphates. In rapid manufacturing processes, such as for frankfurters, a more acidic phosphate blend with acid pyrophosphate (pH 7.2–7.5 vs. pH 8–9) may be

CHAPTER 8 PRINCIPLES OF MEAT PROCESSING

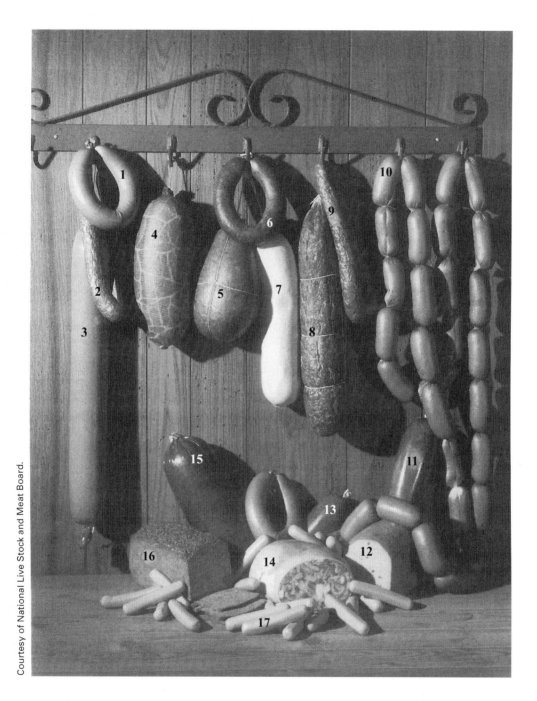

FIGURE 8.11 | Sausages stuffed in either natural or artificial casings. The shape and size of each sausage is dictated in part by the type of natural casing traditionally used for the sausage. The sausages are numbered in the photo and are identified as follows.

FIGURE 8.11 CONTINUED

	Sausage	Casing or Container
1.	Ring bologna	Natural-beef small intestine
2.	Smoked sausage	Natural-hog small intestine
3.	Bologna	Artificial-cellulose tube
4.	Bung bologna	Natural-beef bung
5.	Beer salami	Natural-beef bladder
6.	Mettwurst	Natural-beef small intestine
7.	Liver sausage	Natural-sewn hog bung
8.	B.C. salami	Natural-beef large intestine
9.	Smoked sausage	Natural-hog small intestine
10.	Knackwurst	Natural-beef small intestine
11.	Luncheon salami	Artificial-cellulose tube
12.	Pickle and pimento loaf	Artificial-metal loaf pan
13.	Luncheon salami	Artificial-cellulose tube
14.	Head cheese	Natural-hog stomach
15.	Salami	Artificial-fibrous cellulose
16.	Pepper loaf	Artificial-metal loaf pan
17.	Frankfurter	Artificial-cellulose (removed)

used to gain water binding benefits while providing a lower pH which is more conducive to rapid cure color development.

Antioxidants

Meat inspection regulations in the United States allow the use of BHA, BHT, TBHQ, propyl gallate and tocopherols to retard rancidity in dry sausage, fresh and precooked sausage and meatballs, precooked beef patties, and dried meats. Their use is increasingly common with greatly increased production of precooked, brown-and-serve products that are subject to lipid oxidation. Di- or pyrophosphates also provide antioxidant activity by chelating metal ion catalysts. Consumer interest in natural products has led to increased popularity of ingredients containing natural antioxidants such as rosemary or green tea extracts. These natural phenolic antioxidants may be more stable under repeated heating than are the chemical antioxidants. See a more detailed discussion on antioxidants in Chapter 10.

Forming Processed Meat Products

Most processed meat products are formed at some point in processing, in order to give each product a uniform or characteristic shape. Because sausages and fresh ground meats are comminuted products, they must be placed in some type of forming device or covering to give them shape and to hold them together during further processing. The product shape, provided by the forming device, becomes set permanently when the product is heated. The protein matrix gels and soluble proteins denature and coagulate upon heating to convert the product from semi-fluid to elastic-solid.

Manufacture of meat patties, of which ground beef represents the greatest volume, is itself a well-developed industry. Patty formation is accomplished by portioning and forming machines that produce large numbers of patties with precisely controlled weight and dimensions. After formation, the patties are frozen, usually in a CO_2 or nitrogen cryogenic tunnel or in a spiral type blast freezer. Frozen patties are mechanically stacked before packaging.

Some restructured fresh meat products are formed by freezing, then pressing to produce the desired shape. For example logs of chunked or coarse ground pork might be frozen in round plastic tubes. The frozen log would then be tempered to about -2° to -3° C before pressing it into the oval shape of a boneless pork loin. The loin shaped log would be sliced into chops which would be cooked from the frozen state to make restructured pork chops. The resulting, portion-controlled, restructured pork chops would be exceptionally uniform in size and shape.

Certain smoked meat products are formed in either **molds** or **casings**. Metal molds are often used for loaf-type items that are sliced prior to merchandising. Products are placed in molds after comminution and blending, and then cooked to set their shape. The formed loaves are removed from the molds before slicing. Baked loaves are usually formed in metal pans.

Casings are widely used as forms and containers for sausages. The shape or form of particular types of sausage is dictated by tradition, after having been in use for many years (Figure 8.11).

The process of placing meat products, either comminuted or non-comminuted, into casings is referred to as **stuffing**. Two types of casings are in general use: natural and manufactured.

Natural Casings

Before the development of manufactured casings, only natural casings were available to meat processors. They are derived almost exclusively from the gastrointestinal tracts of swine, cattle, and sheep. Hog casings are prepared from the stomach, small intestine (smalls), large intestine (middles), and terminal end (colon) of the large intestine (bungs). The parts of cattle used for beef casings are the esophagus (weasands), small intestine (rounds), large intestine (middles), bung, and bladder. Only the intestines of sheep are used to produce sheep casings. Natural casings are prepared by first thoroughly rinsing the contents from the gastrointestinal tract usually with cold water. The epithelial lining of the tract is stripped free and removed. Surrounding smooth muscle layers may be removed as for hog and sheep casings or left intact as for bladders or hog stomachs. Natural casings are highly contaminated with bacteria and are often preserved for later use by packing in salt or saturated salt brine. They are somewhat irregular is shape and strength making them difficult to use with high speed stuffing equipment.

Natural casings are very permeable to moisture and smoke. One of their most important characteristics is that they shrink and thereby remain in close contact with the surface of a sausage as it loses moisture. Thus, they are often used in dry sausage manufacture. Most natural casings are digestible and can be eaten. Their presence on sausage imparts resistance to chewing or "bite," a characteristic considered desirable by many consumers.

Manufactured Casings

Four classes of manufactured casings are available:

- cellulose
- inedible collagen
- edible collagen, and
- plastic.

Cellulose casings are manufactured from high quality cellulose obtained from wood pulp. The cellulose arrives at the casing plant as sheets of paper. The cellulose is first dissolved in alkali and then regenerated into casings. Cellulose casings are manufactured in sizes ranging from 1.2 centimeters in diameter for small sausages up to 15 cm for large sausages such as bologna. They are manufactured with stretch and shrink characteristics similar to those of natural casings. Inner surfaces of casings are sometimes coated with an edible, water-soluble dye or natural smoke that transfers to the sausage surface and colors and/or flavors the product. Protein coatings also are used to aid the adherence of casings to dry sausages. Other coatings may be used to improve the ease of removal of the casing as for hot dogs. Or, anti-microbial materials such as bacteriocins may be incorporated into the casing to help control bacterial growth. The advantages of cellulose casings include ease of use, variety of sizes available, uniformity of size, strength, and low microbial loads. Their strength is especially important in view of the widespread use of automated processing procedures. Fibrous cellulose casings include intact cellulose fibers that give strength allowing the casing to hold large sausages (bologna) and hams.

Both edible and inedible collagen casings are regenerated from collagen extracted from the corium layer of skins and hides. Inedible collagen casings combine some of the advantages of both cellulose and natural casings, especially their strength, uniformity, and shrink characteristics. Because of their thickness and strength they are not easy to chew and are considered inedible. So called "inedible" collagen casings are intended to be removed before consumption of the products, as are non-digestible, cellulose casings. Edible collagen casings are used largely for fresh pork sausage links, specialty frankfurters, and smoked sausage. They are very uniform in physical characteristics and may have greater strength than natural casings, but degrade easily upon chewing.

Plastic tubes or bags are used as meat or sausage containers in certain applications. They are impermeable to smoke and moisture. Therefore, they are used with unsmoked products or those that are heat processed in hot water, such as liver sausage and a number of cooked deli meat items such as cooked ham. Plastic chub packages have found widespread use as retail packages for a variety of products such as ground beef and fresh pork sausage. They afford excellent protection against oxidative changes and growth of aerobic organisms because they are impermeable to light and oxygen.

Casings, nets or plastic tubes may be used to apply certain materials to the surface of the meat product. They can be pre-coated on the inner surface with flavorings (liquid smoke, seasonings and spice mixtures) or color transfer products. When they are filled with meat the inner coating is transferred to the surface of the meat.

Not all sausage links are stuffed in casings. Small breakfast sausage links may be extruded to the desired size and shape without casings. This is most common for "brown and serve" type products that are immediately cooked in the plant. Additional treatments are needed to improve the strength of these extruded links if they are not cooked immediately. One approach is to apply a food grade acid to denature surface proteins on the link. This can be done during extrusion using a scintered metal section in the extrusion tube. Another variation involves a process known as **coextrusion**. A product stream is coated with a thin layer of collagen batter using a special extrusion nozzle. The collagen coating is set into a casing-like skin with an acid bath and by drying. This process may be used for fresh sausage links or even for frankfurter type sausages.

Forming Non-comminuted Meat

Forming of non-comminuted meat products differs in several ways from forming of sausage products. Natural and collagen casings generally are not used as containers for non-comminuted meat, but in certain instances fibrous cellulose casings, cloth and elastic nettings, and plastic tubes and films are used. For example, boneless ham muscles are placed in fibrous cellulose or plastic containers after boning and curing. The enclosed product can be flattened by a spring-loaded frame into a shape that is retained after heat processing. Bone-in hams and picnics are placed in netting prior to smoking and heat processing to provide a means of suspension in smoke ovens. After smoking and cooking the netting is removed leaving an attractive diamond grid pattern on the ham surface. Partially frozen (−3°C, 26°F) bacon slabs are formed in high-pressure presses after smoking and before slicing, in order to shape or "square" them to uniform dimensions. This facilitates high speed mechanical slicing, and results in a more uniform slice shape as well as a more uniform number of slices per unit of weight.

Restructured Meat Products

The constant search for new meat products to satisfy changing consumer needs has resulted in many product forms that do not fall into conventional categories of sausage, smoked meat, or anatomical cuts. Products that have undergone **desinewing** and particle size reduction such as sectioning, chunking, slicing, blade tenderization, flaking, macerating, or chopping followed by forming into steaks, roasts, nuggets, patties, or other shapes are called **restructured meats**. These meats may include raw materials that lack tenderness without particle size reduction but, with successful restructuring, possess acceptable tenderness as well as texture resembling intact cuts. In addition, the restructuring process provides for uniform portion size and shape. Even very high value items such as beef tenderloins may be restructured or reformed to improve uniformity of shape and size.

Several variations in manufacturing processes are used to prepare restructured meats. However, one key element in all processes and products is binding of meat pieces together with optimum strength. Several different approaches may be used. The most common approach uses heat gelled proteins from the meat pieces to bind the restructured product. Salt and phosphate are commonly used along with a small amount of water to help dissolve these proteins. Mechanical action from mixing or massaging brings the dissolved proteins to the surface to form a sticky exudate. Upon heating, this exudate gels forming a strong bond between the meat pieces.

Another approach relies on calcium alginate to bind the restructured product together. In this patented process sodium alginate and calcium carbonate are mixed with the meat pieces. The meat is formed into the desired shape and allowed to set under refrigeration. As the reaction proceeds, calcium alginate is formed binding the meat pieces together. Restructured meat products also may be bound using fibrinogen and thrombin derived from beef blood. This patented Fibrimex™ process involves combining fibrinogen and thrombin and quickly mixing with meat pieces. The meat is formed into the desired shape and allowed to set under refrigeration. In the presence of thrombin, fibrinogen is converted to fibrin, which binds the meat pieces together.

Finally, application of molecular biology and fermentation technology has made possible the production of commercially significant quantities of transglutaminase, an enzyme that forms covalent cross-links between glutamine and lysine amino acid side chains. These cross-links effectively bind proteins together. Activa™, containing the transglutaminase enzyme, is coated onto the surface of meat pieces by mixing. The pieces are formed to the desired shape and allowed to set under refrigeration. The enzyme forms cross-links that hold the meat pieces together. Meat products restructured without heating offer increased marketing flexibility but often suffer color and shelf-life problems, which prevent them from being sold along side other fresh meat products. Most restructured products are sold through the frozen meat case and in the restaurant and institutional trades.

Smoking and Heat Processing

Smoking and oven heating of processed meats can be considered as two separate processing steps. They are discussed together since, in most products, the two processes occur simultaneously or in immediate succession, so that variables in one process affect the other. In modern processing methods, the same facilities are used to accomplish both processes. Most products are both smoked and cooked (frankfurters, bologna, and many hams). However, a few products are only smoked with a minimum of heating (mettwurst, some Polish sausage, and bacon) while others are cooked but not smoked (liver sausage, cooked ham, precooked roasts).

Consumer desires for convenience in all forms of meat have resulted in precooking of individual entrees by meat processors. Many products, such as poultry parts and fish fillets, are coated with batter and breading and heated first in frying oil, then in continuous ovens. Hard-shelled crabs are heated under steam pressure before removal of edible components for subsequent packaging. Portioned meat products are components of precooked meals designed to be reheated quickly in the home with conventional or microwave ovens.

Heat Processing

Typical heat-processed meat products are cooked until internal temperatures of 65° C to 77° C (150° F to 170° F) are reached. This is sufficient to kill most of the microorganisms present, including the trichinae occasionally found in pork. Thus, the heating step will typically be identified as a Critical Control Point in the HACCP plan for a process. Pasteurization is one important function of heat processing even though it does not render products sterile or stable at room temperature. Thermal processing as a preservation method is discussed in Chapter 9.

In addition to pasteurization, other important changes result from heat processing. Of special significance is the firming of product texture resulting from protein gelation. The proteins that were solubilized during meat batter formation and those remaining in myofibrillar structures undergo a dynamic process of unfolding (denaturation) and aggregation (coagulation) during heating. These actions initially produce pressure within the meat leading to movement of liquids (fat and water). However, as gelation proceeds fat particles in the product are ultimately entrapped and water is immobilized. The result is a firming that occurs during cooking which sets the structure so that upon casing removal, the product's shape and form are retained. This stabilization of structure is best achieved when constant rates of temperature increase are employed. Preferred heating schedules are those that maintain a constant temperature differential (ΔT) between oven temperature and product temperature. The optimum ΔT for a particular meat product is best determined empirically. If the ΔT or heating rate is too great, fluid release is a common result. In higher fat products, fat melts and coalesces at the surface. In low fat, high added-water products, some of the water will be forced from the product. Such occurrences are more common when temperatures increase in a stepwise manner with large temperature differentials at the start of each step or when high relative humidity is applied in early stages of heating.

Another important purpose of heat processing is to develop and fix the cured meat pigment by development of reducing conditions (release of sulfhydryl groups) and denaturation of nitric oxide myoglobin, as previously discussed. The degree of heating used in processes is dependent on the label specifications and standards of identity under which products are manufactured. Some products, such as bacon, are heated only to an internal temperature of 52° C to 54° C (126°–129° F), at which temperature color fixation occurs (by formation of nitrosylhemochromogen). Such products are not labeled as cooked meat products and, in most cases, require further cooking before they are consumed. Additional heating to a fully-cooked state further stabilizes cured color by continued formation of nitrosylhemochromogen. Meat inspection regulations specify various final internal temperatures for ready-to-eat products based on species (beef, pork, or poultry) and added ingredients (with or without sodium nitrite).

Smoking

Smoking of meat is the process of exposing products to wood smoke at some point during manufacture. Smoking methods originated simply as a result of meat being dried over wood fires. Development of specific flavors and improvement of appearance are the main reasons for smoking meat today, even though smoking provides a preservative effect. More than 700 compounds have been identified in wood smoke. Some classes of chemical compounds present include aliphatic alcohols, ketones, carboxylic acids, heterocyclic hydrocarbons, sulfur-containing organic compounds, phenols, and terpenoids. Although many of these compounds exhibit either bacteriostatic or bactericidal properties, phenolic and acid components account for most of the antimicrobial action of smoke. In addition, phenols have an antioxidant activity. Various combinations of these compounds in wood smoke contribute to the flavor and aroma of smoked meat. Carbonyls inherent in smoke react with amines in meat protein in a Maillard-type browning reaction which contributes to the surface color and sensory properties of smoked meat products.

In most processed meats, smoking contributes little if any preservative action. Smoke components are absorbed by surface and interstitial water in the product, but most penetrate no more than a few millimeters. In products where the surface remains intact, a preservative effect may persist. But if the surface is disrupted, as in slicing or casing removal, the bacteriostatic effect is essentially lost. Few products today reach the consumer with the surface intact and less smoke is applied than in earlier times, so the purpose of smoking meat is mainly to develop a distinctive flavor and aroma that are pleasing to the consumer. However, a few other advantages do accrue from the smoking of meats. For example, the acids in smoke cause denaturation of proteins aiding in the development of a smooth surface or skin beneath the cellulose casing of frankfurters. That protein skin facilitates removal of the casing and contributes to the typical texture of the frankfurter.

Several methods are used to apply smoke. The traditional and most widely used method is the smokehouse. The product is hung from racks or "trees" that, in turn, are placed in the essentially sealed house. Smoke is generated outside the house by controlled combustion (between 650° C and 700° C) of moist sawdust. Sensory properties imparted to meat products by wood smoke are influenced largely by the combustion temperature of the moist sawdust and to a lesser extent by the type of wood used. Smoke is carried into the house by a system of ducts. Because of their low resin content, hardwoods, usually oak or hickory, are most commonly used to generate the smoke. Softwoods sometimes are used to achieve special flavor effects, but care to avoid strong bitter flavor should be exercised.

The modern smokehouse is equipped to heat-process meat products as well as to apply smoke. In order to accomplish both processes, the temperature, density of smoke and relative humidity are carefully controlled. In many cases, smoke is applied during initial phases of cooking, or even before cooking the product. The condition of the product surface (temperature and moistness) and smoke density in the smokehouse determine the length of time that products must be smoked to achieve desired levels of smoke deposition. This is a very important factor in continuous-process frankfurter production, where franks may be smoked and heat processed in less than 45 minutes. For this application processors often rely on **liquid smoke**, a concentrate derived from natural smoke. Frankfurters may be dipped in the liquid smoke or it can be sprayed in the smoke house for rapid development of smoked color and flavor.

Liquid smoke preparations have been developed as an attempt to eliminate the natural wood smoking process and the air pollution to which it contributes. Liquid smoke is prepared by cold condensation and fractional distillation of natural wood smoke. The resulting compounds are chemically filtered and refined to provide desired processing and flavor qualities before being suspended, usually in acetic acid. The resulting liquid smoke is usually applied to meat surfaces as an aqueous spray or an atomized fog. When used with a food grade emulsifier, the liquid smoke may be used as a dip or drench solution. Finally, neutralized liquid smoke may be added directly into the formulation of processed meats. Neutralization is required because the acid carrier of liquid smoke may denature meat proteins or cause emulsion break. To minimize these problems a soy oil carrier may be used or smoke components may be deposited on a malto-dextrin, salt, or other dry granular carrier for direct addition in meat products. Liquid smoke application gives the processor better control of surface color and product flavor than is possible with natural smoke. In addition liquid smoke preparations are free of carcinogenic compounds, such as **3,4-benzopyrene**, that have been discovered in low concentrations in natural wood smoke. The presence and amount of carcinogens in natural smoke depend somewhat on the temperature at which the smoke is generated. Combustion temperatures in the lower ranges produce very few of these compounds.

Humidity control in smokehouses is a critical factor influencing smoke deposition, product color, surface skin formation and product yield. Relative humidity in the smokehouse is influenced by many variables including temperature in the house, relative humidity and temperature of air outside the house, volume of make-up air entering the house, product surface temperature and moistness and the quantity of steam entering the house. Relative humidity of the smokehouse is usually lowered by increasing the volume of make-up air entering the house (and thus increasing the exhaust of moist air from the house) or by raising the house temperature. To raise relative humidity the amount of make-up air is reduced or steam is injected into the house. At the start of a cooking schedule, relative humidity in the smokehouse is often high due to rapid evaporation of moisture from product surfaces. However, as the temperature increases and moist air is exhausted from the house, product surfaces become drier and the humidity drops slowly to the relative humidity control set point. At that point the exhaust dampers close or steam injection begins to maintain humidity at the set point. If relative humidity is controlled at too low a set point, significant product weight loss will occur and a heavy, leathery skin will be formed on the product surface. Acceptable weight losses range from 5–15 percent, depending on product characteristics. If the relative humidity is set too high product yields will improve but surface color may be very pale or dull brown and unattractive. High relative humidity is often used near the end of the smoke/cook cycle for frankfurters in order to improve the ease of casing removal. High relative humidity early in the cooking process effectively increases product surface temperature, due to reduced evaporative cooling, and may lead to emulsion break or fluid loss in sensitive product.

Fermentation, Dehydration, and Aging

Earlier reference was made to dry-cured hams and bacons and dry sausage products, which have unique palatability characteristics. Many of these products are cured and smoked as already discussed, but unlike other processed meats, they rely on microbial fermentation and/or dehydration to develop their special flavor and texture. The combination of lowered pH and dehydration causes extensive protein denaturation, which resembles the changes caused by heating. Consequently, many of these products are consumed without cooking. Processing end points are usually pH, moisture content, or water activity. To assure safety of such uncooked products the manufacturer must conduct a safety validation test to demonstrate that the combination of added ingredients, dehydration, and fermentation of each product is adequate to provide a 5 log reduction for salmonella. When a new product or modification of an existing uncooked ready-to-eat product is considered, a new safety validation test must be conducted. Water activity and pH conditions that destroy salmonella are also effective for destruction of other vegetative pathogens. As result such products are safe to eat even without cooking.

Development of a distinctive, tangy flavor results from acid produced during microbial fermentation. Organisms responsible for fermentation may be lactic acid-producing bacteria that enter the product from the plant environment and processing equipment. If this were the only source, then chance would dictate whether the product would have the correct number and type of organisms. To eliminate the element of chance and to achieve uniform quality in various comminuted fermented products, most processors add specific lactic acid-producing microorganisms as starter cultures. Examples of such starter culture organisms are *Lactobacillus plantarum* and *Pediococcus acidilactici*. These organisms may produce sufficient lactic acid to reduce product pH to 4.5 or lower.

Meat contains very little carbohydrate so added carbohydrate is required for fermentation. The preferred carbohydrate source for meat fermentations is dextrose since it is an easily fermented monosaccharide. Disaccharides such as sucrose can also be fermented by many lactic acid bacteria especially those chosen for inclusion in meat product starter cultures. Successful fermentation is accomplished at 18°–24° C (64°–75° F) for dry sausage products and at 32°–46° C (90°–115° F) for semidry sausage products. Fermentation at lower temperatures requires more time to reach the desired pH and may result in a more complex flavor profile. Higher temperature fermentations are used when the main objective is to quickly reach the desired pH.

Dehydration of meat products is one of several basic processing steps. However, few meat products are dehydrated as a separate process. In those cases where drying is a separate step, the objectives are preservation and product identity, as in jerky and dried beef. Drying to preserve the product can be accomplished by freeze dehydration or conventional high temperature drying. In fermented and dry-cured meat products, drying occurs simultaneously with curing, fermentation and aging.

Dehydration and aging may begin during the fermentation process and continue for varying periods under controlled temperature and humidity conditions. Several objectives are achieved during dehydration and aging, including:

- flavor development
- textural changes
- completion of the various curing reactions
- drying and hardening of the product.

For best product safety, it is essential for curing and dehydration to occur initially at refrigerator temperatures until pH or water activity drop to a point that spores will not germinate. Subsequent dehydration and aging may occur at higher temperatures.

The aging period also is necessary for proper cure development. This is especially true when the curing mixture contains only nitrate, as in some country-cured products, because time must be allowed for growth of nitrate-reducing bacteria, and for conversion of nitrate to nitrite. Bacterial starter cultures used in such products should be selected for their ability to reduce nitrate to nitrite. Two types of textural changes may occur. Tenderization may occur due to the action of autolytic enzymes in muscle (Chapter 5). More commonly, products become firmer or harder due to loss of moisture during drying. Varying amounts of moisture loss are desired, depending on product characteristics. However, in all cases, the rate of drying is closely controlled. Too rapid a drying rate, especially initially, results in loss of moisture primarily from the surface, and development of a hard exterior, called **case hardening**, that retards or prevents proper drying of the interior. Theoretically, the drying rate at the surface should be only slightly greater than that required to remove moisture that migrates to the surface from the interior of the product. Conversely, if the drying rate is too slow, surfaces will be moist enough to support excessive mold, bacterial, or yeast growth.

Aging can either follow or precede the smoking process, depending on the particular product. The length of the aging period for dry or fermented products varies from several hours to several months, depending largely on amount of moisture loss desired. Country-cured hams are cured, salt equalized, and aged for a total of 2.5 to 9 months, during which time they shrink 18-26 percent due to moisture loss. Genoa salami is aged for an average of 90 days, and will sustain an average moisture loss of 25 percent during that time. Summer sausage may be fermented in a period of about 18 hours and experiences a weight loss of about 15 percent during that time.

9

MICROBIOLOGY OF MEAT

OBJECTIVES: *Discuss the principles of microbiology and the sources and types of microorganisms that determine meat safety, spoilage, and shelf life; describe symptoms and conditions contributing to food borne infections and intoxications; discuss measures for prevention and control of microbial contamination.*

Key Terms

- Intervention treatments
- Probiotic
- Spoilage bacteria
- Pathogenic bacteria
- Hazard Analysis Critical Control Point
- Total microbial load
- Viruses
- Fungi
- Molds
- Yeasts
- Parasites
- Lag phase
- Logarithmic growth phase
- Stationary phase
- Death phase
- Psychrophiles
- Mesophiles
- Thermophiles
- Generation interval
- Water activity
- Aerobic
- Anaerobic
- Facultative
- Microaerophilic
- Oxidation-reduction potential
- Lactic acid bacteria
- Bacterioins
- Food borne infection
- Food borne intoxication
- Mycotoxins
- Botulinal toxin
- Enterotoxin
- Endotoxin
- Hemolytic uremic syndrome
- Hemorrhagic colitis
- Bovine spongiform encephalopathy
- Scrapie
- Chronic wasting disease
- Creutzfeldt-Jakob disease
- Kuru
- Prion
- Growth medium
- Countable plates
- Total plate count
- Enzyme-linked immunosorbent assays
- Polymerase chain reaction
- Pulsed field gel electrophoresis
- Bioluminescence
- Sanitation standard operating procedures
- Carcass intervention
- Steam Pasteurization
- Organic acid wash

The muscle of healthy animals is considered to be free of bacteria or other contaminating microorganisms. Natural barriers such as the animal's skin and the walls of the gastrointestinal and respiratory tracts prevent bacterial access while the functioning immune system destroys those microbes that may penetrate the barriers. Contamination of the carcass with bacteria from the animal and its surroundings begins during the slaughter process. Later on, bacteria on the carcass surface are transferred to cuts and products during fabrication and processing. Thus, control measures must be applied at slaughter and continued through fabrication, distribution, and preparation of products for consumption. Common control measures, among others, include sanitary operating procedures that minimize contamination and carcass **intervention treatments** such as lactic acid spraying and low temperature storage to minimize bacterial growth.

Efforts to control bacterial contamination begin on the farm or in the feed lot or rearing barns where vaccines or dietary supplementation with beneficial **probiotic** microorganisms may control specific pathogens or improvements in housing conditions may be used to reduce the presence of manure on the animal's feathers or skin. Withdrawal of feed several hours before slaughter is one practice commonly used to lower the contents in the gastrointestinal tract and thus reduce the likelihood of contamination during evisceration. However, feed withdrawal can lead to increased shedding of certain pathogens in the feces.

Concerns resulting from bacterial contamination of muscle foods include both shelf life and safety. Most of the bacteria that populate the surfaces of carcasses and fresh meat cuts are **spoilage bacteria** that may cause objectionable odors, flavors, and discoloration, leading to disposal of the product. However, spoilage bacteria do not cause disease or illness in people consuming meat products. Meat products may also be populated by pathogenic bacteria that can cause illness in people consuming those products. In fresh meat products, **pathogenic bacteria** are typically greatly outnumbered by spoilage bacteria. However, certain types of pathogens can cause serious illness when only a few of the bacteria are present. Since they are microscopic in size, the number and type of bacteria populating a meat product cannot be easily observed. Bacteria on meat products are detected and enumerated using various procedures, most of which require many hours or days before results are available. Effective sanitation and refrigeration reduce total bacterial numbers and delay spoilage but cannot assure the absence of pathogens.

Heightened interest in bacterial control measures has resulted from several highly publicized food-borne illness outbreaks associated with meat products. The USDA's implementation of **Hazard Analysis Critical Control Point** (HACCP) as the basis for meat plant inspection along with specific testing programs for the pathogens *Escherichia coli* O157:H7, *Salmonella*, and *Listeria monocytogenes* is forcing packers and processors to focus their efforts on pathogens. Since control measures affect all types of bacteria, both pathogenic and spoilage bacteria are reduced when sanitation and operating procedures are improved. Economic benefits accrue in extended shelf life as well as increased food safety.

Thermal destruction of bacteria during cooking is an effective way to assure that pathogens do not infect consumers. Ready-to-eat products such as luncheon meats and hot dogs are cooked during manufacture, then packaged and distributed to consumers. Meat processors take great care to assure that ready-to-eat products are properly cooked and that they are not re-contaminated after cooking. Since these products may be consumed without further heating it is imperative that pathogens be excluded. For most fresh meat products cooking is done at the point of consumption, either in the home or restaurant. Consumer and food handler education is required to assure adequate cooking and avoid cross contamination during preparation.

MICROBIOLOGICAL PRINCIPLES

Sources of Microbial Contamination

Initial microbial contamination of meat results from the introduction of microorganisms into the vascular system by captive bolt stunning or the use of non-sterile knives for exsanguination (that is, sticking and bleeding). Because blood continues to circulate for a short time, microorganisms may be disseminated throughout the animal body. Subsequent contamination occurs by transfer of microorganisms to meat surfaces in almost every operation performed during slaughtering, cutting, processing, storage, and distribution of meat. Carcasses may be contaminated by contact with hides, feet, manure, and viscera (especially the esophagus, rectum, and other openings resulting from punctures) during slaughter. For beef carcasses, the hide has been identified as the major source of bacteria transferred to the carcass. Microbial contamination of meat may also occur when equipment, clothing, hands of personnel, and various surfaces in the processing plant such as doors, walls, floors, and drains become contaminated with bacteria that are then transferred to the meat. Proper sanitation and good operating procedures limit but do not completely eliminate microbial contamination. Thus, **total microbial load** (total amount of microbial contamination) and presence of specific pathogens are important factors in determining shelf life and safety of all meat products.

Microorganisms in Meat

The microorganisms found on or in meat and poultry may consist of **viruses**, **fungi (molds and yeasts)**, **parasites**, and **bacteria**. Viruses are very small organisms that do not usually contribute to meat spoilage but may use meat as a carrier to move from infected workers to other meat plant workers or consumers. Avian or swine flu viruses sometimes infect humans, though transfer is more common for workers handling live animals than for those handling carcasses or meat. Food is not usually a vector for the flu virus. Molds are multi-cellular organisms characterized by their mycelial (filamentous) morphology. They display a variety of colors and are generally recognized by their mildewy or fuzzy, cottonlike appearance. Molds develop numerous tiny spores that are spread by air currents and other means. These spores produce new mold growths if they alight at locations where conditions are favorable for germination. Yeasts are generally unicellular and can be differentiated from common bacteria by their larger cell sizes, unique morphology, and production of buds during division. Yeasts, as mold spores, can be spread through the air or by other means, and will contaminate meat and equipment surfaces wherever they settle. Parasites may be distinguished from other biological contaminants in that they are often multi-cellular and always spend a part of their life cycle outside the primary host. Parasites mature and reproduce in their primary host, often causing temporary illness symptoms. They may remain dormant in the host for an extended time. However, in order to complete their life cycle they must enter a new host following shedding into the environment (usually in feces) or via consumption of infected tissue by the new host. Parasites are easily destroyed by heating or other antimicrobial agents. Bacteria are unicellular and vary in morphology from elongated or short rods to spherical or spiral forms. Some bacteria exist in clusters; in others the rods or spheroids are linked together to form chains. Pigments are produced by some bacteria, whose colors range from shades of yellow to brown or black. Pigmented bacteria having intermediate colors, such as orange, red, pink, blue, green, and purple, also are found. Some bacteria also produce spores which can allow the organism to survive severe conditions such as the cooking temperatures employed in making ready-to-eat meat products. Certain spores are also extremely resistant to chemicals, desiccation, and other agents. When the offending condition is removed the spore may germinate and develop into a viable bacterial cell.

Increases in numbers of microorganisms occur in phases, as shown in the growth curve of Figure 9.1. When favorable conditions exist, cells

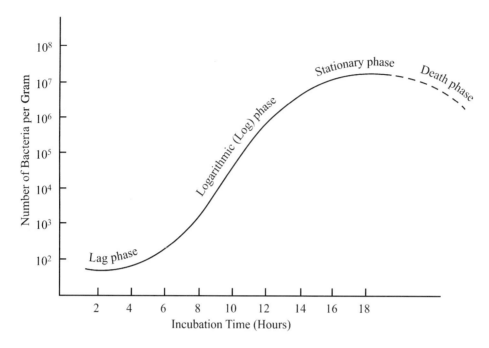

FIGURE 9.1 | A growth curve for a pure culture of bacteria under ideal growth conditions.

synthesize enzymes necessary for reproduction but do not increase in number during the **lag phase**. This is followed by a marked increase in number of microorganisms in the **logarithmic growth** (or exponential) phase, and it continues until some environmental factor becomes limiting. Rate of growth then slows and reaches an equilibrium point, where cell replication and cell death are in balance. In this **stationary phase**, nutrients continue to be depleted and by-products accumulate so that the rate of cell death eventually exceeds the replication rate. At this time the population is said to be in the **death phase**. For practical purposes, evidence of spoilage (slime formation and off odor) becomes apparent during the logrithmic phase of growth. Thus, control measures such as refrigeration are applied in order to extend the lag phase.

Type and number of microorganisms present are important factors contributing to the rate of meat spoilage. Even though meat always contains populations of bacteria, molds, and yeasts, a number of properties inherent to specific products and their surrounding environments markedly affect the kind, rate, and degree of spoilage. Organisms that eventually grow enough to cause spoilage are those that find existing conditions most favorable. Even though many species of microorganisms may originally be present, usually only one, and seldom more than three, multiply rapidly enough to cause spoilage. Predominant microorganisms causing spoilage originally may have been present in lower numbers than many others that did not grow because of selective growth conditions.

Factors Affecting Microbial Activity in Meat

Microbial survival and growth in meat products depend on a variety of factors in the meat and the surrounding environment. Some factors that affect microbial growth in meat are moisture content, pH, oxidation-reduction potential, nutrients available, and presence or absence of inhibitory substances. In addition, factors such as temperature, presence or absence of oxygen, and physical form of meat (carcass, wholesale or retail cuts, or comminuted forms) affect growth rates of microorganisms. However, the factors having greatest influence on growth of microorganisms in fresh meat are storage temperature, moisture, and oxygen.

Temperature

Each microorganism has an optimum (as well as a minimum and maximum) temperature for growth. Consequently, the temperature at which meat is stored markedly influences the kind, rate, and extent of microbial growth. A temperature change of only a few degrees may favor the growth of entirely different organisms and result in different types of spoilage. These characteristics provide the basis for use of temperature as a means of controlling microbial activity.

The optimum temperature for growth of most microorganisms is 15° to 40° C. However, some organisms grow well at refrigeration temperatures; some even grow well at subzero temperatures, and others grow at temperatures exceeding 100° C. Microorganisms that have their optimal growth at temperatures lower than 20° C are called **psychrophiles**. Those that have optimal growth at temperatures higher than 45° C are called **thermophiles**. Microorganisms with optimal growth between the psychrophiles and thermophiles are called **mesophiles**. Bacteria, molds, and yeasts each have some genera that are thermophiles, mesophiles, and psychrophiles. Thus, at normal refrigeration temperatures of -1° C to 3° C, meat spoilage may involve bacteria, molds and yeasts. However, for fresh meat products with plentiful moisture, psychrophilic bacteria tend to be the predominant contributors to spoilage due to their shorter generation time versus molds and yeasts.

The effects of temperature on **generation interval** (the time required for one bacterial cell to become two) of psychrophilic bacteria in hamburger are shown in Figure 9.2. Marked increases in generation interval below 5° C emphasize the extreme importance of continually maintaining proper refrigeration temperatures during meat storage. Likewise, holding meat at ambient temperatures during transit and in the home, even for a few hours, may shorten the lag phase or decrease the generation interval in the log phase. Effects of

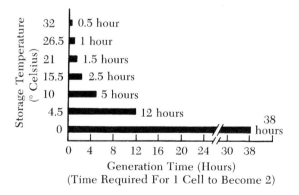

FIGURE 9.2 | The generation intervals for one species of psychrophilic bacterium at different storage temperatures. Adapted from E.A. Zottola, *Introduction to Meat Microbiology*, American Meat Institute.

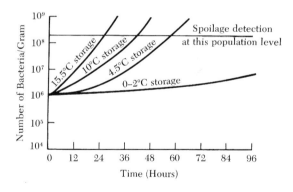

FIGURE 9.3 | The time required at different storage temperatures for the spoilage of hamburger initially contaminated with 1 million psychrophilic bacteria/gram. The spoilage was detected by odor and slime formation at a population level of 300 million bacteria/gram. From E. H. Zottula, *Introduction to Meat Microbiology*, American Meat Institute.

temperature on shelf life or storage life of hamburger are illustrated in Figure 9.3. Freshly ground hamburger ordinarily contains about 1 million bacteria/gram, but the number required to cause abnormal odor is approximately 10 million/gram and slime development is observed at about 300 million/gram. From Figure 9.3 it is evident that storage life of hamburger containing 1 million bacteria/gram is only approximately 24 hours at 15.5° C, at which time bacterial numbers reach 300 million/gram and detectable spoilage has developed. However, at normal refrigerated storage temperatures, about -1° C to 3° C, the storage life exceeds 96 hours.

As temperatures approach 0° C, fewer microorganisms can grow and their proliferation is slower. Temperatures below approximately 5° C greatly retard growth of more prevalent spoilage microorganisms and also prevent growth of nearly all pathogens except *Listeria monocytogenes*. Consequently, 5° C is considered the critical temperature during meat handling and storage, and it cannot be exceeded for a significant time without substantial reduction in quality and appearance of meat.

Moisture

Water is an absolute requirement for microbial growth. The availability of water determines which types of microorganisms can grow. **Water activity** (a_w) is the term used to describe the availability of water for bacterial growth. Water activity is defined as the vapor pressure of the solution divided by the vapor pressure of the pure solvent, ($a_w = p/p_o$, where p equals the vapor pressure of the solution and p_o equals the vapor pressure of pure water). Water activity may be reduced by removal of water or by addition of solutes (ie.,salt) that make the water unavailable to microorganisms. Bacteria generally require highest a_w (>.90). Most yeasts and molds also prefer a_w greater than .90 but continue to grow at a_w of .87 or less (yeasts) and .84 or less (molds). Water for microbial growth is in excellent supply in meat animal carcasses and fresh meat products, which typically have a_w in excess of .98. Nevertheless, manipulation of a_w may be used to control microbial growth. Reduced relative humidity in carcass coolers leads to surface drying and reduced a_w inhibiting surface bacteria. Fat covered surfaces of the carcass are especially affected since limited moisture is available in the fat. The application of a water shower in the carcass cooler keeps the a_w of the surface high. Also, wide swings in relative humidity when warm air or hot carcasses enter the cooler may lead to condensation on previously dry meat surfaces. Such conditions often lead to increased microbial growth.

Oxygen

Availability of oxygen determines the type of microorganisms that are active. Some microorganisms have an absolute requirement for oxygen, whereas others grow only in the complete absence of oxygen, and still others grow either with or without oxygen. Microorganisms requiring free oxygen are called **aerobic** organisms, and those growing only in its absence are called **anaerobic** organisms. Microorganisms capable of growing with or without free oxygen are called **facultative** organisms. **Microaerophilic** is a term used to describe microorganisms growing in the presence of a reduced concentration of oxygen (less than the 21 percent typically present in the atmosphere). All molds that grow in meat are aerobic, and yeasts also grow best when aerobic conditions prevail. On the other hand, bacteria found in meat may be aerobic, anaerobic, microaerophilic, or facultative organisms. Aerobic conditions prevail in meat stored in air, but only on or near the surface because oxygen diffusion into tissues is very limited. Thus, microbial growth occurring on meat surfaces is largely aerobic with some facultative organisms, and interior portions of meat support growth of anaerobic, microaerophilic, and facultative bacteria. Some processes such as grinding lead to incorporation of oxygen into the meat while vacuum packaging involves removal of oxygen. Transient or long term changes in microbial populations may result from such procedures.

Acidity

Microorganisms are capable of growth over a range of pH values, with the optimum pH for growth generally near neutrality (pH 7.0). As a group, molds have the widest range of pH tolerance (pH 2.0 to 8.0), although their growth is generally favored by acid pH. Molds can thrive in media that are too acidic for either bacteria or yeasts. Although yeasts also can grow in acid environments, they grow best in intermediate acid (pH 4.0 to 4.5). On the other hand, bacterial growth is generally favored by near-neutral pH. Nevertheless, at normal ultimate meat pH of approximately 5.4 to 5.6 many types of bacteria are still capable of growth if other environmental factors are suitable. In meat with low pH values (5.2 or lower), microbial growth is markedly reduced from that at normal pH. On the other hand, meat with high ultimate pH (as in dark cutting beef) is generally very susceptible to microbial growth, even under excellent sanitary conditions.

Oxidation-Reduction Potential

Oxidation-reduction potential of meat indicates its relative oxidizing or reducing power. To attain optimal growth, some microorganisms require reducing conditions and others, oxidizing conditions. Aerobic microorganisms are strongly favored by high oxidation-reduction potential (oxidizing activity). Low potential (reducing activity) largely favors growth of anaerobic organisms. Facultative microorganisms are capable of growth under either condition. Microorganisms are capable of altering the oxidation-reduction potential of meat to the extent that activity of other organisms is affected. Aerobes, for instance, may decrease oxidation-reduction potential to such a low level that their own growth is inhibited while growth of anaerobes is promoted.

Following exsanguination, during conversion of muscle to meat, the oxidation-reduction potential falls. In postmortem muscle, reducing conditions generally prevail, oxygen penetration is markedly inhibited, and many reducing groups are available. Because of oxygen from the air, the oxidation-reduction potential is highest at surfaces and lowest in interior portions. Comminution of meat markedly increases its oxidation-reduction potential by incorporating more oxygen.

Other Nutrients

In addition to water and oxygen, aerobes have other nutrient requirements. Most microorganisms need external sources of nitrogen, energy, minerals, and B vitamins to support growth. They generally obtain nitrogen from amino acids and other non-protein nitrogen sources, but some use peptides and proteins. *Pseudomonas* and other organisms common on fresh meat use carbohydrate for energy until the supply becomes limited. Subsequently, they rely more heavily on amino acids for energy and release increasing amounts of sulfur containing compounds (hydrogen sulphide, methylsulphide and dimethylsulphide) that contribute to the putrid sulphury odor of spoiled meat. All microorganisms need minerals, but their requirements for vitamins and other growth factors vary. Molds (and some bacteria) can synthesize enough B vitamins to meet their needs, but other microorganisms require ready-made supplies, even though some are needed only in very small amounts. Meat has an abundance of each of these nutrients; consequently, it is an excellent medium for microbial growth.

Microbial Inhibitors

Meat does not naturally contain significant quantities of materials that inhibit microbial growth. However, such materials may be part of many meat products. Chapter 8 describes the addition or application of various non-meat ingredients that control microbial growth. Inhibitory substances may also be produced by bacteria growing on the product. Production of such substances probably provides a competitive advantage for the microorganism allowing it to establish itself in the presence of many other organisms. Production of organic acids is a common mechanism exhibited by some bacteria. **Lactic acid bacteria** ferment carbohydrate and produce lactic acid that lowers pH and limits growth of other bacteria. Hydrogen peroxide or carbon monoxide are inhibitory substances that may be produced under certain conditions. **Bacteriocins** are potent antimicrobial peptides produced by many bacteria, especially lactic acid bacteria.

Many are heat stable even at 100° C. Bacteriocins are unique in that they recognize the target cell by means of specific receptors on the cell wall.

Physical Form
Nearly all microbial contamination of meat and poultry products occurs at the surface. Since muscle is not a fluid material the microorganisms are not able to penetrate far into the surface. However, each time a cut is made new surfaces are exposed and simultaneously contaminated with microorganisms. Hence, small cuts and comminuted meat support rapid growth of microorganisms and are more susceptible to spoilage. Factors that contribute to increased microbial load of comminuted or small cuts of meat are greater exposed meat surface areas, additional processing time, and contact with more sources of contamination such as knives, saws, grinders, and choppers. In addition, contaminating microorganisms are distributed throughout meat products during comminution. Thus, both surface and interior of comminuted products must be considered when applying control measures to reduce spoilage or assure safety.

Interactions
Effects of such factors as temperature, oxygen, pH, and a_w or microbial activity are not independent. At temperatures near growth minima or maxima, microorganisms generally become more sensitive to a_w, oxygen availability, and pH. Under anaerobic conditions, for example, facultative bacteria may require higher pH, a_w, and temperature for growth than when aerobic conditions prevail. In fact, psychrophilic microorganisms are usually aerobic and generally require high a_w in order to grow. Consequently, lowering a_w or excluding oxygen from meat held at low temperatures significantly reduces rates of microbial spoilage. Usually, some microbial growth continues when any one of the factors controlling growth rate is at a limiting level. But if two or more factors become limiting, growth is drastically curtailed or prevented entirely.

Spoilage and Pathogenic Bacteria

Spoilage Bacteria
Following slaughter, the surfaces of carcasses usually have high proportions of bacteria of fecal and soil origin. While some of these may be pathogenic the overwhelming majority are non-pathogenic. These are commonly referred to as spoilage bacteria since their growth and metabolic products eventually lead to objectionable appearance and odor. As the carcass temperature declines growth of psychrophilic bacteria is favored. Eventually as few as 2-4 genera of bacteria predominate on the meat surface. *Pseudomonas, Moraxella, Psychrobacter,* and *Acinetobacter* are the predominant organisms populating the aerobic surface of refrigerated meat and poultry carcasses and cuts. Other bacteria that may be present in smaller numbers include *Aeromonas, Alteromonas, Shewanella, Micrococcus, Lactobacillus, Streptococcus, Leuconostoc, Pediococcus, Flavobacterium,* and *Proteus*. As storage conditions change the predominant species may also change. For example, vacuum packaging leads to dramatic changes with lactic acid bacteria becoming predominant. Addition of salt, especially in fermented meat products, inhibits Gram-negative spoilage bacteria such as *Pseudomonas* while the more salt tolerant lactic acid bacteria thrive and produce lactic acid. Freezing kills or damages many bacteria on meat, and the number of viable organisms continues to decrease during subsequent freezer storage. However, many bacteria survive and resume growth upon thawing.

Pathogenic Bacteria
Development of illness due to ingestion of food can result from any of several causes. These include allergies, overeating, poisoning from chemical contaminants, toxic plants or animals, bacterial toxins, infection by microorganisms, or infestation by animal parasites. Although each of these factors is recognized as a potential source of illness in humans, the subsequent discussion is confined to illnesses caused by microorganisms and the parasites *Toxoplasma gondii* and *Trichinella spiralis*.

Food borne infection occurs following ingestion of viable pathogenic (disease-producing)

CHAPTER 9 MICROBIOLOGY OF MEAT 221

TABLE 9.1 Characteristics of Some Common Food Poisonings and Infections

Illness	Causative agent	Symptoms	Average time before onset of symptoms	Food usually involved	Preventative measures
Botulism (food poisoning)	Toxins produced by *Clostridium botulinum*	Impaired swallowing, speaking, respiration, coordination, dizziness and double vision	12–48 hours	Canned low-acid foods including canned vegetables, meat, seafood, smoked and processed fish.	Proper canning, smoking, and processing procedures. Cooking to destroy toxins, proper refrigeration and sanitation. Curing wih nitrite.
Staphylococcal (food poisoning)	Enterotoxin produced by *Staphylococcus aureus*	Nausea, vomiting, abdominal cramps and diarrhea	3–6 hours	Custard and cream-filled pastries, potato salad, dairy products, cooked ham, tongue, and poultry.	Pasteurization of susceptible foods, proper refrigeration and sanitation.
Clostridium perfringens (food infection)	Toxin produced in the body by *Clostridium perfringens* (live cell intoxication)	Nausea, occasional vomiting, diarrhea and abdominal pain and flatulence	8–24 hours	Cooked meat, poultry and fish cooked and held at non-refrigerated temperatures for long periods of time.	Prompt refrigeration of cooked meat, poultry, or fish; maintain proper refrigeration and sanitation. Curing with nitrite.
Salmonellosis (food infection)	Infectin produced by ingestion of any of more than 1200 species of *Salmonella* that can grow in the human gastrointestinal tract	Nausea, vomiting, diarrhea, fever, abdominal pain; may be preceded by chills and headache	6–24 hours	Insufficiently cooked or warmed-over meat, poultry, eggs, and dairy products; these products are especially susceptible when kept unrefrigerated for a long time.	Cleanliness and sanitation of handlers and equipment; pasteurization, proper refrigeration and packaging.
Campylobacteriosis (food infection)	*Campylobacter jejuni* and *Campylobacter coli*	Headache, fever, abdominal pains and diarrea.	1–7 days	Insufficiently cooked poultry and meat products, unpasturized milk and dairy products.	Cleanliness and sanitation of handlers and equipment, pasteurization and proper cooking.

Continued...

CHAPTER 9 MICROBIOLOGY OF MEAT

Disease	Organism	Symptoms	Incubation	Source	Prevention
Listeriosis (food infection)	*Listeria monocytogenes*	Diarrhea, fever, abortion	Up to 8 weeks	Refrigerated, ready to eat foods, luncheon meat, hot dogs, milk, soft cheese.	Avoid recontamination after heat processing, effective sanitation during packaging.
Hemorrhagic colitis (food infection)	*Escherichia coli*	Fever, bloody diarrhea, hemolytic uremic syndrome, kidney failure	2–5 days	Undercooked ground beef, low-temperature fermented products (meat or cheese), contaminated water.	Cook ground beef to 71°C and avoid recontamination with raw meat juices, use only validated fermentation and aging protocols for low temperature fermented products.
Toxoplasmosis (infection)	*Toxoplasma gondii*	Often no symptoms, some flu-like symptoms, severe cases-eye redness and light sensitivity	2–28 days	Insufficiently cooked pork, lamb or venison. Exposure to cat feces.	Thorough cooking of meat (to an internal temperature of 63°C or higher); avoid cross contamination between raw and cooked meat.
Trichinosis (infection)	*Trichinella spiralis* (nematode worm) found in muscle	Nausea, vomiting, diarrhea, profuse sweating, fever, and muscle soreness	2–28 days	Insufficiently cooked pork and products containing pork or wild carnivores.	Thorough cooking of pork (to an internal temperature of 59° C or higher); freezing and storage of uncooked meat at −15°C, or lower, for a minimum of 20 days; avoid feeding raw garbage.

organisms that grow and cause illness in the host. Food infections commonly occur when raw or under-cooked foods are eaten or when ready-to-eat foods come in direct or indirect contact with raw foods. Indirect contact occurs when raw foods are handled or prepared and then ready-to-eat foods are handled or prepared using the same facilities without effective sanitation procedures between uses. Consumption of foods such as unpasteurized milk, rare hamburger or raw seafood leads to high risk of food borne infections.

Food borne intoxication is an illness caused by ingestion of microbial toxins already present in a food. Toxins causing food borne intoxication are produced by bacteria and fungi (molds and yeasts). Toxins produced by fungi are referred to collectively as **mycotoxins**. Because they are common in mold-infested grains and legumes, mycotoxins often constitute a health problem for livestock. Fresh meat, especially pork, and organ meat such as liver and kidney frequently contain animal feed derived mycotoxins at low part-per-billion concentrations. Mycotoxin production within meat is generally limited to dried or fermented meat products on which some mold growth occurs during an extended aging period. However, no evidence indicates that ingestion of meat containing low levels of mycotoxins causes disease or illness in humans. In contrast, illnesses following ingestion of foods containing bacterial toxins are well known. These bacterial toxins are relatively odorless and tasteless and are readily consumed by unsuspecting victims.

Food Borne Infections and Intoxications

Botulism

Botulism is a food borne intoxication resulting from ingestion of a toxin produced by the bacterium *Clostridium botulinum* during its growth in food. This bacterium is anaerobic, spore forming, gas forming, and found primarily in soil. Toxin produced by this organism is extremely potent and exceedingly dangerous. It affects the central nervous system of victims and in a large percentage of cases, results in death from respiratory failure.

Some characteristics of botulism, as well as those of other common food poisonings and infections are presented in Table 9.1.

Because of the presence of *Clostridium botulinum* in soil, it follows logically that it is also present in surface water. Hence, seafood is a greater potential source of botulism than are other muscle foods. Honey is another common source of botulism spores. Historically, however, improperly home-canned vegetables and fruits, with low to medium acid content, constitute the greatest potential source of botulism. Because the organism is anaerobic, canning and vacuum-packaging foods including meat, poultry, and seafood enables production of **botulinal toxin**. Fortunately, distribution of botulinum spores in meat is usually very low and great care is taken by the industry to either destroy the spores (canning) or prevent growth of C. botulinum (curing, salting, refrigeration), so botulism poisoning from canned or vacuum-packaged cured meat products is very rare.

Protecting people from possible food-borne botulism outbreaks requires proper refrigeration, thorough cooking, or other means of control. The botulinum toxin itself is relatively heat labile, but the bacterial spores are very heat resistant, and severe heat treatment is required to destroy them. Thermal processing at 85° C for 15 minutes inactivates the toxin. Destruction of spores requires much more heating. Conditions such as reduced a_w, near neutral pH, or presence of protein or fat lead to even greater spore resistance to thermal destruction. The following time/temperature combinations are needed to produce a minimum of 12 log reduction in spores required in a low acid, hermetically sealed canned food:

Temperature (° Celsius)	Time (minutes)
100	360
105	120
110	36
115	12
120	4

In addition to high and low temperatures, several other factors suppress the outgrowth of botulism spores. These include a$_w$ below 0.93, pH below 4.6, percent brine greater than 10 percent [percent brine = (percent salt/percent salt + percent water) × 100], and preservatives such as nitrite. Some synergism exists among these barriers to spore outgrowth. Canned foods showing evidence of swelling (swollen cans result from gas produced by organisms) should never be eaten. Processed, precooked-frozen, or raw-frozen foods should not be stored at room temperature for extended periods of time.

Staphylococcal Food Poisoning

Staphylococcal food poisoning is caused by ingesting the **enterotoxin** produced by *Staphylococcus aureus*. Toxin produced by this facultative, non-spore forming organism is called an enterotoxin because it causes an inflammation of stomach and intestinal linings (gastroenteritis). It also affects the central nervous system, but seldom results in death. When mortality does occur, it is usually due to added distress in people already suffering from other illnesses. *Staphlococcus aureus* organisms are widely distributed in nature, and have been isolated from many healthy individuals. Thus, handling of food by infected individuals is probably the greatest source of contamination. Foods most commonly associated with staphylococcal poisoning are cream- and custard-filled pastries, potato salad, dairy products, and cooked meats, particularly ham and poultry. Under favorable conditions, organisms multiply to extremely high numbers in foods, including meat products, without significantly changing the color, flavor, or odor of the food. Although the bacterium grows over a temperature range of approximately 7° C to 45° C and a pH of 4.0 to 9.8, growth rate and toxin production are most rapid above 20° C and in foods having little acidity. *Staphylococcus aureus* organisms are quite easily destroyed by heat (66° C for 12 minutes), but destruction of the enterotoxin requires severe heat treatment beyond the capability of any typical kitchen. Ordinary cooking temperatures do not destroy the enterotoxin.

Clostridium perfringens Food Infection

This food infection is referred to by the name of the causative bacterium, *Clostridium perfringens*. It is an anaerobic spore-former that produces a variety of toxins and abundant amounts of gas during growth. *C. perfringens* organisms, as well as their spores, are found in many foods including fresh and processed beef, lamb, pork, poultry, veal, and fish. The greatest number of organisms have been observed in cooked meat items that were allowed to cool slowly and then held for extended time before serving. Large numbers of active organisms must be ingested for this type food poisoning to occur. Symptoms include a variety of digestive disturbances.

The spores from various strains of the organism differ in their resistance to heat; 100° C temperatures kill some in minutes, but others require from 1 to 4 hours at this temperature for destruction. *Clostridium perfringens* food infection can be controlled by rapidly cooling cooked and heat-processed foods. Proper refrigeration of foods at all times, especially leftovers, and good sanitation are essential for preventing outbreaks. When foods are held on steam tables, temperatures should exceed 60° C to prevent food poisoning by this organism. When leftover foods are reheated, they require thorough heating to destroy living organisms.

Salmonellosis

Salmonellosis is a food infection resulting from ingestion of any one of numerous species of living Salmonella organisms. These organisms grow in the consumer and produce an **endotoxin** (toxin retained within the bacterial cell) that is the causative factor of illness. The usual symptoms of salmonellosis are nausea, vomiting, and diarrhea caused by irritation of intestinal walls. In healthy individuals with normal immune capability, up to one million organisms must be ingested for infection to occur. Mortality from salmonellosis is generally low, with most deaths occurring in infants, the elderly, or victims who are already debilitated from other illnesses.

Principally of intestinal origin, salmonellae are non-spore forming, facultative bacteria. They are frequently present in intestinal tracts and other tissues of meat animals and poultry without producing apparent symptoms of infection. While Salmonella may be present in animal tissues, a major source of infection results from cross-contamination of carcasses and meat during slaughter operations. Many cases of salmonellosis result from cooked or prepared foods contacting raw meat or its juices. Some hazard also exists when undercooked, rare meat is consumed, especially when bacterial resistance is enhanced by dry-heat cookery. Thermal processing conditions normally used to cook meat are sufficient to destroy most species of Salmonella, but their resistance to heat increases as water activity decreases. The organisms are so widespread that they may be considered a component of normal meat microflora. The following combinations of time and temperature are required to produce at least 7 log reduction of salmonella in beef:

Temperature (° Celsius)	Time (minutes)
53.3	121
57.2	37
60.0	12
62.2	5

Campylobacter Enterocolitis

Campylobacter jejuni and *Campylobacter coli* cause a food infection when as few as 500 organisms have been ingested. Diarrhea and severe abdominal pain are typical symptoms, but victims may experience several other types of distress. Poultry is often contaminated, but organisms are not difficult to isolate from other meat animal species as well. Carcasses and meat cuts are frequently cross-contaminated during processing and handling. These bacteria may be controlled by temperature since normal cooking temperatures are sufficient to kill them and they only multiply at temperatures of about 30° – 42° C.

Listeriosis

Listeria monocytogenes is a low temperature pathogen which thrives in cold, damp environments, especially damp floors, pooled water, or drains in meat plants. It is widely distributed and may be found in milk, meat, and poultry products. It can survive various conditions including low water activity, high salt, or low pH. While healthy adults are usually not affected, *Listeria monocytogenes* can cause a food infection when ingested by susceptible individuals including young children, elderly, immune compromised, or pregnant women. Symptoms include diarrhea, fever, and in severe cases, abortion and meningitis. *Listeria* is more heat tolerant than *Salmonella* but is normally killed during heat processing of meat products. Special care must be exercised to avoid recontamination of ready-to-eat products such as sliced luncheon meats or frankfurters.

Hemorrhagic Colitis

Escherichia coli strains are common inhabitants of the gastrointestinal tracts of birds and mammals. While most strains are not harmful, a few shiga toxin producing strains (STEC) are quite virulent. One strain, *E. coli* O157:H7, has been associated with numerous food infection outbreaks resulting from consumption of under cooked ground beef and other foods. *E. coli* O157:H7 may be carried in the gastrointestinal tract of a healthy steer or heifer. In properly refrigerated meat products the *E. coli* is not likely to grow or increase in number. However, the infectious dose for *E. coli* O157:H7 is very low, possibly less than 10 cells. Thus, even minimal contamination can lead to illness if the pathogen is not killed in a heating step before consumption. Following infection, the organisms produce shiga toxin which may attack the kidney and cause **hemolytic uremic syndrome** (HUS). The individuals most susceptible to infection are young children, the elderly, and immune compromised individuals. Symptoms include **hemorrhagic colitis**, bloody diarrhea, and in severe cases, HUS and kidney failure. In addition to *E. coli* O157:H7 other STEC's belonging to serogroups O26, O45, O103, O111, O113, O121, and O145 have emerged as important food-borne pathogens of public health significance. Several of these have been isolated from beef, pork, chicken, and game meats.

Other Bacterial Infections

Infrequently, several other bacterial infections occurring in humans cause illnesses with typical food poisoning symptoms. One common infection is that caused by the *Streptococcus faecalis* bacterium. Whether it is a food-borne pathogen is unclear, but it has been isolated from meat and dairy products. Food infections have been caused by *Yersinia enterocolitica* in individuals of all ages. It is associated with milk, but is present in a variety of foods, including meat. It grows easily at refrigeration temperatures and usually is killed by temperatures of 60° C or higher.

Bovine Spongiform Encephalopathy

An outbreak of **bovine spongiform encephalopathy** (BSE) in cattle in Great Britain in the early 1990's led to greatly elevated awareness of this and related illnesses. BSE is one of several related transmissible spongiform encephalopathies (TSE) that occur in various species. **Scrapie** (in sheep), **Chronic Wasting Disease** (in deer and elk), and **Creutzfeldt-Jakob Disease** and **Kuru** (in humans) are TSE's that cause similar symptoms. The disease attacks the central nervous system leading to various neurological symptoms. It is thought to be fatal without exception but can only be positively diagnosed by a postmortem histopathological examination. The causative agent is thought to be a **prion**—a protein particle smaller than a virus and lacking any DNA or RNA. Prion proteins are normally produced by various cells and are thought to be involved in cellular communication during embryonic development. Abnormal prions involved in TSE's may be able to transmit the disease by modifying the structure of normal prions. In general these TSE's were thought to be species specific until a BSE outbreak in Great Britain was traced to offal from scrapie infected sheep that was fed to cattle. The rendering process in place at the time was not adequate to destroy the infectious agent. More than 500,000 cattle in Great Britain were thought to be infected with BSE with many of those entering the food chain before the danger was well understood. In an effort to stop the spread of BSE in Great Britain, more than 4 million head of cattle were destroyed and a ban on feeding of beef by-products was implemented. At the same time, diagnosis of a number of cases of new variant Creutzfeldt-Jakob Disease (vCJD) in British citizens fueled speculation that the BSE infectious agent may have crossed another species barrier to infect humans. Near the peak of the BSE outbreak in Great Britain in the year 2000, the number of vCJD cases peaked at 28. Since then the number of vCJD cases has declined consistently, with only one case reported in 2008. Through 2009 a total of 164 vCJD cases occurred in Great Britain.

In the United States, through the year 2010, four cases of BSE have been diagnosed with one animal originating in Canada (where 15 BSE cases have been diagnosed). The remaining three US cases diagnosed in 2004, 2006, and 2012 are listed as atypical-BSE since their etiology does not fit the pattern of common BSE found in Great Britain. The very low incidence of BSE in US cattle is attributed in part to a rendering process that is much more effective than the procedure previously used in Great Britain. In addition, FDA regulations now preclude the feeding of rendered offal from sheep and cattle to cattle. Other precautions limiting consumer risk include removal of "Specified Risk Materials" (brain, trigeminal ganglia, spinal cord, and dorsal root ganglia) from carcasses of cattle over 30 months of age and elimination from the food supply of all beef from non-ambulatory cattle.

Infection from Parasites

The list of possible food-borne infections includes several caused by parasites originally present in animals and transmitted to humans following the ingestion of improperly cooked meat. Among these are Toxoplasmosis, a very commonly occurring infection, and Trichinosis, a very uncommon infection but one which continues to have great impact on the pork industry.

Toxoplasmosis is considered by the US Centers for Disease Control to be the third leading cause of death attributed to food-borne illness in the US. More than 60 million men, women, and children in the US carry the *Toxoplasma gondii* parasite, though very few have symptoms because the immune system usually prevents the parasite

CHAPTER 9 MICROBIOLOGY OF MEAT 227

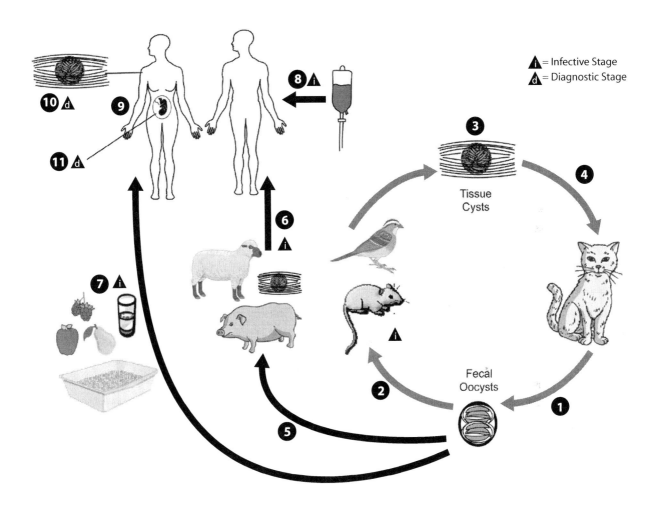

FIGURE 9.4 | The life cycle of Toxoplasma gondii. Source: http://www.dpd.cdc.gov/dpdx

from causing illness. *Toxoplasma gondii* is a protozoan parasite that is commonly acquired from raw or undercooked meats, especially pork, lamb, or venison containing tissue cysts. The primary reservoir for *Toxoplasma gondii* is actually cats (Figure 9.4). Infected cats shed huge numbers of oocysts in their feces. Intermediate hosts including birds, rodents, livestock, and many others consume the oocysts and subsequently become infected. After a few days or weeks of active infection the *Toxoplasma gondii* organisms establish cysts in muscle and nervous tissue. These cysts typically remain dormant throughout the life of the animal. Consumption of under cooked muscle tissue from infected animals starts the life cycle again.

Oocysts of *Toxoplasma gondii* in contaminated soil or water may remain infectious for 6 months or more. Contamination of fresh vegetables, soil or surface water is a common route of human infection. Tissue cysts of *Toxoplasma gondii* in meat are easily inactivated. Frozen storage of meat is generally effective for inactivating tissue cysts. They are

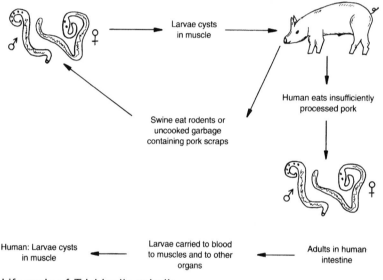

FIGURE 9.5 | Life cycle of *Trichinella spiralis*.

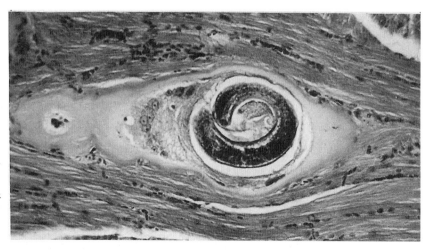

FIGURE 9.6 | A photomicrograph showing an encysted *Trichinella spiralis* larva in skeletal muscle (x 450).

also destroyed by normal cooking procedures. The US Centers for Disease Control suggests heating to 63° C to inactivate tissue cysts. Lower temperatures may be sufficient if heating time is extended.

In healthy adults toxoplasmosis symptoms are usually mild or nonexistent. If symptoms are observed, "flu" is likely the diagnosis. Following the primary infection a heightened immune response generally assures that subsequent exposures cause no illness. For people with weakened immune systems toxoplasmosis may be quite serious or even fatal. Preventive measures include proper cooking of meat products and good sanitation practices in the kitchen to avoid cross contamination from raw to cooked foods. Care should be taken to avoid possible fecal contamination from cats such as indoor litter boxes, outdoor sand boxes, or vegetable gardens.

Trichinosis (Trichinellosis), another parasitic infection, is caused by consuming the dormant *Trichinella spiralis* organisms found in insufficiently cooked meat from infected animals, especially pork (Figure 9.5). These nematodes, small worm-like organisms, are present as encysted larvae in muscles of infected animals. Pork is the greatest single source of human infection, but wild game including bear, boar, cougar, and walrus have also been implicated along with horse meat. Ingested dormant larvae that were not killed during cooking are released within the host when digestive enzymes hydrolyze cyst walls. Liberated larvae invade the upper portion of intestinal walls, where they mature within 5 to 7 days. Mature nematodes reproduce in intestines, and newly hatched larvae migrate into the circulatory system where they are carried to muscles of the new host. Larvae then complete their life cycle by becoming encysted in muscles of the host, as shown in Figure 9.6. Symptoms of infection occur as a result of irritation of intestinal linings caused by immature larvae burrowing through intestinal walls. These symptoms are followed by fever and generalized weakness that accompanies migration of larvae into muscles. Muscle pain usually occurs during encystment.

Trichinosis can be most easily prevented by adequately cooking infected meat. The thermal death temperature for *Trichinella spiralis* cysts in pork muscle is considered to be 58.5° C. The USDA advises a margin of safety for cooking fresh pork at home and recommends heating to 63°C internal temperature with a three minute hold time at that temperature. For commercial meat processors the USDA requires that all pork products either be heated to a safe temperature, cured with salt, or frozen and held at specific temperatures for a prescribed time to destroy Trichinella spiralis cysts. The US Food and Drug Administration and USDA permit low-dose ionizing irradiation of pork for trichina destruction (Chapter 10). Figure 9.7 shows combinations of time and temperature required to inactivate Trichinella larvae in country-cured hams. Frozen storage time required to inactivate

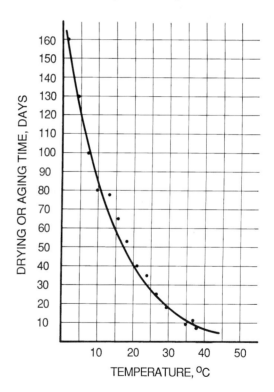

FIGURE 9.7 | Time-temperature relationship for *Trichinella spiralis* control in country hams. From *Federal Register* Vol. 50, No. 26, 1985 and modified from Christian, "Curing Georgia Hams Country Style," Univ. Georgia Cooperative Extension Service Bulletin 627, 1982.

TABLE 9.2 The Storage Time Required, at Various Freezer Temperatures, to Destroy Trichinella Larvae in Various Thicknesses of Pork

Freezer temperature (° Celsius)	Days of storage required	
	<6 inches thick	>6 inches thick
−15	20	30
−23.3	10	20
−29	6	12

Adapted from Brandley, P. J., G. Migaki, and K. E. Taylor. Meat Hygiene, 3rd ed. Lea & Febiger, Philadelphia, 1966. p. 741, Appendix Table 1.

these organisms depends on temperatures employed and thickness of specific pork products (Table 9.2). Fresh pork held at these prescribed temperatures and times, under supervision of a USDA meat inspector, may be identified as "Certified" pork. This meat may be used in meat products, such as summer sausage, that are processed to internal temperatures less than 58.5° C or in fast food products where final cooking temperature may not be well controlled.

Hogs may become infected with trichinae when they eat infected muscle tissue. Feeding of raw kitchen scraps and garbage to hogs has been largely discontinued, in part, because of concerns for trichinae infection. Hogs, especially those kept outdoors, may consume rodents or wild animals leading to trichinae infection. Thus, infected pork may occasionally find its way into the food supply. As explained in the preceding paragraph, the US regulatory approach to trichinae control involves treating all pork products by heat or other means to assure safety. Pork in the US is not regularly tested for trichinae so current incidence is not accurately known. In 1996 USDA APHIS tested more than 220,000 pork carcasses from mid-western confinement operations and found none with trichinae infection. However, with increasing interest in open or free-range pork production a low level of trichina infection will likely continue. In European countries where samples from all pork carcass are pooled and tested the incidence is very low (generally less than 0.0001%) with countries such as Denmark often finding no positive samples. In the US some pork producers work with USDA to establish and monitor a Trichina Herd Certification program. Herd certification involves audit of Good Production Practices, bio-security, feed production and storage, rodent control and general hygiene. Pork from Certified herds would not have to be subjected to trichina control procedures required of other pork.

Assessing Microbial Numbers, Growth, and Activity in Meat

Several methods are available to assess microbial growth and activity in meat. The method of choice depends on information needed, specific products in question, and nature of the microorganisms. Representative sampling of products or surfaces to be evaluated is essential for obtaining useful information. Additionally, microbiological analyses are less objective than are chemical analyses. Thus, considerable subjective judgment is required for interpretation of microbiological data. Regardless of limitations, microbiological analyses provide meat and poultry processors with much useful information. They provide indications of sanitary conditions, specific pathogens, microbiological load, spoilage problems, and expected shelf life.

Microbial Testing Methods

Whatever methods are selected, there are some common elements in microbial testing: sampling, dilution, plating, incubation, and counting. The purpose of sampling is to collect a representative sample of the bacteria present and keep them alive in liquid diluent for plating and counting. Sampling methods include swabbing, sponging, or rinsing of surfaces as well as collection of tissue

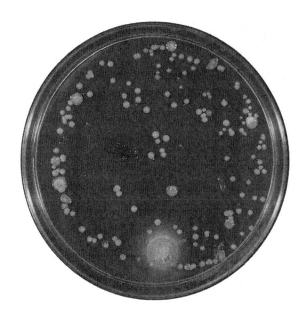

FIGURE 9.8 | A typical culture plate, showing bacterial and mold colonies.

such as ground meat. Tissue samples require further processing to disintegrate the tissue and free the bacteria into the liquid diluent.

For many materials such as fresh meat or carcass surfaces, the number of bacteria in the initial sample is too large to produce separate colonies on the plate. Thus, it is necessary to dilute the samples before plating. Sample dilution is usually achieved by transferring 1 ml of the liquid sample to 99 ml of sterile diluent. The result is a 1:100 dilution. Additional transfers may be made to further dilute the sample to the desired concentration for plating.

Plating usually involves the transfer of 1 ml of the diluted sample to a petri plate along with nutrient media that allows the bacteria to multiply. The **growth medium** is gelled due to addition of agar so that as a single bacterium multiplies it remains in place and ultimately produces a visible colony. It is very important that the sample be dispersed evenly over the whole area of the plate so that each bacterium produces a separate colony. Counting the number of colonies thus gives the number of bacteria present in the diluted sample.

The plates containing sample and nutrient media are incubated for a period of time, usually 1–2 days, at a specific constant temperature. During incubation, individual bacterial cells multiply to produce visible colonies on the plate. Since bacteria are sensitive to temperature extremes, either high or low, it is essential that the incubator chamber be capable of maintaining constant temperature. The temperature needed for incubation depends on the type of test being conducted. In general a temperature around 35–37° C results in the most rapid growth and formation of colonies. However, some bacteria, called psychrotrophs, can also grow slowly at refrigerator temperatures. Thus, incubation at a lower temperature may be used for psychrotrophic bacteria recovered from a cold area, such as a meat cooler or refrigerated meats.

Following incubation, visible colonies on the plate are counted. At this point the importance of the dilutions becomes obvious. If too many colonies are present they grow together, making it impossible to count individual colonies. It is generally recommended that **countable plates** should have between 25 and 250 colonies. A typical plate, showing various microbial colonies, is shown in Figure 9.8. A colony counter is quite helpful for this process. It shines a light up through the plate and has a magnifying glass to make colonies easier to see and count. The number of viable bacteria in the original sample may be estimated by multiplying the number of colonies by the dilution factor used in preparing the sample.

The most commonly reported microbial count is the **Total Plate Count**. For this measurement, a non-selective medium is used to support the growth of most types of bacteria. A Total Plate Count provides an indication of total populations of microorganisms. It can be used to assess microbial loads on or in meat products, in air and water, and on equipment and facilities. Enumeration of specific types of bacteria can be accomplished using similar methods but incorporating selective media that inhibit the growth of all but the desired types of microorganisms.

Rapid Microbiological Methods

Emphasis on microbiological monitoring and standards in the meat and poultry industries has increased the demand for rapid or automated bacterial testing methods. Traditional microbial methods can be accelerated or simplified using various innovations. A spiral plating system may be used to avoid the need for successive sample dilutions. Convenient plating systems are available that already contain the gelled growth medium. At the present time most methods still require a lengthy (1–2 day) incubation period in order to enumerate viable cells.

In the interest of speeding up the availability of results, indirect methods may be used to give an indication of microbial populations. *Pseudomonas*, common meat spoilage organisms, produce catalase enzyme. By assaying for microbial catalase the laboratory technician can quickly estimate the progress of meat spoilage. Triphenyltetrazolium chloride and various other dyes may be used to indicate changes in pH or redox potential resulting from increasing microbial growth. Such dyes can be used to assess total microbial population and can be useful in indicating the presence of specific types of bacteria. **Enzyme-linked immunosorbent assays** (ELISA), utilizing antibody recognition systems, have been developed for quick detection of some pathogens such as Salmonella and Listeria.

The field of molecular genetic technology is producing many rapid and precise techniques for microbial detection and enumeration. **Polymerase Chain Reaction** (PCR) is used to quickly amplify microbial DNA producing a large quantity suitable for further investigation. The resulting DNA may be hybridized to identify the microbial species or it may be subjected to **pulsed field gel electrophoresis** (PFGE) to identify the specific serotype. PFGE identification (finger printing) is widely used to connect disparate infections when a food borne outbreak occurs in multiple locations. For example, the Salmonella outbreak associated with peanut butter in 2008 and 2009 involved hundreds of infections in 43 states. Each candidate infection was tested by PFGE finger printing to determine if it was part of the nationwide outbreak.

Other rapid monitoring systems are designed to assess the overall cleanliness of food contact surfaces such as table tops, tools or equipment. ATP hygiene-monitoring systems do not give a direct indication of bacterial contamination but are sensitive to the presence of adenosine triphosphate (ATP) which is present in most biological materials including meat and bacteria. Surfaces are swabbed and ATP picked up by the swab is then analyzed using a very rapid **bioluminescence** procedure (luciferin/luciferase). A high level of ATP indicates that the surface is not clean and may be harboring bacteria.

MICROBIAL CONTAMINATION OF MEAT ANIMAL CARCASSES AND PRODUCTS

A combination of control procedures is required in order to produce meat and poultry carcasses with minimal numbers of spoilage and pathogenic bacteria. For ease of discussion, those control procedures may be grouped into prevention and correction procedures. Prevention means avoiding contamination or bacterial growth on the carcass while corrective procedures are applied in order to remove or destroy bacteria on the carcass.

Preventing Microbial Contamination

Bacterial contamination of the meat animal carcass occurs as plant workers and machinery repeatedly touch contaminated surfaces and move microorganisms onto the carcass surface. At the start of operations, following preoperational sanitation, machinery and tools are essentially sterile. However, the animal carries a huge quantity of microorganisms into the plant. The hide is the main source of contaminants that are transferred onto the carcass. Thoughtful design of operating procedures, especially for hide removal and evisceration, can greatly reduce contamination. Frequent washing and sanitizing of hands, tools, and machinery is required in order to remove bacteria before they reach the exposed surface of the carcass. United States meat plants are required to maintain written **Sanitation Standard Operating Procedures**

(SSOP) describing pre-operational cleaning and sanitizing procedures for equipment and facilities. In addition, the plant's SSOP must describe procedures used during normal operations to prevent contamination. Another required component of the SSOP is monitoring and record keeping to demonstrate adherence to the written procedures. Sanitation requirements for US Meat Inspection are discussed in Chapter 13.

Pre-operational Sanitation

Pre-operational cleaning and sanitizing procedures often take place soon after completion of manufacturing operations but must be completed before the start of the next day's operations. Cleaning procedures include an initial dry cleaning step (removal of large pieces of debris) followed by an initial rinse, detergent foam, hot water rinse, and sanitizer application. The dry cleaning step is beneficial in improving the total cleaning process while also reducing the quantity of solids in the wastewater. The initial rinse with warm water wets soil on the surfaces to be cleaned. For predominantly protein soils, the water temperature should not be so hot (>60° C) that it coagulates protein as this leads to increased adhesion to surfaces. The detergent acts to loosen grease and protein. Alkaline detergents are commonly used for high fat containing materials such as meat. Application of the detergent as a foam provides good coverage and permeates all areas of equipment and facilities. Water temperatures for foaming may be 50°–60° C. Higher temperatures improve detergent penetration into higher fat soils but may lead to the release of chlorine gas from chlorinated detergents. The final rinse is completed using hot water, 60°–70° C. After the final rinse, various sanitizers may be used, most commonly chlorine and quaternary ammonium. In some plants, no chemical sanitizer is applied but a final rinse temperature of 80° C or higher is used to kill temperature-sensitive microorganisms.

Operational Sanitation

Operational sanitation procedures usually involve personal hygiene, tool and equipment sterilization, and carefully designed operating procedures. Personal hygiene requires hair covering and clean, washable clothing that is used only in processing operations. Frequent hand washing and properly designed personal equipment (knives, steels, gloves, etc.) are absolute requirements for all personnel. Equipment sanitation must be maintained constantly. Processing must be interrupted at regular intervals (maybe after every cut) to permit thorough cleaning and sanitizing of equipment. Knife and saw dip-sterilizers with 82° C water are commonly located conveniently near workers for easy access. In addition to processing equipment, all walls, doors, rails, ceilings, light fixtures, air handlers, and other installations must be free of contaminants such as rust, grease, paint flakes, and other materials. Sanitary slaughter procedures are prescribed methods for slaughtering animals that minimize opportunities for product contamination. These include proper sequence of operations, use of approved techniques, and minimal carcass-to-carcass contact. For example, abdominal cavities may not be opened before feather, hide or hair removal is complete. To avoid spreading contamination, blood stains may be removed by washing with water only after any contaminated areas are removed by trimming. Carcasses of the same species are not permitted to contact each other until postmortem inspection is complete, and carcasses of different species are generally stored and handled in separate areas.

Even with the best procedures, avoidance or removal of all bacteria is not practical in the dressing and fabrication of fresh meat products. Thus, it is important to further protect the meat by slowing bacterial growth or destroying the bacteria present. Slowing bacterial growth is usually achieved by chilling the carcass or meat quickly to 4° C or less.

Corrective Procedures for Removing or Destroying Bacteria

Trimming is the first response when visible contamination is present on a carcass in the form of fecal material or other contaminants. While bacteria are not visible, the presence of visible stains or residue is a good indication that bacteria are present. Trimming is preferable to washing since a water spray may spread the contamination to other areas. Some frequently contaminated parts of the carcass may be routinely subjected to corrective/preventive measures even when no visible contamination is present. A steam/vacuum process is commonly used for this purpose. A specially designed device sprays saturated steam on the carcass surface while at the same time drawing a vacuum in the surrounding area.

A whole **carcass intervention** treatment is applied at the completion of the slaughter process after all visible contamination has been trimmed away. Commonly used intervention treatments include washing the carcass with 82° C water, spraying with an organic acid solution, or **steam pasteurization**. **Organic acid washes** typically utilize a 2–2.5 percent solution of lactic or acetic acid. The acid, which is sprayed on and left in place, sometimes leads to surface lean color fading. In the steam pasteurization process the carcass enters a chamber where saturated steam heats the surface quickly; a cold water shower then cools it down again. The steam pasteurization process is used in many beef processing plants and gives a 2–4 log reduction of bacterial numbers on the carcass. (One log reduction is a 90 percent reduction.)

Irradiation may be used to reduce bacterial numbers on fresh and frozen meat and poultry carcasses or products. It can reduce spoilage and pathogenic bacteria to undetectable numbers with little or no effect on flavor or appearance. Irradiation is discussed in more detail in Chapter 10.

10

DETERIORATION, PRESERVATION, AND STORAGE OF MEAT

OBJECTIVES: *Explain mechanisms involved in deterioration of meat. Describe refrigeration and freezing of meat and how these processes affect meat functional properties. Discuss thermal processing, drying, irradiation, and non-meat ingredients used for microbial control and meat preservation.*

Bacteriocidal
Bacteriostatic
Halophilic
Non-ionizing radiation
Ultraviolet radiation
Irradiation odor
Rad units
Kilogray
Cold pasteurization
Gamma radiation
Ionizing radiation
Sublimation
Case hardening
Water activity
Heat shock proteins
Specific heat

Thermal conductivity
Sterilization
Commercially sterile
Pasteurization
Bone darkening
Recrystallization
Individual quick frozen
Eutectic point
Nucleation sites
Freezing boundary
Super-cooled
Cryogenic
Retained water
Chill tunnels
Blast coolers
Drip pan
Temperature cycling
Freezer burn
Drip

Bloom
Warmed-over-flavor
Antioxidants
Free radical scavengers
Autoxidation
Oxidative rancidity
Hydrolytic rancidity
Lipase
Modified atmosphere packages
Round sour
Ham sour
Bone sour
Microaerophilic
Spoilage
Aerobic
Anaerobic
Putrefaction

Food products of all types are subject to deterioration due to microorganisms, enzymes, chemical reactions such as oxidation and physical changes such as dehydration. Preservation of food products involves a variety of processes and ingredients that prevent, delay, or slow the progress of destruction or deterioration. Refrigeration is a process widely used to preserve high moisture foods such as meats. The term "refrigeration" in this textbook refers to storage at temperatures between –2° and 5° C where microbial growth is slowed and water remains liquid. Almost all fresh meat is stored under refrigeration. Refrigeration usually begins with chilling of carcasses shortly after slaughter. It continues through subsequent storage, breaking, transit, fabrication, retail display, and storage by consumers. Most processed meat products also are handled under refrigeration temperatures from the time of final processing until consumption. In addition to simple refrigeration some fresh or processed meats are frozen at -10° C or less for extended storage. Thermal processing, dehydration, and irradiation contribute to preservation by destroying microorganisms and certain enzymes. Some chemicals used in curing, pickling, and as sausage ingredients, and some compounds present in wood smoke, provide limited preservative effects to meat products. Ultimately, effective preservation prevents or delays product spoilage so the useful life of the product (shelf life) is extended. Shelf life extension leads to improved efficiency of food utilization by increasing the opportunity for distribution and consumption prior to spoilage.

DETERIORATION OF MEAT

It is not possible to make a definitive determination of when meat is spoiled because that which one individual considers spoiled might well be considered edible or even delectable by another. The usual characterization of spoiled meat, or any food, is the point at which it becomes unfit for human consumption. In its usual connotation, spoilage is equated with decomposition and **putrefaction** resulting from microbial growth or from chemical reactions such as lipid oxidation. There is seldom any doubt about fitness of meat for consumption when it shows evidence of such spoilage. But the presence of pathogens such as *Listeria* or *Salmonella* or chemical contaminants such as drug residues may render a product unfit for consumption without decomposition or putrefaction. Nevertheless, the explanations in this section deal with meat spoilage from the perspective of deterioration of organoleptic properties.

Food products of all types are subject to deterioration due to microorganisms, enzymes, chemical reactions such as oxidation and physical changes such as dehydration. Immediately following harvest, changes are due mainly to endogenous enzymes which degrade protein, carbohydrate, and lipid into various forms. Initially, endogenous enzymes are responsible for degradation of complex molecules. But, as numbers of microorganisms and their activity increase, they contribute to, and may eventually account for nearly all, subsequent reactions. Complex molecules are degraded into simpler compounds that are used as nutrient and energy sources for microbial growth and activity. End products of microbial action vary with microbial species and available nutrients. Fresh meat normally has enough carbohydrate to temporarily support aerobic bacteria that produce CO_2 and water or anaerobic bacteria that produce lactic acid. When the carbohydrate is depleted the microbes depend more on protein for energy. Under aerobic conditions products of protein hydrolysis are simple peptides and amino acids. Under anaerobic conditions, proteins are degraded to a variety of sulfur-containing compounds, many of which are extremely odorous and generally obnoxious. End products of non-protein nitrogen compounds usually include ammonia.

Lipases (lipid-hydrolyzing enzymes) are naturally present in milk and may be secreted by microorganisms growing in other products such as meat. These enzymes hydrolyze triglycerides and phospholipids into glycerol and fatty acids, and, in the case of phospholipids, into nitrogenous bases and phosphorus as well. The resulting free fatty acids impart odors and flavors in the product. Extensive lipolysis can accelerate lipid oxidation making fatty acids more accessible to free radicals.

Meat is generally very low in carbohydrates, except when they are added during processing. However, microorganisms prefer carbohydrates to other compounds as sources of energy and therefore, readily use them when available. Carbohydrate use by microorganisms results in a variety of end products, including alcohols and organic acids. In some sausage products, microbial fermentation of added sugar leads to production of organic acids (primarily lactic acid), which contribute distinctive flavors. While this is desirable in fermented sausage, the same reaction leading to accumulation of lactic acid in vacuum packaged smoked ham or other cured meat is considered spoilage.

Deterioration Caused by Microorganisms

Growth of microorganisms in meat leads to a combination of physical and chemical changes. Meat color, appearance, odor, flavor, tenderness, and processing properties may all be affected to some degree. Meat **spoilage** may be classified as being either **aerobic** or **anaerobic**, depending on conditions under which it occurs as well as principal organisms causing spoilage: bacteria, molds, or yeasts.

Aerobic spoilage by bacteria and yeasts usually results in slime formation, undesirable odors and flavors, and color changes. Specific bacteria or yeast species causing slime formation vary, depending on environmental conditions such as temperature and water activity. As a result of production of oxidizing compounds (peroxides) by bacteria, myoglobin and oxymyoglobin may be changed to metmyoglobin and other oxidized pigment forms, resulting in gray, brown, or green colors. Some species of bacteria cause "greening" in sausage, and pigmented bacteria and yeasts cause various other surface colorations. Hydrolysis of lipids is enhanced by lipolytic bacteria and yeasts with development of oxidative rancidity and undesirable odors and flavors.

Aerobic spoilage by molds results in sticky meat surfaces. The formation of "whiskers" by molds is commonly observed on aged meat products such as dry-aged beef or Genoa salami. Because of colors associated with growth of specific mold colonies, various surface colorations, such as creamy, black, or green, are seen. As in bacterial and yeast spoilage, that caused by molds also results in lipolysis and enhances oxidative rancidity as well as the production of musty odors and flavors.

Aerobic spoilage is essentially limited to meat surfaces, where oxygen is readily available. Therefore, affected areas can be trimmed off and meat that remains generally is acceptable for consumption. The unique flavor developed in dry aged meat is largely due to microbial enzyme action at or near the product surface. When surface molds are trimmed off of aged meat, lean tissues below may have little or no microbial growth. However, if extensive bacterial growth occurs on surfaces, organisms may penetrate deeply into tissues, particularly along bones and connective tissue septa.

Anaerobic and **microaerophilic** spoilage occurs in vacuum packaged products, in sealed containers or sometimes deep inside whole muscle cuts where oxygen is either absent or present in limited quantities. This type of spoilage is caused by facultative and anaerobic bacteria and usually is described by the terms taint, souring, and putrefaction. Taint is a nonspecific term used to describe off odors and flavors. Souring (development of a sour odor or flavor), results mainly from accumulation of organic acids during bacterial enzymatic degradation of complex molecules. Occasionally, souring is accompanied by production of various gases. **Round sour**, **ham sour**, and **bone sour** are terms commonly used to describe sour or putrid odors occasionally found in tissues surrounding bones, particularly in rounds or hams. These sours, or taints, are caused by anaerobic bacteria that originally may have been present in lymph nodes or bone joints, or may have gained entrance along bones during storage and processing. If chill room temperatures are inadequate for rapid dissipation of body heat, or if temperatures are improperly maintained during storage and processing, growth of these microorganisms occurs and souring may develop.

Vacuum-packaged, fresh meat cuts sometimes develop off odors at meat surfaces resembling those of bone souring. The extent of this problem is greater in meats having unusually high pH, for example dark cutting beef, because low acidity allows for faster microbial growth. Such products may also exhibit putrid or sulfury odors since low carbohydrate leads to increased protein utilization by bacteria. Hydrogen sulfide or sulfur dioxide-producing bacteria are generally responsible for these odor problems.

Modified atmosphere packages (MAP) with head space (gas space above the product) exhibit some unique deterioration problems. The head space is usually filled with nitrogen and carbon dioxide or, in the case of fresh meat, oxygen and carbon dioxide. However, volatile compounds from the product also accumulate in the head space and, though the concentration may be low, the consumer gets full exposure when the package is opened. Fresh meat in MAP, especially packages containing a small amount of carbon monoxide to stabilize color, may develop offensive odors in the head space gas before the meat shows obvious evidence of spoilage. The product looks fine but upon opening the consumer detects the offensive aroma. In spite of the initial odor (which dissipates quickly) the product may look and taste just fine. Another problem associated with fresh meat in MAP results when oxygen in the package is used in chemical reactions within the meat. If oxygen is the predominant head space gas (ie., 80 percent oxygen) the package will collapse as that oxygen becomes depleted. Lipid oxidation is one common reaction that uses oxygen so a collapsed package may contain meat with oxidized lipids and off flavors.

Deteriorative Changes Caused by Insects

Under sanitary meat handling conditions, insects seldom constitute a problem in meat deterioration because sanitation and insect control measures are sufficiently stringent to prevent infestations. Insect infestation not only reduces product acceptability, but constitutes a health hazard because of the disease transmission potential of insects.

Insect control is a particularly important requirement in processing country-cured hams and similar dried and aged products. During aging and storage these products are held at relatively high temperatures (21°–38° C), generally for many weeks or months. Thus, if aging and storage rooms are not constructed to prevent entry of insects, or if products are not protected by some other means, infestation by one or more species of insects may result. Among commonly found insects in these products are skippers, or leaping larvae that feed on muscle and other soft tissues. Other insects found on or in country-cured pork products include larder beetles, cheese or ham mites, and blowflies. In addition to structural damage, infestations lead to discoloration and weight loss. In the US, methyl bromide has been approved for use as a fumigant for insect control in certain meat drying and aging facilities.

Lipolysis and Lipid Oxidation

Lipids in foods are susceptible to degradation by enzymatic (lipase) and oxidative (autoxidation) pathways. Lipase releases free fatty acids from triglycerides by hydrolysis of the ester bond attaching the fatty acid to the glycerol backbone. On the other hand, lipid oxidation involves nonenzymatic degradation of unsaturated fatty acids at the double bond. This reaction is catalyzed by metal ions such as iron or copper among others. This type of lipid degradation requires molecular oxygen and produces free radicals which trigger oxidation of other double bonds leading to a cascade known as autoxidation. These pathways lead to off flavors associated with free fatty acids (**hydrolytic rancidity**) or fatty acid fragments (**oxidative rancidity**) in the product. This is usually undesirable but, in certain aged meat or cheese products the rancid flavor and aroma notes are desired.

For the most part, efforts focus on limiting the occurrence of lipid oxidation and rancid flavor in meat products. The characteristic flavor and odor of oxidized fat is caused by the presence of low molecular weight aldehydes, acids, and ketones that form during oxidation and decomposition of unsaturated fatty acids. Polyunsaturated fatty acids including those of phospholipids are much more susceptible to autoxidation than are mono-unsaturated or saturated fatty acids. The fatty acids associated with phospholipids are more polyunsaturated than fatty acids in adipose tissues. Thus, polyunsaturated fatty acids are focal points for understanding autoxidation of fats.

The **autoxidation** process includes three distinct stages: initiation, propagation, and termination. The initiation of autoxidation occurs as activated molecular oxygen attacks a fatty acid double bond, cleaving it to produce two free radicals. The initiation step is enhanced by pro-oxidants, such as metal ions (i.e., copper, iron, manganese, and cobalt) which help to activate oxygen, heat, ultraviolet light, low pH, and numerous other substances or agents.

The free radicals made during initiation quickly produce additional activated oxygen species which migrate to other unsaturated fatty acids where they each attack and cleave another double bond producing still more free radicals. This so called propagation stage continues to gain momentum with each subsequent cycle doubling the number of free radicals. Propagation continues as long as oxygen and double bonds are available to participate in the reactions.

The lipid peroxide radicals formed during lipid oxidation carry an unpaired electron that may react in a variety of ways. During propagation it commonly reacts with molecular oxygen and directly or indirectly with fatty acid double bonds. As propagation proceeds, the availability of double bonds and, possibly, oxygen concentration decline at the same time that free radical concentration increases. This leads to an increasing frequency of one free radical reacting with another to produce a non-reactive product. This represents a natural termination of the oxidation process. However, it only occurs after considerable oxidation has occurred and intense rancid flavor has developed. Termination may be achieved much earlier if oxygen is excluded or a free radical scavenger is added.

Thus, lipid oxidation may be controlled or largely prevented by several actions that slow the initiation and speed the termination steps. Eliminate light, heat and oxygen exposure, and eliminate or chelate pro-oxidant metal ions to slow initiation. Eliminate oxygen and add **free radical scavengers** to accelerate termination.

The storage life of postrigor ground products is shortened because of incorporation of oxygen during grinding. Addition of salt to processed meat products also accelerates oxidation by supplying metal ions present as contaminants. However, some ingredients added during processing possess lipid antioxidant properties and tend to retard development of rancidity (free radical scavengers such as nitrite, ascorbate, sage, majoram, rosemary, thyme and wood smoke phenolic fraction, and metal chelators such as phosphate, EDTA, and citric acid).

For products such as lard and fresh pork sausage, rancidity may be inhibited by addition of **antioxidants** during processing. Two widely used antioxidants are butylated hydroxytoluene (BHT) and butylated hydroxyanisole (BHA). Antioxidants used commercially generally contain combinations of several ingredients to take advantage of desirable properties of each. Combinations of BHT, BHA, N-propyl gallate, and citric acid are commonly used in edible fats, and in foods with moderate to high fat content.

Warmed-over-flavor (WOF) is an off flavor due to lipid oxidation which may develop in cooked meat during storage. It is thought to result from a combination of factors during and subsequent to cooking. Heat denatures myoglobin and releases free iron and possibly other catalysts. Heat changes membrane lipid structure making unsaturated fatty acids in membrane phospholipids more accessible to oxygen and free radicals. At the completion of heating, as the product cools, surrounding air including oxygen is drawn into spaces in the meat and dissolves into the water phase of the meat. WOF becomes most apparent after cooked meat has been chilled and stored for some time (hours or days) before reheating. For some consumers WOF is quite objectionable while others consider it to be acceptable or even desirable. In cured meat products, which are cooked, chilled and stored, WOF is commonly controlled by nitrite, phosphate and vacuum packaging each of which blocks a separate aspect of the lipid oxidation reactions.

Protein Oxidation

Proteins in food are susceptible to oxidation especially in modified atmosphere environments with high oxygen concentration i.e. 70–80 percent oxygen. Initiation of oxidation of meat proteins is not well understood. Following death of the animal native antioxidant systems within muscle begin to fail. Reactive oxygen species produced by lipid oxidation may contribute to protein oxidation. However, some protein oxidation may be observed separate from lipid oxidation. Reactive radical intermediates formed during protein oxidation include amino acid derivatives. Cysteine is among the first to be oxidized along with methionine. Further reaction of these and other intermediates leads to cross-linking of proteins with loss of water binding capacity and enzymatic activity. Myosin, constituting more than 50 percent of myofibrillar protein, is very susceptible to oxidation of its cysteine, methionine and tyrosine residues. Disulfide and dityrosine cross-links lead to polymerization and aggregation of myofibrillar protein. These structural changes along with possible loss of calpain activity are thought to account for the loss of tenderness observed in fresh meat held in a high oxygen MAP environment. Control of protein oxidation may require different strategies than those used for lipid oxidation. Increased dietary vitamin E for turkeys is reported to reduce susceptibility to protein oxidation. However, antioxidants capable of inhibiting lipid oxidation (free radical chain terminators such as phenolic compounds in rosemary extract) are not effective for stopping protein oxidation. Improved understanding of protein oxidation is needed in order to better manage its effects on meat palatability and functional properties.

Chemical Contamination

Chemical residues may result from materials consumed, inspired, and absorbed by animals. Residues derived from the feed include antibiotics, hormones, and hormone-like compounds, minerals (such as selenium from feeds grown in high-selenium soils), and residual chemicals in feeds resulting from use of herbicides and fungicides for disease and insect control in crop production. Mercury residues in fish and polychlorinated biphenyls in other animals are further examples. Residues may accumulate when animals are reared near industrial environments where gaseous pollutants, such as sulfur dioxide or dioxins, contaminate the atmosphere or collect on forage. Skin absorption of chemical compounds used to control external parasites may contribute to tissue residues. Tolerance limits for chemical residues vary, depending on their specific effects in humans. In the US, the Food and Drug Administration has established pre-slaughter withdrawal times for drugs permitted for use in food-producing animals. If a drug residue is present in meat, the livestock producer may have failed to observe proper withdrawal time requirements or used a drug not permitted for use in food-producing animals.

Physical Deteriorative Changes

Dehydration

Surface discoloration may become noticeable when meat is held at relative humidity levels low enough to result in dehydration. Moisture loss from meat surfaces concentrates pigments and, due to loss of intracellular water, reduces light reflection, both of which cause meat to appear dark. Such cuts of meat are described as having lost their bloom. Frozen meat generally possesses a darker color than fresh meat, particularly after slow freezing. On thawing, meat color is improved, but it usually does not correspond to the color of its fresh counterpart.

Loss of moisture from meat surfaces during storage produces a dried, stale, coarse appearance that adversely affects eye appeal and acceptability. Unless severe dehydration has occurred, the problem is confined largely to meat surfaces. Although the trimming of dried surfaces essentially restores consumer appeal, the removal of trimmings, and the labor required, are serious economic losses.

Severe dehydration usually results in very dry products following cooking and thus affects palatability.

Freezer Burn and Drip

During frozen storage of properly packaged food, minimal **sublimation** (evaporation) of moisture from the surface occurs even after several months. However, without an effective barrier, sublimation occurs leading eventually to surface dehydration and a quality defect called **freezer burn**. Freezer burn in meat is characterized by corklike texture and gray to tan surface color. It can result when wrapping materials have been punctured, are not in contact with meat surface, or when moisture-proof wrapping is not used. Improper maintenance of temperature, with frequent up and down temperature cycles during storage, also contributes to freezer burn. During development of freezer burn moisture evaporates, proteins become denatured and do not rehydrate properly upon thawing. Meat with severe freezer burn exhibits reduced juiciness and tenderness and has off flavors from lipid oxidation.

The common practice of freezing fresh meat in its retail package is a leading contributor to occurrence of freezer burn. Fresh meat is often packaged in a low barrier film with considerable void space inside the package. After freezing the film is easily damaged or moisture may evaporate through the intact film. Additionally, the void spaces inside the package allow moisture to evaporate from the product surface then freeze as frost on the packaging film. Repeated **temperature cycling** causes further evaporation and frost formation so that the meat surface becomes dehydrated. Freeze dehydration (like freeze drying) leads to protein denaturation and the resulting porous structure exposes the interior of the meat to oxygen and eventual lipid oxidation.

In addition to freezer burn, freezing and thawing may cause damage to meat structure leading to loss of fluid or drip. This is especially true for meat that is frozen slowly. As meat freezes ice crystals form first in the extracellular spaces within the tissue. These ice crystals grow with addition of water which diffuses from the adjacent muscle cells. As the ice crystals grow they eventually contain much of the cellular water and may encroach on cell membranes causing physical damage to the membranes. When the product is thawed the ice crystals change back to liquid. Since the ice crystals were located in the extracellular space the liquid water is now in that space and is free for migrate through the lymph vessels to the cut surface of the meat. Additionally, cell membranes damaged by the large ice crystals may allow some remaining cellular fluid to escape. Fluid released during thawing collects in the package and is called **drip**. Excessive drip results in unattractive packages, losses of nutrients, and reduced juiciness of cooked meat.

Absorption of Off-Flavors

The texture, composition, and consistency of meat render it highly susceptible to the absorption of volatile materials. The water in the meat accommodates hydrophilic materials while hydrophobic materials may be absorbed by the fat. Thus, many aromatic compounds from other foods, such as apples or onions, are readily absorbed by meat tissues. Consequently, off-flavors may occur when unprotected meat is stored in the presence of such products.

REFRIGERATED STORAGE

Throughout the food industry temperature reduction is applied almost universally for food preservation during manufacture, storage, distribution, or retail display. Chilling or freezing of food may have significant impact on its functional and palatability properties. In the cooling system, air is the most common heat transfer medium though water, salt brine or glycol chillers are also used. As warm air moves across the cold surface of the refrigeration unit the air is cooled and water vapor in the air may condense on the surface. The condensed water must be collected to avoid dripping on carcasses or products. A **drip pan** is placed under the refrigeration unit for this purpose. In freezers the condensed water freezes on the refrigeration unit fins. If the ice is allowed to accumulate it will block the air flow across the unit and prevent further cooling.

The water vapor which condenses on the surface of the evaporator may be from warm moist air which entered the cooler or freezer through an open door. However, when the doors remain closed water may evaporate from exposed or improperly packaged meat. Such evaporation helps cool the meat but also represents loss of salable weight. Carcasses, meat cuts or products remaining uncovered in the cooler lose weight due to evaporation. Hot carcasses from slaughter may lose as much as 2.5 percent of their weight in the first 24 hours. After the first day the rate of loss slows considerably but continues as long as the carcass or product remains uncovered. Similar problems may occur in a freezer where water evaporates by sublimation from uncovered or improperly packaged products leading to freezer burn.

Initial Chill

At the completion of the slaughter process, internal temperatures of animal carcasses generally range between 30° and 39° C. This body heat must be removed during initial chilling, and the internal temperature of the thickest portion of the carcass should be reduced to 5° C or less. Beef, pork, lamb, veal, and calf carcasses are initially chilled in **blast coolers** or **chill tunnels** at temperatures ranging from -10° C to 0° C; poultry and fish are generally chilled by immersion in ice water. Major factors affecting chilling rates include specific heat of the carcass, carcass size, amount of external fat, and temperature of the chilling environment. Fat reduces efficiency of heat dissipation. In conventional chill coolers, beef carcasses may require 48 hours or longer to reach an internal temperature of 5° C or lower (Figure 10.1). In coolers equipped for high-velocity air movement, chilling time can be reduced by 25 to 35 percent. The relative humidity in chill coolers usually is kept high to prevent excessive carcass shrinkage from moisture evaporation. Many plants use intermittent cold water showers to keep chilled carcass weights close to hot carcass weights. However, if carcasses gain weight during chilling, as is common for poultry chilled in ice water, the fresh products from those carcasses must include a statement on the retail label reporting the **Percent Retained Water**. Other factors affecting chilling rates include number of carcasses in chill coolers and their spacing.

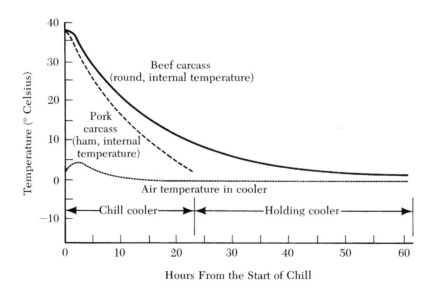

FIGURE 10.1 | Typical cooling curves for beef (270 kg) and pork (75 kg) carcasses. Temperatures represent the interior of the round and ham, respectively. Source: Retrum, R., *Beef Carcass Chilling and Holding*. Am. Soc. Refrig. Eng. Refrigeration and Meat Packing Conference (1957).

Sufficient spacing should be allowed for thorough air circulation to assure rapid heat dissipation. After approximately 6 to 24 hours of chilling, carcasses are moved to holding coolers at 0° to 3° C until fabricated or shipped to users. Pork, lamb, and veal carcasses are usually fabricated or shipped to processors or consumers after 24 hours. Beef carcasses may be held an additional 24 hours before grading and fabrication.

Immersion in ice water is a relatively rapid method of chilling. For poultry, US Department of Agriculture regulations require that internal muscle temperatures be decreased to 5° C or lower within 8 hours of slaughter or before the product is packaged. Although cold shortening may occur in chicken and turkey muscles at these low temperatures, rapid onset of rigor mortis and some elapsed time between slaughter and chilling usually eliminate measurable toughness problems. Likewise, rapid chilling of fish in ice water has no appreciable effect on its tenderness. In contrast, rapid chilling treatments may cause some toughening in beef, pork, lamb, and veal carcasses. (See Chapter 5 for discussion of rigor mortis and cold shortening.)

Duration of Refrigerated Storage

Refrigerated storage of meat is generally limited to relatively short periods, because deteriorative changes continue to occur and the rate of many of these changes accelerates with time. Major factors influencing storage life of meat under refrigeration include initial microbial load, temperature, and humidity conditions during storage, type of packaging, if any, species, and type of product. To achieve maximum storage duration while maintaining acceptable quality, all variables influencing storage life must be optimized.

Initial microbial loads have profound effects on storage life of fresh and processed meat products, but prevention of further contamination during subsequent handling, processing, packaging, and storage is essential to maintain optimum qualitative properties and prolong shelf life. Maintenance of constant temperature storage conditions of 3° C or less is essential to preservation of meat quality. Because worker comfort is important, temperatures of most cutting, fabrication, and shipping rooms exceed 5° C, so the length of time meat is exposed to these elevated temperatures should be minimized. Maintenance of proper refrigeration temperatures of 3° C or less sometimes fails during transit, especially during loading and unloading.

Effective refrigeration along with various protective coverings allow longer storage or shipping of cuts and carcasses. Fat cover, skin, or scales offer some protection against microbial contamination, dehydration, and discoloration of meat surfaces. Some lamb and veal carcasses and fresh pork loins are protected from contamination in transit by loosely wrapped paper or plastic. However, most beef, veal, and lamb carcasses are fabricated into sub-primal cuts that are vacuum packaged and boxed for storage and shipping. When stored at less than 3° C, such packages may have a shelf life of up to 45 days. Proper packaging materials for fresh retail cuts and processed meat products are important to maintain desirable color and prevent dehydration and contamination during retail display. Fresh cuts in oxygen permeable wrap may have 5 to 7 days of storage life, and vacuum packaged cured meat may be safely stored up to 90–120 days.

Meat storage duration depends on product characteristics. For example, pork, poultry, and fish possess more highly unsaturated fats than beef or lamb, are more susceptible to development of oxidative rancidity, and therefore are seldom stored for more than a few days without vacuum packaging. In addition, storage time depends on the amount of contamination carcasses receive during slaughter, which varies among species. Poultry carcasses, for example, are subjected to more microbial cross-contamination during cold water chilling than are other species that are air chilled.

The length of time that consumers may keep meat safely under refrigerated storage in the home is determined by previous handling conditions and refrigerator temperature. However, even under ideal home refrigeration conditions, fresh meat should be consumed within 4 days of purchase. Fresh meat that will not be consumed within this time should be frozen immediately. Some deterioration occurs during slow freezing in home refrigerators, but such losses in quality are preferable to bacterial spoilage and discoloration that could develop when unfrozen meat is stored in home refrigerators for extended periods.

Freezer Storage

Freezing is recognized as an excellent method for preservation of meat. It results in fewer undesirable changes in qualitative and organoleptic properties than other methods of preservation. In addition, most of the nutritive value of meat is maintained during freezing and frozen storage. Some loss in nutritive value occurs when water-soluble nutrients are lost during thawing, but the amount of drip and nutrient loss varies with freezing and thawing conditions. Nutrients found in drip include salts, proteins, peptides, amino acids, and water-soluble vitamins. None of the nutrients present in meat are destroyed or rendered indigestible by freezing. For properly packaged meat, color does not change appreciably due to freezing and thawing. If meat is frozen prior to bloom it will generally bloom when exposed to oxygen after thawing. Meat that has a bright, oxymyoglobin color before freezing will generally keep that color for some time during frozen storage. Thus, quality properties of frozen meat approximate those of fresh meat.

The length of time meat is held in refrigerated storage before freezing affects ultimate frozen meat quality. To preserve optimum quality, meat to be frozen must be handled with the same care as refrigerated meat, especially if products are going to be in freezer storage for several months. Some deterioration continues to occur in meat even at freezer storage temperatures especially at temperatures above -10° C. In addition, quality of frozen meat is influenced by freezing rate, length of freezer storage, and freezer storage conditions such as temperature, temperature variation, and packaging materials used. Changes that may occur during frozen storage are rancidity and discoloration, with the latter change being due to surface dehydration. At temperatures below about -10°C, most deterioration due to microbial and enzymatic activity is essentially curtailed. On the other hand, in meat with relatively high microbial loads or in improperly chilled meat, slow freezing rates may allow considerable microbial growth before the temperature is reduced.

Freezing Rate

Freezing rates affect physical and chemical properties of meat. They may be influenced by temperature of the freezing medium, type and movement of the freezing medium, packaging materials used, and composition of meat products to be frozen. In the latter case, tissues containing fat have lower thermal capacity than lean tissues, and therefore may freeze more rapidly. However, when fat is present at the surface it may freeze quickly then slow the transfer of heat from the interior of the product.

Rapid heat transfer is required for fast freezing and may be accomplished by extremely low temperatures (-40° C), rapid air movement, or direct contact with a freezing medium (direct immersion or spray application) at intermediate temperatures (-20° C). Air is a poor heat transfer medium, so freezing in air is a relatively slow process, even when high-velocity air movement is used. More rapid freezing rates are obtained when condensed gases such as liquid nitrogen, dry ice (solid carbon dioxide) or liquefied nitrous oxide are used to freeze meat. Because of the extremely low boiling or sublimation temperatures of these condensed gases (liquid nitrogen, -195° C; dry ice, -98° C; and liquid nitrous oxide, -78° C) freezing methods in which they are used are called **cryogenic**.

Undesirable physical and chemical effects occurring in meat during freezing are associated with one or more of the following factors:

- formation of large ice crystals in extracellular locations
- mechanical damage to cellular structures resulting from volume changes, and
- chemical damage caused by increased concentration of solutes, such as salts and sugars during ice crystal formation.

Nonvolatile solutes lower the freezing point of the fluid in meat. Thus, meat freezes at approximately −2° C to −3° C. Ice crystals formed within the meat are composed of pure water so the solute concentration increases in the remaining liquid water. The degree of tissue damage attributable to freezing is a function of ice crystal size and increased solute concentration in the remaining liquid water.

Slow Freezing

During slow freezing, the meat surface is initially **super-cooled** to below its freezing point until nucleation of ice crystals begins. Subsequently the temperature rises slightly and remains near the freezing point for an extended time (Figure 10.5). As a result, a continuous **freezing boundary** forms and proceeds slowly from exterior to interior. Extracellular water freezes more readily than intracellular water because it has a lower solute concentration. Slow freezing also favors formation of pure ice crystals and increased concentration of solutes in remaining, unfrozen solutions. Additionally, intracellular solutions may be deficient in **nucleation sites** (suspended microscopic particles) necessary for formation of many small ice crystals. These conditions favor gradual migration of water out of muscle fibers, resulting in both collection of large extracellular pools at sites of ice crystal formation, and intracellular concentration of solutes. Consequently, intracellular freezing temperatures are lowered even further. As ice crystal formation continues, remaining solute concentration increases and freezing point decreases until reaching the saturation point, called the **eutectic point** or eutectic temperature. As freezing continues below this temperature solutes begin to crystallize in parallel with ice crystal formation. The total contribution of this process to freezing damage is not completely understood, but it apparently causes chemical alterations, including losses in protein solubility and elasticity of thawed muscle tissue.

During slow freezing, long periods of crystallization time exist, producing numerous large extracellular masses of ice crystals that are easily lost as drip during thawing. Freezing curves and periods of crystallization for various freezing temperatures are shown in Figure 10.2. Mechanical damage due to volume changes is more likely to occur during slow freezing because of expansion associated with formation of large ice masses, as well as concomitant shrinkage of muscle fibers that have lost water to extracellular pools. Such muscle tissue has distorted structure in frozen form that completely obliterates normal striations (Figure 10.3).

Fast Freezing

Fast freezing is achieved when small portions such as patties are individually frozen in a blast or cryogenic freezer. During cryogenic fast freezing, meat product temperatures fall rapidly below the initial freezing point. Numerous small ice crystals with filament-like appearance are formed both intra- and extracellular at approximately the same speed. Because of rapid rates of heat transfer, small ice crystals formed have little opportunity to grow in size. Thus, fast freezing causes spontaneous formation of many small ice crystals, resulting in discontinuous freezing boundaries and very little translocation of water. Most of the water inside muscle fibers freezes intracellular, so drip losses during thawing are considerably lower than in thawing of slowly frozen meat. In

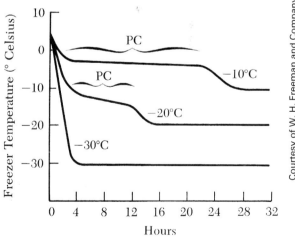

FIGURE 10.2 | Freezing curves, showing the relative rates of freezing at various freezer temperatures. (Key: PC=period of crystallization)

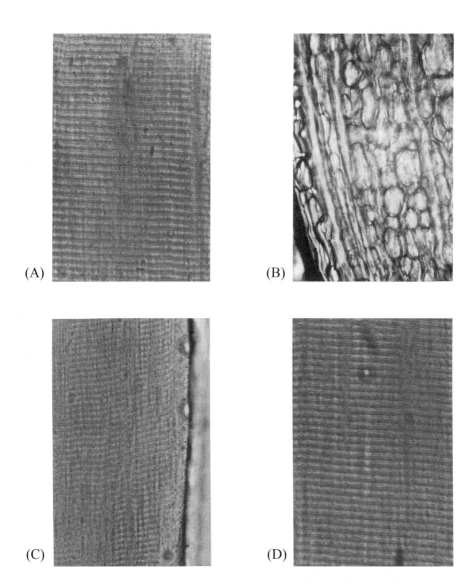

FIGURE 10.3 | The ultrastructural appearance of an unfrozen muscle fiber is shown in photograph (A). Photograph (B) shows a muscle fiber that was slowly frozen at -2.5° C. Photograph (C) shows a fiber that was rapidly frozen at -40° C. Photograph (D) shows a thawed fiber that was frozen slowly at -2.5° C (x 525). From Rapatz, G. L., and B. J. Luyet, Biodynamica, 8:295, 1959. Photographs courtesy of Dr. G. L. Rapatz.

addition, muscle fiber shrinkage and distortion are minimized during fast freezing, resulting in near normal ultrastructure in the frozen state (Figure 10.3). Volume changes are less and periods of crystallization are shorter (Figure 10.2) than in slowly frozen muscle and, consequently, mechanical damage is correspondingly less. Filament-like ice crystals entrap solutes and thus minimize the ion concentration effect. In addition, smaller and more numerous ice crystals in rapidly frozen meat reflect more light from meat surfaces, resulting in lighter color than that of slowly frozen meat.

Freezing Methods

Several commercial methods are used to freeze meat products, including:

- still air
- plate freezing
- high velocity air—blast freezing
- liquid immersion and liquid sprays
- cryogenic freezing.

The rate at which meat freezes varies with freezing method and size of the cut. For larger cuts or products ice crystal formation inside the cut may continue for many hours even in severe blast freezing conditions. Some important features of each method, as applied to freezing meat products, are discussed in the following sections.

Still Air

In still-air freezing, air is the heat transfer medium. This method is entirely dependent on natural convection and meat freezes very slowly, as shown in Figure 10.4. Home freezer units, as well as refrigerator freezers, may operate on the principle of still-air freezing. Temperatures commonly used in commercial still-air freezing range from about -10° C to -30° C. Temperatures in most home freezers also fall within this range, but their capacity to maintain these temperatures when loaded with large quantities of unfrozen meat is limited and freezing rate may be further reduced.

Plate Freezer

The heat transfer medium in this freezing method is metal, rather than air. Trays containing meat products or flat surfaces of products themselves are placed directly in contact with metal freezer plates or shelves. Plate freezer temperatures usually range from about -10° C to -30° C in commercial practice. Plate freezing is generally limited to thin pieces of meat such as steaks, chops, fillets, and patties or thin blocks of boneless meat or fish. Conduction is the heat transfer mechanism in this method and, consequently, freezing rates are slightly faster than in still air (Figure 10.4). Although plate freezing is relatively slow, it can be accelerated by circulating cold air over freezing products. Several modifications of conventional plate freezers are used in industry to accommodate batch, semicontinuous, or continuous production operations. Other modifications include systems in which moveable plates are in direct contact with both top and bottom product surfaces.

Blast Freezing

The most commonly used commercial method for freezing meat products is cold air blast freezing in rooms or tunnels equipped with fans that provide rapid air movement. Air is the heat transfer medium, but because of its rapid movement, rate of heat transfer is greatly increased over that in still air, and rate of freezing is markedly increased (Figure 10.4). High air velocity increases both cost of freezing and severity of freezer burn in unpackaged meat products. Because of the severity of this freezing process, products are often identified by one of several names: quick frozen, fast frozen,

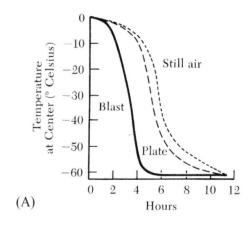

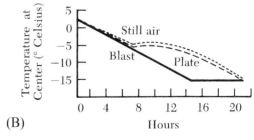

FIGURE 10.4 Freezing curves for 6-inch meat cubes at -62° C (A), and at -17° C (B). Reprinted from *Journal of Food Science* 1953, 7:505. © Institute of Food Technologists.

sharp frozen, or blast frozen. However, these terms may not be descriptive of the true freezing rate if products are large (greater than 5 pound) or contained in corrugated boxes which delay heat transfer.

In blast freezing air velocities range from about 30 to 1070 meters/minute (mpm), and temperatures range from about -10° C to -40° C in blast freezers. However, air velocity of about 760 mpm and temperature of -30° C are probably the most practical and economical conditions used in industry.

Proper product spacing and stacking on pallets or shelved racks in blast freezers is important for rapid and efficient freezing. Boxed product on pallets should be stacked so air columns can move through the stack. Anything which obstructs air flow will lead to uneven freezing. In blast tunnels, meat products are placed on moving metal mesh belts or similar conveyor systems. Conveyor speed depends on time necessary to freeze products. Sometimes meat products are frozen or partially frozen in blast freezer tunnels prior to wrapping, then packaged at a later time. Alternatively, meat products are passed through blast freezer tunnels just long enough to produce surface freezing and hardening. They are then wrapped, and the partially frozen packaged products are transferred to still-air freezers for completion of freezing and subsequent storage. This procedure may lead to quality defects due to greatly extended freezing time in the still-air freezer.

Liquid Immersion and Liquid Sprays

Liquid immersion or spray is the most widely used commercial method for freezing poultry. However, other meat and fish products also are frozen by this method. Because of rapid heat transfer, higher temperatures are generally used than in blast freezing but freezing rates are comparable.

Products to be frozen are placed in plastic bags, stacked on pallets or shelved racks, and either immersed in freezing liquid by forklift trucks, or moved through liquid by conveyors. In another application, products are conveyed through enclosed freezing cabinets while cold liquid is continuously sprayed on product surfaces. After products are removed from immersion tanks or freezing cabinets, the freezing liquid is rinsed from product surfaces with cold water. The length of time products are immersed or sprayed determines the extent of freezing. When surfaces are frozen (crusted), products are usually transferred to freezer rooms for completion of freezing and subsequent storage.

Liquids used for freezing must be nontoxic, relatively inexpensive, have low viscosity, low freezing point, and high heat conductivity. Sodium chloride brine, glycerol, and glycols such as propylene glycol are widely used. A problem associated with salt brines is corrosion of metal tanks and equipment. Another problem associated with immersion freezing is the presence of holes in protective packages. Seepage of liquid into packages necessitates washing and repackaging of products.

Cryogenic Freezing

Cryogenic freezing is very low temperature freezing as accomplished with condensed or liquefied gases. Any one of three systems may be used: direct immersion, liquid spray, or circulation of cryogenic agent vapor over products. Most commonly used cryogenic agents are nitrogen, either as liquid or vapor, and carbon dioxide, either as liquid under high pressure, solid (dry ice block or snow) or vapor. Occasionally, liquefied nitrous oxide also is used. Large pieces of meat are rarely immersed directly into liquid nitrogen because of extensive shattering or cracking of products. Therefore, systems generally evaporate liquid nitrogen in freezing chambers and use its great cooling capacity as it changes into nitrogen gas. Liquid nitrogen spray or liquid carbon dioxide spray released as snow, combined with conveyor systems, are used to rapidly freeze meat products of relatively small size, such as patties, diced meat, fish, and shellfish. Meat products frozen by these methods suffer minimal damage due to freezing and thawing.

Individual Quick Frozen (IQF) products are commonly prepared using a cryogenic freezing system. IQF products are usually individual servings such as meat patties, chicken legs or drums, nuggets, meat balls, or other consumer size pieces. The pieces are conveyed through a cryogenic tunnel as separate units then packaged individually or in bulk. Irregular shape of most IQF products leads to excess void space in bulk packages and increased likelihood of freezer burn. To reduce the problem IQF products may receive a water mist spray following freezing in order to provide a thin coat of ice. During storage the ice may undergo sublimation without damaging the frozen muscle.

Length and Conditions of Frozen Storage

Previous discussion has emphasized effects of freezing rates on retention of quality in frozen meat. However, conditions under which frozen meat is stored are also important for maintaining quality. The length of time frozen meat may be successfully stored varies with species, product, freezer temperature, temperature fluctuations, and wrapping materials.

In general, storage time of frozen meat may be extended by lowering storage temperatures. Rates of chemical deterioration are greatly reduced by freezing, but reactions such as lipid oxidation continue slowly even in the frozen state. Most chemical changes could be eliminated by reducing temperature to −80° C, but this temperature is not economically feasible in most storage facilities. Growth of pathogenic and spoilage microorganisms, and most enzymatic reactions, are greatly reduced, if not entirely curtailed, at temperatures below −10° C. In general, storage temperatures of less than −18° C are recommended for both commercial and home freezer units. Most of these units operate at temperatures between −18° C and −30° C. Although it is more expensive to maintain the lower temperatures of this range, length of meat storage life may be significantly extended.

Temperature fluctuations during frozen storage should be avoided as much as possible to minimize ice crystal growth, formation of large crystals, and associated drip losses. Nearly all water in meat is frozen at about −18° C (Figure 10.5). However, as temperature increases, liquid proportions increase, and become especially marked above −10° C. Small ice crystals are thermodynamically less stable than large crystals and may disappear completely as the product warms to −10° C. As the temperature again declines no new ice crystals are formed but remaining large ice crystals grow even larger. Ice crystal growth is enhanced by relatively high storage temperatures and by temperature fluctuations. Fluctuating storage temperatures also may cause excessive frost accumulation in air voids inside packages, much of which is lost as drip on thawing.

Acceptable quality may be maintained in frozen meat products for several months only if certain critical packaging requirements are met. These requirements include use of vapor-proof packaging material with low temperature stability to retain moisture and exclude oxygen from packages. Moisture losses due to improper packaging materials or techniques result in dehydration, freezer burn, and oxidative rancidity. Another requirement for extended frozen storage is to eliminate all void space from packages. Tight fitting bags, such as vacuum pouches, are ideal. Other requirements include use of odorless, grease-proof packaging materials that are strong when wet and

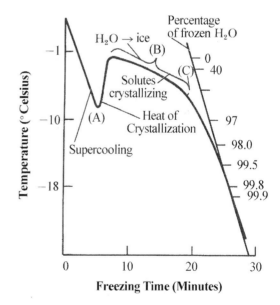

FIGURE 10.5 | Freezing curve of a thin section of beef showing the period of crystallization and percentage of frozen water. From N. Desrosier, *The Technology of Food Preservation*, p. 37, 1959, AVI Publishing Company.

TABLE 10.1 Maximum Recommended Length of Storage of Certain Meat Items at Various Temperatures for the Preservation of Optimum Quality

Item	-12°C	-18°C	-24°C	-30°C
	\multicolumn{4}{c}{Months}			
Beef	4	6	12	12
Lamb	3	6	12	12
Veal	3	4	8	10
Pork (fresh)	2	4	6	8
(cured)*	0.5	1.5	2	2
Variety meats (liver, heart, tongue)**	2	3	4	4
Poultry	2	4	8	10
Ground beef and lamb	3	6	8	10
Seasoned sausage* (pork, bulk)	0.5	2	3	4
Fish		6		12

*Salted products should be vacuum packaged before freezing.
**It is not recommended that brains and sweetbreads be frozen.

resist scuffing, tearing, and puncturing under low temperature conditions.

The amount of time meat may be held in frozen storage, while maintaining acceptable quality, depends in part on degree of saturation of meat fats. Fish, poultry, and pork fats are more unsaturated than beef and lamb fats, so they are more susceptible to oxidative changes. Hence, recommended frozen storage times differ for various species (Table 10.1). Gradual decreases in flavor and odor acceptability during frozen storage are primarily due to oxidation of lipids.

Further, length of frozen storage is influenced by the processing state of meat products—fresh, seasoned, cured, smoked, precooked, or comminuted. Salt usually contains impurities which enhance development of rancidity, and processed meat products containing salt and other ingredients have limited frozen storage life. It is generally recommended that cured and smoked products not be stored frozen or stored only for very limited time. Sliced meat products such as bacon and luncheon meat should not be frozen unless vacuum packaged because air incorporated during slicing, together with salt and curing ingredients, lead to rancid flavor development in a matter of weeks. Precooked frozen meat and poultry products lose their fresh-cooked flavor during frozen storage and develop warmed-over flavor due to lipid oxidation. These oxidative changes increase with cooking time and temperature but are minimized to some degree by treatment with antioxidants such as alkaline polyphosphates, vegetable or seed flour based sauces or gravies. Oxidation is inhibited by pH increases and metal chelation caused by alkaline polyphosphates and by natural antioxidants present in many vegetables and seeds. Frozen storage life of precooked meat products also may be extended by packaging in inert gas, such as nitrogen, in which case elimination of oxygen from packages is responsible for inhibition of oxidation. However, void space in such packages may still permit surface desiccation of the meat product.

Recommended durations of frozen storage for some fresh meats, at various storage temperatures, are shown in Table 10.1. Recommended storage times indicated are applicable only when near optimum handling, packaging, and storage conditions have been provided. If recommended maximum times are exceeded, meat products will remain edible and safe for consumption, but probably will have decreased quality of flavor, odor, and juiciness as compared to fresh products.

Thawing and Refreezing

Meat product damage during thawing may be on par with the damage during the freezing process. As with the freezing process, quality changes associated with thawing are influenced by both product and process variables. Due to low temperature differential and poor heat transfer, thawing occurs more slowly than freezing. Thawing under refrigeration at about 4° C provides a very low temperature differential compared to the 20° C differential commonly applied during freezing. During thawing, temperatures rise rapidly to the freezing point, then remain there throughout the entire course of thawing (Figure 10.6). The increased duration of thawing compared to freezing provides greater opportunity for enlargement of ice crystals (**recrystallization**), for increased microbial growth, and for chemical changes. Conversely, the extended thawing time is thought to allow more opportunity for fluid migration and re-hydration of intracellular proteins which may have become dehydrated during freezing. This could be most beneficial for products which experienced slow freezing.

Meat products may be thawed in several ways:

- in cold air
- in warm air
- in water, or
- during cooking.

Unless meat products are being cooked directly from the frozen state, they should be thawed with packaging material intact to prevent dehydration. It is generally believed that rapid thawing reduces drip, but for large cuts, exterior areas may reach excessively high temperatures. Time required for thawing frozen meat depends on a number of factors:

- temperature of meat
- thermal capacity of meat (lean products have higher thermal capacities and therefore thaw more slowly than fat products)
- size of meat products
- nature of thawing medium (water provides faster heat transfer than air)
- temperature of thawing medium, and
- movement of thawing medium.

It is generally recommended that meat be thawed at refrigerator temperatures. Admittedly, this is a slow process and might require several days for items as large as whole turkeys or hams. A slow thawing process has been reported to improve meat texture for beef and chicken as compared with cuts cooked from the frozen state. It is thought that during extended thawing greater time is allowed for diffusion of moisture back to its original location. Rapid thawing at room temperature or in warm water may also increase the opportunity for microbial growth and spoilage of the product. Thawed meat products are no more susceptible to microbial growth than fresh meat, but when surfaces reach temperatures above 0° C, microbial growth readily occurs. Thus, frozen meat products should not be

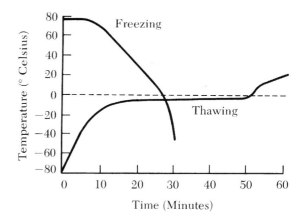

FIGURE 10.6 Graph of the average temperature at the geometric center of cylindrical specimens of a starch gel during freezing and thawing under equal temperature differentials. Can size 303 X 407 (3-$\frac{3}{16}$" diameter by 4-$\frac{7}{16}$" tall). From Fennema, D. and W. D. Powrie, *Advances in Food Research*, 13:219, 1964, Academic Press.

thawed long before cooking. In fact, cooking from frozen or partially thawed states largely eliminates the opportunity for appreciable microbial growth to occur and has no known detrimental effects on cooked meat quality. Cooking frozen ground beef patties may lead to improved palatability compared with pre-thawed patties. However, for most chops and steaks tested there was no difference in palatability for pre-thawed versus cooked from frozen cuts.

Refreezing thawed meat produces some additional loss of quality with each freeze/thaw cycle. However, refreezing may still be preferable to discarding excess thawed product if optimum procedures are followed. Temperature of defrosted products should not exceed 7° C, and microbial status of the product should not have been compromised. Products with high microbial loads should not be refrozen in an effort to restore quality because refreezing will not improve microbial status. In addition, all undesirable physical and chemical effects associated with less than optimum freezing conditions will occur again during refreezing. Therefore, repeated freezing is not recommended unless thawing times and temperatures are such that no appreciable microbial contamination or growth has been permitted to take place. Under practical conditions, meat that has been frozen and thawed several times experiences loss of flavor and juiciness.

Bone Darkening

Bone discoloration or **bone darkening** is a condition induced by freezing and thawing young chickens, particularly fryers. It rarely occurs in older chickens and is not a problem in turkeys, ducks, or geese. After freezing and thawing, muscle areas adjacent to bones exhibit an intense red appearance in the uncooked state. During cooking, the red color changes to dark gray or brown, and in severe cases, to black. The red color is due to hemoglobin that has leached from marrow of the relatively porous bones of young chickens. During cooking, the leached hemoglobin is denatured and converted to a hemichrome pigment, causing the dark discoloration. In general, discoloration occurs around leg and thigh bones, knee joints, and wing second joints, but is occasionally present in backs and breasts. Bone darkening is undesirable in appearance but does not affect flavor, odor, or texture of cooked meat. Bone darkening is greatly reduced by precooking before freezing.

THERMAL PROCESSING

Heating, as a method of preservation, is used to kill spoilage and pathogenic microorganisms in food products, and to inactivate endogenous enzymes that cause deteriorative changes. Two levels of heat processing are employed in meat preservation. Moderate heating, during which products reach temperatures of 58° C to 75° C, is employed in preparation for serving fresh meat products and in cooking most processed meat items. The process may be referred to as **pasteurization**. This heat treatment kills most vegetative cells but, does not destroy spores so products must be kept hot until serving or chilled rapidly to avoid spore germination. Shelf life of meat products is extended by pasteurization, but they must be stored under refrigeration. More extreme heating to temperatures above 121° C is used to prepare **commercially sterile** meat products that are stable at room temperature for one or more years. This process, called **sterilization**, kills vegetative cells as well as spores. Texture of meat generally changes as it is heated to temperatures above those used for pasteurization. Myofibrillar proteins become compact and hard and connective tissues begin to disintegrate. Shelf-stable canned meat products typically have strong sulfhydryl flavor due to extensive protein denaturation and hydrolysis during heat processing. In addition, the texture of these products is modified due to protein hardening and breakdown of connective tissues.

Subsequent sections of this chapter contain information on thermal processing for commercial sterilization. The effects of heating meat to moderate temperatures as required in processing are discussed in Chapter 8, and as applied in cookery are discussed in Chapter 11.

CHAPTER 10 DETERIORATION, PRESERVATION, AND STORAGE OF MEAT

Principles of Thermal Processing

Heat Transfer

Conventional methods of thermal processing involve heat transfer by conduction, convection, and/or radiation. Heating by conduction involves direct transfer of heat from particle to particle without use of a medium other than the product itself. For example, transfer of heat from the surface of a griddle to a meat patty during grilling is almost entirely by conduction. Convection heating involves heat transfer by mass movement of heated particles in fluids such as air, steam, oil or water. Radiant heating involves transfer of energy through space by means of electromagnetic radiation including infrared, radio or microwave frequencies. One or more of these heat transfer mechanisms is involved in all procedures used to thermally process meat products.

Several variables affect rates of heat transfer to products and extent of heating. Of course, rate and extent of temperature increase are directly proportional to temperature differentials between products and the heat source, and to the length of time the heat is applied. Products differ in specific heat and thermal conductivity. The **specific heat** of substances is the amount of heat energy required to change the temperature of 1 gram of product by 1° C. **Thermal conductivity** expresses the rate of heat movement by conduction through substances. A block of metal will warm more rapidly than a piece of meat of the same size because metal has both higher thermal conductivity and lower specific heat. Meat with high fat content has lower specific heat than does very lean meat, so less heat energy is required for a given temperature increase. However, rate of temperature increase in higher fat product may be slower than expected because fat has lower thermal conductivity. Product consistency and movement (as accomplished by agitation of containers) affects the amount of heat transferred by conduction versus convection. Fluid or semifluid products transfer heat energy faster than solid products, because heat transfer in the former occurs by both conduction and convection. Finally, products with larger surface area per unit weight pick up heat from the surroundings faster than those with smaller surface area.

When heating in an oven with air as the primary heat transfer medium moisture evaporates from the meat surface. This evaporation effectively cools the surface, lowering the surface temperature and slowing heat transfer to the product. As heating proceeds surface moisture in the product becomes depleted and the surface temperature rises toward the temperature of the surrounding air. However, moisture within the product migrates slowly to the surface so some evaporative cooling continues even near the end of the heating process. Products sealed in an impervious package such as a Cook-in-Bag or heated in moisture-saturated air (steam cook or pressure cooker) experience minimal evaporation. Similarly, no evaporation or evaporative cooling occurs in meat when it is boiled in water.

It is important to consider factors affecting the rate and extent of heating, as well as fundamental heating principles, when specific thermal processing schedules are developed for meat products. However, meat products vary widely in fat content, water content, consistency, and homogeneity; hence, calculation of exact processing schedules is virtually impossible. Thermal processing schedules usually are developed by trial and error for individual products. Sometimes individual plants, processing the same product, require different processing schedules due to variations in equipment performance or prevailing atmospheric conditions.

Heat Resistance of Microorganisms

Cells and spores of microorganisms differ widely in resistance to high temperatures. These differences result from both controllable and uncontrollable factors in organisms. Factors known to affect heat resistance of cells and spores must be considered when thermal processes for destruction of microorganisms are developed for, or applied to, production of meat products.

The concentration of cells or spores present in products has important effects on required heat treatment, i.e., greater numbers of cells or spores require greater heat treatment for a complete kill.

The previous history of cells or spores also affects their heat resistance. For example, bacterial cells are most susceptible to thermal processing during the logarithmic phase of their growth. Also, bacterial cells previously exposed to sub-lethal heat or other severe stress such as high salt or desiccation are more resistant to subsequent heat treatments. The initial stress triggers production of **heat shock proteins**. Heat shock proteins are common to animal, plant, insect, and microbial species. These proteins serve a protective role by associating with other biologically functional proteins in the cell and preventing them from denaturing. Immature spores are less heat resistant than mature ones. The composition of the medium in which cells or spores are heated has major effects on heat resistance. Moist heat is much more effective for killing microorganisms and spores than dry heat; thus products with low moisture content require a higher temperature for sterilization than do foods with high moisture content. In general, microorganisms are most heat resistant at or near neutral pH. An increase in either acidity or alkalinity hastens killing by heat, but changes in pH toward increased acidity are more effective than similar changes toward alkalinity.

DEHYDRATION

Drying of meat in the sun or over a fire dates back to prehistoric times. Preservative effects of dehydration are due to reduction of **water activity** (a_w) to such a level that microbial growth is slowed or eliminated, making some meat products stable without refrigeration. Other products, such as dry and semidry sausages, are partially dried in air during aging. The combination of reduced a_w and acidity developed by fermentation leads to exceptional resistance to microbial spoilage.

Dried meat products are subject to several quality problems depending on the drying and storage procedures used. High temperature drying processes cause protein denaturation which leads to poor rehydration and loss of desirable texture. Such products are often used as ingredients in soups or entrées where texture is less important. Oxidation of fat and resulting rancidity may develop during storage of dried products. Addition of antioxidants or vacuum packaging of dehydrated product may be used to control this problem. Dried products will not support microbial growth but they may become contaminated if not properly protected during storage. In addition, dried meat products may become partially rehydrated if they are exposed to humid air during storage. Protective packaging is critical to the quality and safety of dried meat products.

Dehydration Methods

Hot Air Drying

The oldest and simplest method is hot air drying. Product surface temperature is raised above the wet bulb temperature of the surrounding air and water begins to evaporate. Evaporation is rapid while free water is available at the product surface and heat is available to overcome evaporative cooling. During this time evaporation rate can be increased by increasing the rate of air movement across the product surface. As drying proceeds, water lost from the surface is replaced by water migrating out from the interior. This process is relatively slow and quickly becomes the limiting factor determining the rate of water removal. Increasing temperature or air flow rate during this stage of drying may lead to **case hardening**—a shell of very dry, hardened protein at the surface. Case hardening further limits the release of moisture to the surrounding air and inhibits subsequent rehydration. Hot air drying is commonly used for comminuted meat products with large surface area. The wet product is spread in thin layers on drying trays. Temperature, particle size, and rate of air movement must be carefully controlled to minimize case hardening and assure acceptable quality.

Freeze Drying

In conventional freeze drying, meat remains frozen throughout drying while its ice is removed by the application of heat under vacuum to transform it directly to water vapor without going through the liquid state. The transformation of substances directly from solid to vapor, without passing through an intermediate liquid state, is called **sublimation**.

Rapid sublimation of water from meat cools it sufficiently to prevent thawing and, as drying proceeds from outside inward, the low heat exchange coefficient of dehydrated outer portions prevents frozen inner portions from reaching temperatures high enough to cause thawing. Freeze drying is carried out in vacuum chambers with pressures of 1.0 to 1.5 mm of mercury and chamber temperatures as high as 43° C. Rate of freeze drying is limited by rate of heat transfer into meat for continuance of sublimation. Rates of heat transfer and drying may be increased by placing meat directly on heated shelves, or by using microwave or infrared heating. Infrared heating is most commonly used for heating in commercial facilities. The residual moisture content of freeze-dried meat products generally is below 2 percent.

Meat cooked before freeze drying generally is more stable than uncooked freeze-dried meat, and has two to four times greater shelf life. The shelf life of cooked, freeze-dried beef with vacuum packaging is approximately 24 to 28 months. Stability of freeze-dried meat depends on the method of drying, residual moisture content, packaging method, storage temperature, and quality of products prior to drying. The principal requirement for maintaining product stability is the exclusion of oxygen and moisture and necessitates use of impermeable packaging materials. Prominent physical and chemical factors affecting the stability, rehydration, and palatability of freeze-dried meat products are rancidity development, non-enzymatic browning, and protein denaturation. Most nutrients in meat, except thiamin, are relatively stable in freeze drying. Protein denaturation occurring during drying has little effect on biological value.

Freeze-dried non-comminuted meat products retain their original shape and size. Consequently, they are very porous and are much more readily rehydrated than hot air dried meat products. However, the texture and flavor of freeze-dried raw and precooked meat are greatly affected by the method of rehydration. If 1 to 2 percent sodium chloride and 0.1–0.15 percent alkaline pyrophosphate are added to the rehydration water, palatability characteristics are greatly improved. The extent of rehydration is greatest if rehydrating solutions are maintained at temperatures just below boiling. However, even thinly sliced freeze-dried meat generally does not rehydrate to original moisture content. This characteristic is responsible for reduction in flavor, texture, and juiciness. The dehydrated flavor characteristic of freeze-dried meat largely may be masked by addition of seasonings, such as those used in freeze-dried chili.

Comminuted meat may be freeze dried and stored for future use as raw material for further processed meats if it has been properly prepared for freeze drying. For maximum rehydration and subsequent performance in formulations, raw materials should be ground and cured prerigor. Such addition of salt leads to solubilization of meat proteins and improves rehydration after freeze drying. Addition of nitrite suppresses lipid oxidation and improves the stability of freeze-dried products during storage. Trials have shown that batter-type sausages manufactured from freeze-dried raw materials are equal in quality to those made from fresh materials. Of course, shelf-stable raw ingredients for sausage production offer processors great flexibility in meeting varied production requirements. However, economic feasibility of such systems has not been proven.

The most common freeze-dried meat products are dehydrated soup mixes. Because of lipid rancidity problems, meat items that are low in fat are preferred for dehydrated soups. Other dehydrated meat products include those used by the military, and a growing number of products available for campers.

IRRADIATION

Ionizing radiation has been the subject of meat preservation research since the 1950's. It is currently used to pasteurize a variety of food products including spices and flours and is widely used to sterilize surgical and personal health care items. Ionizing radiation is defined as radiation having energy sufficient to cause loss of electrons from atoms to produce ions. This includes high-speed electrons produced by a variety of electron accelerators, x-rays generated by electrons when they strike heavy metal, and electrons and **gamma**

radiation emitted from radioactive isotopes such as Cobalt-60 (radioactive cobalt) or Cesium-137 (radioactive cesium). Ionizing radiation kills microorganisms in and on meat with minimal temperature rise in the product. It is therefore sometimes referred to as **cold pasteurization**.

The amount of radiation energy absorbed by meat being irradiated is expressed in **rad units**. One million rads equal 1 megarad, and 100,000 rads equal 1 **kilogray** (kGy). To sterilize products so they are stable during subsequent unrefrigerated storage, radiation dosage of about 4.5 megarads is required. This amount of irradiation ensures destruction of vegetative cells and the most resistant spores but, imparts objectionable flavor and odor. Currently approved dosages for meat and poultry products in the US range from 1.5–3.0 kGy for poultry products and up to 7.0 kGy for red meats. These lower levels of irradiation are used to destroy most vegetative cells but not spores. Individual irradiation applications have been approved at various times leading to a maze of approved dosages for various products and processes. Irradiation is approved for pathogen destruction and shelf life extension in beef, pork, lamb, goat, and horse. Dosages of 1.5–4.5 kGy are approved for chilled products and 1.5–7.0 kGy for frozen products. An earlier approval for poultry for control of pathogenic bacteria (no mention of shelf life extension) listed 1.5–3.0 kGy. Low-dose irradiation is effective in destroying *Trichinella spiralis* parasites that may be present in pork. US Food and Drug Administration regulations allow use of up to 1 kGy for pork irradiation for the purpose of destroying *Trichinella spiralis*. The Radura symbol (Figure 10.7) is recommended by the FDA for labeling of irradiated food products.

High (sterilizing) doses of ionizing radiation (4 to 10 megarads) cause undesirable chemical and physical changes in meat products, including discoloration and the production of intense objectionable odors and flavors. The lower doses currently approved for meat and poultry products cause only minimal off odors or flavors. With properly controlled procedures, off odors and flavors may be further reduced or eliminated. Vacuum packaging to eliminate the presence of oxygen or freezing during irradiation reduce the occurrence of detectable **irradiation odor** in fresh meat. MAP packaging is generally not suitable since the head space in the package allows offensive odors to accumulate. Cooking further reduces the irradiation odor sometimes detected in raw products.

In order to reduce the likelihood of subsequent contamination it is desirable to package products before irradiation. However, not all of the packaging films currently in use for meat products have received approval for irradiation applications. Films and film components such as plasticizers or laminate binders may be affected by irradiation at doses used for food applications. Low molecular weight hydrocarbons and halogenated polymers could be evolved in the film and migrate into the product during irradiation. It is important to check the approval status of the packaging film and tray before deciding to use irradiation.

Non-ionizing radiation also can have lethal effects on microorganisms and is useful in meat preservation. **Ultraviolet radiation** can cause death when absorbed by microorganisms; thus, giving a germicidal effect. However, practical value of ultraviolet light is limited since it possesses very low penetrating power and is effective only at the surfaces of carcasses and meat products. Ultraviolet irradiation also accelerates discoloration of meat and development of rancidity

FIGURE 10.7 | The Radura symbol that is recommended for labeling irradiated food products.

in fat. Historically, ultraviolet radiation was used in the meat industry to control surface microbial growth on beef carcasses during accelerated high-temperature aging or during extended dry aging.

PRESERVATION BY CHEMICAL INGREDIENTS

A number of ingredients added to meat products during processing impart varying degrees of preservative properties. Sodium chloride's preservative action in cured meat has long been known, and probably has the longest history of usefulness of any preservative. Preservation by salt is achieved by lowering water activity and by chloride ion toxicity. However, effective preservation by salt alone requires up to 11 percent salt in finished products. This is much higher than the 1 to 2 percent salt commonly found in commercially cured meat products. While some microorganisms are inhibited by these lower salt concentrations, water activity is usually high enough to support growth of molds, yeasts, and **halophilic** (salt-loving) bacteria. Thus, salt in processed meat products provides only limited preservative effects, and other methods of preservation are necessary to prolong shelf life.

Addition of sodium or potassium nitrite to processed meat provides marked bacteriostatic and antioxidant properties. Nitrite effectively inhibits growth of several types of bacteria including pathogens, most notable of which is *Clostridium botulinum*. Nitrite in thermally processed, cured, canned meat products inhibits germination of surviving spores. Even at low concentrations, nitrite functions synergistically with salt to provide certain cured meat products, such as canned hams and canned luncheon meats, with effective preservative and **bacteriostatic** properties. In addition to its antimicrobial properties, nitrite in meat is reduced to nitric oxide which is a very effective free radical scavenger and autoxidation chain reaction terminator.

Use of sugar as a preservative agent would require levels well above those normally used in cured meat or other processed meat products. However, sugar added to fermented sausage products indirectly serves as a preservative because of lactic acid formed by microorganisms. In addition, sugar reacts with protein in the Maillard browning reaction producing reaction products which impart brown color and flavor and serve as antioxidants in the cooked meat. The antioxidant function of the browning products is most notable in gravies and sauces where conditions favor their extensive production.

A number of seasonings and spices, especially those containing essential oils such as mustard and garlic, contribute some preservative effects to processed meat products through their bacteriostatic or antioxidant actions. Rosemary extract is commonly used as a natural anitioxidant. Other spices such as clove, cinnamon or oregano may be used as antioxidants as well but their use is limited by their aromatic character.

Some aldehydes, ketones, phenols, and organic acids in wood smoke impart bacteriostatic and **bacteriocidal** effects to the surface of smoked meat products. Even though most smoked meat products have greater stability and shelf life than their fresh meat counterparts, they also depend on means other than smoke for preservation.

Addition of acids to meat products, such as pickled sausage products and pigs' feet, provides them with limited bacteriostatic properties. A decrease in pH by one unit increases bacteriostatic effects about tenfold. Acetic acid (vinegar) is most commonly used in meat preservation for pickling. Citric and lactic acids are added to certain semidry sausage products as a substitute for the much slower lactic acid fermentation.

Carbon dioxide and ozone gases have received limited use for preservation of meat. Use of these gases in holds of ships with prolonged shipping schedules selectively curtails aerobic growth, and favors proliferation of anaerobic microorganisms. However, use of these gases may have deleterious effects on meat quality. Carbon dioxide may cause discoloration in concentrations greater than 25 percent. Prolonged storage of sausages in carbon dioxide gas at concentrations greater than 50 percent causes souring to occur, because the gas dissolves in the product resulting in formation of carbonic acid. Ozone enhances development of rancidity and, at bacteriostatic levels, it causes loss of bloom in meat. Ozonation is a commonly used procedure for destruction of bacteria in water used to chill poultry carcasses or processed meat products. During recirculation in the chiller the water is subjected to ozone gas. Since the gas does not stay in the water the meat product is not exposed directly to ozone. Thus, one can take advantage of ozone's antibacterial properties without degrading product quality.

PRESERVATION BY PACKAGING

Principal preservative functions of meat packages are to provide protection against damage, physical and chemical changes, and further microbial contamination. Thus, packages are designed to maintain quality of products, but they cannot improve quality in any way. Specific requirements for packaging materials used in storage and retail display of fresh, frozen, and processed meat are described in Chapter 8.

Packaging materials for both fresh and cured meat must protect products from further microbial contamination during subsequent storage, handling, and merchandising. Packaging materials must not impart odors or flavors, but should retain natural odors and flavors inherent to products. They should have sufficient tensile strength and resistance to tearing and scuffing to withstand normal handling. Packaging materials also should be grease-proof so that meat fat will not be absorbed and reduce strength or vapor barrier properties.

Packaging requirements for frozen meat include low oxygen and moisture vapor transmission, pliability, strength, and grease resistance. Materials used for heat shrink packaging of meat products should take the shape of products and yet retain their strength and barrier characteristics.

Several materials are available for packaging meat. These include glass and metal containers, aluminum foil, paper and paper board, cellophane, films manufactured from polyethylene, polypropylene, polyesters, nylon, polystyrene, polyvinyl chloride, Surlyn™, and chemically treated rubber. Combinations of these materials, including laminates and copolymers, produce packaging materials with a wide variety of functional properties.

11

PALATABILITY AND COOKERY OF MEAT

OBJECTIVES: *Explore the palatability characteristics of meat and the factors that affect palatability. Describe the effects of heat and cookery methods on meat constituents and meat palatability characteristics.*

Key Terms

Palatability
Softness to tongue and cheek
Resistance to tooth pressure
Ease of fragmentation
Mealiness
Adhesion
Residue after chewing
Background toughness
Actomyosin (rigor) toughening
Enzyme tenderizers
Papain
Bromelin
Ficin
Mechanical treatment

Bone sour
Oxidative rancidity
Sugar-amine browning
Denaturation
Coagulation
Protein hardening
Thermal shrinkage
Gelatin
Warmed over flavor
Dry heat
Moist heat
Conduction
Convection
Radiant
Microwave radiation

People eat meat because of tradition, nutritive value, availability, wholesomeness, variety, satiety value, and social or religious custom. Of even greater consequence is the fact that meat is a delectable and nutritious food and is the central item of most meals in many countries.

Ultimately, meat value is based on its degree of acceptability to consumers. Satisfaction derived from meat consumption depends on psychological and sensory responses unique among individuals. Such factors as appearance, purchase price, aroma during cooking, cooking losses, ease of preparation for serving, edible portion, tenderness, juiciness, flavor, and perceived nutritive value govern composite reactions of individuals. Among individuals, there is wide variation in importance attributed to such factors.

PALATABILITY CHARACTERISTICS

The term **palatability** is only slightly more precise than the broad spectrum of acceptability factors listed above. Characteristics of meat that contribute to palatability are those that are agreeable to the eyes, nose, and palate. Consequently, assessing palatability, or its absence, begins with appearance of meat items.

Appearance

Consumers expect raw meat to have attractive color. With the exception of fish and some poultry parts, meat is normally some shade of red (Chapter 6). Dark color is often associated with lack of freshness, even though it usually indicates an old animal, or one that was slaughtered under stress. Such an impression reduces expectations for, or prejudices perception of, flavor when meat is consumed. With respect to fat, the most desired color is a creamy white. Yellow fat, as found in some grass-fed cattle, is less appealing to most consumers.

Color of cooked products has impact on consumer enjoyment. Brown surfaces of meat cooked with dry heat are associated with crispness and unique flavor. Golden-brown roasts or steaks readily stimulate salivary glands and thus, usually are considered desirable even before they are tasted. Interior color of many cuts also influences palatability reactions, depending on consumer preference for rare (pink), medium (light pink to gray), or well done (uniform gray) products (Figure 11.1, see color section).

Textural properties of cooked meat affect its appearance and impart sensory impressions related to adhesion, mealiness, or fragmentation. Overcooked meat may be stringy in appearance and is associated by previous experience with dryness and lack of flavor.

Ratios of muscle to bone and fat have pronounced effects on appearance and, consequently, on palatability of meat. Less pleasure is derived from consuming products that are excessively fat, because of dissection efforts required and reduction of edible portion.

Tenderness

No palatability factor has received more research study than tenderness. Many experiments have been conducted to identify constituents of muscle responsible for tenderness, yet much variation in this attribute cannot be explained. The sensation of tenderness has several components of varying importance, and perception of tenderness by humans is very difficult to duplicate by scientific instrumentation. Therefore, it is difficult to describe the experience of eating tender meat in a few simple terms.

Perception of Tenderness

Perception of tenderness is described by the following conditions within meat during mastication:

- **Softness to tongue and cheek** is the tactile sensation resulting from contact of meat with the tongue and cheeks. There is wide variation in softness of meat, ranging from mushy to woody consistency.
- **Resistance to tooth pressure** relates to the force needed to sink the teeth into meat. Some meat samples are so hard that little indention

can be made with the teeth, whereas other samples are so soft that they offer practically no resistance to such force.

- **Ease of fragmentation** is an expression of the ability of the teeth to cut across meat fibers. Rupture of the sarcolemma and myofibrils is required to accomplish fragmentation.
- **Mealiness is an exaggerated** type of fragmentation. Small particles cling to the tongue, gums, and cheeks and give the sensation of dryness. This condition apparently exists when fibers fragment too easily.
- **Adhesion** denotes the degree to which fibers are held together. Strength of connective tissues surrounding fibers and bundles influences this characteristic.
- **Residue after chewing** is detected as that connective tissue remaining after most of the sample has been masticated. Coarse strands of connective tissue of perimysial or epimysial origin are responsible for this component.

Three major components of meat contribute to tenderness, or lack of it: the connective tissues collagen and elastin, muscle fibers, and adipose tissue. However, meat lipids are of only minor importance in this respect.

Connective Tissue

Many of the meat tenderness differences associated with animal age, muscle location, and sex (Chapter 5) result from differences in connective tissues. Estimates of amount and nature of collagen (white fibrous connective tissue protein) in meat cuts generally indicate that high amounts of collagen and chemically mature collagen are associated with toughness. Figure 11.2a,b illustrates that, in tough muscle fibers, abundant collagen surrounding the fibers facilitates formation of many interfiber connections per unit of fiber surface area. Many heat-stable cross-links, as found in chemically mature collagen, retain high residual strength in collagen after cooking. Although most connective tissue fibers in muscles are collagen, elastin fibers also are present and may contribute to toughness.

The connective tissue content of muscle changes with muscle growth and functional demand. Connective tissue is deposited before contractile proteins so connective tissue concentration is higher in muscles of very young animals. In more mature animals differences in connective tissue content of muscle is probably a reflection of functional demands during life. For example, powerful muscles of the legs must develop more strength than some muscles surrounding the backbone. Supporting and connecting tissues therefore must undergo extensive development. Quantities of collagenous connective tissue fibers are great in such muscle, and chemical cross-linking (Chapter 2) in these fibers is extensive. The result is a strong, web-like structure surrounding muscle fibers (endomysium), muscle fiber bundles (perimysium), and muscles themselves (epimysium). While epimysium is often trimmed from the muscle surface before consumption, endomysium, and perimysium are integral parts of meat as consumed. When heat stable crosslinks are present the connective tissues remain largely intact following cooking and impart a **background toughness** to meat. Although connective tissue concentration is slightly higher, the background toughness often is less in meat from young animals compared to that of old animals because of fewer heat stable crosslinks. Also, animals that have been allowed to grow rapidly prior to slaughter have reduced background toughness derived from collagen.

Contractile Protein Interactions and Myofibril Integrity

Degree of tenderness of muscles is determined partially by rigor state and structural integrity of the myofibrils. The formation of stable crossbridges between actin and myosin filaments during development of rigor mortis (Chapter 5) causes a marked increase in muscle toughness. This loss of tenderness in the first few hours postmortem has been referred to as **actomyosin** or **rigor toughening**. Degree of tenderness of muscles also is determined partially by postrigor contraction state and sarcomere length, which are controlled in part

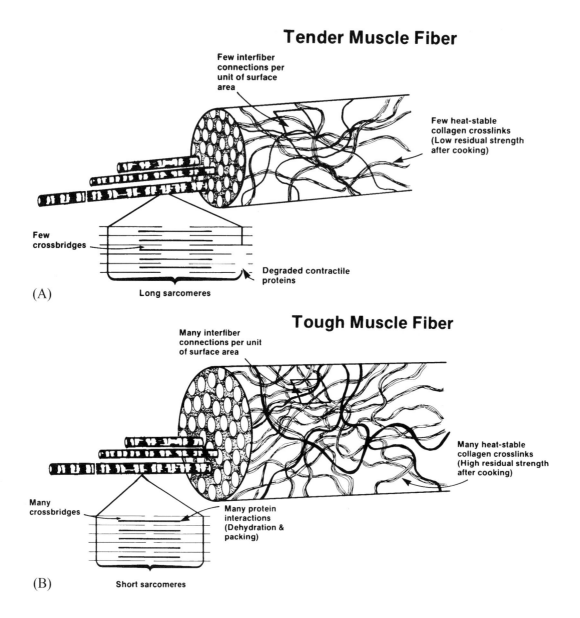

FIGURE 11.2 | Diagram of a (A) tender and a (B) tough muscle fiber.

by skeletal attachments and amount of tension on muscles during rigor onset. As discussed in Chapter 5, contraction state also is regulated by temperature during rigor mortis onset and by electrical stimulation. Even though some muscle toughening may be prevented by avoidance of shortening during rigor onset, formation of cross-bridges is a toughening process that occurs even in the absence of shortening.

The combination of cross-bridge formation and shortening causes muscle fibers to toughen because many contractile proteins interact and form stable structures (Figure 11.2a,b) and muscle fibers become firm and inextensible. These proteins become tightly packed and partially dehydrated as a result of protein-to-protein interactions making fewer sites available for water binding. Thus, water-holding capacity and tenderness are reduced after muscles undergo rigor-associated changes. Tenderness and water-holding capacity are well correlated. Characteristic toughness found immediately after completion of rigor mortis may be partially due to the fact that water is free to escape from muscle fibers during cooking. According to this theory, fibers so shrunken do not rupture easily during chewing, and an impression of toughness is given.

Formation of actin/myosin cross-bridges during rigor mortis development and the normal integrity of ultrastructural components of muscle fibers reduce tenderness. During postmortem aging after rigor mortis development, proteolytic enzymes begin to degrade several of the myofibrillar proteins other than actin and myosin (Chapter 5). These subtle changes occurring naturally within the tissues have profound effects on tenderness. Action of the calpains is responsible for degradation of the myofibrillar proteins titin, nebulin, and troponin T and several cytoskeletal proteins. Progressive degradation of Z disks that occurs in muscles during aging may facilitate some fragmentation of myofibrils and slight lengthening of sarcomeres. The overall effects of these changes are progressive increases in tenderness during postmortem aging.

Changes also may occur in membrane permeability or in its general structural integrity, permitting redistribution of ions, and shifts in water-binding activity. Thus, some tenderization accompanying aging may be attributed to cationic shifts and resultant improvements in ability of proteins to retain water.

Adipose Tissue

Intramuscular lipids have been credited by meat industry personnel with making meat more tender. However, little research can be cited to show a strong positive influence of lipid components in tissues (marbling) on tenderness. It is more likely that some lipid acts as a lubricant in mastication of less tender meat, thus improving perceived tenderness, and easing the process of swallowing. However, animals normally slaughtered for retail meat cuts are youthful and relatively tender, making any tenderness due to marbling relatively unimportant. The persistence of marbling as an indicator of tenderness in industry practice may indicate some usefulness of the trait as an indicator of animal production practices. When animals are reared in comfortable surroundings on high nutritional planes, their rapid growth may ensure meat tenderness and marbling may be deposited merely as a product of high energy production systems.

The relationship of marbling to overall palatability of meat is shown in Figure 11.3. The rectangle in the figure represents the window of acceptable palatability of beef. The range in fat content for the window is from 3.0 to 7.3 percent. This corresponds to the degrees of marbling in the lower end of the USDA Select grade through the USDA Choice grade. Thus, some marbling is necessary to ensure acceptable palatability, but palatability does not increase incrementally with each additional degree of marbling. As can be seen by the shape of the palatability curve, only small increases occur within the window beyond the Select grade and even smaller increases occur with the higher degrees of marbling characteristic of the USDA Prime grade.

Juiciness

Meat juices play important roles in conveying overall impressions of palatability to consumers. They contain many important flavor components and assist in fragmenting and softening meat during chewing. Regardless of other virtues of meat, absence of juiciness severely limits its acceptability and destroys its unique palatability characteristics.

The principal sources of juiciness in meat, as detected by consumers, are intramuscular lipids and water. In combination with water, melted lipids constitute a broth that, when retained in meat, is released upon chewing. This broth also may stimulate flow of saliva and thus improve apparent juiciness.

Marbling in meat serves to enhance juiciness in an indirect way. During cooking, melted fat apparently becomes translocated along bands of perimysial connective tissue. This uniform distribution of lipid throughout the muscle may act as a barrier to moisture loss during cooking. Consequently,

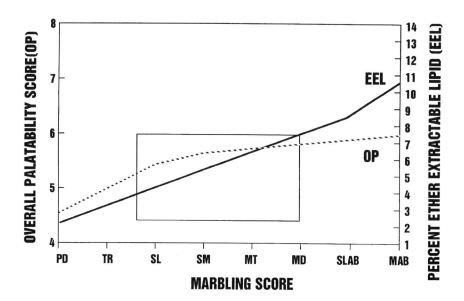

FIGURE 11.3 | Relationship of marbling to overall palatability (broken line) and ether-extractable lipid or fat (solid line) of beef longissimus muscle. Degrees of marbling are presented along the horizontal axis; overall palatability (OP) scores are on the left vertical axia and percent ether-extractable lipid (EEL percent) on the right vertical axis. The rectangle in the figure represents the window of acceptable palatability. The wondow includes high traces through a moderate degree of marbling and EEL from 3.0 to 7.3 percent. Redrawn from Savell, J. W. and H. R. Cross, "The Role of Fat in the Palatability of Beef, Pork and Lamb," in *Designing Foods,* National Academy Press, Washington, DC, 1988.

meat with some marbling shrinks less during cooking and remains juicier. Subcutaneous fat also minimizes drying and moisture loss during dry heat roasting.

In spite of influences of fat, the major contributor to the sensation of juiciness is water remaining in cooked product. Because fat-free water content of meat is relatively uniform, differences in juiciness relate primarily to ability of muscles to retain water during cooking. Because aging improves water-binding properties, aged meat is juicier than un-aged meat.

Flavor and Aroma

Many of the psychological and physiological responses experienced when meat is eaten are elicited by product flavor and aroma. Meaty flavor and aroma stimulate flow of saliva and gastric juices, thus aiding in digestion.

Flavor and aroma sensations result from a combination of factors that are difficult to separate. Physiologically, perception of flavor involves detection of five basic sensations (salty, sweet, sour, bitter, and umami) by nerve endings in taste buds imbedded in the surface of the tongue. Aroma is detected when numerous volatile compounds stimulate nerve endings in the lining of the nasal passages. The total sensation is a combination of gustatory (taste) and olfactory (smell) stimuli.

Components of meat responsible for flavor and aroma have not been completely identified. It is likely that many constituents of muscle, connective and adipose tissue become flavor compounds upon being heated. Some evidence shows that inosine monophosphate (IMP) and hypoxanthine enhance flavor or aroma. Because IMP and hypoxanthine are breakdown products of ATP, it is obvious that muscles with large energy stores would have more pronounced flavor. This may explain the intensity of flavor found in frequently used muscles of the carcass, and the stronger flavor of some game animals. In addition, flavor developed during carcass aging results partially from destruction of mononucleotides.

There is general acceptance of the view that most constituents of meat responsible for meaty flavor are water soluble components of muscle tissue. Several sulfur-containing compounds have been identified as impact compounds relative to meat flavor. Species flavor and aroma are thought to arise from water soluble compounds in adipose tissue, many of which are volatilized when heated.

Palatability Interrelationships

The experience of eating meat does not consist of separate impressions of individual palatability factors of tenderness, juiciness, and flavor. Rather, consumers get overall impressions of satisfaction or dissatisfaction, unless there is one particularly outstanding characteristic. Consequently, products that are juicy and flavorful might seem more tender than shearing tests would verify. Palatability tests of meat by laboratory or consumer taste panels usually show high correlations among various factors evaluated. These close relationships are partially due to psychological factors mentioned previously. However, some tissue properties, such as water-binding capacity or level of energy metabolites, may affect more than one palatability attribute, and thus provide a basis for palatability interrelationships.

Overall palatability of meat cooked by dry heat methods is influenced to a greater extent by temperature, both cooking temperature and end-point temperature (degree of doneness), than by marbling (quality grade). Cooking losses are increased as cooking temperature and end-point temperature increase. Greater cooking losses typically decrease meat juiciness, but the lack of juiciness also tends to reduce meat flavor. In addition, high cooking losses along with the protein hardening and toughening induced by high cooking temperatures (72° C to 74° C) markedly reduce meat tenderness. Meat cooks so rapidly at these high temperatures that cooking time may not be long enough to allow a sufficient amount of solubilization of collagen to take place to ensure acceptable tenderness.

FACTORS ASSOCIATED WITH MEAT PALATABILITY

Antemortem and immediate postmortem factors associated with development of meat quality and palatability are discussed in Chapter 5. Yet, several other factors are either associated with or influence meat palatability.

Carcass Grade

Beef and lamb carcass quality grades estimate meat eating characteristics based on evidence observable in carcass tissues (Chapter 15). In general, meat tenderness is indicated by youthfulness, which, in turn, indicates minimal cross-linking of collagen. Juicy cuts of meat come from carcasses which have some intramuscular fat (marbling) and whose muscles do not exude excessive amounts of moisture. Flavor intensity becomes greater as carcass maturity and marbling increase.

Postmortem Aging

Postmortem aging is required to develop maximum tenderness of most species of meat. Meat tenderness is least at the time rigor mortis is fully developed and usually increases progressively during postmortem aging. Chemical changes in muscle during postmortem aging and factors that affect these changes are described in Chapter 5. Tenderness increases most rapidly during the early portions of the postrigor period, for example, during the first 72 hours postmortem in beef muscle. But, tenderness increases progressively less as aging time progresses. The rate of postrigor tenderization varies by muscle and is affected by temperature, postmortem electrical stimulation, genetic differences among animals, and other factors (See Chapter 5). Aging-based tenderization proceeds most rapidly in poultry muscle and generally least rapidly in beef. For most fresh meat products, aging occurs only during the time meat is being distributed from the slaughter plant to retail or food service establishments. Intentional aging of meat separate from that occurring in the normal distribution system is mostly limited to specialty products such dry-aged beef. In dry aging, selected cuts are held under refrigeration in open air for up to several weeks. During this time meat tenderness improves due to endogenous enzymes and a unique dry-aged flavor develops due to microbial growth and other changes occurring at the meat surface. Wet aging, on the other hand, normally occurs after carcasses have been fabricated and primal or retail cuts have been placed in long-term packaging materials. Though not mandatory, often times the choice is vacuum packaging. Occasionally, high-end restaurants will offer both "wet-" and "dry aged" steaks on their menus. Understanding the difference in these methods of aging fresh meat helps consumers frame their eating expectations.

Tenderizers

Use of weak acids such as vinegar or lemon juice is a traditional method for overcoming connective tissue toughness in meat. These marinades promote swelling of collagen, which causes some disruption of hydrogen bonds within collagen fibrils. **Enzyme tenderizers** of plant origin also are used on meat to reduce both connective tissue and muscle fiber toughness. The most commonly used proteolytic enzyme is **papain**, an extract of papaya plants. **Bromelin** from the pineapple and **ficin** from the fig along with preparations derived from *Aspergillus oryzae* and *Bacillus subtillus* are other proteolytic enzymes used in meat tenderizers. Enzyme preparations may be applied to meat cut surfaces or injected. They are activated by warming during cooking. However, when sufficiently high temperatures are reached, enzymes are denatured and lose their activity. Of these enzymes, papain has the least activity on collagen. Bromelin has strong effects on collagen and the least on the myofibril and elastin. Ficin has the greatest degradative action on collagen, elastin and myofibrils. Care must be taken when using these enzyme tenderizers to avoid over tenderization and development of mushy texture.

Tenderizing effects of **mechanical treatments**, such as pounding, cubing, or blade tenderization depend on mechanical destruction of connective tissues and muscle fibers. Individual boneless meat pieces may be fused together to form cubed steaks by rotary blade tenderizers. Boneless subprimal cuts may be tenderized with tenderizers that convey cuts under banks of blades or needles, which move up and down.

Processing Methods

Virtually all commercial processes used on meat affect palatability, and most of these produce desirable effects. As example, meat curing and smoking produce unique texture and flavor because of combined effects of curing ingredients, water introduction, heating, and smoking. Meat batters form finished products that are very uniform in all palatability attributes even though many different meat raw materials are utilized. Restructured meat is uniformly tender even though less tender raw materials may have been used.

Microbiological

End products of microbial growth may produce undesirable flavor if they are present in large quantities. Undesirable flavors may result from growth of aerobic microorganisms confined to free surfaces and anaerobic microorganisms in deep portions of meat. The latter type causes putrid odors and is known as **bone sour** when incomplete chilling or inadequate cooking rates permit such organisms to grow and break down muscle proteins. On the other hand, desirable flavor results from lactic acid produced during controlled microbial fermentation in some dry and semidry products. Microbes contribute to flavor of long-term aged meat; yeasts produce characteristic aged meat flavor after several days of aging.

Chemical Changes

Flavor changes occur after extended storage due to chemical breakdown of certain constituents, escape of volatile materials, and oxidation of certain components. Desirable flavors develop in aged meat from breakdown of nucleotides. Undesirable flavors develop due to **oxidative rancidity** (Chapter 8) when fatty acid chains are broken at points of unsaturation (double bonds) by chemical addition of oxygen. Formation of carbonyls, particularly low molecular weight volatile aldehydes, is directly responsible for rancid flavor and sharp aroma.

COOKERY

The art of meat cookery exists because of shared culinary experience from generation to generation. Certain principles of meat preparation must be observed to increase the likelihood of palatability in cooked product. These principles include knowledge of time and temperature combinations that assure preparation of meat having maximum eating satisfaction. Knowledge of various characteristics of meat cuts and probable response to heat also is vital to successful meat cookery.

Effects of Heat on Meat Constituents

Appearance

Cooked meat color is relied on by consumers to assess degree of doneness. External meat color of cuts heated with dry heat results from a combination of surface dehydration and **sugar-amine browning** reactions. Amine groups on muscle proteins react with any available reducing sugars, such as free glucose, in tissues. Browning occurs at high (approximately 90° C) temperatures, such as those found at cut surfaces during roasting or broiling. Internal colors and meat temperatures associated with various degrees of doneness are illustrated in Figure 11.1 (see color section). Myoglobin begins to denature at about 60° C, thus the first changes in color are observable at this temperature. Progressive loss of redness with increasing temperature results from progressive denaturation of pigments. Cooked meat pigments show the brown color of hemichrome because of oxidation and denaturation by heat (Figure 8.3).

Visible textural properties of meat are affected by cookery. Cooked muscle may appear soft, hard, coarse, or smooth. Adhering adipose tissue may appear soft or crisp. Juices may exude or products may appear dry.

Structural Changes

When proteins of muscle are exposed to heat, they lose their native structure and undergo several changes in configuration. In general, even though specific changes are unique to each protein, **denaturation** (alteration of structure due to nonproteolytic changes) occurs. This may be accompanied or followed by aggregation, or clumping, of protein

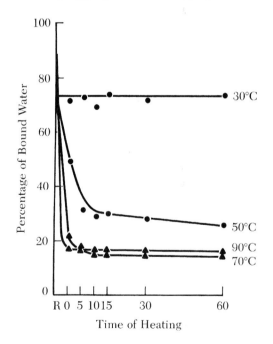

FIGURE 11.4 | Influence of time and temperature of heating on the water-holding capacity of beef muscle. R-O is the decrease of water-holding capacity during heating from 20° C up to the given temperature. From Hamm, R.A. Lebensm. Untersuch, Forsch, 116, p. 120 (1962).

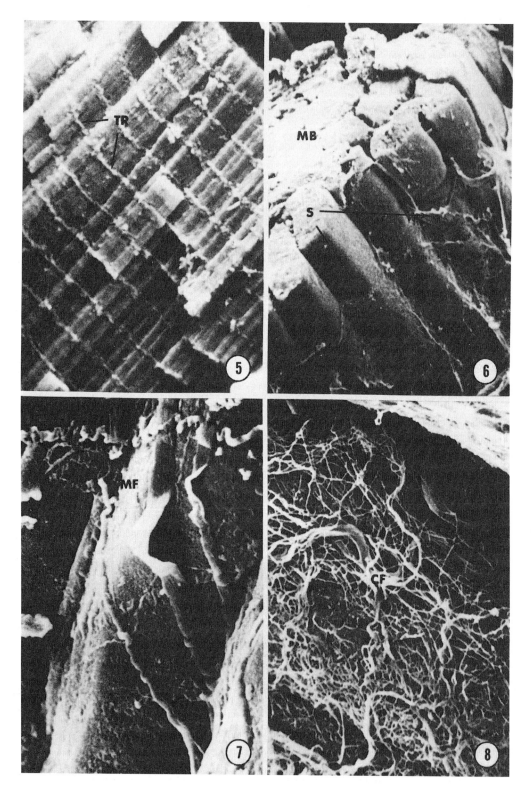

FIGURE 11.5A | Scanning electron micrographs of unheated beef muscle: 5-myofibrils, 6-fiber bundle, 7-muscle fiber with endomysial fibers, 8-perimysial connective tissue. From Leander et al, *J. Food Science* 45:1 (1980).

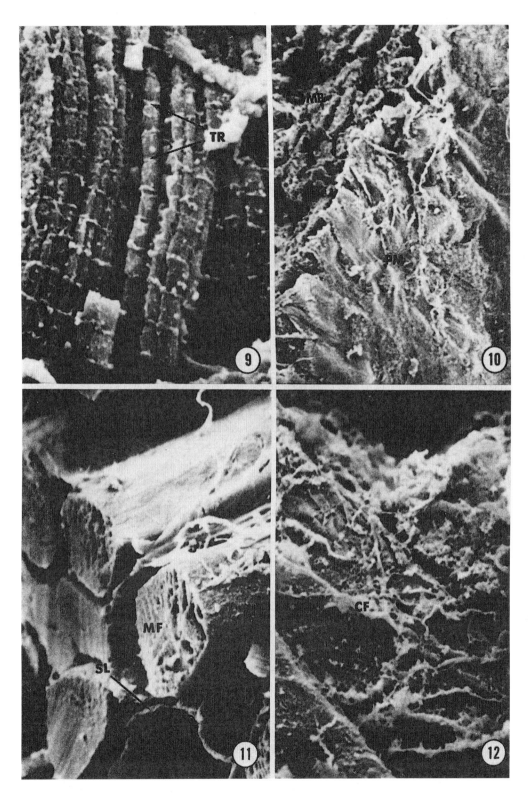

FIGURE 11.5B | Scanning electron micrographs of beef muscle heated to 63° C. 9–myofibrils, 10–perimysium, 11–muscle fibers, 12–endomysium. From Leander et al, *J. Food Science* 45:1 (1980).

molecules (**coagulation**), the presence of which indicates a loss in protein solubility.

Muscle proteins do not all denature at the same time as meat is cooked. For example, myosin in the thick filament, α-actinin in the Z line and the sarcoplasmic proteins denature in the temperature range of 40°–50° C, titin and myoglobin begin to denature at 60° C and actin begins to denature at 70° C. Collagen fibers denature around 60° C.

Loss of water holding capacity is one of the first physical changes during cooking and may begin at temperatures as low as 40° C. It is directly related to denaturation of myosin. Myofibrils begin to shrink in their transverse dimension at about 45° C and water is translocated from the intramyofibrillar to the intermyofibrillar space. Water holding capacity is markedly reduced by 50° C and reaches is lowest level by 70° C (Figure 11.4). Figure 11.4 also indicates that water holding capacity falls more rapidly with higher cooking temperatures and that duration of heating is important at temperatures between 50° C and 70° C.

Coagulation of myofibrillar proteins is associated with easily observed changes in gross physical character of tissues. Fiber microstructure is recognizable after heating (Figure 11.5) but disintegration of filaments and, ultimately, Z disks, occurs progressively with higher temperatures. Even though the microscopic striated appearance of muscle fibers persists after heating, in well-cooked meat an increased rigidity typically occurs. This phenomenon sometimes is referred to as **protein hardening**. Temperatures associated with various phases of protein hardening are incompletely identified, but some research indicates that it does not occur below approximately 63° C (Figure 11.7).

Coagulation and loss of solubility occur in most proteins of muscle fibers with heating. On the other hand, when collagen of the surrounding connective tissue is heated, it undergoes some physical changes that cause increased solubility. The first change taking place is physical shortening of collagen fibrils to one-third of their original length. This may occur at temperatures as low as 56° C in some fibrils, and is usually complete in half the muscle collagen fibrils of postrigor muscle at 61°–62° C. Effects of heat on collagen fibrils are evident in Figure 11.5B. These changes, called **collagen shrinkage** or **melting**, are accompanied by marked decreases in collagen fibril strength and increased collagen solubility. On extended heating in the presence of moisture, collagen is further hydrated and hydrolyzed and forms **gelatin** on cooling. Thus, collagen becomes more tender with heating.

The connective tissue protein elastin is not susceptible to effects of heat. Even though properly cooked, some meat has persistent toughness because of high elastin content. The only feasible method of degrading elastin is application of plant proteolytic enzymes as previously discussed.

Several other alterations occur upon heating muscle proteins, which may affect ultrastructure. These include changes in pH, reducing activity, ion-binding properties, and enzyme activity. Slight upward shifts in pH (approximately 0.3 unit) result from exposure of reactive groups on the amino acid histidine. Increased reducing activity develops as a result of unfolded protein chains and exposed sulfhydryl groups. Likewise, conformational changes in proteins alter their ability to bind various ions, such as Mg^{2+} and Ca^{2+}. With regard to enzyme activity, heat generally inactivates enzymes, but varying degrees of heat resistance are shown by several muscle enzymes.

Fat translocation in cooked meat was discussed earlier in this chapter in relation to palatability. Although extent of fat migration is uncertain, the action is initiated by rising muscle temperature. Solubilization of collagenous connective tissue provides channels through which melted fat may diffuse. Thus, cooking action results in movement, and possibly emulsification, of fat with soluble protein.

Tenderness

Heat may cause both toughening and tenderization of meat, depending on factors such as rate of temperature increase, time at specific temperatures, amount of collagen in meat, and the prevalence of heat-stable collagen crosslinks. In general, those heat-induced changes in myofibril proteins that result in denaturation, coagulation, and dehydration reduce tenderness. Toughening of the myofibrils

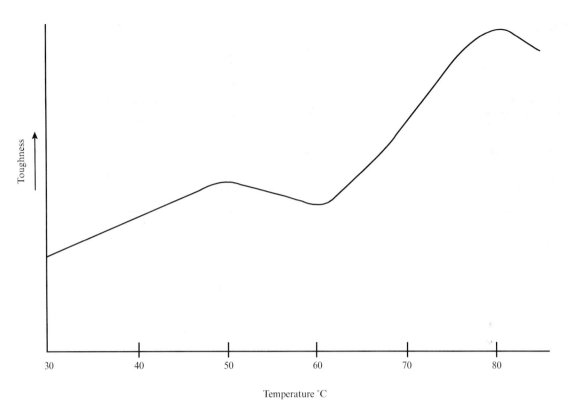

FIGURE 11.6 | Illustration of the effect of heating on toughening of beef *semitendinosus* muscle. Modified from Christensen, M., P. P. Purslow and L. M. Larsen. The effect of cooking temperature on mechanical properties of whole muscle, single muscle fibers and perimysial connective tissue. *Meat Science* 55:301 (2000).

begins at about 60° C and progresses as the temperature increases. Conversely, those heat-induced changes in collagen that result in shrinkage, hydration and solubilization increase tenderness. This effect begins at about 50° C as collagen shrinks (melts) and becomes increasingly hydrated. It continues above 50° C as collagen shrinkage is completed and hydration continues, eventually leading to solubilization of some collagen molecules. However, as is explained below, the initial effect below 50° C is toughening as collagen shrinkage begins.

Figure 11.6 illustrates the generalized effect of heating on tenderness of beef *semitendinosus* muscle from animals about 48 months of age. Toughening occurs in two stages, first between 30° C and 50° C and then again between 60° C and 80° C. One explanation for the toughening that occurs between 30° C and 50° C is straightening of the crimped collagen fibers in the perimysium due to partial denaturation as shrinkage begins resulting in increased numbers of collagen fibers in a given cross-section. Although myosin, α-actinin and sarcoplasmic proteins begin to denature in this temperature range, evidence indicates this does not contribute to toughening. Reduced rate of toughening or even reduced toughness between 50° C and 60° C probably is caused by continued hydration, shrinkage and possibly solubilization of collagen molecules in the perimysium. While proteins of the myofibril continue to denature at these temperatures, their associated coagulation is limited until temperatures reach 60° C and higher. Any toughening effect of myofibril protein denaturation is masked by the marked weakening of collagen fibers that occurs between 50° C and 60° C. Thus, the length of time during cookery that meat is at temperatures near 60° C is important for softening of intramuscular connective tissues without undue coagulation of myofibrillar proteins. A beef ribeye roast ("prime rib") cooked slowly to a rare degree of doneness (60° C) is an example of cookery to accomplish collagen shrinkage without extensive myofibril coagulation and toughening.

Toughening between 60° C and 80° C (Figure 11.6) occurs at a greater rate than between 30° C and 50° C and is believed due to further denaturation, coagulation and dehydration of the myofibrils. Actin denatures in this temperature range. Coagulation and dehydration of the myofibril proteins produces a firmer myofibril structure that is less tender. The strength of perimysial collagen fibers, however, continues to decrease. Above about 80° C, meat with high collagen content may soften and become more tender due to increasing hydration of collagen and its conversion to gelatin, particularly when heated in the presence of adequate moisture at these temperatures. For example, in beef chuck roasts that are cooked with moist heat in a covered skillet to temperatures above 80° C, the myofibrils and myofibril bundles may be quite firm but are easily separated from each other because the perimysial connective tissues have shrunk, hydrated and lost their tensile strength, allowing the fiber bundles to be easily fragmented and giving the perception of increased tenderness.

Figure 11.7 shows the effect of temperature on the rate of tenderization caused by collagen shrinkage. At very low meat temperatures of 56°–58° C, tenderization is relatively slow. On raising temperatures slightly to 62°–64° C, collagen shrinkage occurs more quickly. At higher temperatures of 72°–74° C, rapid shrinkage of collagen is followed by protein hardening and toughening. However, continued heating at this temperature results in substantial hydrolysis of collagen and eventually, this change causes meat tenderization.

Juiciness

Obviously, severe cooking procedures causing extensive dehydration of meat also render it less juicy to consumers. The greater the degree of doneness, the less juicy a meat cut will be. Meat containing 68–75 percent moisture in the raw state contains about 70, 65, and 60 percent moisture after being dry-heat roasted to 60° C, 70° C, and 80° C, respectively. Losses occurring are due to evaporation and drip losses. Of course, the extent of losses also depends on the water-holding capacity of tissues and the degree of protection afforded by lipid translocation (as discussed earlier in this chapter) and subcutaneous fat.

Flavor and Aroma

Certain volatile materials are driven off during cooking of meat that contribute to unique flavor and aroma. These include various sulfur and nitrogen compounds, as well as certain hydrocarbons, aldehydes, ketones, alcohols, and acids. Changes in quantity and type of volatile materials present are extensive but poorly understood. Yet, the effects of heat are so unique that the type and conditions of cookery often may be identified solely by flavor and aroma of products. Dry heat cookery imparts certain flavors, particularly at exposed surfaces where temperatures become very high. On the other hand, cooking with moisture under pressure causes development of pronounced and

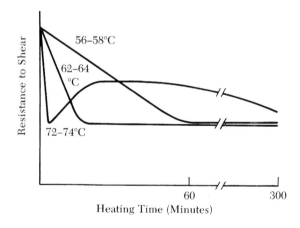

FIGURE 11.7 | Effect of heating temperature and time on the tenderness of 1.27cm diameter cylinders of beef (semitendinosus) muscle. Summarized from Machlik, S. M. and H. N. Draudt, "The Effect of Heating Time and Temperature on the Shear of Beef Semitendinosus Muscle," *J. Food Science*, 28:711 (1963).

unique flavor changes in deep tissues of meat cuts. The aroma released from fat during cooking differentiates pork, lamb, beef, and poultry.

Storage of precooked uncured meat for a short period results in the development of **warmed-over flavor**, characterized as "old, rancid, stale, and painty" odor and flavor. This objectionable flavor apparently is caused by the iron-catalyzed oxidation of unsaturated fatty acids and becomes more noticeable when refrigerated precooked meats are reheated.

Cookery Methods and Recommendations

There are several systems by which temperature increases are accomplished in meat. Although the major objective in meat cookery is to achieve a particular internal temperature, surface color development and cooked flavor development are also important. The rate of heating, equipment used, and many other factors influence characteristics of cooked products.

Of particular importance in determining cooked characteristics of meat is the amount of moisture present during heating. Water is a good conductor of heat, and its presence aids in penetration of heat into the deepest parts of meat cuts. On the other hand, moist surfaces of meat also may delay heating due to evaporative cooling. Water also is necessary for developing tenderness and the final texture of cooked meat because of hydrolysis of collagen.

One of the essentials for success in meat cookery is knowledge of proper duration of heating. For thin cuts, such as steaks or chops, criteria are subjective. Experienced cooks often determine the cooking endpoint from the color and rigidity of cuts. However, differences in product pH and certain ingredients my influence these subjective endpoints. Consistent control of endpoint temperature requires the use of a meat thermometer. The thermometer is inserted into the thickest part of a cut, avoiding pockets of fat and bones, so temperatures in the coolest region will be detected. For thick cuts, this is the only method that will completely assure the desired degree of doneness.

Practical guides to meat cookery have been developed on the basis of time and temperature influences on palatability. For example, unnecessary toughening of most meat cuts may be avoided by prevention of internal temperatures that cause protein hardening. However, it is necessary to cook some meat to this temperature range to develop other desirable palatability characteristics. To develop flavor as desired by most consumers and completely convert pigments to brownish-gray denatured forms, pork, poultry, and lamb usually are cooked to a well-done degree. Other meats, notably beef, are cooked to a degree consistent with consumer preferences. Internal temperatures required for these degrees of doneness are shown in Figure 11.1.

Heating Systems

Cooking outcomes are largely dependent on the heating system selected for a specific product. Heating systems are categorized as either **dry-heat** or **moist-heat** depending on the conditions at the surface of the product. It should be noted, however, that the interior of a meat product is almost always moist even following dry-heat cooking.

Heating systems are also categorized on the means whereby heat is transferred to the product, including **conduction, convection, radiant** and **microwave radiation**. Each heating system relies on one or a combination of these heat transfer methods.

Dry Heat

Cooking with dry heat can be accomplished by any method that surrounds meat with hot, dry air. Broiling and roasting are the best examples of dry heat cookery.

Broiling is appropriate only for tender cuts, such as steaks or chops, because the heating period is usually rather short and there is inadequate time to achieve connective tissue degradation. High surface temperatures result in development of unique flavor and extensive browning. Duration of heating is extremely critical, since cuts are relatively thin and temperatures used are quite high.

Charcoal broiling is a very popular method of cooking that also imparts unique flavor. Many products, such as chops, steaks, chicken, ribs, kabobs, sausage, or roasts, are cooked by this method. Products acquire a pronounced smoked flavor from combustion of charcoal and melted fat drippings. Roasting with dry heat is appropriate for tender roasts. It is usually accomplished in an oven at temperatures of 150° C to 175° C. It imparts unique flavor by the sugar-amine browning reaction on meat surfaces as mentioned earlier. However, cuts should be protected during roasting by a layer of external fat or aluminum foil to prevent excessive moisture losses. If cuts are large, such as intact rounds, hams, or turkeys, it is possible to use relatively low (120° C) oven temperatures for extended cooking. Limited surface area per unit of meat mass in these large cuts prevents extensive moisture losses.

Recommended for dry heat cookery are: all seafoods; all cuts of young poultry; all cuts of pork except thin shoulder cuts and hocks; all cuts of lamb except breast and shank cuts; all veal roasts; all youthful maturity beef steaks and roasts from ribs, short loins, sirloins, and selected areas of rounds; and all comminuted meats.

Moist Heat

If meat cuts contain relatively large amounts of connective tissue, it is desirable to provide extra water during cooking. This provides all the water necessary for complete hydrolysis of collagen even when the cooking time is lengthy. Low temperatures are prescribed, over long cooking times, to allow this conversion to occur without hardening myofibrillar proteins. It is emphasized that this action occurs in collagen, but does not occur in elastin. Thus, cuts that do not tenderize in spite of extensive cooking are probably high in elastin connective tissue fibers.

Braising, cooking in water, or pot roasting are examples of moist heat cookery. Heating is accomplished in closed containers with added water. Seasoning, sauces, or flour also may be added to enhance development of desired flavor and texture in cooked products.

Tenderizing action in moist heat cookery may be achieved by wrapping cuts in moisture-proof materials and heating them in a dry oven. Natural juices are trapped, and moisture loss is minimized. Consequently, heating may be extended to allow collagen to become hydrolyzed. Temperatures are usually in the range of 95° C to 100° C for moist heat cookery.

Recommended for moist heat cookery are some seafoods; stewing poultry; pork shoulder cuts and hocks; lamb breast and shank cuts; veal chops, cutlets, steaks, shoulder and round roasts, and shank and breast cuts; and beef chuck, round, fore shank, brisket, short plate, flank, and tip cuts.

Conduction Heating

Conduction heating relies on direct contact between the food and a heating surface. Food service units rely heavily on conduction heating for many meat products. Thin cuts and patties often are cooked by grilling, heating action of which occurs from contact with hot metal surfaces. Meat stews usually are cooked in open kettles with direct conduction of heat from the kettle to the stew. Cooking in liquid relies partially on conduction for heat transfer. Examples are deep fat frying, open kettle cooking in water or pressure cooking.

Convection Heating

Convection heating involves circulation of a heat transfer medium (air, water or oil) within a local environment moving thermal energy from the heat source to the product. Convection ovens circulate dry or moist air around products, thereby increasing rate and uniformity of cooking. Convection heating also occurs in products cooked in liquids. In this process the heat transfer medium may be water or oil. Examples of this method are deep fat frying and open kettle boiling. Most breaded chicken and fish products are cooked by deep fat frying.

Radiant Heating

Radiant heat transfer depends on a hot radiation source (usually a flame or an electric heating element) to produce radiation which is absorbed by the product surface. Radiant heating may produce very high surface temperatures with considerable browning or charring. Most tender steaks and chops and many meat patties are cooked by radiant heat from gas-fired or electric broilers. When charcoal or heated briquets are the sources of heat, radiant heating occurs but smoked flavor also develops from fat drippings. Infrared heat generated by infrared lamps is used infrequently for warming and holding hot foods in food service establishments.

Microwave Radiation Heating

Heat transfer by microwave radiation requires a microwave generator (915 or 2450 megahertz magnetron tube) to produce radiation which is absorbed by the product. Unlike infrared radiation, microwave radiation penetrates several millimeters or centimeters into the product as it imparts its thermal energy. Thus, the surface of the microwave heated product remains comparatively cool until heat from the interior makes its way to the surface.

Use of microwave heating can be an extremely rapid method of cooking especially when small units of product are heated. Variations in product shape and non-uniformity in the microwave field lead to variation in doneness and meat textural properties. Variability is diminished when external heating coils are used to brown outer surfaces of meat, while interior areas are being cooked by microwave radiation. In addition, use of low power (30 percent) cooking for large meat cuts improves uniformity of cooking as compared to 100 percent power. Low power ranges are slower than high ranges but still faster than conventional cookery. Microwave cookery is not recommended for less tender cuts requiring long cooking times to achieve tenderization. Such cuts also may be less juicy because of greater cooking losses and less flavorful because of limited surface browning after microwave cookery. Microwave heating is particularly useful for reheating precooked or processed meats. Vending machines often use microwaves for meat pies and sandwiches. Food service units use microwaves for heating frankfurters, bacon, and other precooked products.

SERVING COOKED MEAT

Satisfaction gained from consuming meat is enhanced if cooked meat is portioned properly and served graciously. Several artistic and scientific principles underlie these components of the meat distribution system.

Carving

Carving skills are required to portion meat roasts and large steaks. Knowledge of anatomy and sharp carving knives are basic to successful accomplishment of tasks. However, carving meat at the dinner table adds a dimension of pleasure and prior preparation is well rewarded by satisfaction received. Large cuts such as roasts are firmer and easier to carve and lose fewer meat juices if allowed to stand for a few minutes after cookery before carving. In addition, flavor and aroma of the meat is at its optimum when carved and served.

Carving should yield uniform servings of meat cut perpendicular to the grain (across the muscle fibers). In loin or rib cuts, this may be facilitated by prior removal of the vertebral column (chine bone). In leg cuts, it is accomplished by cutting slices perpendicular to the femur. In poultry, muscle arrangements do not generally permit carving across muscle fibers.

Appropriate cuts for carving:

- beef rib roasts
- pork and lamb crown roasts
- pork, beef, or lamb loin roasts
- hams
- lamb legs
- and whole turkeys.

12

NUTRITIVE VALUE OF MEAT

OBJECTIVES: *Describe the major nutrient contributions of meat and other animal products in the human diet and present some of the most prominent health concerns associated with the consumption of animal products.*

Recommended dietary allowances
Dietary reference intakes
Dietary guidelines for Americans
High quality protein
Essential amino acids
Biological value
Complementarity
Triglycerides
Unsaturated
Saturated
Low density lipoprotein
Chlolesterol
Trans fatty acid
Conjugated linoleic acid
Non-heme iron
Heme iron
Vitamin b12
Variety meats

CHAPTER 12 NUTRITIVE VALUE OF MEAT

The human diet, especially that in the United States, is complex and varied. Consumers have almost unlimited food selection available year round. Less than 10 percent of disposable income is spent on food. For those with limited income, food assistance programs such as Food Stamps, Child Nutrition Programs and the Special Supplemental Nutrition Program for Women, Infants, and Children, better known as "WIC" currently provide assistance to more than 9 million households in the US. As a result consumer food choices are based largely on taste preference while nutrient content and dietary balance are of secondary importance. In addition, some consumers avoid certain foods for religious or ethical reasons or because of real or perceived health concerns.

Government agencies including the US Department of Agriculture (USDA) and the Food and Drug Administration (FDA), an agency of the Department of Health and Human Services (DHHS), attempt to monitor eating habits in the US and provide dietary advice based on sound nutritional science. USDA's Nationwide Food Consumption Survey (NFCS), conducted every 10 years, and the Continuing Survey of Food Intake by Individuals (CSFII), conducted every 1–3 years, use consumer recall information to assess food consumption. USDA's Food Consumption, Prices and Expenditures report, updated annually, uses retail food disappearance information to estimate food consumption. These surveys and others provide comprehensive information about what Americans are eating. Dietary recommendations from various government and health organizations continue to evolve. **Recommended Dietary Allowances** (RDA) for key nutrients, updated every 5–10 years by the Food and Nutrition Board of the National Academy of Sciences, provide a basis for most recommendations. RDA's are set to meet known nutrient needs for practically all healthy persons. However, increased interest in dietary influences on chronic diseases such as cancer, osteoporosis and cardiovascular disease have lead to development of a confusing variety of **Dietary Reference Intakes** (DRIs). Along with RDA, other DRIs used by nutrition professionals include Adequate Intake (AI), Tolerable Upper Intake Level (UL),

and Estimated Average Requirement (EAR). Daily Values (DV) for key nutrients are used in FDA mandated nutritional labeling (Figure 12.1) and USDA and DHHS translate the DRIs into specific food group serving recommendations in the **Dietary Guidelines for Americans** (Table 12.1) and the MyPlate logo (Figure 12.2).

Nutrition Facts

Serving Size 1 cup (228g)
Servings Per Container about 2

Amount Per Serving

Calories 250   Calories from Fat 110

	% Daily Value*
Total Fat 12g	18%
Saturated Fat 3g	15%
Trans Fat 3g	
Cholesterol 30mg	10%
Sodium 470mg	20%
Total Carbohydrate 31g	10%
Dietary Fiber 0g	0%
Sugars 5g	
Proteins 5g	
Vitamin A	4%
Vitamin C	2%
Calcium	20%
Iron	4%

* Percent Daily Values are based on a 2,000 calorie diet. Your Daily Values may be higher or lower depending on your calorie needs:

	Calories:	2,000	2,500
Total Fat	Less than	65g	80g
Saturated Fat	Less than	20g	25g
Cholesterol	Less than	300mg	300mg
Sodium	Less than	2,400mg	2,400mg
Total Carbohydrate		300g	375g
Dietary Fiber		25g	30g

For educational purposes only. This label does not meet the labeling requirements described in 21 CFR 101.9.

FIGURE 12.1 | Nutritional Facts Label. Source: US Department of Health and Human Services.

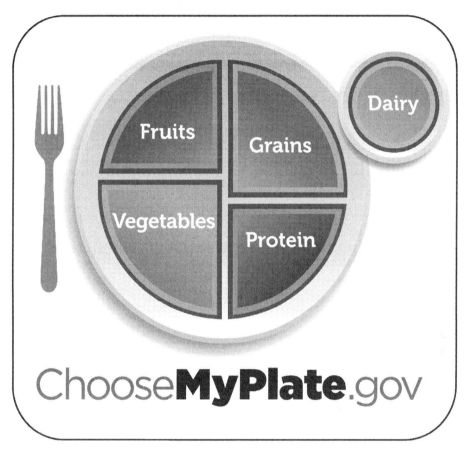

FIGURE 12.2 | Myplate. Source: US Department of Health and Human Services.

Animal products account for less than 26 percent of calories but more than 62 percent of protein in the American diet (Table 12.2). Muscle foods, including meat, fish and poultry products, represent a significant portion of the US diet accounting for more than 14 percent of the calories, 39 percent of the protein and 24 percent of the fat. Dairy products, including butter, contribute 10.3 percent of calories, 19.7 percent of protein and 15 percent of fat in the diet. Current consumption of lean meats and fat free dairy products fall somewhat short of USDA's Dietary Guidelines for Americans but, a small change in consumer food choices could have a great impact on demand for animal products. Against this backdrop of numerous consumer choices and various dietary recommendations the following discussion of the nutritive value of animal products is intended to provide insight on some key issues relating to meat, dairy and eggs in the diet.

PROTEINS

Dietary proteins provide amino acids needed to build and maintain body tissues as well as contribute to many regulatory processes in the body. Excess protein may be utilized for energy, supplying 4.5 Calories per gram. Animal products supply most of the protein in the American diet. In addition the protein supplied by animal products is **high quality protein**. A high quality protein is one that supplies all **essential amino acids** in amounts equivalent to human requirements (high **biological value**), is highly digestible, and is easily

CHAPTER 12 NUTRITIVE VALUE OF MEAT

TABLE 12.1 Suggested Servings[a] of Food from the Basic Food Groups. USDA Food Patterns (Appendix 7. Dietary Guidelines for Americans 2010) US Department of Health and Human Services and US Department of Agriculture.

Calorie level	1,000	1,200	1,400	1,600	1,800	2,000	2,200	2,400	2,600	2,800	3,000
Fruits	1 c	1 c	1½ c	1½ c	1½ c	2 c	2 c	2 c	2 c	2½ c	2½ c
Vegetables	1 c	1½ c	1½ c	2 c	2½ c	2½ c	3 c	3 c	3½ c	3½ c	4 c
Dark-green vegetables	½ c/wk	1 c/wk	1 c/wk	1½ c/wk	1½ c/wk	1½ c/wk	2 c/wk	2 c/wk	2½ c/wk	2½ c/wk	2½ c/wk
Red and orange vegetables	2½ c/wk	3 c/wk	3 c/wk	4 c/wk	5½ c/wk	5½ c/wk	6 c/wk	6 c/wk	7 c/wk	7 c/wk	7½ c/wk
Beans and peas (legumes)	½ c/wk	½ c/wk	½ c/wk	1 c/wk	1½ c/wk	1½ c/wk	2 c/wk	2 c/wk	2½ c/wk	2½ c/wk	3 c/wk
Starchy vegetables	2 c/wk	3½ c/wk	3½ c/wk	4 c/wk	5 c/wk	5 c/wk	6 c/wk	6 c/wk	7 c/wk	7 c/wk	8 c/wk
Other vegetables	1½ c/wk	2½ c/wk	2½ c/wk	3½ c/wk	4 c/wk	4 c/wk	5 c/wk	5 c/wk	5½ c/wk	5½ c/wk	7 c/wk
Grains	3 oz-eq	4 oz-eq	5 oz-eq	5 oz-eq	6 oz-eq	6 oz-eq	7 oz-eq	8 oz-eq	9 oz-eq	10 oz-eq	10 oz-eq
Whole grains	1½ oz-eq	2 oz-eq	2½ oz-eq	3 oz-eq	3 oz-eq	3 oz-eq	3½ oz-eq	4 oz-eq	4½ oz-eq	5 oz-eq	5 oz-eq
Enriched grains	1½ oz-eq	2 oz-eq	2½ oz-eq	2 oz-eq	3 oz-eq	3 oz-eq	3½ oz-eq	4 oz-eq	4½ oz-eq	5 oz-eq	5 oz-eq
Protein foods	2 oz-eq	3 oz-eq	4 oz-eq	5 oz-eq	5 oz-eq	5½ oz-eq	6 oz-eq	6½ oz-eq	6½ oz-eq	7 oz-eq	7 oz-eq
Seafood	3 oz/wk	5 oz/wk	6 oz/wk	8 oz/wk	8 oz/wk	8 oz/wk	9 oz/wk	10 oz/wk	10 oz/wk	11 oz/wk	11 oz/wk
Meat, poultry, eggs	10 oz/wk	14 oz/wk	19 oz/wk	24 oz/wk	24 oz/wk	26 oz/wk	29 oz/wk	31 oz/wk	31 oz/wk	34 oz/wk	34 oz/wk
Nuts, seeds, soy products	1 oz/wk	2 oz/wk	3 oz/wk	4 oz/wk	4 oz/wk	4 oz/wk	4 oz/wk	5 oz/wk	5 oz/wk	5 oz/wk	5 oz/wk
Dairy	2 c	2½ c	2½ c	3 c	3 c	3 c	3 c	3 c	3 c	3 c	3 c
Oils	15 g	17 g	17 g	22 g	24 g	27 g	29 g	31 g	34 g	36 g	44 g
Maximum sofas[b], calories (% of calories)	137 (14%)	121 (10%)	121 (9%)	121 (8%)	161 (9%)	258 (13%)	266 (12%)	330 (14%)	362 (14%)	395 (14%)	459 (15%)

[a] For each food group or subgroup, recommended average daily intake amounts at all calorie levels. Recommended intakes from vegetable and protein foods subgroups are per week. For more information and tools for application, go to MyPyramid.gov.

[b] Solid Fats and Added Sugars

TABLE 12.2 Contribution of Animal Products to the Diet (in percent of total)

Nutrient	Red Meat	Poultry	Milk products except butter	Eggs	Fish and Seafood	All Animal Products except butter	Butter (salted)	Total per day from all foods
Calories	9.3	4.2	9.3	1.3	0.7	24.8	1.0	3800 Kcal
Protein	21.7	13.4	19.3	3.8	4.2	62.4	0.04	110g
Total Fat	17.4	6.6	12.3	2.1	0.4	38.9	2.7	159g
Saturated FA's	20.3	5.8	23.6	2.0	0.3	52.0	5.2	52g
Monounsaturated	18.7	6.6	8.6	2.0	0.3	36.3	1.7	65g
Polyunsaturated	6.7	7.4	2.1	1.5	0.8	18.5	0.5	31g
Cholesterol	25.9	14.4	16.1	34.4	3.6	94.3	2.7	410mg
Niacin	18.1	16.5	1.4	0.1	3.6	39.7	0.0	29mg
Riboflavin	13.1	4.3	30.7	6.5	0.8	55.3	0.1	2.6mg
Thiamin	16.4	1.8	6.2	0.8	0.5	25.7	0.0	2.7mg
Folate	4.3	2.5	7.3	4.7	0.6	19.4	0.0	331ug
Vitamin A	17.5	3.4	17.4	4.2	0.5	43.0	2.3	1520ug[1]
Vitamin E	2.0	1.5	2.8	2.1	1.3	9.7	0.7	16.9mg[2]
Vitamin C	0.9	0.9	2.7	0.0	0.2	4.7	0.0	124mg
Vitamin B6	22.0	12.3	9.7	2.0	2.2	48.1	0.0	2.3mg
Vitamin B12	58.5	5.0	21.0	4.1	9.4	98.0	0.1	8.1ug
Calcium	1.3	1.0	72.8	1.7	1.0	77.8	0.1	960mg
Phosphorus	14.6	7.2	32.8	3.5	2.9	61.0	0.0	1680mg
Magnesium	6.5	4.2	16.4	0.9	2.1	30.0	0.0	380mg
Iron	10.6	4.1	2.1	2.3	1.6	20.7	0.0	21.2mg
Zinc	30.4	9.1	18.9	2.8	1.9	63.2	0.0	13.2mg
Copper	9.8	2.7	2.8	0.2	1.5	17.0	0.0	1.9mg
Potassium	10.5	4.3	18.5	1.1	1.9	36.3	0.0	3780mg

[1] Retinol equivalents
[2] Alpha tocopherol
Source: USDA, Economic Research Service. 1999. Food Consumption, Prices and Expenditures, 1970–97 (1994 data). Statistical Bulletin, No. 965.

absorbable. Amino acids are the basic building blocks of which all proteins are composed, and essential amino acids are those that cannot be synthesized by the body in amounts sufficient to meet requirements. Adult humans need nine essential amino acids—phenylalanine, valine, tryptophan, threonine, methionine, leucine, isoleucine, lysine, and histidine. Most animal products such as meat, milk, and eggs have high biological value since they supply a proper balance of these amino acids. The biological value of some human foods are: mother's milk—100 percent, whole egg—88–94 percent, cow's milk—88-90 percent, meat—75–85 percent and whole wheat—64 percent. When animal products serve as the principal dietary protein source, the RDA for protein for adult males is 47 gm per day versus 63 gm per day when the dietary protein source is not specified. On the other hand, most vegetable, fruit, grain, and animal connective tissue proteins are somewhat low in biological value. High amounts of the nonessential amino acids glycine, proline, and hydroxyproline in collagen are responsible for the lower biological value of some variety meats containing connective tissue as the primary source of protein. Small additions of skeletal muscle to connective tissue, cereal, or legume proteins markedly improves the biological value of those proteins because of **complementarity** of amino acids. Availability of amino acids following digestion is high in animal proteins. Meat, dairy and egg proteins are 95–100 percent digestible whereas some plant proteins are only 65–75 percent digestible.

Meat proteins are largely those of muscle and connective tissues. The largest proportion of total muscle proteins are those of the myofibrils. Sarcoplasmic proteins, consisting of muscle enzymes and myoglobin, make up the next largest fraction. This is followed in abundance by connective tissue proteins, consisting largely of collagen and some elastin. Lean portions of raw meat contain 19–23 percent protein (Table 12.3). This content varies inversely with amount of fat present and, because of moisture and fat losses during cooking, increases to 25–30 percent in cooked meat. Meat, poultry and fish products account for 43 of the 110 grams of protein Americans consume daily.

The average US protein consumption is at least 175 percent of the recommended dietary allowance (RDA) for protein. As prescribed by the Food and Nutrition Board of the National Research Council, the RDA for protein for a grown man, for example, is 63 gm per day. Large amounts of free amino acids or peptides cannot be stored in the body, so it is essential that protein be consumed every day. Because a normal-sized serving of cooked lean meat is approximately 100 gm, and its protein content is 25–30 percent, this serving furnishes 25–30 gm of protein, or about 40–48 percent of the RDA. Based on USDA CSFII figures the average daily consumption in the US of cooked edible red meat, poultry, fish, and processed meat is about 185 gm.

In addition to proteins, meat also contains some non-protein nitrogen compounds, such as free amino acids, simple peptides, amines, amides, and creatine. Although these compounds do not contribute significantly to nutritive value, they are potential sources of nitrogen for amino acid and protein synthesis.

LIPIDS AND CALORIES

Dietary lipids supply energy and essential fatty acids. They also transport fat soluble vitamins in the diet. Lipid content of meat or milk is generally its most variable component. The amount of lipid in meat cuts depends on amount of untrimmed fat within and between muscles and the amount of external fat remaining after cutting and trimming (Table 12.3). In processing of fluid milk, lipid is standardized to concentrations ranging from 3.25 percent (whole milk) to nearly 0 percent (skim milk). In whole eggs the fat is essentially all found in the yolk and may be easily separated from the albumen if desired. The caloric value of lipids (9 Calories per gram) is derived from fatty acids in triglycerides and phospholipids, of which triglycerides constitute the bulk (Table 2.3).

Triglycerides are composed of three fatty acids attached to a glycerol backbone. The fatty acids vary in carbon chain length and presence or absence of double bonds. Each type of fatty acid has different physical and biological properties. Longer chain fatty acids generally produce harder fats such a beef tallow while presence of double bonds often leads to softer more oily fats.

Fatty acids in triglycerides may be **saturated** (no double bonds) or **unsaturated** (one or more double bonds in each fatty acid). Subcutaneous fats from chicken, pork, beef, and lamb are composed of about 33, 45, 54, and 58 percent saturated fatty acids, respectively and are progressively harder fats. On the other hand, milk fat has about 62 percent saturated fatty acids but many of those have shorter carbon chains so that milk fat (butter) is rather soft and oily. In meat fats and egg yolk, the most abundant fatty acid is the monounsaturated oleic acid (18 carbons with one double bond at the ninth carbon). However, other fatty acids present in high proportions are saturated and include palmitic and stearic acids. Thus, animal fats, as all fats in nature, include a mixture of saturated and unsaturated fatty acids.

TABLE 12.3 Proximate Composition and Caloric Content of Separable Lean of Raw and Cooked[a] Retail Cuts[b] of Beef, Chicken, Halibut, Lamb, Pork and Veal and of Selected Milk and Egg Products

Food source and physical state	Percent					Calories per 100 gm
	Protein	Moisture	Fat	Carbohydrate	Ash	
Beef, separable lean with 1/4 in. trim						
Raw	20.78	70.62	6.16	0	1.02	144
Cooked	29.58	59.25	9.91	0	1.20	222
Chicken, light meat, without skin						
Raw	23.20	74.86	1.65	0	0.98	114
Cooked	30.91	64.76	4.51	0	1.02	173
Chicken, light meat, with skin						
Raw	20.27	68.60	11.07	0	0.86	186
Cooked	29.02	60.51	10.85	0	0.93	222
Chicken, dark meat, without skin						
Raw	20.08	75.99	4.31	0	0.94	125
Cooked	27.37	63.06	9.73	0	1.02	205
Chicken, dark meat, with skin						
Raw	16.69	65.42	18.34	0	0.76	237
Cooked	25.97	58.63	15.78	0	0.92	253
Halibut						
Raw	20.9	76.5	1.2	0	1.4	100
Cooked	25.2	66.6	7.0	0	1.7	171
Lamb						
Raw	19.5	71.5	7.0	0	1.5	145
Cooked	27.0	61.5	8.5	0	2.0	200
Pork						
Raw	20.22	71.95	6.75	0	1.04	147
Cooked	27.04	58.97	13.04	0	1.10	233
Veal						
Raw	20.0	75.0	3.5	0	1.0	130
Cooked	29.0	63.0	5.5	0	1.6	175
Milk, pasteurized, fluid						
Whole, 3.25%	3.22	88.3	3.25	4.52	0.69	60
Reduced fat, 2%	3.30	89.3	1.97	4.68	0.71	50
Reduced fat, 1%	3.37	89.92	0.97	4.99	0.75	42
Skim	3.37	90.84	0.08	4.96	0.75	34
Egg, without shell						
Whole, fresh	12.58	75.84	9.94	0.77	0.86	143
Whole, fried	13.63	69.13	15.31	0.88	1.05	196

[a]Roasted; [b]Composite of cuts; [c]Values of less than 1 percent are listed as 0

*Source: US Department of Agriculture Handbook Nos. 8, 8–5, 8–10, and 8–13 and USDA National Nutrient Database for Standard Reference, Release 20 (2007).

High consumption of certain saturated fatty acids has been implicated as a contributing factor in cardiovascular disease risk especially elevated **low density lipoprotein** (LDL) **chlolesterol**. Many vegetable oils have a lower proportion of saturated fatty acids than animal fats and consequently are considered by some medical authorities to be safer for consumption. Current USDA Dietary Guidelines recommend that no more than 10 percent of calories should come from saturated fats. However, processors of unsaturated and polyunsaturated vegetable oils (such as margarine and cooking oils) hydrogenate these oils, a process that increases the proportion of saturated fatty acids and transforms some of the double bonds in unsaturated fatty acids from cis to trans forms. Unsaturated fatty acids with isolated (i.e., nonconjugated) trans double bonds behave much like saturated fatty acids in the body. As a result, hydrogenated vegetable fats, containing trans fatty acids, are nutritionally similar to saturated fats. USDA Dietary Guidelines recommend that trans fatty acids be kept a low as possible (less the 1% of calories) in the diet. FDA now requires specific labeling of **trans fatty acid** content on the Nutrition Facts panel of food labels (Figure 12.1). The focus on health concerns associated with trans fatty acids has led to many food product reformulations and modifications in hydrogenation procedures in efforts to reduce the occurrence of trans fatty acids.

Conjugated linoleic acid (CLA) is a unique fatty acid found in the lipid of meat and dairy products. With a conjugated configuration (two double bonds on adjacent carbons) its trans isomers are functionally quite different than the isolated trans fatty acids discussed above. In fact, this naturally occurring trans fatty acid is considered to have numerous nutritional benefits. In various animal and human research trials CLA is reported to be anti-carcinogenic, anti-obesity (alters nutrient partitioning), anti-atherogenic (reduces cholesterol) and enhances immune function and bone mineralization. CLA is produced primarily in the rumen of cattle and sheep. Mammary production of CLA has also been demonstrated in cattle. The amount of CLA deposited in muscle adipose tissue and milk fat (3–10 mg/g of fat) may not be enough to produce detectable biological effects in people consuming typical amounts of meat or milk. However, genetic selection and dietary changes such as feeding more grass and less grain may lead to considerable increases in CLA in meat and milk.

Cholesterol is a component of animal product lipids. In the body cholesterol is utilized as a precursor for steroid hormone synthesis and is an essential component of cell membranes. Dietary cholesterol may be an important factor contributing to exercise induced muscle growth or maintenance especially in older individuals. In humans, high serum concentrations of cholesterol, especially LDL cholesterol have been shown to increase the risk of cardiovascular disease. On the other hand, dietary intake of cholesterol, particularly at levels found in meat, has little impact on serum cholesterol levels in normal individuals. Cholesterol is synthesized by the body at the rate of 600 to 1500 mg per day. When cholesterol is consumed, the amount synthesized by the body decreases. USDA Dietary Guidelines recommend no more than 300 mg cholesterol intake per day. Red meat products account for about 26 percent of the cholesterol consumed by Americans (Table 12.2).

Research on dietary fats and cardiovascular disease has shown mixed results. Diet is one of several risk factors including age, sex, genetics, blood pressure, tobacco smoke and stress which contribute to cardiovascular disease. Also, the high caloric content of dietary fats (9 Calories per gram) contributes to increased obesity. Obesity and relative inactivity are additional factors cited as contributing to the high to incidence of cardiovascular diseases in the US. Thus, discretion should be exercised in amount of fat consumed, whether of animal or vegetable origin, to curtail caloric intake. Nevertheless, a 100-gram serving of cooked lean meat with separable fat removed contains only about 165–230 Calories (Table 12.3), which represent 8 to 12 percent of a 2000-Calorie diet. About 100 to 150 Calories of such a serving are from protein, and the remainder are from lipids.

Obesity and high caloric intake have been associated with increased risk of cancer. Each 1 percent of excess body weight increases the risk of death from cancer by approximately 1 percent.

Beyond this, dietary fat itself may increase chances of cancer. Correlations between amount of fat consumed and incidence of cancer in human populations are significant for all types of fat but are slightly higher for animal fats than fats of other origins.

Meat fats contain ample quantities of fatty acids that are essential in diets of humans. Because daily requirements for essential fatty acids are relatively small, the RDA is met easily from intramuscular fat, even when most external fat is cut away. Fatty acids known to be essential are linoleic and arachidonic and, perhaps, linolenic.

CARBOHYDRATES

There is no specific dietary requirement for carbohydrates. Nevertheless, the US Dietary Guidelines recommend that 45–65 percent of total calories come from carbohydrates especially those found in fiber-rich fruits, vegetables and whole grains. Carbohydrates serve mainly as an energy source (4.5 Calories per gram) and, in the case of certain complex carbohydrates, as dietary fiber. Animal products generally contain limited amounts of carbohydrates and essentially no dietary fiber.

TABLE 12.4 Cholesterol Content of Common Measures of Selected Foods

Food	Amount	Cholesterol (mg)
Milk, skim, fluid or reconstituted dry	1 cup	4
Cottage cheese, uncreamed	1/2 cup	8
Mayonnaise	1 Tbsp.	10
Butter	1 pat	11
Lard	1 Tbsp.	12
Cottage cheese, creamed	1/2 cup	17
Milk, reduced fat, 2%	1 cup	18
Half and half	1/4 cup	23
Ice cream, approx. 10 percent fat	1/2 cup	30
Cheese, cheddar	1 oz	30
Milk, whole	1 cup	34
Oysters, salmon	3 oz	40
Clams, halibut, tuna	3 oz	55
Chicken, turkey, light meat	3 oz	67
Beef, pork, chicken, turkey	3 oz	75
Lamb, veal, crab	3 oz	85
Shrimp	3 oz	130
Lobster	3 oz	170
Heart, beef	3 oz	230
Egg or egg yolk	1 each	250
Liver (beef, calf, pork, lamb)	3 oz	370
Kidney	3 oz	680
Brain	3 oz	1700

From "Diet and Coronary Heart Disease," Council for Agricultural Science and Technology Report No. 107 (1985).

Carbohydrates constitute less than 1 percent of the weight of fresh meat, prompting USDA's National Nutrient Database to list values of zero for these products (Table 12.3). Most of the carbohydrate in meat is present as glycogen with a significant portion of that being converted to lactic acid in postmortem muscle. The liver is the principal storage site for glycogen, so most carbohydrate in the animal body is in the liver. Thus, most fresh meats are poor sources of carbohydrates. Processed meat products such as sausages and cured meat products typically have sugar or other carbohydrates added during manufacture. Added carbohydrates are usually limited to 0.5–1.0 percent but some products such as Sweet Lebanon Bologna may contain as much as 12 percent.

Milk contains about 4.5–5 percent carbohydrate with most being lactose, a disaccharide of glucose and galactose. Many people are unable to adequately digest lactose and sometimes exhibit symptoms of lactose intolerance after consuming a meal containing this disaccharide. Lactose intolerance is caused by inadequate production lactase in the small intestine. Lactase is an enzyme which breaks the disaccharide bond of lactose releasing glucose and galactose, which are then absorbed into the bloodstream. Lactose intolerance is most common among African and Asian Americans and American Indians. It is least common among people of northern European descent. Lactose intolerance becomes more apparent in older individuals as lactase production decreases with age. Dairy products such as cheese, cottage cheese and yogurt have reduced lactose due to microbial fermentation or aging.

Whole egg contains about 0.77 percent carbohydrate including roughly equal amounts (0.11 percent) of sucrose, fructose, lactose, maltose and galactose. Glucose is present at about 0.21 percent. The carbohydrate in egg is located in both the yolk and the albumen.

MINERALS

Animal products in the diet combine to provide adequate amounts of most minerals. Meat is generally a significant source of all minerals except calcium while milk is an excellent source of calcium. Animal products provide most of the calcium, phosphorus and zinc in the diet (Table 12.2). Minerals in meat are associated mostly with lean tissue. In eggs the minerals are distributed between the yolk and albumen. The mineral composition of selected animal products is presented in Table 12.5.

Meat is an especially good source of iron, an essential nutrient for maintaining good health (Table 12.5). Iron is required for synthesis of hemoglobin, myoglobin, and certain enzymes. Because little free iron is stored in the body, regular intake of dietary iron is important. Red meat and poultry account for nearly 15 percent of the dietary iron consumed daily in the US. Also, the iron from meat is in a readily absorbable form. Absorbable iron is the proportion of dietary iron the body can absorb. It varies widely, depending on iron status of the consumer and type of meal consumed. Meals that include meat contain greater absorbable iron regardless of iron source. That is, **heme iron**, which accounts for about 40–60 percent of the iron in meat, is several times more absorbable than **non-heme iron**, and meat contains an unidentified factor that increases absorbable iron from all non-heme sources.

Meal planners should consider the amount of absorbable iron contributed by meat and the enhancement of iron absorption provided by meat. For example, servings of lean beef and navy beans contain approximately equal quantities of iron but beef contains nearly four times the amount of absorbable iron as navy beans. Further, individuals must consume meat at the same time as other iron-containing foods to benefit from the meat factor that increases iron absorption from those foods. Adult men and women require 10 and 15 mg of iron per day, respectively. A serving of liver provides 100 percent or more of these requirements. A 100-gram serving of meat provides about 32 and 21 percent of the RDA for men and women, respectively, plus the enhancement effect of iron absorption from other sources. However, American women consume on average only about 90 percent of the RDA for iron.

TABLE 12.5	Mineral Content of Meat and Poultry Retail Cuts,[a,b] Milk and Eggs								
	Mg/100 gm								
	Calcium	Iron	Magnesium	Phosphorus	Potassium	Sodium	Zinc	Copper	Manganese
Beef	9	2.99	26	233	360	67	6.93	0.125	0.017
Chicken, light meat	15	1.06	27	216	247	77	1.23	0.050	0.017
Chicken, dark meat	15	1.33	23	179	240	93	2.80	0.080	0.021
Lamb	15	2.05	26	210	344	76	5.27	0.128	0.028
Pork	21	1.10	26	237	375	59	2.97	0.61	0.018
Veal	24	1.16	28	250	338	89	5.10	0.120	0.038
Whole Milk	113	0.03	10	91	143	40	0.41	0.011	0.003
Egg, Fried	59	1.98	13	208	147	204	1.20	0.111	0.041

[a]Roasted
[b]Composite of cuts, separable lean only
Source: USDA National Nutrient Database for Standard References, Release 20 (2007)

Zinc is another mineral that is more available to the body when provided from meat rather than plant sources. In spite of this, most Americans consume less than the 15 mg RDA for zinc. However, more than 40 percent of dietary intake of zinc comes from red meat, poultry, and fish (Table 12.2). When plant foods replace meat in the diet, high levels of ingested phytate may chemically bind zinc and reduce its availability for uptake by intestinal cells. Zinc is essential for growth, wound healing, immunity, taste acuity, and DNA synthesis.

Meat contains significant amounts of other minerals required in the diet. Diets that contain sufficient meat to fulfill protein requirements are likely to provide sufficient sodium, potassium, and magnesium. Copper also is found in liver and some seafoods.

VITAMINS

Meat is generally an excellent source of B complex vitamins, but is a poor source of vitamin C, and except for liver is a poor source of vitamins A, D, E, and K. All B complex vitamins are present in meat, but thiamine, riboflavin, niacin, vitamin B6, and **vitamin B12** are present in highest quantities (Table 12.6). Abundant quantities of thiamin are found in pork, niacin and vitamin B6 in chicken, and vitamins B6 and B12 in beef. Red meat and poultry products account for over 63 percent of vitamin B12 in the diet. Liver is rich in B complex vitamins, especially riboflavin and niacin. Unless supplements are used, proper B vitamin nutrition is extremely difficult to achieve without meat in the diet.

Liver is the only meat product rich in vitamin A. Its concentration of the vitamin is quite high and far exceeds the RDA. All meat is a poor source of vitamin C, except when ascorbate has been added to processed meat products.

VARIETY AND PROCESSED MEATS

Variety meats have variable content of protein as compared to skeletal meats (Table 12.7). Yet these items are often more economical sources of protein and vitamins than conventional retail cuts of meat. In addition, liver provides the richest source of iron, vitamin A, niacin, and riboflavin of any ordinary food. Nutritionists recommend that liver be included in the diet on a regular basis.

TABLE 12.6   Vitamin Content of Meat and Poultry Retail Cuts,[a,b] Milk and Eggs

	mg/100 gm						IU/100 gm	microgram/100 gm	
	Thiamin	Riboflavin	Niacin	Pantothenic Acid	B6	E	A	Folacin	B12
Beef	0.100	0.240	4.130	0.400	0.370	0.14	0	8	2.64
Chicken light meat	0.065	0.116	12.42	0.972	0.600	0.27	29	4	0.34
Chicken dark meat	0.073	0.227	6.548	1.210	0.360	0.27	72	8	0.32
Lamb	0.100	0.280	6.320	0.690	0.160	0.19	0	23	2.61
Pork	0.846	0.345	5.172	0.684	0.434	0.18	7	1	0.75
Veal	0.060	0.340	8.420	1.330	0.330	0.42	0	16	1.65
Whole milk	0.044	0.183	0.107	0.362	0.036	0.06	102	5	0.44
Egg, fried	0.075	0.518	0.077	1.558	0.155	1.22	729	51	1.39

[a]Roasted, separable lean only
[b]Composite of cuts
Source: USDA National Database for Standard References, Release 20 (2007)

Processed meat products are to a greater or lesser extent formulated to predetermined composition targets during manufacture. Since most processed products are ready-to-eat, there is little opportunity for consumers to modify the composition before eating, as they might do by trimming the fat from the surface of a cooked steak. Many of the processed products currently offered contain less protein and more fat than the consumed portions of fresh meat (Table 12.8). However, there is great opportunity to change the composition of processed products in response to consumer demand. Low fat and fat free versions of traditional products (Table 12.8) generally are available. Such products often contain added starch, hydrocolloid or non-meat protein along with additional water so that the caloric content may not be greatly reduced. Percentages of some minerals in processed meats are higher than in fresh meat because of added salt and seasonings (Table 12.9). Of particular concern to some consumers is the amount of sodium in processed meat because of its association with hypertension. In recent years, as consumers have become increasingly sensitive to salt, meat processors have reduced the amount used in various smoked and luncheon meats. In addition, low sodium versions of many meat products are offered with potassium salts used in place of sodium salts.

NUTRIENT RETENTION DURING HEATING

Nutritive value of meat proteins remains essentially unchanged during heat processing or cooking. Mineral content of meat is likewise unchanged, but some B complex vitamins may escape in drippings. Because these vitamins are water soluble, moist cookery methods result in greater losses than dry methods. In addition, thiamin may be partially destroyed by heat and other processes. For most complete use of nutrients in meat, drippings should be salvaged and the water portion used in preparation of other foods.

NUTRIENT LABELING

Nutrition labeling of retail meat products provides useful information to consumers who are concerned about diet composition. Prepackaged items including fresh meats and luncheon meat products

TABLE 12.7 — Protein and Fat Composition Percentages for Cooked[a] Variety Meats

	Beef		Pork		Veal		Lamb		Chicken	
	Protein	Fat	Protein	Fat	Protein	Fat	Protein	Fat	Protein	Fat
Brain	11.07	12.53	12.14	9.51	10.5	7.4	12.7	9.2		
Heart	28.79	5.62	23.60	5.05	26.3	4.5	21.7	5.2	26.41	7.92
Kidney	25.48	3.44	25.40	4.70	26.3	5.9	23.1	3.4		
Liver	24.38	4.89	26.02	4.40	21.5	7.62	3.7	10.9	24.36	5.45
Lung	20.40	3.70	16.60	3.10	18.8	2.6	20.9	3.0		
Pancreas	27.10	17.20	28.50	10.80	29.1	14.6	23.3	9.6		
Spleen	25.10	4.20	28.20	3.20	23.9	2.6	27.3	3.8		
Thymus	21.85	24.98			18.4	2.9				
Tongue	22.11	20.74	24.10	18.60	26.2	8.3	21.5	20.5		
Gizzard									27.15	3.66

[a]Simmered or braised

Source: US Department of Agriculture Handbook Nos. 8, 8–5, 8–10, 8–13 and 8–17 and National Live Stock and Meat Board, "The Nutritive Value of Meat," (1976).

TABLE 12.8 — Proximate Composition and Caloric Content of Selected Sausage, Cooked, and Smoked Meats

	Protein	Percent moisture	Fat	Calories per 100 gm
Bacon, raw	8.66	31.58	57.54	556
Bacon, cooked	30.45	12.94	49.24	576
Bologna, beef, and pork	11.69	54.30	28.26	316
Bratwurst, cooked	14.08	56.13	25.87	301
Braunschweiger	13.50	48.01	32.09	359
Dutch loaf	13.42	59.41	17.82	240
Frankfurter, beef & pork	11.28	53.87	29.15	320
Frankfurter, chicken	12.93	57.53	19.48	257
Frankfurter, light	11.40	66.75	14.90	193
Frankfurter, fat free	12.60	78.80	0.60	73
Headcheese	16.00	64.75	15.78	212
Ham, sliced, extra lean	19.35	70.52	4.96	131
Ham, sliced, regular	17.56	64.64	10.57	182
Luncheon meat, pork & beef	12.59	49.28	32.16	353
Pate', goose liver	11.40	37.04	43.84	462
Pepperoni, pork and beef	20.97	27.06	43.97	497
Polish sausage, pork	14.10	53.15	28.72	326
Pork sausage, fresh	11.69	44.52	40.29	417

Continued...

Pork sausage, cooked	19.65	44.57	31.16	369
Salami, cooked, beef & pork	13.92	60.40	20.11	250
Salami, dry, beef & pork	22.86	34.70	34.39	418
Turkey ham, thigh meat	18.93	71.38	5.08	128
Turkey roll, light meat	18.70	71.55	7.22	147

Source: US Department of Agriculture Handbook Nos. 8 and 8–7.

TABLE 12.9 Mineral Content of Selected Sausage, Cooked and Smoked Meats

	Mg/100 gm					
	Iron	Magnesium	Potassium	Sodium	Zinc	Copper
Bacon, raw	0.60	9	139	685	1.15	0.064
Bacon, cooked	1.61	24	486	1596	3.26	0.17
Bologna, beef and pork	1.51	11	180	1019	1.94	0.08
Frankfurter, beef and pork	1.15	10	167	1120	1.84	0.08
Ham, sliced, extra lean	0.76	17	350	1429	1.93	0.07
Ham, sliced, regular	0.99	19	332	1317	2.14	0.10
Luncheon meat, pork and beef	0.86	14	202	1293	1.66	0.04
Pepperoni, pork and beef	1.40	16	347	2040	2.50	0.07
Pork sausage, cooked	1.25	17	361	1294	2.50	0.14
Turkey ham, thigh meat	2.76		325	996		
Turkey roll, light meat	1.28	16	251	489	1.56	0.04

Source: US Department of Agriculture Handbook Nos. 8 and 8–7.

must carry a full Nutrition Facts panel (Figure 12.1) showing the serving size and servings per container along with total calories, calories from fat, total fat, saturated fat, cholesterol, sodium, total carbohydrates, dietary fiber, sugars, protein, vitamin A, vitamin C, calcium, and iron. Percentages of RDA per serving are required for vitamin A, vitamin C, calcium, and iron. Both quantity and percent of Daily Value are given for each item.

Certain additional items may be added to the Nutrition Facts panel if the processor wishes. The composition information provided in the Nutrition Facts panel may be based on chemical analysis or estimates calculated from published database values and product formulation.

13

MEAT INSPECTION AND FOOD SAFETY

OBJECTIVES: *Explore the history and development of meat inspection and the application of meat inspection laws and regulations to insure safe, wholesome muscle food products.*

- Antemortem inspection
- Postmortem inspection
- Product inspection
- Sanitation standard operating procedures (SSOP)
- Food inspectors
- Laboratory inspectors
- Veterinary medical officers
- Official establishment
- Horse meat or horse meat product
- Sanitation
- Condemned
- HACCP or hazard analysis critical control point
- Hazard analysis
- Critical control point
- Critical limit
- Corrective action
- Record keeping
- Verification
- US condemned
- Bovine spongiform encephalitis or BSE
- Non-ambulatory
- US retained
- US inspected & condemned
- US inspected & passed
- Tanking
- Approved ingredients
- Reinspection
- Kosher
- Halal

HISTORY

Since earliest recorded history, people have recognized the importance of wholesome sources and proper processing of their meat supply. For public health, the earliest Mediterranean civilizations regulated and supervised slaughter and handling of meat animals. For example, Mosaic laws proclaimed those animals that were suitable and unsuitable for human food. These food laws (Leviticus 11, and Deuteronomy 14) forbidding consumption of pork and many other types of meat, are strictly observed by Orthodox Jews. Islamic laws in the Quran forbid consumption of pork, products containing pork, or blood by Muslims.

Several noted Greek and Roman writers, including Aristotle, Hippocrates, and Virgil, recognized the similarity between diseases of animals and humans. However, the medieval Roman Catholic Church voiced displeasure toward any speculation about possible relationships between them. As a consequence, meat inspection in medieval Europe often was carried out in opposition to the Church, and in a sporadic and superficial manner. Nevertheless, meat inspection was practiced in France as early as 1162, in England by 1319, and in Germany by 1385.

Meat inspection in the United States was carried out in a rudimentary manner before passage of the Meat Inspection Act of 1906. In early colonial times, production and slaughter of livestock were entirely local enterprises. Animals were slaughtered by local butchers and sold to customers who generally were able to identify products closely with the butcher and even, in many cases, with the farmer who produced the animal. As a rule, local butchers were prominent members of the community and, in attempts to eliminate diseased animals, usually scrutinized every animal slaughtered. But because of a general lack of training in detecting diseased conditions, much unwholesome meat was inevitably processed. Later, as population increased and transportation systems developed, the livestock and meat industry expanded from local to national enterprises. More animals were slaughtered and processed by a few large slaughter plants and fewer animals by local butchers.

With concentration of livestock marketing and slaughtering at central locations, needs developed for improved sanitation and inspection procedures to safeguard consumers from sale of unwholesome meat and meat products. In the early 1880s, the press focused public attention on the problem of quality and purity of food products sold to the public. Meat was the most suspected product, and consequently, attention was focused on the growing packing industry. Chicago newspapers published charges of unsanitary conditions and diseased animals slaughtered for human consumption. These reports adversely affected meat exports and created a growing public suspicion of packers and their products. Chicago packers and processors of meat products were aroused to action and, in efforts to correct deplorable situations, they cooperated with health authorities to create an improved inspection system that would detect diseased animals. However, this improved system did not entirely correct existing situations.

In 1890, in response to growing demand for action, Congress passed a law that provided for inspection of meat "in the piece," but only when it was intended for export. Obviously, this initial move by Congress was designed to assist foreign trade, rather than to protect the American public. Provisions of this law provided for inspection to determine the manner in which products such as salted pork and bacon were packed, and the condition of these products at time of shipment. Since the law did not authorize inspection of animals at the time of slaughter, diseased animals were not detected. Foreign governments consequently refused to remove the prohibition they had placed against American pork, because inspection required by law was deemed inadequate.

In 1891 Congress passed a law that extended the inspection system. This law stipulated that the Secretary of Agriculture would establish a pre-slaughter inspection system for all cattle, hogs, and sheep that were intended for interstate commerce. The success of this law led to a further extension of the inspection system, with passage of an amended law in 1895. This act conferred power upon the Secretary of Agriculture to make such rules and provisions as deemed necessary to prevent transportation of condemned carcasses of cattle, hogs, or sheep from one state or territory to another state or territory or to any foreign country. However, these laws and their administration did not satisfy demands of the American public for an adequate national system of meat inspection. No provision had been made to control sanitation of surroundings in which animals were slaughtered and their meat processed. The press continually focused public attention on the unsanitary manner in which packing houses were being operated. Upton Sinclair's book, *The Jungle*, depicted existing conditions in the packing industry and aroused public anger against the industry. These reports led to appointment of committees by the Secretary of Agriculture and the President to investigate conditions of the packing industry in the Chicago area. A message to Congress, entitled "Conditions in the Chicago Stockyards," from President Theodore Roosevelt brought about enactment and passage of a comprehensive Meat Inspection Act on June 30, 1906. This act applied not only to inspection of live cattle, hogs, sheep, and goats before slaughter (**antemortem inspection**), and their carcasses following slaughter (**postmortem inspection**), but to meat and meat products in all stages of processing as well (**product inspection**). It also applied to surroundings in which livestock were slaughtered and meat was processed. The Horse Meat Act of 1919 extended provisions of the Meat Inspection Act to cover horse slaughter and processing of horse meat products.

Inspection of poultry was not covered in the Meat Inspection Act of 1906. Until the middle of the twentieth century, the bulk of poultry was bought by consumers either alive or "New York dressed" (only blood and feathers removed). Housewives dressed and prepared the birds and determined whether the meat was wholesome. By the 1920s, with development of the poultry processing industry, needs for legislation covering inspection of poultry had become apparent. In 1926, a voluntary Poultry Inspection Program evolved, patterned after an agreement among the US Department of Agriculture, the New York Live Poultry Commission Association, and the Greater New York Live Poultry Chamber of Commerce.

Voluntary inspection was paid for by poultry slaughter and processing plants. Subsequently, other cities passed ordinances requiring inspection of poultry and poultry products shipped into these cities. A Poultry Products Inspection Act was signed into law in 1957. Provisions of this law were similar to those contained in the Meat Inspection Act of 1906.

The Humane Slaughter Act of 1958 required plants desiring to sell meat products to federal agencies to slaughter animals in a humane manner, as described by the Secretary of Agriculture. Approved methods of immobilizing animals prior to exsanguination include chemical, mechanical, and electrical methods. Exposure of animals to carbon dioxide gas, in a way that accomplishes anesthesia with minimum excitement and discomfort, is an approved chemical method. Captive bolt stunners, compressed air concussion devices, and firearms are examples of approved mechanical devices. Administration of electrical current to produce surgical anesthesia is approved as an electrical method. Kosher slaughtered animals are exempt from provisions of this act.

The Meat Inspection Act of 1906 and the Poultry Products Inspection Act of 1957 applied to animals slaughtered and products processed for interstate and foreign commerce, but did not apply to intrastate commerce. Small slaughterhouses and processors selling meat in strictly local areas were not required to comply with provisions of these acts. They needed only to comply with non-uniform and often poorly enforced city or state laws. To improve on this situation, the Wholesome Meat Act and the Wholesome Poultry Products Act were enacted by Congress and signed into law in 1967 and 1968, respectively. These laws provided for a state-federal cooperative program, under which all intrastate slaughter and processing plants were placed under state supervision. The act provided for federal financial, technical, and laboratory assistance in setting up state meat and poultry inspection programs equivalent to federal standards. Individual states were given two years in which to bring their meat inspection programs up to these standards. Some states failed to act, and the federal government assumed all meat inspection responsibilities within them.

When the Wholesome Meat Act and Wholesome Poultry Products Act were passed, provisions of these acts and those of the Federal Meat Inspection Act of 1906, and its amendments, and of the Poultry Products Inspection Act of 1957, were consolidated. Under provisions of that legislation, antemortem inspection is now mandatory for all cattle, sheep, swine, goats, horses, mules, and other equine animals and chickens, turkeys, ducks, geese, and guineas. Under provisions of the act of 1906, antemortem inspection was not mandatory, but the Secretary of Agriculture, at his discretion, could require an antemortem inspection of all cattle, sheep, swine, and goats. The current legislation provides for exemptions from antemortem and postmortem inspection when the meat will be used by the owner of the animal (meat is not sold) or, in the case of poultry, when the operation raises and slaughters less than 20,000 birds per year.

The Processed Products Inspection Improvement Act of 1986 was passed in response to a need for improved use of meat inspection resources. A new concept of discretionary inspection was incorporated into law, which allows processing plants (not slaughter plants) with total quality control programs to receive less than continuous inspection. The shifting of more inspection responsibility to processors having excellent records of compliance allowed reallocation of inspection resources for improvement of marginally acceptable plants, development of new analytical tests, training of highly skilled inspection teams, and other uses. Even though meat inspector presence was lessened in some plants by this act, the purpose of the change was to increase overall effectiveness of product inspection by better use of personnel and resources.

Studies and reports by the National Academy of Science, the Government Accounting Office and the USDA between 1980 and 1995 addressed the adequacy of the meat inspection system to assure safe, wholesome meat and poultry products. These reports uniformly recommended implementation of more science-based inspection procedures. In the early 1990s, several serious and highly publicized food poisoning outbreaks caused by the organism *Escherichi coli O157:H7* and several other food poisoning organisms further heightened concern about the safety of meat and poultry products and the inspection procedures used for their production. In 1996, USDA issued sweeping new regulations, the **Pathogen Reduction and Hazard Analysis Critical Control Point (HACCP)** systems, to improve food safety of meat and poultry products. These systems were designed to reduce the occurrence and numbers of pathogenic microorganisms on meat and poultry products, reduce the incidence of foodborne illnesses caused by consumption of these products, and modernize the meat and poultry inspection system.

APPLICATION AND ENFORCEMENT OF INSPECTION LAWS

The Secretary of the US Department of Agriculture is given responsibility of administering federal meat and poultry inspection laws and is authorized to promulgate regulations that will implement these laws and enable effective administration of provisions. The federal inspection program, as conducted by the Food Safety and Inspection Service (FSIS), consists of

- the examination of animals and their carcasses at slaughter
- inspection at all stages of preparation of meat and meat food products to assure sanitary handling and sanitation of equipment, buildings, and grounds
- verification of adequacy and effectiveness of each facility's or plant's **sanitation standard operating procedures (SSOP)** and HACCP plans
- product sampling and analysis to assure compliance with performance standards
- condemnation of unfit products to prevent their use for human food
- examination of all ingredients used in preparation of meat food products to assure their fitness for food

- prescription and application of identification standards for inspected meat food products
- enforcement of measures that ensure informative labeling
- prohibition of false and deceptive labeling
- inspection of foreign meat and meat food products offered for importation
- administration of a system of certification to assure acceptance of domestic meats and meat food products in foreign commerce.

Federal and state governments hire, train, pay, and assign meat inspectors. Cost of meat inspection to the federal government amounts to less than 2 dollars per person per year. The packer or processor is required to compensate the government only for the cost of overtime inspection.

FIGURE 13.1 — A federal inspection stamp similar to those used on inspected fresh meat cuts.

Meat Inspectors

Federal inspection personnel are classified into three groups: food inspectors, laboratory inspectors, and veterinary medical officers. **Food inspectors** are experienced laymen who are trained to assist in antemortem and postmortem inspections, monitor SSOP and HACCP reports and compliance, and report to **veterinary medical officers**, who are graduate veterinarians. These veterinary medical officers are in charge of inspection, wherever slaughtering is conducted. **Laboratory inspectors** are trained generally in food or meat science and specifically in bacteriological and chemical examination of meat and meat products. All permanent employees engaged in federal meat inspection are civil service employees.

FIGURE 13.2 — A federal inspection stamp similar to those used on inspected meat products.

Identification of Inspected Products

The presence of federal or state meat inspection stamps on large cuts of meat (Figure 13.1) or meat products (Figure 13.2) assures consumers the product was wholesome when it was shipped from the plant where it was inspected. There are specific legal provisions controlling use of inspection stamps. The circular stamp inscribed with the legend "US INSP'D & P'S'D" must have a number on it to identify the **official establishment**. For example,

FIGURE 13.3 — Imported meat bears the inspection stamp of the country exporting the product and that of the US. The latter stamp includes the name of the port of entry.

the number 38 is reserved by the US Department of Agriculture for illustrative purposes. Each plant under federal inspection has a different number. In some cases, letters appear with numbers to indicate specific plants in multiple plant operations. Special edible inks approved for use by the Food and Drug Administration are used to apply inspection legends on fresh meats. Poultry products are stamped with a round inspection brand identical to that used for red meat species. However, the establishment number is preceded by a P to indicate poultry.

Meat imported into the US from foreign countries is subjected to the same controls as federally inspected meat. Countries that desire to export meat to the US request acceptance of their system of inspection and certification as equivalent to that maintained by the US. Investigations of foreign plants then are made to determine whether inspection is comparable to that in the US. If foreign inspection is satisfactory, arrangements are made to provide a system of certification whereby each shipment of meat, upon arrival at a port of entry in the US, is identified by source. If, upon inspection by FSIS, the meat is found to be wholesome, free from adulteration, and properly labeled, it is stamped (Figure 13.3), permitted to enter the US, and moved freely in interstate commerce.

Meat exported from the US carries certificates of inspection recognized in foreign countries. Consignments of meat for foreign countries are not permitted to leave the US unless they are identified by export stamps or certificates indicating products were prepared under FSIS supervision.

REQUIREMENTS FOR GRANTING INSPECTION SERVICE

Owners or operators of meat processing plants who plan to engage in interstate or foreign meat trade, or to furnish meat to federal agencies, must send detailed information relative to the nature and volume of proposed operations to FSIS. Operators then will be informed of acceptance or rejection of proposed facilities for federal inspection. If so, formal application forms are furnished. Along with this application, operators are required to furnish detailed plans and specifications of proposed plants. Applications must be accompanied by drawings showing the character of plant construction, arrangement of all essential equipment, and location of the establishment. Included in federal requirements for inspection are potable and ample water supplies, approved sewage systems, adequate natural or artificial light, adequate ventilation, adequate refrigeration, and ample supplies of pressurized hot water for cleaning. In addition, establishments must have developed written SSOPs and have conducted a hazard analysis and developed and validated a HACCP plan. If these requirements are met, inspectors are assigned and the system of inspection is applied in accordance with provisions of inspection laws.

Federal meat inspection is conducted only at establishments that meet sanitary standards and are equipped with facilities that assure sanitary preparation, handling, and storing of meat. If a plant does not comply with these standards, inspection is withdrawn and the plant cannot sell meat. Any anticipated change in facilities and equipment under federal inspection must be in compliance with federal regulations. States having cooperative meat inspection programs in compliance with the Wholesome Meat Act or the Wholesome Poultry Products Act have requirements and procedures similar to the federal system for granting inspection service.

US Department of Agriculture regulations require that slaughter of horses and processing of horse meat products be conducted in plants that are separate, by location or time, from those processing other meat animals. Any meat or food products from horses must be plainly and conspicuously labeled, marked, branded, or tagged **Horse meat** or **Horse meat products** as the case may be. A hexagonal brand in green ink is used for horse meat (Figure 13.4). Labels on horse meat products must be printed with black ink on light green paper.

US Department of Agriculture regulations require that poultry be slaughtered and processed in facilities that are separate, by location or time, from those used for other meat animals. The basis for this regulation is the recognized

 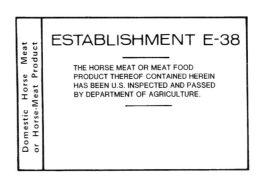

FIGURE 13.4 A federal inspection stamp similar to those on inspected fresh horse meat, and a label similar to those used on inspected horse meat products.

difference between large animals and poultry in populations of normal and pathogenic microflora carried by each.

ELEMENTS OF INSPECTION

Elements of inspection consist of the following essential parts: **sanitation**, hazard analysis and HACCP, antemortem and postmortem inspections, control and disposition of **condemned** materials, product inspection, laboratory inspection, and marking and labeling. These elements are applicable to all meat animals, poultry, and horses.

Sanitation and SSOP

For purposes of this text, sanitation includes all procedures and processes used to maintain or restore cleanliness and insure hygiene in meat and poultry processing in order to prevent foodborne illnesses. An excellent sanitation program is the first key to reducing the incidence of pathogenic microorganisms and insuring food safety in meat products. USDA meat inspection regulations require each establishment to develop and implement sanitation standard operating procedures (SSOP). Requirements for and enforcement of sanitary standards and SSOPs begin with livestock pens or cages and extend to every subsequent operation. Attention is given to every component of the environment in which animals are slaughtered and meat is handled. This includes structural aspects of premises, water supply, equipment used, sewage disposal, and locker and rest-room facilities for personnel. No establishment is allowed to employ, in any department where meat products are handled or prepared, any person affected with a serious communicable disease.

SSOPs are developed to describe all procedures that are followed daily and during operations to prevent the direct contamination or adulteration of products. SSOPs must describe steps taken before beginning operation to insure clean facilities, equipment, and utensils, the steps taken during operation to maintain a sanitary environment such as cleaning, sanitizing, and disinfecting during breaks or between shifts, and the steps taken to clean and sanitize facilities, equipment, and utensils at the completion of operations. Employee hygiene is described, including cleanliness of frocks, aprons, and gloves, washing of hands, use of hair coverings, and health and personal hygiene. Finally, SSOPs specify how implementation and effectiveness will be monitored. Personnel responsible for monitoring SSOPs are designated and a record keeping system is established to document results. Effectiveness of SSOPs may be monitored in several ways including sensory (e.g., sight, smell or feel), chemical (e.g., check of chlorine level), or microbiological (e.g., swabbing surfaces for total plate count determination) measurements. Corrective actions must be taken and documented whenever there is a deviation from established procedures.

For slaughter operations, SSOP effectiveness is monitored by sampling selected carcasses for the presence of E. coli, biotype 1, as an indicator of fecal contamination. USDA regulations mandate this monitoring by plant personnel and specify the acceptable performance criteria. If performance criteria are not met, corrective actions must be implemented to reduce the occurrence of fecal contamination and lower the E. coli count. Nevertheless, these test results are not a basis for carcass condemnation. Current regulations specify zero tolerance for visible fecal contamination. Corrective procedures for visible fecal contamination include trimming of the contaminated area and modification of slaughter procedures to prevent contamination of other carcasses.

Inspectors are responsible for verifying the effectiveness and adequacy of SSOPs and for identification of deficiencies. Verification may include review of SSOPs, review of daily records maintained by the establishment and of any corrective actions, direct observation of procedures and corrective actions, and observation or testing to assess sanitary conditions. When deficiencies occur, the seriousness of the deficiency determines the nature of the response. For example, if it is certain that a deficiency will result in an adulterated or contaminated product that will reach and have a detrimental effect on consumers, the response is immediate cessation of the practice or process in question and implementation of corrective action. Distribution of the product affected by the deficiency would be prohibited and it may be condemned indicating that it is unfit for human consumption. When determined by inspectors that any equipment, utensil, room, or compartment of an establishment is unclean, or its use would be in violation of sanitary requirements, the item or area may not be used until it has been made acceptable. In addition, the discovery of unsanitary conditions is taken as an indication that the SSOPs are inadequate or ineffective. This may lead to withdrawal of inspection, effectively closing the plant.

Hazard Analysis Critical Control Points (HACCP)

Revised inspection regulations issued in 1996 established **Hazard Analysis Critical Control Points (HACCP)** as the organizing structure for FSIS meat food safety programs. HACCP principles provide the framework to develop science-based process control systems that minimize safety hazards in food production systems. HACCP is designed to identify safety hazards, establish controls, and monitor the effectiveness of the controls. These principles were first applied in food production in 1959 by the Pillsbury Company to assure safe food for NASA astronauts during space travel. Today, HACCP is accepted and used worldwide in the design and implementation of safe food processing systems.

HACCP as defined in meat inspection is based on the application of seven principles to all operations in slaughter and processing, with the objectives of making products as safe as possible, and documenting that product was processed in as safe a manner as possible. These principles are:

- Conduct a hazard analysis. A **hazard analysis** is the process of identification and listing of all the food safety hazards that are reasonably likely to occur during the production of a product and of the preventative measures needed to control the hazards. Hazards may be biological, chemical, or physical factors that may adulterate a product or make it unsafe for human consumption. An identified hazard should be something whose prevention, elimination, or reduction to acceptable levels is essential to produce a safe product.
- Identify critical control points. A **critical control point** (CCP) is a point in the process, a step, or a procedure where control can be applied and a food safety hazard can be prevented, eliminated, or reduced to acceptable levels. Critical control points are identified by asking a series of questions about all of the hazards identified during the hazard analysis: (1) Do preventative measures exist for the hazard? (2) Will the preventative measure eliminate or reduce the occurrence of the hazard to an

acceptable level? (3) Could contamination occur or increase to greater than acceptable levels? (4) Will subsequent steps in production eliminate the hazard or reduce it to acceptable levels? A decision tree for identifying critical control points is illustrated in Figure 13.5.

- Establish critical limits. A **critical limit** is the maximum or minimum value to which a variable must be controlled at a critical control point to prevent, eliminate, or reduce to an acceptable level a food safety hazard. Critical limits could be established for variables such as temperature, time, humidity, moisture level, water activity, pH, salt concentration, preservatives, or survival of pathogens. They are based on FSIS regulations, scientific and technical data, experimental studies, or the

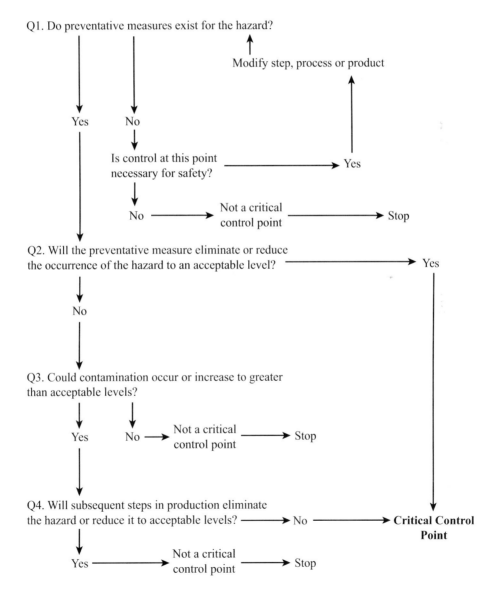

FIGURE 13.5 | Decision tree to identify critical control points.

recommendations of recognized experts. Critical limits must be established for every critical control point.
- Monitor critical control points. Monitoring is an integral part of HACCP. It consists of observations and measurements to insure that the critical control point is within the critical limits. Responsibility for monitoring must be assigned and personnel must be adequately trained to record results and initiate corrective action immediately if critical limits are exceeded.
- Establish corrective actions. HACCP plans must define the corrective actions that will be taken when there is a deviation from a critical limit. **Corrective actions** should define what will be done to bring the process back within the critical limits and the disposition of potentially unsafe products or those that do not meet standards.
- Establish effective record keeping. **Record keeping** procedures that document the entire HACCP process must be developed and maintained. The availability of objective, relevant data is one of the principal benefits of HACCP systems to both industry and inspection personnel.
- Verify the system. **Verification** involves a series of steps to establish that the HACCP system is in compliance with the HACCP plan and/or whether the plan needs to be modified or revalidated to achieve food safety objectives. The establishment is responsible for developing the scientific validation that the critical control points and critical limits are adequate to control potential hazards, for insuring that the entire HACCP system functions properly, and for documentation and reassessment of the HACCP plan. The FSIS also verifies the HACCP system by reviewing the HACCP plan, critical control point records, corrective actions taken when deviations occur, critical limits, and other records and by direct observations or measurements at a critical control point, by sample collection and analysis, or by on-site observations.

TABLE 13.1 A HACCP Plan Checklist

A. Describe the Product

1. Does the HACCP plan include:

    a. The producer/establishment and the product name?

    b. The ingredients and raw materials used along with the product recipe or formulation?

    c. The packaging used?

    d. The temperature at which the product is intended to be held, distributed, and sold?

    e. The manner in which the product will be prepared for consumption?

2. Has a flow diagram for manufacture of the product been developed that is clear, simple, and descriptive of the steps in the process?

3. Has the flow diagram been verified for accuracy and completeness against the actual operating process?

B. Conduct a Hazard Analysis

1. Have all steps in the process been identified and listed where hazards of potential significance occur?

    Have all hazards associated with each identified step been listed?

    Have safety concerns been differentiated from quality concerns?

    Have preventative measures to control the identified hazards been identified, if they exist, and listed?

C. Identify Critical Control Points

1. Has the CCP Decision Tree been used to help determine if a particular step is a CCP for a previously identified hazard?
2. Have the CCPs been entered on the forms?
3. Have all significant hazards identified during the hazard analysis been addressed?

D. Establish Critical Limits

1. Have critical limits been established for each preventative measure at each CCP?
2. Has the validity of the critical limits to control the identified hazard been established?
3. Were critical limits obtained from the regulations, processing authority, etc.? Is documentation attesting to the adequacy of the critical limits maintained on file at the establishment?

E. Establish Monitoring Procedures

1. Have monitoring procedures been developed to assure that preventative measures necessary for control at each CCP are maintained within the established critical limits?
2. Are the monitoring procedures continuous or, where continuous monitoring is not possible, is the frequency of monitoring sufficiently reliable to indicate that the hazard is under control?
3. Have procedures been developed for systematically recording the monitoring data?
4. Have employees responsible for monitoring been identified and trained?
5. Have employees responsible for reviewing monitoring records been identified and trained?
6. Have signatures of responsible individuals been required on the monitoring records?
7. Have procedures been developed for using the results of monitoring to adjust the process and maintain control?

F. Establish Corrective Actions

1. Have specific corrective actions been developed for each CCP?
2. Do the corrective actions address re-establishment of process control, disposition of affected product, and procedures to correct the cause of non-compliance and to prevent the deviation from recurring?
3. Have procedures been established to record the corrective action?
4. Have procedures been established for reviewing the corrective action records?

G. Establish Record Keeping Procedures

1. Have procedures been established to maintain the HACCP plan on file at the establishment?
2. Do the HACCP records include description of the product and its intended use, flow diagram for the process with CCPs indicated, preventative measures, critical limits, monitoring system with corrective action plans for deviations from critical limits and record keeping procedures for monitoring, and procedures for verification of the HACCP system?

H. Establish Verification Procedures

1. Have procedures been included to verify that all significant hazards were identified in the HACCP plan when it was developed?
2. Have procedures been included to verify that the critical limits are adequate to control the identified hazards?
3. Are procedures in place to verify that the HACCP system is functioning properly?
4. Are procedures in place to reassess the HACCP plan and system on a regular basis or whenever significant product, process, or packaging changes occur?

Source: Modified from Burson, D. E. and E. Dormedy, 1998. Hazard Analysis Critical Control Point (HACCP) Model for Frankfurters. University of Nebraska-Lincoln, Lincoln, NE. Copyright © Dennis Burson. Reprinted with permission.

A HACCP plan is required for every product produced by an establishment if the hazard analysis reveals one or more food safety hazards that are reasonably likely to occur. A HACCP plan checklist is presented in Table 13.1 and the steps to develop a HACCP plan consistent with the checklist are presented in Tables 13.2–13.9 and Figure 13.9. These are discussed in detail in the Product Inspection section of this chapter.

Pathogen reduction is an additional focus of the regulations implemented in 1996. Plants may establish microbial monitoring programs as part of their SSOP and HACCP plans. In order to encourage the meat industry to increase its efforts to reduce pathogens in meat products, USDA, FSIS has implemented its own testing of selected raw and ready-to-eat meat and poultry products for specific pathogens: salmonella, *E. coli O157:H7* and *Listeria monocytogenes*.

Salmonella is used as an indicator organism for the presence of a number of other food poisoning organisms because

- it is a very common bacterial cause of foodborne illness
- it colonizes most species of meat animals and occurs at frequencies that permit changes to be detected
- reliable methods are available to recover Salmonella from a variety of raw products
- strategies that reduce fecal contamination and Salmonella are effective in reducing other pathogens.

The specific performance standards for Salmonella vary according to the species and class of carcass and various types of ground raw products. Test results that exceed the performance standards may lead to withdrawal of inspection and effective closure of the plant.

Eschericia coli O157:H7 is a particularly troublesome pathogen often associated with fresh ground beef products. USDA, FSIS regularly samples ground beef in manufacturing plants and distribution channels. The performance standard for this pathogen is zero. Tens of thousands of samples have been analyzed. Approximately 1 sample in 600 is found to contain *Eschericia coli O157:H7*. A positive test leads to condemnation of the contaminated product and recall of product in commerce.

Listeria monocytogenes is a unique pathogen since it can grow in vacuum packages under refrigeration. Thus, it is of particular concern when recontamination after heat processing leads to its presence in packaged, ready-to-eat meat or poultry products. USDA, FSIS samples ready-to-eat products for *Listeria monocytogenes*. The performance standard is zero. A positive test result generally leads to a product recall since packaged products are typically moved quickly into distribution.

Antemortem Inspection

Antemortem examination is made of all cattle, sheep, swine, goats, poultry, and horses, for the purpose of eliminating unfit animals from the food supply. Such inspection is made on the day of slaughter (Figure 13.6). Animals plainly showing symptoms of disease or condition that would cause condemnation of their carcasses on postmortem inspection are marked "**US Condemned**" and are killed by the establishment and disposed of in a prescribed manner. These include dying animals and those showing symptoms of such diseases as rabies, tetanus, hog cholera, anaplasmosis, parturient paresis, anthrax and **bovine spongiform encephalitis** (BSE). Such animals may not be taken into any establishment to be slaughtered or dressed, nor conveyed into any department connected in any way to production of edible products. Animals that do not show distinct symptoms of such diseases, but which are suspected of having a disease or condition that may cause condemnation of part or all of the carcass upon postmortem inspection, are marked "US Suspect". Their identity is maintained during slaughter until final postmortem inspection is performed. With the discovery of BSE in several cattle in Washington State in December, 2003, USDA banned the slaughter of **non-ambulatory** disabled (downer) cattle and their entry into the food supply. This regulation is intended to lessen the risk of meat from BSE-infected cattle entering the food supply.

FIGURE 13.6 | All animals are examined by inspectors before slaughter.

Postmortem Inspection

Postmortem inspection in livestock and poultry slaughter plants has two distinct components—carcass-by-carcass visual inspection and HACCP. Visual inspection is made of carcasses and noncarcass components simultaneously with slaughter and dressing operations (Figure 13.7). Postmortem inspection is conducted under supervision of a veterinary medical officer. Inspectors are located at specific stations along dressing lines so that carcasses, heads, and viscera of all animals may be inspected. Lungs, livers, lymph glands, spleens, and hearts receive particular attention, because symptoms of disease are readily detectable in these organs. Positive identity of all parts of each animal is maintained through use of duplicate numbered tags, or by synchronized chains and moving tables containing the carcass and corresponding viscera.

When inspectors observe lesions or other conditions in carcasses, internal organs, or detached parts that might render them unfit for food purposes, and will therefore require subsequent inspection, the carcasses, organs, or parts are retained by attachment of "**US Retained**" tags. They are completely isolated from other carcasses until complete final examinations are made by a veterinarian. In some instances laboratory examinations are necessary before final decisions may be made.

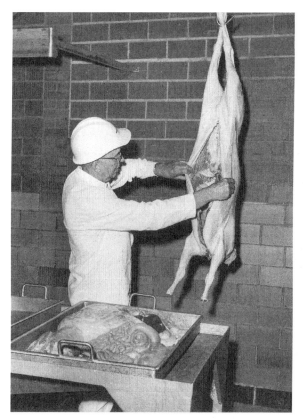

FIGURE 13.7 | All parts of the carcass receive a thorough postmortem examination by an inspector.

If carcasses or parts are found to be unwholesome and thereby unfit for human food, the veterinarian conspicuously marks the surface of the items "**US Inspected and Condemned**." If they are found to be wholesome and fit for human food, the veterinarian allows them to reenter the production line and to be marked "**US Inspected and Passed**." In cases where lesions or conditions are localized, inspectors condemn affected parts, and pass the remainder of such carcasses. Bruises are trimmed out of otherwise healthy carcasses and the remainder passed, provided only limited areas are involved.

Discovery of a BSE infected animal in 2003 led to implementation of regulations intended to minimize human exposure to materials that have been demonstrated to contain the BSE infective agent in cattle that are infected with the disease. Such materials are designated as "Specified Risk Materials" (SRMs). SRMs are inedible and their use for human food is prohibited. They include the tonsils and distal illium of the small intestine of all cattle. They also include the brain, skull, eyes, trigeminal ganglia, spinal cord, vertebral column (except for the tail, transverse processes of the thoracic and lumbar vertebrae, and the wings of the sacrum), and dorsal root ganglia of all cattle 30 months of age or older.

FIGURE 13.8 | Inspection is conducted at all stages of processing, in order to assure that consumer that the product has been processed in compliance with US Department of Agriculture approved regulations.

TABLE 13.2 | Generic Product Description for a HACCP Plan

Process category: raw product	Product: ground beef
1. Common name?	Ground beef
2. How is it to be used?	Cooked and consumed
3. Type of package?	Bulk packed (e.g. plastic bag, vacuum packed)
4. Length of shelf life, at what temperature	12 months at 0° F or below, 7 days at 40° F
5. Where will it be sold? Consumers? Intended use?	Retail and HRI, wholesale General public, no distribution to schools or hospitals.
6. Labeling instructions?	Keep refrigerated
	Cooking instructions (minimum internal temperature for cooking)
	Safe food handling label
7. Is special distribution control needed?	Keep refrigerated

Source: Modified from Generic Model for Raw, Ground Meat and Poultry Products. US Department of Agriculture, Food Safety and Inspection Service. September, 1998. HACCP-3. http://www.fsis.usda.gov/OPPDE/nis/outreach/models/HACCP-3.pdf.

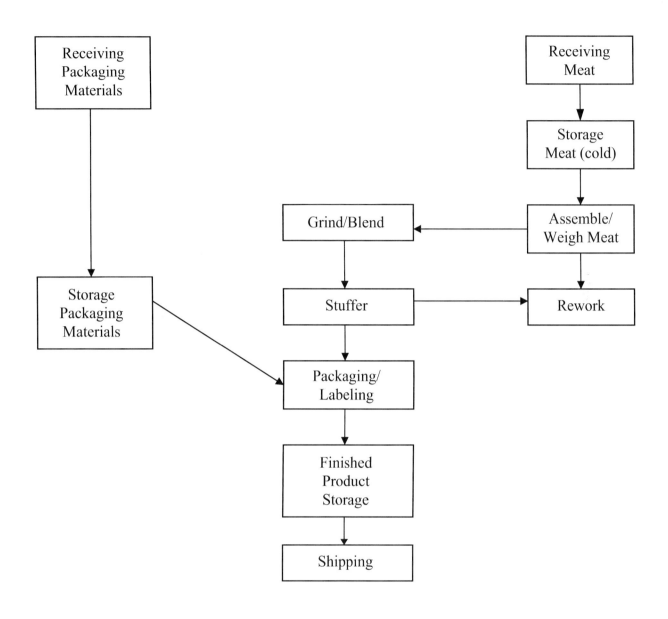

FIGURE 13.9 | Generic example of a process flow diagram for raw ground beef for a HACCP plan. Modified from Generic Model for Raw, Ground Meat and Poultry Products. US Department of Agriculture, Food Safety and Inspection Service. September, 1998. HACCP-3. http://www.fsis.usda.gov/OPPDE/nis/outreach/models/HACCP-3.pdf.

Control and Restriction of Condemned Material

When carcasses, parts of carcasses, or organs are condemned, they must be marked "US Inspected and Condemned." All condemned material remains in the custody of inspectors until disposal by approved methods. The most commonly used method for disposal is called **tanking**. Condemned materials are cooked at temperatures sufficiently high to destroy disease-producing organisms. Products from the tanking process are used only for inedible purposes. All tanks and equipment used for rendering, preparing, or storing inedible products must be in rooms separated from those used for rendering, preparing, or storing edible products.

Product Inspection

Processed meat product inspection requires development, implementation, validation, and verification of HACCP plans for every product manufactured in an establishment. In addition, each step in manufacture of processed meat products may be subject to scrutiny of food inspectors (Figure 13.8). Examples of such products are sausages, canned meat, cured and smoked meats, and edible fats.

A model or generic example of how a hazard analysis might be conducted and a HACCP plan constructed for raw ground beef manufacture is shown in Tables 13.2–13.9. The specific steps begin with preparation of a complete product description (Table 13.2), which describes the product in detail, how it is used, type of packaging, shelf life, distribution, and labeling. All ingredients, casings, and packaging materials used in the manufacture and packaging should be listed The next step is to develop a production flow chart for the product (Figure 13.9). The flow chart lists in sequence all of the operations required to produce the product beginning with raw meat ingredients and non-meat ingredients and ending with shipping and distribution. The production flow chart is then used to conduct the hazard analysis, which is shown in Table 13.3. A critical control point decision tree similar to that illustrated in Figure 13.5 is used to identify the critical control points (CCP). This process is shown in Table 13.3. Next, critical limits are established for each CCP based on existing regulations, scientific literature, or validation studies conducted by the company. Monitoring and recording procedures are established for each CCP along with designation of individuals who are responsible for monitoring, recording, and reviewing monitoring records. Corrective actions are described for responding when any critical limit is exceeded and individuals responsible for making corrective actions and recording them are designated. The specific HACCP records that will be maintained to document monitoring and corrective actions are described. Finally, verification procedures are established and responsible persons are designated to assure that the HACCP plan is followed and to validate the effectiveness of the plan for controlling hazards. The results of these steps are presented in the final model HACCP plan for raw ground beef manufacture (Table 13.4). Tables 13.5 though 13.9 illustrate generic room temperature log, thermometer calibration log, metal detection log, corrective actions log, and pre-shipment review log for a ground beef HACCP plan.

In addition to verified HACCP plans, all formulas used in processed meat products must be filed with and approved by FSIS. Only **approved ingredients** in controlled amounts are permitted in processed meat products. Process temperatures, product yields, and finished product composition are monitored and regulated. Processors as well as inspectors must be thoroughly familiar with and abide by all published regulations governing manufacturing, storage, and handling of processed meat products. This supervision guards against adulteration, contamination, and misrepresentation of meat products before they enter distribution channels.

Reinspection is performed in operations such as cutting, boning, trimming, curing, smoking, rendering, canning, sausage manufacturing, packaging, handling, or storing of meats. All products, even though previously inspected and passed, are reinspected as often as necessary to

TABLE 13.3 Generic Example of Hazard Analysis for a HACCP Plan

Process category: raw, ground
Product: ground beef

Process step	Food safety hazard	Reasonably likely to occur?	Basis	If YES in Colum 3, What measures could be applied to prevent, eliminate or reduce the hazard to an acceptable level?	Critical control point
Receiving—meat	Biological: pathogens *Salmonella*, *E.coli* O157:H7, other pathogens and BSE infective agent in beef trimmings	Yes	*Salmonella* and *E. coli* may be present in incoming raw product. Few BSE infected cattle have been found in the US	Certification from suppliers that product has been sampled for *salmonella* and *E. coli* and meets performance standards. Certification from suppliers that product meets USDA standards for "specified risk materials".	1B
	Chemical	No			
	Physical—foreign materials	No	Plant records show no incidence of foreign materials in products received		
Receiving—packaging materials	Biological—none				
	Chemical—not acceptable for intended use	No	Letters of guarantee received from all suppliers of packaging materials		
	Physical—metal, glass wood.		Records show that foreign material contamination has not occurred for several years		
Storage (cold)—meat	Biological—pathogens *Salmonella* and *E. coli*	Yes	Pathogens are likely to grow in product if temperature is not maintained at or below a level sufficient to prevent growth	Maintain product temperature at or below a level sufficient to prevent pathogen growth.	2B
	Chemical—none				
	Physical—none				
Assemble/Weigh meat	Biological—none				
	Chemical—none				
	Physical—none				

Continued...

Process Step	Hazard	Is hazard a food safety hazard?	Justification	Control Measures	CCP #
Grind/Blend	Biological—none				
	Chemical—none				
	Physical—metal contamination	Yes	Records show that metal contamination may occur during the grinding process	Maintain grinder blades and plates to preclude metal contamination. Routine examination during equipment break-down. There will be a metal detector at packaging.	
Stuffer	Biological—none				
	Chemical—none				
	Physical—none				
Rework	Biological—Pathogens	Yes	Use of rework can provide a medium for pathogen growth	Rework left at the end of the day will be condemned or used in a cooked product at the plant.	
	Chemical—none				
	Physical—none				
Packaging/Labeling	Biological—Pathogens	Yes	Pathogens have historically occurred in raw beef trimmings	Labels that clearly indicate this is a raw product, along with cooking instructions and the safe food handling statement.	3B
	Chemical—none				
	Physical—metal contamination	Yes	Metal contamination may have come in the raw product or occurred during the grinding and stuffing process	A functional metal detector is on-line in the packaging-labeling area to remove product with metal contamination.	4B
Finished product Storage (cold)	Biological—pathogens	Yes	Pathogens are likely to grow in this product if temperature is not maintained below a level sufficient to prevent their growth	Maintain product temperature at or below a level sufficient to preclude pathogen growth.	5B
	Chemical—none				
	Physical—none				
Shipping	Biological—none				
	Chemical—none				
	Physical—none				

Source: Modified from Generic Model for Raw, Ground Meat and Poultry Products. US Department of Agriculture, Food Safety and Inspection Service. September, 1998. HACCP-3. http://www.fsis.usda.gov/OPPDE/nis/outreach/models/HACCP-3.pdf.

TABLE 13.4 Generic Example of a HACCP Plan

Process category: raw product, ground
Product: fresh ground beef

CCP# and location	Critical limits	Monitoring procedures and frequency	HACCP records	Verification procedures and frequency	Corrective actions
1B Receiving—meat Pathogens in raw beef trimmings	Supplier certification that product meets FSIS performance standards for *Salmonella*, *E. coli* and "specified risk materials" must accompany shipment.	Receiving personnel will check each shipment for certifications.	Receiving log	Every two months, QA will request FSIS *Salmonella* and *E. coli* data results from the company for at least 2 suppliers.	Will not receive product unaccompanied by certifications. If product does not have certifications, it is rejected or returned. Assure that procedures for guaranteed supplier list are kept current and guaranty is on file in shipping/receiving log. If supplier does not meet FSIS performance standards, product will not be purchased from them until they can meet performance standards.
2B Storage—(cold) meat	Raw product storage area shall not exceed 40° F.	Maintenance personnel will record raw product storage area temperature every 2 hours, initial/sign and date log.	Room temperature log Corrective action log Thermometer calibration log	Maintenance supervisor will verify accuracy of the Room Temperature Log once per shift and observe plant employee performing monitoring.	QA will reject or hold meat until time/temperature deviation and its implications are reviewed. Product disposition will depend on this expert review. QA will identify the cause of the deviation and devise measures to prevent reoccurrence.

Continued...

CHAPTER 13 MEAT INSPECTION AND FOOD SAFETY

CCP	Critical Limits	Monitoring	Records	Verification	Corrective Actions
(continued)				QA will check all thermometers used for monitoring devices for their accuracy and verify to within 2° F on a daily basis.	Maintenance will adjust scheduled cooler upkeep and review if necessary or repair any malfunctioning equipment.
3B Packaging/ Labeling	Product must clearly be labeled as Raw-Cook Before Eating. Cooking instructions that state "recommend cooking to 160° F" must be on package. Safe food handling statement must be part of label.	Packaging line supervisor will select 2 packages of product hourly and ensure labeling requirements are met.	Corrective action log; Labeling log	QA will observe packaging line supervisor perform monitoring activity once per shift. QA will select 3 labels intended for use from label storage area twice weekly to ensure label accuracy. QA will check labels once a day on packaged product to ensure accurate labels are place on packaged product.	QA will segregate and hold all affected product. QA will ensure that proper labeling is applied to all affected product prior to shipment. QA will determine cause of deviation and institute preventative actions.
4P Packaging/ Labeling	No metal particles larger to exceed 1/32 inch. All contaminated product is removed from system by functioning kick-out mechanism. All kick-out product will be visually examined and any metal removed.	Packaging line supervisor will check the metal detector using a seeded sample every two hours to determine limits are not exceeded.	Metal detection control log; Corrective action log	QA, outside the packaging unit, will verify that the metal detector is functioning as intended by running the seeded sample through the metal detector twice per shift (once AM and once PM).	Packaging supervisor will control and segregate affected product. Maintenance personnel will identify and eliminate any problems with the metal detection or kick-out mechanism. Preventive maintenance program will be implemented. QA will run seeded sample through the detector after repair.

CHAPTER 13 MEAT INSPECTION AND FOOD SAFETY

CCP# and location	Critical limits	Monitoring procedures and frequency	HACCP records	Verification procedures and frequency	Corrective actions
					All potentially contaminated product will be run through functional metal detector prior to shipment. All product rejected by detector will be reworked.
5B Finished product—storage (cold)	Finished product storage area shall not exceed 40° F.	Maintenance personnel will check finished product storage area temperature every two hours.	Room temperature log Corrective action log Thermometer calibration log	Maintenance supervisor will verify the accuracy of the room temperature log once per shift. QA will check all thermometers used for monitoring and verification activities for accuracy daily and calibrate to within 2° F accuracy as necessary. QA will observe maintenance personnel check finished product storage area once per shift.	If a deviation from a critical limit occurs, the following corrective actions will be taken: 1. Product which may not have met CL will be identified and held. 2. The cause of the temperature exceeding 40° F will be identified and eliminated. 3. The CCP will be monitored hourly after the corrective action is taken to ensure that it is under control. 4. When the cause of the deviation is identified, measures will be taken to prevent it from recurring.

5. If room temperature exceeds the critical limit, the processing authority will evaluate the product temperature to ensure that it is sufficient to preclude pathogen growth prior to shipment. If temperature is not sufficient to preclude pathogen growth, product will be cooked in the establishment to ensure destruction of pathogens.

Signature: _____ Date: _____

Source: Modified from Generic Model for Raw, Ground Meat and Poultry Products. US Department of Agriculture, Food Safety and Inspection Service. September, 1998. HACCP-3. http://www.fsis.usda.gov/OPPDE/nis/outreach/models/HACCP-3.pdf.

TABLE 13.5 Generic Example of a Room Temperature Log for a HACCP Plan

Room:			Date:		
Time	Temperature	Deviation from critical limit (check if yes)	If Yes in previous column, action?	Monitored by:	Verified by:

Time/Temperature Critical Limit—40° F

Source: Modified from Generic Model for Raw, Ground Meat and Poultry Products. US Department of Agriculture, Food Safety and Inspection Service. September, 1998. HACCP-3. http://www.fsis.usda.gov/OPPDE/nis/outreach/models/HACCP-3.pdf.

TABLE 13.6 Generic Example of a Thermometer Calibration Log for a HACCP Plan

Criteria: within + or - 1° F of control thermometer								
Date	Time	Department or area	Thermometer ID#	Control thermometer reading	Personal thermometer reading	Adjustment required (Yes or No)	Initials	Comments

• **If a thermometer is broken or taken out of service, document this in the comment column.**

Verified by: _____   *Date/Time:* _____

Source: Modified from Generic Model for Raw, Ground Meat and Poultry Products. US Department of Agriculture, Food Safety and Inspection Service. September, 1998. HACCP-3. http://www.fsis.usda.gov/OPPDE/nis/outreach/models/HACCP-3.pdf.

TABLE 13.7 Generic Example of a Metal Detection Log for a HACCP Plan

Date	Product	Lot #	Results	Seeded sample	Time	Monitored by:	Verified by:

Source: Modified from Generic Model for Raw, Ground Meat and Poultry Products. US Department of Agriculture, Food Safety and Inspection Service. September, 1998. HACCP-3. http://www.fsis.usda.gov/OPPDE/nis/outreach/models/HACCP-3.pdf.

TABLE 13.8 Generic Example of a Corrective Actions Log for a HACCP Plan

Product: raw ground beef   Lot #

Critical control point	Deviation/problem	Corrective action procedures/explain	Disposition of product	Responsible person	Time

Signature: _____   Date _____

Source: Modified from Generic Model for Raw, Ground Meat and Poultry Products. US Department of Agriculture, Food Safety and Inspection Service. September, 1998. HACCP-3. http://www.fsis.usda.gov/OPPDE/nis/outreach/models/HACCP-3.pdf.

TABLE 13.9 Generic Example of a Pre-Shipment Review Log for a HACCP Plan

Product: raw ground beef   Date:

Lot ID	Time records were reviewed	By whom	Lot released for shipment? Signature	Comments

Source: Modified from Generic Model for Raw, Ground Meat and Poultry Products. US Department of Agriculture, Food Safety and Inspection Service. September, 1998. HACCP-3. http://www.fsis.usda.gov/OPPDE/nis/outreach/models/HACCP-3.pdf.

assure continued wholesomeness as they leave the plant (official establishment) where inspection is performed. If items are found to be unfit for human food upon reinspection, original inspection marks are removed or defaced and the items are condemned.

If FSIS inspectors find unwholesome products bearing an inspection mark outside an official establishment, they have authority to condemn them under provisions of the Wholesome Meat Act of 1967 and the Wholesome Poultry Products Act of 1968. Food and Drug Administration officials, under provisions of the Pure Food, Drug and Cosmetic Act of 1906, also have authority to condemn foods in wholesale and retail outlets judged to be unwholesome.

Laboratory Inspection

The FSIS has several regional inspection laboratories equipped to make chemical, microbiological, and other technical determinations. Samples of ingredients used in manufacture of products are submitted to these laboratories for analysis. Likewise, samples of finished products are submitted by inspectors for analysis to ascertain compliance with specific regulations governing production. Products or tissues suspected of containing non-approved ingredients or drug residues are submitted to these laboratories for analysis. Government pathology laboratories also assist in diagnosis of diseased and abnormal tissues submitted to them by inspectors.

Marking and Labeling

Brands and labels applied to carcasses, wholesale cuts, processed meats, edible meat by-products, and containers holding meat, are controlled by FSIS. Inspection legends are applied under supervision of meat inspectors. No edible products may be removed legally from official establishments without being marked or labeled in accordance with regulations.

Inspection legends appear on labels of prepared meat products. In addition, labels must contain the common or usual name of product, name and address of the processor or distributor, handling instructions, nutritional content panel, and statement of ingredients used. Ingredients are listed in descending order of their content, if applicable (Figure 13.10). Inspected plants are permitted to use only those labels and markings which meet all FSIS criteria including restrictions on lettering size and appearance. Pictures used on labels must accurately represent products contained and must not convey false impressions of origin or quality. Meat products originating from state-inspected plants also must bear state labels, requirements for which are essentially the same as for federally inspected products.

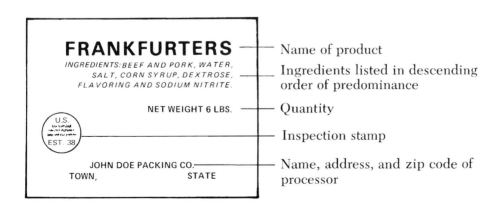

FIGURE 13.10 | Sample of a US Department of Agriculture approved label.

RELIGIOUS INSPECTION AND CERTIFICATION

In addition to federal or state inspection, there are religious forms of inspection. The term kosher is derived from a Hebrew word meaning "properly prepared" and, when applied to meat, it means that products meet requirements of Mosaic and Talmudic laws. The term halal when applied to meat and other food products means they meet the requirements of Islamic food laws.

Kosher slaughter is performed according to prescribed rabbinical procedures, under supervision of authorized representatives of the Jewish faith. Actual bleeding, examination for fitness, and removal of certain organs is usually performed by the "shohet," a specially trained scholar of dietary laws, who is the slaughterer. If carcasses and parts thereof pass this examination, they are marked in Hebrew with the date of slaughter and name of the shohet or rabbi.

Halal is a Quranic word meaning "permitted" or "lawful." Halal slaughter is performed by a Muslim of proper age, the name of Allah must be pronounced at the time of slaughter and the slaughter must be done by cutting the throat in a manner to insure rapid, complete bleeding and death of the animal.

Neither kosher nor halal inspection meet requirements of federal meat inspection laws. In plants subject to federal or state inspection, usual inspection is performed by federal or state inspectors, in addition to that performed by representatives of the Jewish or Islamic faith.

SEAFOOD INSPECTION

Inspection of seafood is under the regulatory authority of the Food and Drug Administration (FDA). For regulatory purposes, seafood is defined to include all fresh and saltwater finfish, molluscan shellfish, crustaceans, and products manufactured from these species. Regulations are established under the authority of the Food, Drug and Cosmetic Act, as amended, the Fair Packaging and Labeling Act, and the Low-Acid Canned Food program.

The essential elements of seafood inspection include:

- inspections of seafood processors and related commercial entities
- sampling and analysis of fish and fishery products for the presence of toxins, chemicals, and other potential hazards
- examination of imported seafood offered for entry into the US
- standard setting for contaminants.

Continuous or piece-by-piece inspection of seafood processing is not practiced. Inspections of processors are unannounced and focus on compliance with good manufacturing practices and sanitation, labeling, use of additives, and verification of HACCP plans. The FDA published its final rule in December, 1995, requiring that all seafood in interstate commerce in the US must be processed under HACCP controls. This includes seafood that is domestically processed and seafood that is imported. HACCP principles applied to seafood are the same as those applied to other muscle foods and their implementation is similar to that discussed previously in this chapter.

FDA works cooperatively with other agencies to implement seafood inspection. One such example is the Federal/State National Shellfish Sanitation Program. This program exercises control of all sanitary aspects of growing, harvesting, shucking, packing, and interstate transportation of molluscan shellfish. Another is the National Oceanic and Atmospheric Administration's voluntary, fee-for-service inspection program for seafood processors that utilizes FDA's safety and other standards for seafood.

14

MEAT GRADING AND EVALUATION

OBJECTIVES: *Develop an understanding of the various forms of meat grades and learn to interpret and utilize information supplied by such grades. Distinguish between federal meat grading and private sector evaluation and sorting of meat products.*

Key Terms

- Meat grading
- Meat evaluation
- Quality grades
- Yield grades
- Brands
- Kind
- Class
- Maturity
- Physiological age
- Ossifies
- Buttons
- Break joint
- Spool joint
- Texture
- Marbling
- Flank streaking
- Feathering
- Firmness
- Conformation
- Fleshing
- Muscling
- Finish
- Cutability
- Backfat thickness
- Fat depth
- Fat thickness
- Kidney, pelvic and heart fat
- Rib-eye
- Loin eye
- Fat free lean
- Dressing percent
- Beef carcass data service
- Beef carcass information service
- Beef carcass evaluation service
- Certification of carcasses eligible for approved breed programs
- Process verification program

CHAPTER 14 MEAT GRADING AND EVALUATION

Meat grading is a procedure by which carcasses, meat, or meat products are segregated on the basis of expected palatability or yield attributes, or other economically important traits. **Meat evaluation** is an extension of grading that identifies carcass composition and value differences with greater precision than grades alone. Grading serves to segregate products into standardized groups with common characteristics, such as appearance, physical properties, or edible portion. Assigned grades within species determine subsequent processing and form of finished meat products. Grades have specified minimum levels for traits used in grading, but they encompass variability above these minimum requirements. The purpose of meat grading and evaluation is to facilitate marketing and merchandising through standardization of product characteristics that are valuable to consumers.

TYPES OF GRADING AND EVALUATION

Federal Grades

The Federal Meat Grading Service was established by an act of Congress on February 10, 1925. The Bureau of Agricultural Economics was designated to provide this service. Standards for beef carcass grades were soon published, and official grading of beef carcasses began in May 1927. However, tentative US standards for dressed beef had been formulated in 1916 to provide a basis for national reporting of dressed beef markets. These tentative dressed beef standards also formed the basis for the first official grading, begun in 1927. Tentative standards for pork carcass grades were published first in 1931 and subsequently made official in 1933. Grading of lamb and mutton was initiated in 1931. Standards and grades for poultry products were developed independently of those for beef, pork, and lamb, and were administered in separate bureaus within the Department of Agriculture. However, historical development of poultry grades paralleled development of beef, pork, and lamb grading systems. Grades for poultry meat and carcasses were first proposed in October 1927. Since their initial institution, official standards for meat and poultry grades have been amended periodically, both to improve the grading systems and to accomplish their intended purposes more efficiently. In 1946, the Agricultural Marketing Act was enacted, reemphasizing and extending responsibilities and provisions of the Federal Meat Grading Act of 1925.

Federal grades for meat were developed in response to demands of livestock producers for market reforms. Until the last half of the nineteenth century, both livestock production and the meat industry were highly decentralized; small farmers raised livestock mainly to meet their own needs. But they also sold some of their livestock to local butchers, who in turn sold meat directly to consumers. Following the Civil War, a widespread economic reorganization occurred in the US, producing very complex marketing processes for all agricultural products. Factors that contributed to this economic reorganization included transition from subsistence agriculture to commercial agriculture, movement of processing from farms to factories, growth of urban markets, extension of transportation facilities, and expansion of foreign and domestic trade. Livestock producers did not understand the roles of distributors and processors in the new marketing system. In general, they felt these groups of middlemen were exploiting both producers and consumers. Their first effort toward understanding meat marketing was to obtain price information pertaining to meat trade, so that livestock market conditions could be interpreted. Uniformity of meat classification and terminology was necessary before effective market reporting could occur. Thus, initial grade standards were specifically designed to facilitate a national price reporting service for livestock.

At present, grading of beef, calf, veal, pork, lamb and mutton, chicken, turkey, duck, geese, pigeon, and rabbit is administered by the US Department of Agriculture, Agricultural Marketing Service, Livestock and Seed Program, Meat Grading and Certification Branch. Graded products or carcasses must be prepared under supervision of either federal meat inspection or equivalent state inspection. Federal grading is a service that is available, on voluntary request, to any company, plant, or individual with financial interest in products to be graded. In contrast to federal meat inspection, cost of federal grading is borne by companies or individuals that request it. Charges for service are based on standard rates established by the Agricultural Marketing Service. In general, two types of services are available:

- grading of meat for sale in commercial channels on a graded basis, and
- examination of contract purchased meat delivered to federal, state, county, or municipal institutions, to ensure conformance with grade specifications.

There are two general types of federal grades for meat. Grades for quality are intended to categorize meat on the basis of its acceptability for consumer cuts. Variables used in assigning **quality grades** include factors related to palatability and acceptability of meat products to consumers. Grades for quantity (**yield grades**) are designed to categorize carcasses on the basis of expected yield of trimmed retail cuts, and are established only for beef, pork, and lamb carcasses.

Brands

Brands are grades designed and applied to meat or meat products by the processing industry, including individual packers, processors, distributors, or retailers.

Since development of federal grade standards was not supported enthusiastically by the meat packing industry, many national packers immediately developed their own branding systems for beef carcasses following institution of federal grading in 1927. Many slaughter companies subsequently retained their own grading or branding system, but now use it to supplement rather than compete with the federal system. There is little uniformity of standards for specific products among companies, nor is there uniformity in grade names. Initially, a much larger volume of beef was branded than was federally graded. That situation then reversed, so the proportion of fresh meat branded by packers is now much smaller than that which is federally graded. On the other hand, some large national packers and national chain retailers have developed and promoted retail brands for fresh meat, which have lessened their reliance on federal grades. Also, brands have been developed that promote certain breeds of livestock or specific production practices. An example of a breed designation is Certified Angus Beef. Other brands might promote grass-fed beef from cattle that are produced without grain feeding. Usually, these brands specify certain minimum USDA grade characteristics to qualify for the brand. Fresh poultry and fish are seldom merchandized with a government grade designation. For these products, company specific brands predominate in the marketplace.

Even though private brands are applied to carcasses infrequently, they are widely used for all types of processed meats. Figure 14.1 illustrates use of brands on two processed meat products, bacon and bologna. Such brands reflect differences in ingredients, formulation costs, product image, and retail price. Brands play important roles in merchandising and promotion of fresh and processed meat products.

Commodity Evaluations

Some livestock commodity groups use carcass evaluation systems that are independent of federal grades. These systems are useful in carcass evaluation for genetic improvement and provide more detailed information than federal grades or brands. Even though these evaluations are generally more detailed and may be applied to limited numbers of carcasses, they represent a type of grading and perform many functions of grading.

FUNCTIONS OF GRADES

All functions of grading are related in one way or another to some phase of the marketing process, from livestock producer to ultimate meat consumer. For livestock producers, carcass grades and evaluation data form a portion of the basis on which animals are bred, fed, bought, and sold. Grades established for live animals are designed to correspond to respective federal grades for carcasses. Producers estimate, with varying accuracy, grades of animals they plan to produce, prices they expect to receive for those animals, and the market segment their production will serve. For packers, processors, and retailers, grades provide means for segregating and pricing animals, carcasses, and meat into uniform groups. For consumers, grade or brand stamps on products represent assurance that those products conform to established standards or their personal standards. This provides convenient means for selection of products that satisfy their needs.

FEDERAL QUALITY GRADES

For the most part, quality grades for various species are intended to classify meat on the basis of palatability and cooking traits. However, palatability is broadly defined to include factors that affect overall acceptability of products to consumers. These factors include the shape of carcasses or cuts, amount of fleshing and fat cover, and overall appearance including physical defects. Identifying marks placed on carcasses or packages to indicate US Department of Agriculture quality grades are shown in Figure 14.2.

(A)

(B)

(C)

(D)

(E)

(F)

Courtesy of Emge Packing Co., Inc., Fort Branch, Indiana.

FIGURE 14.1 | Brands used by a meat processor. (A, B, C) Brands of bacon. (D, E, F) Brands of frankfurters.

Factors Used to Establish Quality Grades

Kind and Class

In grading meat, each species is referred to as a **kind**. Each kind of meat is divided into classes that are quite similar in physical characteristics. Primary factors in establishing the **class** of carcasses are sex and maturity. Maturity is especially important in establishing classes of poultry, beef, pork, and lamb or mutton carcasses.

Maturity

As used in grade standards, **maturity** is defined as the physiological age of animals or birds from which carcasses are produced. **Physiological age** is an expression of degree of aging visible in animal tissues. Of all grade factors used, maturity is most closely related to meat tenderness. In general, meat from physiologically mature animals is less tender than meat from immature animals. This difference largely reflects qualitative changes occurring in muscle connective tissue (collagen) as it matures. These changes are discussed in Chapter 2.

In all species, the most useful indicators of carcass maturity are characteristics of bone and cartilage (Figure 14.3). In young animals, cartilage is abundant, particularly in the epiphyseal plates, between vertebrae, at the ends of dorsal vertebral processes, and in the breast bone. As animals mature, cartilage **ossifies** from infiltration with bone salts. Therefore, during aging, a progression from soft cartilage, to hard cartilage, to bone is evident. In birds, pliability of the keel bone cartilage is evaluated. In beef carcasses, characteristics of **buttons**, the cartilage caps located on the ends of dorsal processes of all vertebrae, may be evaluated in carcasses split along the dorsal midline. Shapes and colors of ribs change during maturation of all species. The presence of a **break joint**, the epiphyseal cartilage at the distal end of metacarpal bones, is used to classify carcasses as lamb. As animals mature, this cartilage is converted to bone, and the foot is removed during slaughter at the **spool joint**

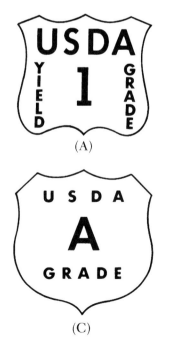

FIGURE 14.2 | Identifying marks used for the US Department of Agriculture grades for meat: (A) beef, lamb, and mutton yield grades; (B) beef, lamb, and mutton quality grades; (C) poultry grades; and (D) pork carcass grades.

322 CHAPTER 14 MEAT GRADING AND EVALUATION

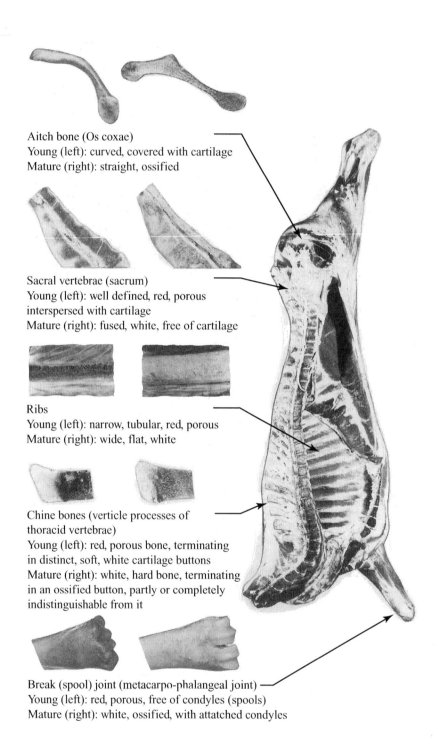

FIGURE 14.3 | Indices of maturity in meat animal carcasses. A comparison of immature and mature characteristics. From Briskey, E. J., and R. G. Kauffman, "Quality Characteristics of Muscle as a Food," in *The Science of Meat and Meat Products*, 2nd ed. J. F. Price and B.S. Schweigert, eds. W. H. Freeman and Co., San Francisco, 1971.

rather than the break joint. The presence of a spool joint is a criterion for designation of mutton.

Meat from young animals is very light red in color. As they mature, myoglobin concentrations increase in muscles, and they become darker red. Therefore, color changes of lean meat are used, to some extent, in assessing carcass maturity. Changes in color of lean are more pronounced in species such as beef that normally have very red muscles. In contrast, less apparent color changes occur in the white breast meat from chickens or turkeys.

The term **texture** relates to size of muscle fiber bundles and to thickness of perimysial connective tissues between fiber bundles. Meat surfaces that have been cut across the fibers have a smooth, shiny appearance in fine-textured muscles, whereas they appear dull and slightly rough in coarse-textured muscles. As animals mature, texture of lean changes from fine to coarse because of prominence of perimysial connective tissues resulting from muscle fiber degeneration without replacement. On the other hand, fine and coarse texture in animals of similar age result from small and large fiber bundles, respectively.

Marbling

Visible intramuscular fat, located in perimysial connective tissues between muscle fiber bundles, is called **marbling**. The amount of marbling present in meat has been considered an important meat quality characteristic for many years. As stated in Chapter 11, marbling has been widely credited with making meat tender, even though there is little research evidence to indicate that its presence has a strong positive influence on meat tenderness. Marbling has a stronger beneficial effect on juiciness and flavor of meat than on tenderness. Modest amounts of marbling, uniformly distributed throughout meat (Figure 14.4a), provide optimum flavor and juiciness. Meat that is nearly devoid of marbling (Figure 14.4b) may be dry and flavorless, but excessive amounts of marbling (Figure 14.4c) do not give proportional increases in palatability.

With regard to establishment of quality grades, ten degrees of marbling are recognized. The US Department of Agriculture maintains photographic reference standards that illustrate each marbling degree for evaluation of marbling. Amounts of marbling are subjectively determined based on these photographs of the longissimus muscle cross-section at the twelfth rib in beef and lamb and tenth rib in pork. Marbling is a major factor used in establishing beef carcass quality grades. In pork and lamb carcasses, which normally are

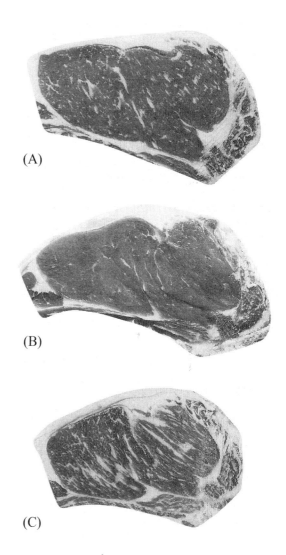

FIGURE 14.4 | Marbling in beef longissimus muscle. (A) Ideal marbling; a modest amount uniformly distributed and finely dispersed in the muscle. (B) A muscle deficient in marbling. (C) A muscle containing excessive marbling.

not ribbed (a cross-section of longissimus muscle is not available), it is customary to evaluate other quality indicating characteristics. Marbling in lamb carcasses is indicated by amounts of flank streaking. **Flank streaking** refers to the streaks of fat located beneath the epimysium of the flank muscles, as viewed from the inside of the abdominal cavity. Marbling in pork carcasses is indicated by fine streaks of fat visible in the intercostal muscles between the ribs, known as **feathering**. Both feathering and flank streaking are used as indicators of marbling, but their relationship to amount of marbling is not close. Thus, in view of a relatively low relationship of marbling to meat palatability, it is apparent that feathering and flank streaking are of even more doubtful value.

Firmness

As used in quality grading, the term **firmness** refers to firmness of the flank area or a lean cut surface. Carcasses with soft, oily fat lack firmness and are discriminated against. Firmness also is influenced by amount of fat present because carcass chilling makes fat much firmer than muscle. Therefore, carcass fat, regardless of location, contributes to overall firmness. Firmness makes no direct contribution to meat palatability, but it is a quality trait because firm retail cuts retain their shape and are more attractive than soft cuts.

Firmness is important in some meat products such as pork bellies, which are subjected to extensive processing. Manufacture of bacon requires several mechanical steps including cure pumping, forming, and slicing that are difficult to accomplish on soft tissues. Carcass firmness and belly thickness are therefore important quality criteria in pork.

Color and Structure of Lean

Wide variation in color and structure of lean (within species) occurs as a result of physical and chemical changes during conversion of muscle to meat (Chapter 5). In both color and structure, extremes are undesirable and diminish quality grades. At one extreme is the pale, soft, exudative lean often observed in pork, but also seen occasionally in meat of other species. At the other extreme is dark, firm, dry lean typical of the dark cutting condition in beef and, occasionally, in other species.

Color and structure of lean are evaluated to establish federal grades of meat, because they affect appearance of cuts, which in turn affects consumer acceptability. In general, consumers object to dark meat because it often is associated with meat from old animals or with deteriorated meat. Research indicates that dark cutting beef or pork are equivalent to typical meat in tenderness, flavor, and juiciness but may have slightly less microbiological storage life. Dark cutting beef may be downgraded as much as one full grade, depending on severity of the condition. Consumers generally do not object seriously to appearance of pale meat, but, because of low water-holding capacity and tendency to lose juices during cooking, pale, soft, exudative meat is less juicy than meat with typical color and structure. Pork color scores as used in carcass evaluation are shown in Figure 14.5 (see color section).

Conformation, Fleshing, and Finish

As used in meat grading, **conformation** refers to the proportionate development of carcass parts or wholesale cuts, and to the ratio of muscle to bone. Thus, conformation is primarily a function of muscular and skeletal system development, but its perception also is affected by amount of subcutaneous and intermuscular fat. **Fleshing** (or **muscling**) refers specifically to development of skeletal musculature. Conformation and fleshing have no direct effects on palatability of meat. These factors are included in determination of quality grades, partly because more desirably shaped retail cuts are obtained from carcasses with acceptable conformation. For example, birds with thinly fleshed thighs and drumsticks, and narrow, tapering, crooked, or concave breasts are not attractive to consumers when selecting meat. A round, full longissimus muscle makes more attractive rib steaks than long, narrow muscles.

Finish refers to the amount, character, and distribution of external, internal, and intermuscular fat, either in carcasses or wholesale cuts. Excess fat detracts from retail cut yield obtainable from carcasses, but a minimum amount of finish is desired for optimum meat quality. Fat beneath the skin of birds serves to baste the meat during dry heat cooking and helps prevent dryness in cooked products. The same advantages may be cited for external finish on other kinds of meat. Small amounts of external finish on beef and lamb carcasses also are necessary to prevent desiccation and shrinkage of muscles during chilling, aging, merchandising, and cooking.

Carcass Defects

In establishing poultry grades, several potential carcass defects are of special importance. Before quality grades may be assigned to ready-to-cook poultry, carcasses must be free of protruding pinfeathers. Exposed flesh from cuts, tears, missing skin, and broken or disjointed bones detract from the appearance of birds. Extent of skin discoloration, bruises, and defects resulting from freezing also are considered. The number and severity of such permitted defects depend on their location. Defects on the breast are more objectionable than those on other carcass areas.

Establishing Quality Grades

Terminology used to designate quality grades varies widely among kinds of meat. There is also variation in grade terminology among classes within kinds of meat. US Department of Agriculture quality grade terms for each kind of meat, and for various classes within each kind, are given in Table 14.1. Not all quality grade factors discussed earlier are used to establish grades for each kind and class of meat. To illustrate, specific factors used to establish quality grades for various kinds of meat are given in Table 14.2.

Beef carcass quality grades are examples of grades developed by combination of various grade factors. Figure 14.6 shows the relationship between

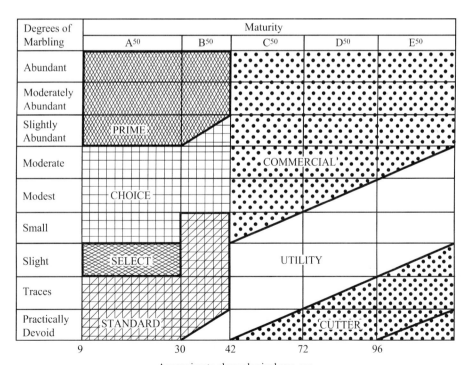

FIGURE 14.6 | Relationships among marbling, maturity, and quality as used to establish beef carcass quality grades.

TABLE 14.1 US Department of Agriculture Grades for Meat

Kind	Class	Grade Names
Beef quality grades	Steer, heifer, cow*	Prime, Choice, Select, Standard, Commercial, Utility, Cutter, Canner
	Bullock	Prime, Choice, Select, Standard, Utility
Beef yield grades	All classes	1, 2, 3, 4, 5
Calf quality grades		Prime, Choice, Good, Standard, Utility
Veal quality grades	Same as calf	
Lamb and mutton quality grades	Lamb, yearling mutton	Prime, Choice, Good, Utility
	Mutton	Choice, Good, Utility, Cull
Lamb yield grades	All classes	1, 2, 3, 4, 5
Pork carcasses	Barrow, gilt	US No. 1, US No. 2, US No. 3, US No. 4, US Utility
	Sows	US No. 1, US No. 2, US No. 3, US No. 4, US Utility
Chicken	Rock Cornish game hens	A, B, C
	Broiler or fryer	A, B, C
	Roaster	A, B, C
	Capon	A, B, C
	Cock or rooster	A, B, C
	Hen, stewing chicken, baking chicken, fowl	A, B, C
Turkey	Fryer or roaster	A, B, C
	Young turkey	A, B, C
	Yearling turkey	A, B, C
	Mature or old turkey	A, B, C
Duck	Duckling	A, B, C
	Roaster	A, B, C
	Mature or old duck	A, B, C
Geese	Young goose	A, B, C
	Mature or old goose	A, B, C

Guineas	Young guinea	A, B, C
	Mature or old guinea	A, B, C
Pigeon	Squab	A, B, C
	Pigeon	A, B, C

*Cows are not eligible for the Prime grade.

TABLE 14.2 Factors Used to Establish US Department of Agriculture Grades for Meat

Beef quality grades	Beef yield grades
Class	Adjusted fat thickness
Maturity	Rib-eye area
Buttons	Percent kidney, heart, and pelvic fat
Sacral vertebrae	Carcass weight
Color of lean	**Lamb yield grades**
Texture of lean	Adjusted fat thickness
Marbling	
Lamb quality grades	**Poultry grades**
Class	Class
Conformation	Condition
Maturity	Conformation and fleshing
Break joint color and porosity	Finish
Shape and color of ribs	Defects
Color of lean	Pinfeathers
Flank streaking	Exposed flesh
Firmness of lean and fat	Discoloration
Pork carcass grades	Bruises, freezing damage
Class	
Backfat thickness	
Degree of muscling	
Color of lean	
Firmness of fat and lean	
Belly thickness	

degree of marbling and carcass maturity as they are combined to establish quality grades. Note that the minimum degree of marbling required for each grade increases with increasing maturity. Beef carcasses differing in quality grade are shown in Figure 14.7.

Quality grades of lamb, yearling mutton, veal, and calf carcasses are established by systems that are similar to that used for beef. However, those kinds of carcasses are not always ribbed, so quality indicators are used instead of marbling degrees. Figure 14.8 shows the relationship between flank fat streaking and maturity in lamb and yearling mutton and corresponding quality grades. Figure 14.9 shows similar relationships for veal and calf carcasses.

Quality is only a minor component of pork carcass grades. It is included in pork carcass grading only to the extent that carcasses are judged to have either acceptable or unacceptable quality.

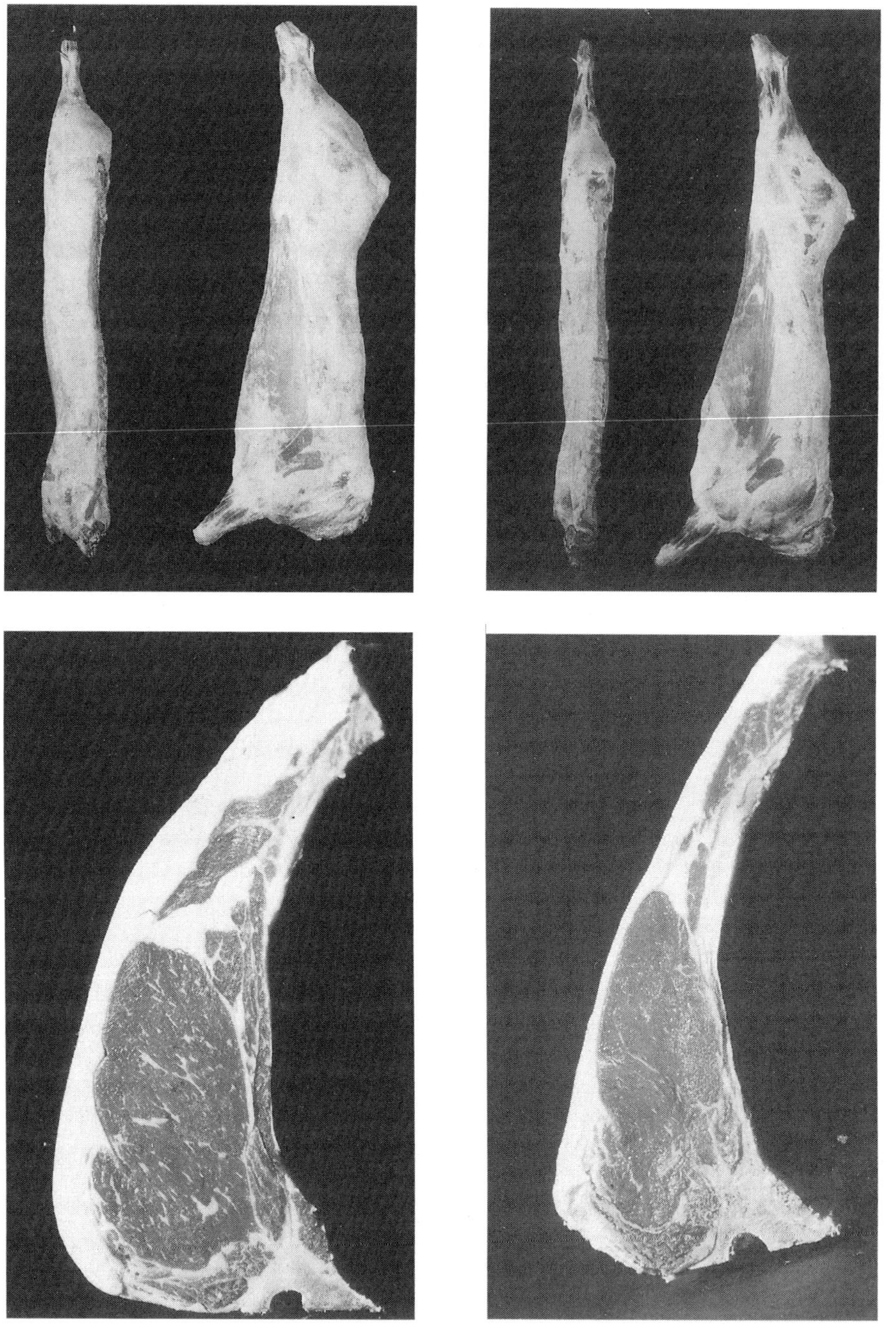

328 **CHAPTER 14** MEAT GRADING AND EVALUATION

(A)

Marbling	Modest
Maturity	A
Quality Grade	Choice

(B)

Marbling	Traces
Maturity	A
Quality Grade	Standard

Courtesy of the National Live Stock and Meat Board.

FIGURE 14.7 | (A) USDA Choice grade beef carcass. (B) USDA Standard grade beef carcass.

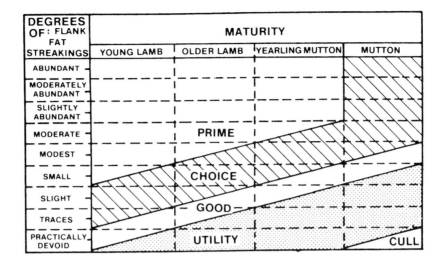

FIGURE 14.8 | Relationships among flank fat streaking, maturity, and quality as used to establish lamb, older lamb, and yearling mutton carcass quality grades.

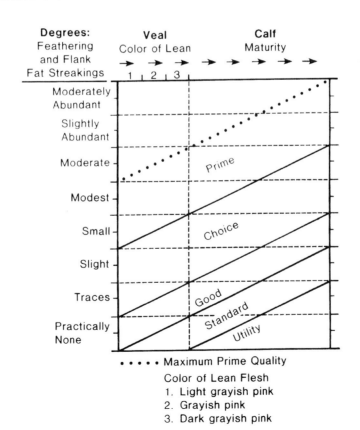

FIGURE 14.9 | Relationships among feathering and flank fat streaking, maturity, and quality as used to establish veal and calf carcass quality grades.

Acceptable quality carcasses must meet minimum standards for muscle color and structure, carcass firmness, fat characteristics, and belly thickness. Pork carcass quality grades are not available to consumers for use in selecting pork, as in other kinds of meat, because quality plays a relatively minor role in establishing carcass grades.

Poultry carcasses are graded on the basis of quality. However, as in lamb carcass quality grades, conformation is a component of the quality grade for poultry. Chicken carcasses differing in conformation are shown in Figure 14.10.

CARCASS CUTABILITY AND FEDERAL YIELD GRADES

In a broad sense, the term **cutability** indicates the proportionate amount of saleable retail cuts that may be obtained from a carcass. Beef, pork, and lamb carcass merchandising includes removal of excess fat and bones, and subdivision of carcasses into retail cuts such as steaks, chops, roasts, and ground meat before presentation to consumers. Yields of such cuts and potential retail value vary widely, even among carcasses that have the same quality grade. Therefore, separate and distinct grading systems have been established to segregate carcasses, based on their expected retail yield. For beef and lamb carcasses, these grades are called yield grades. Beef and lamb yield grades identify carcasses according to differences in expected yields of closely trimmed retail cuts (Table 14.3). Only one grading system is established for pork carcasses. It is fundamentally a yield grading system based on expected yield of the four lean wholesale cuts (ham, loin, picnic shoulder, and Boston shoulder) (Table 14.3). However, pork producers use yet another system to evaluate muscle content of carcasses.

Factors Used to Determine Cutability and Yield Grades

In general, quality grades are subjective grades and yield grades are more objective in nature. Objective measurements are obtained from carcasses and combined in equations based on their relative value in predicting yield grade, percent retail cuts, or percent muscle. Factors closely related to the predicted attribute receive more emphasis in prediction equations than factors less closely related.

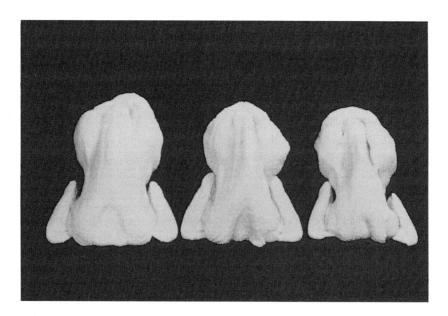

FIGURE 14.10 | Front view of stewing chickens illustrating from left to right A, B, and C quality grades.

TABLE 14.3 | Expected Retail Yield for Beef and Lamb Yield Grades and Expected Percent Lean Cuts for Pork Carcass Grades

Beef or lamb yield grade	Retail yield		Pork grades[c]	Lean cuts
	Beef[a]	Lamb[b]		
1.0	54.6	51.0	US No. 1	60.4 and greater
2.0	52.3	49.7	US No. 2	57.4–60.3
3.0	50.0	48.4	US No. 3	54.4–57.3
4.0	47.7	47.1	US No. 4	Less than 54.4
5.0	45.4	45.8	US utility	Unacceptable quality

[a]Percent boneless, closely trimmed, retail cuts from the round, loin, rib, and chuck. [b]Percent boneless shoulder, semiboneless leg (femur bone remaining), and bone-in rib and loin trimmed to 0.25 cm of external fat.
[c]Percent trimmed ham, loin, picnic shoulder, and Boston shoulder based on chilled carcass weight.

Amount of Fat

The amount of external, internal, and intermuscular fat in carcasses has more effect on percent retail yield or lean cuts than any other single factor. Excess fat must be removed to prepare attractive, easily merchandisable retail cuts. Considerable time and effort are required to measure excess fat on carcasses, but some easily obtainable measurements may be made to estimate such fat with reasonable accuracy. These measurements are used in yield grading and other commodity evaluation systems.

Thickness of the layer of pork backfat (subcutaneous or external fat that covers the pig's back) is an accurate indicator of total fat in carcasses. A single measurement of **backfat thickness** is taken on the carcass midline (split surface) opposite the last rib. Another measurement, known as **fat depth** or **fat thickness**, is located off the midline at a point three-fourths of the width of the longissimus muscle from the medial side and between the tenth and eleventh ribs.

In beef carcasses, thickness of external fat is measured at the three-fourths longissimus width position between the twelfth and thirteenth ribs (location of carcass quartering or ribbing). In lambs, fat thickness is measured over the midpoint of longissimus muscle width between the twelfth and thirteenth ribs. Additional measurements of fat thickness over other carcass areas improve accuracy of fat estimation, but these are not obtained because the improved accuracy does not offset the time and effort required.

Fat surrounding the kidneys, lining the pelvic channel, and surrounding the heart and lungs is left on beef carcasses during slaughter. This fat is removed and discarded when carcasses are cut for retail sale. Thus, amounts of internal fat deposits must be taken into consideration in establishing yield grades. Amounts of **kidney**, **pelvic**, and **heart fat** are estimated as percentages of carcass weight. Experienced graders make these estimates with high degrees of accuracy.

Muscle Development

Degree of skeletal muscle development influences percent retail yield. In general, carcasses with minimal muscle development have lower ratios of muscle to bone, and lower cutability than more muscular carcasses. Several measures of muscle development are used in grading and evaluation. In beef and lamb, **rib-eye** or loin eye area is the cross-sectional area of the longissimus muscle between the twelfth and thirteenth ribs. In pork, the measurement is made between the tenth and eleventh ribs. **Loin eye** area is not used in grading lamb and pork carcasses even though it would contribute somewhat to prediction of retail yield. This is partly because, in industry practice, pork and lamb carcasses are not ribbed. However, pork loin eye area is used to predict percent muscle in carcass improvement programs used by producers.

332 **CHAPTER 14** MEAT GRADING AND EVALUATION

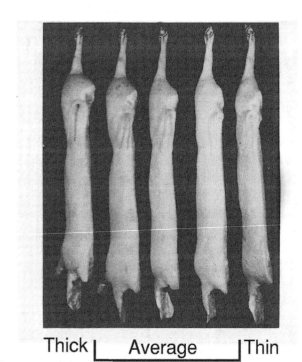

FIGURE 14.11 | Degrees of muscle development for pork carcasses.

For muscularity determination in pork carcass grading, three degrees of muscle development, "thick", "intermediate," and "thin," are recognized (Figure 14.11). Specific degrees of muscle development are assigned, based on subjective visual appraisal of carcass conformation. Measurements of muscle development based on subjective scores for conformation are less accurate as predictors of retail yield than fat thickness measurements and consequently have less influence on assigned yield grades.

Carcass Size

Carcasses differ in cutability due to differences in degrees of fatness and muscle development in relation to skeletal size. For example, large beef carcasses may have larger ribeye areas than small carcasses, especially if they are from cattle with large mature size. Thus, to make accurate predictions of retail yield, rib-eye area must be considered in relation to carcass size. The most widely used indicator of skeletal size is carcass weight. As carcass weight increases, beef carcasses must have larger rib-eye area to maintain the same yield grade. If rib-eye area does not increase with carcass

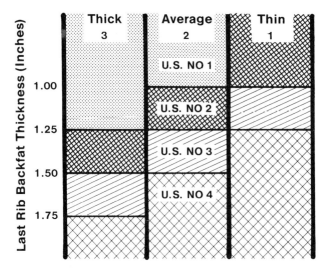

FIGURE 14.12 | Relationships between backfat thickness and muscling score as used to establish pork carcass grades.

TABLE 14.4 Comparison of Yields of Retail Cuts and Retail Values Between US Department of Agriculture Yield Grade 2 and Yield Grade 4 Beef Carcasses, Each Weighing 600 Pounds

Closely trimmed retail cuts	Weight of retail cuts from a 600-lb. Carcass*		Price per pound	Retail value	
	Yield Grade 2	Yield Grade 4		Yield Grade 2	Yield Grade 4
Rump, boneless	21.0 lbs.	18.6 lbs.	$3.29	$69.09	$61.19
Inside round	27.0	22.2	3.49	94.23	77.48
Outside round	27.6	25.2	2.99	79.76	72.83
Round tip	15.6	14.4	3.29	51.32	47.38
Sirloin	52.2	47.4	4.39	229.16	208.09
Short loin	31.2	30.0	5.29	165.05	158.70
Rib, short cut (7in.)	37.2	36.0	5.29	196.79	190.44
Blade chuck	56.4	50.4	1.89	106.60	95.26
Chuck arm, boneless	36.6	33.0	2.59	94.79	85.47
Brisket, boneless	13.8	11.4	2.69	37.12	30.67
Flank steak	3.0	3.0	5.19	15.57	15.57
Ground beef	141.0	115.8	1.59	224.19	184.12
Fat	76.2	137.4	0.10	7.62	13.74
Bone	59.4	53.4	0.05	2.97	2.67
Kidney	1.8	1.8	0.89	1.60	1.60
Totals	600.0	600.0		$1,378.83	$1,245.21

*Based on yields presented in Marketing Bulletin No 45, USDA Yield Grades for Beef, US Department of Agriculture, 1968.

weight, estimated cutability will be lower. Skeletal size measurements are not included in standards for pork grades and lamb carcass yield grades.

Establishing Yield Grades

As evident from discussion of yield grades and cutability, a variety of factors are considered to predict retail yield or other measures of leanness. Specific factors used in grading for each kind of carcass are summarized in Table 14.2, and listed in descending order of importance as predictors of percent retail yield. The following equation may be used to establish yield grades for beef carcasses: Beef Carcass Yield Grade = 2.50 + (2.50 × fat thickness, inches) − (0.32 × rib-eye area, square inches) + (0.20 × percent kidney, heart, and pelvic fat) + (0.0038 × warm carcass weight, pounds).

The following equation may be used to establish yield grades for lamb carcasses: Lamb carcass yield grade = 0.4 + (10 × adjusted fat thickness, inches). For both beef and lamb, carcasses with calculated grades of 1.0 to 1.9 are designated as Yield Grade 1, and those with calculated grades of 2.0 to 2.9 are designated as Yield Grade 2. Yield Grades 3, 4, and 5 are designated in the same manner. Figure 14.12 shows pork carcass grades based on backfat thickness and muscling score. When carcasses display muscling degrees of thick or thin (Figure 14.11), the grade is raised or lowered one grade, respectively. The following equation may be used to establish grades for pork carcasses: Grade = (4.0 × backfat thickness) − (muscle score). Muscle score is coded as follows: Thick = 3; Intermediate = 2.5, 2.0 or 1.5; Thin = 1. Although not used in federal grading, a prediction equation is available

TABLE 14.5 Dressing Percentages of Various Kinds of Livestock by Grades

Grade	Range	Average	Grade	Range	Average
*Cattle			Barrows and gilts*,****		
Prime	62–67	64	US No. 1	68–72	70
Choice	59–65	62	US No. 2	69–73	71
Select	58–62	60	US No. 3	70–74	72
Standard	55–60	57	US No. 4	71–75	73
Commercial	54–62	57	US Utility	67–71	69
Utility	49–57	53	Turkeys, 84–112 days		
Cutter	45–54	49	Males	78–80	79
Canner	40–48	45	Females	77–81	79
Calves and vealers*,**			Chicken broilers, 56 days		
Prime	62–67	64	Males		72
Choice	58–64	60	Females		70
Good	56–60	58	Channel catfish		60
Standard	52–57	55			
Utility	47–54	51			
Cull	40–48	46			
Lambs (Wooled) *, ***					
Prime	47–55	51			
Choice	45–52	49			
Good	43–49	47			
Utility	41–47	45			

*All percentages are based on warm carcass weights.
**Based on hide-off carcass weights.
***Kidney and pelvic fat removed.
****Based on packer style dressing (ham facings, leaf fat, kidneys, and head removed).
Source: US Department of Agriculture, Consumer and Marketing Service, Livestock Division; Lesson and Summers, Poul. Sci. 59:1237 (1980); Hayes and Marion, Poul. Sci. 52:718(1973); Lovell, Sci. of Food and Agric. 3:10 (1985).

for use with pork carcasses that estimates percent muscle, expressed as **Fat Free Lean**, as follows: Pounds of Fat Free Lean = 8.588 + (.465 × warm carcass weight, pounds) + 3.005 × loin eye area, square inches) – (21.896 × fat depth over tenth rib, inches). Percentage of Fat Free Lean = (Pounds of Fat Free Lean/warm carcass weight, pounds) ×100.

To illustrate the importance of cutability, weights of various retail cuts that may be obtained from Yield Grade 2 vs. Yield Grade 4 beef carcasses, both of which weigh 273 kilograms (600 pounds), and the retail value of these cuts, are presented in Table 14.4. Both carcasses are US Department of Agriculture Choice quality grade. Excluding bone and fat trim, 25.1 kilograms (55.2 pounds) more salable retail cuts are obtained from the Yield Grade 2 carcass, and its retail value is $133.62 greater. Retail values given include wholesale cost of the carcasses plus retailer markup for operating expenses and profit. Figure 14.13 illustrates typical examples of Yield Grade 2 and Yield Grade 4 beef carcasses.

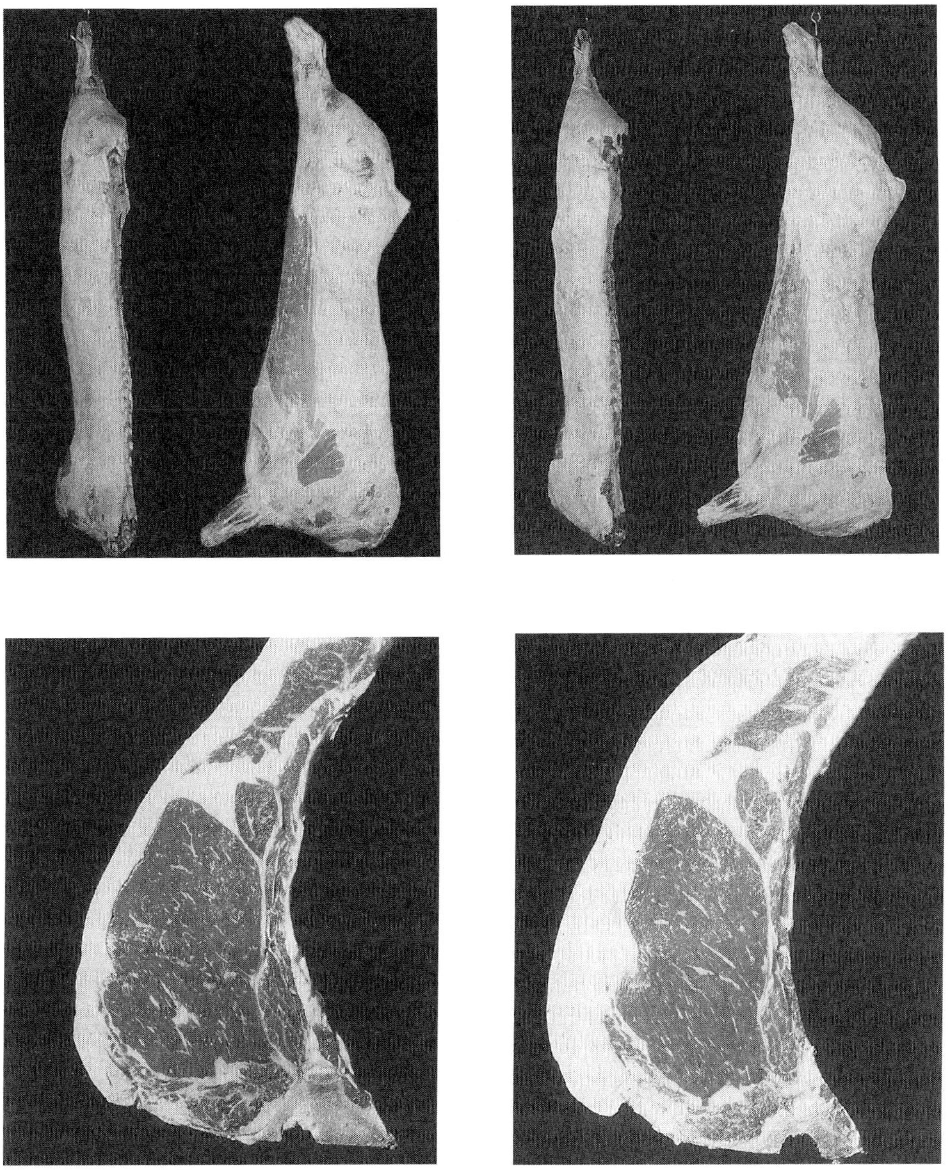

Courtesy of The National Live Stock and Meat Board.

(A)

Yield grade	2
Fat thickness	0.4 inches
Ribeye area	12.3 square inches
Kidney, heart, and pelvic fat	3.0 percent
Carcass weight	605 pounds

(B)

Yield grade	4
Fat thickness	0.9 inches
Ribeye area	10.5 square inches
Kidney, heart, and pelvic fat	3.5 percent
Carcass weight	665 pounds

FIGURE 14.13 Examples of beef carcass yield grades. Both carcasses are USDA Choice quality grade. (A) Yield Grade 2 and (B) Yield Grade 4.

Dressing Percent

The ratio of dressed carcass weight to the weight of the live animal, expressed as a percentage, is known as **dressing percent**. This may be calculated either on a warm or chilled carcass weight basis. Intestinal tract size and contents, muscle bone ratio, and carcass fatness all have influence on dressing percent. As shown in Table 14.5, there is a progressive increase in dressing percent of beef, veal, and lamb from lower to higher quality grades, and in pork from USDA No. 1 to USDA No. 4 grades. These grade-associated trends are caused largely by progressive increases in muscling and to a lesser extent fatness which accompany increases in dressing percent.

SERVICES OFFERED BY THE USDA GRADING SERVICE

Application of USDA Quality and Yield Grades is the primary function of the USDA Grading service. However, related specialized services are offered to meet various needs of the livestock and meat industry. Grading services are offered on a fee basis including the USDA Quality and Yield grading programs outlined above. All of these services provide the industry with a third-party independent evaluation that is often used in the price discovery process.

Carcass Data

The **Beef Carcass Data Service** allows producers and feeders to request data on the individual quality and yield factors on carcasses from animals in which at some point they had a financial interest. A **Beef Carcass Information Service** is available to cattle feeders and others who buy and sell cattle on a formula or carcass grade and yield basis. This information is provided on a lot rather than an individual basis. A **Beef Carcass Evaluation Service** provides researchers at colleges and universities with more detailed evaluation of quality and yield factors such as firmness and color of lean, texture of marbling, bone and muscle maturity, and rib-eye area tracings. Under this program other data collection services can be developed for specific customers.

Certification and Verification Services

Certification of Carcasses Eligible for Approved Breed Programs. Several breed organizations have developed quality and or yield standards for carcasses from their specific breed to differentiate and highlight certain characteristics of their product for consumers. The Meat Grading and Certification Branch provides certification programs to assure consumers that carcasses marketed under the program meet the standards that have been established with regard to breed, quality, and composition.

Audit Services

The USDA **Process Verification Program** provides livestock and meat producers an opportunity to assure customers of their ability to provide consistent quality products by having their written manufacturing processes confirmed through independent, third party audits. The verification program facilitates marketing claims such as breed, feeding, or other production practices to be marketed as "USDA Process Verified." The program uses the International Organization for Standardization's ISO 9000 series standards for documented quality management systems as a format for evaluating documentation to ensure consistent auditing practices and promote international recognition of audit results.

For a complete and up-to-date listing and description of services offered go to the USDA web site at http://www.ams.usda.gov/lsg/mgc.

15

ELECTRONIC ASSESSMENT OF CARCASSES AND FRESH MEAT

OBJECTIVES: *Understand the biological and physical principles underlying electronic technologies used to segregate meat products based on key quality or composition characteristics.*

Colorimetry
Tristimulus method
Spectral reflectance (reflectance spectrum)
Spectrophotometer
Diffraction grating
Fiber optics
Invasive technology
Visible light
Near-infrared
Vision-based systems
Conductivity
Impedance (resistance)

Ultrasound
Pulser
Transducer
Receiver
Display
Memory device
A-mode
B-mode
Real-time ultrasound
Electromagnetic scanning
X-ray scanner

Key Terms

The development of new electronic technologies that can be used to determine or predict product quality and quantity has been driven by a number of developments in the meat animal industry. First, as producers strive for greater production efficiencies and invest greater resources into new production practices, such as superior breeding stock, sophisticated nutrition programs, or use of growth-modifiers, they expect greater returns on their investment when live animals are merchandized. Consequently, processing companies have evaluated closely technologies that would allow them to accurately and rapidly identify animals and products with desired quality or quantity traits and thus to reward producers for producing these animals and products. Accurate, rapid means to predict yield and quality immediately after or during the harvesting process allow for greater rewards to flow through the industry.

Segregation of product based on these characteristics is not simply a reward to the producer. Processors have the potential to make large gains in efficiency as well. For example, if a technology were available so animals of similar yield or quality characteristics could be assigned to a particular group prior to harvesting, animals could be sorted as they arrived at the holding facilities and all animals of a given yield or quality could be harvested at the same time. Likewise, if carcasses could be sorted to particular rails or chill coolers based on predicted quality or composition parameters, then all carcasses with a given classification could be processed together. This could result in more effective fabrication, trimming, and packaging operations and thus improved product uniformity, which is key in any type of manufacturing process.

Advances in a number of electronic technologies in recent years have allowed their adoption and application in a widening array of industries. For example robotics, optics and electronics that are widely used in automotive- and electronic-based manufacturing plants are now becoming common in food processing plants. High-volume, high-speed processing facilities benefit from the adoption of technologies with the capability to sense rapidly and accurately various raw or finished product characteristics, which then can be used to predict ultimate yield and quality parameters. Several of these technologies have been adopted to more accurately establish value of carcasses or their products. As such, producers are often paid for their livestock based on a "grid", or pricing structure, using this objective information. These technologies will be discussed in some detail in the remainder of this chapter.

REQUISITES FOR SUCCESSFUL TECHNOLOGY ADOPTION

Before discussing individual technologies, it is important to understand the criteria used to evaluate the potential effectiveness of any new technology in the meat industry. First and foremost, any technology must be accurate in predicting an economically relevant characteristic. As eluded to above, two of the most important characteristics that producers would like to optimize in their livestock are the amount of meat that will be harvested from a particular animal and the ultimate eating quality of that meat when it is merchandized and consumed. Though the former becomes less important once a carcass is fabricated, the general need to know the amount and quality of the product produced seems to carry throughout harvesting and processing.

One of the inherent hurdles to be overcome in the development of such technologies is that, in many cases, these technologies may be attempting to "predict a predictor". Though seemingly odd at a glance, consider information from Chapter 14 where yield and quality grading of beef carcasses was discussed. During this activity, grades are assigned to a carcass based on specific indicators or physical parameters, which are then used as predictors of a particular yield or quality characteristic. Using these predictors, carcasses are assigned a particular yield or quality grade, which are subsequently used to help segregate carcasses on either of these two valuable attributes. For example, marbling abundance in the rib-eye area, as determined by a trained federal employee, is not directly evaluating eating quality but rather is evaluating a characteristic of the carcass that is useful, at some level, in predicting the eating quality of meat from that carcass. So, development of a technology that attempts to remove the subjectivity of the trained technician by evaluating the marbling objectively, for example using a camera and sophisticated image processing software, is really attempting to predict the amount of marbling in the rib-eye muscle, a predictor of eating quality, or simply predicting a predictor.

Though this concept is a bit advanced for a general textbook, it is important to understand that many technologies fail, not because they lack accuracy but because traditional predictors may be somewhat inaccurate in predicating a trait. As such, the overall inaccuracy is additive and unacceptable. To this end, developing objective techniques to measure the various meat quality and quantity traits presents a major challenge to the meat industry. Typically any electronic means of sensing is limited in the range of traits for which it is useful. Therefore, it is necessary to match the type of sensors with the particular trait of interest. The basic chemical or physical parameters that delineate the trait to be assessed must be identified before an effective technology can be developed.

The following discussion is intended to highlight the principles on which some of the successful technologies are based. Even so, quality is a generic term in the meat industry that implies a degree or grade of excellence and meat is a complex system with many properties that contribute to the overall degree of excellence. The importance of the various properties depends greatly on the specific meat product and the way it is to be utilized by the purchaser, whether it is a processor, food service operator, retailer, or the ultimate consumer. For example, those characteristics that define fresh meat quality in a steak or chop, may be quite different from those used to evaluate processed meat. Thus, quality parameters needing assessment vary within the industry.

PRINCIPLES OF ELECTRONIC ASSESSMENT

A number of electronic instruments are configured with sensors capable of detecting biological or physical parameters of meat that are useful in segregating products. Using differences in these unique tissue properties across different products, electronic signatures are created that can be used to establish quality or quantity predictions. These types of equipment, by and large, use computers and various forms of software to process very complex sensor input data. Given the rather harsh nature of most processing facility environments (moisture, heat, shock, etc.), implementation of such technologies has been challenging, but has been successful in many instances.

Optical-Based Technologies

The visual properties of meat are essentially a collective result of how light is reflected from or transmitted in the tissues. When light interacts with meat, it is not only absorbed by various tissue components such as water, myoglobin, and cytochromes, it is also reflected and scattered by structures inherent to the tissue. Though not completely understood, structures associated with the myofibril and other muscle-specific organelles as well as components of connective tissues such as collagen and reticular fibers may affect the path of light as it travels through meat. A number of technologies use visible or infrared light to determine the optical properties of meat. Depending on the configuration of light sources and detectors, different parameters of meat can be ascertained. The premise that light, regardless of the wavelength, when emitted onto the surface or within meat, is reflected or transmitted in such a manner that it provides information about meat color and structure is the basis of all optical technologies.

Colorimeters are instruments often used to determine lean or fat color, which is a major contributor to fresh meat quality. The term **colorimetry** identifies the process of determining color using an instrument or technology. There are two major technological approaches by which color is measured, the **tristimulus** method or the **spectral reflectance** method. In order to understand how color is measured, it is first important to recognize that the human eye has photoreceptors (cone cells) that are sensitive to light in three different wavelength bands, or color bands. When color is analyzed by a piece of equipment that may be more or less capable than the human eye of sensing the individual wavelengths of light from a particular color, results may be quite different.

In an attempt to define color, the International Commission on Illumination (Commission Internationale de L'Elcairage, CIE) was held in 1931. This collection of experts was charged with developing a system to describe color as viewed by the human eye. As a result of this commission, a system was adopted in which any color could be defined by the combination of three values, X, Y and Z (Figure 15.1 in the color section), which roughly correspond to red, green and blue, respectively. This is known as the **tristimulus method**. In other words, using a three-coordinate system, every color can be defined by the contribution of each of three colors (X, Y and Z). Most other tristimulus-based color coordinate systems are based on the original XYZ color space, such as L*a*b*, L*C*h, and HSI. Each of these color scales relies on three coordinates to define a color within a three-dimensional space. Because of the simplicity of the technology, many colorimeters especially less expensive, hand-held models use this tristimulus method for determining color. Essentially, the light reflected by the object, or piece of lean meat, is reflected through a small opening or aperture on the device. The reflected image (light) is subjected to three electronic sensors. The data from these sensors are fed into a microcomputer and yield numerical values for X, Y and Z. The tristimulus approach for determining color best reflects a human's perception of color because the human eye tends to view images in a red, green and blue, or XYZ color space. A number of hand-held colorimeters are available that use this technology and many have been used to define meat color.

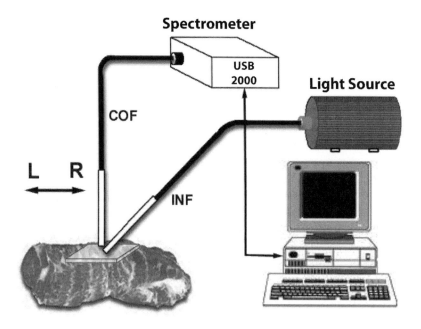

FIGURE 15.2 | Components of most fiber optic-based spectroscopy technologies. Light is provided through fiber optics (INF) to meat and reflected or scattered light is captured by yet other fiber optics (COF) connected to a spectrometer, which provides data to a computer. Often, source and detecting fiber optics are contained within a single lead. Source: Xia, et al., 2006, *J Biomed Opt.* 11(4):040504.

The **spectral reflectance** (or **reflectance spectrum**) method for determining color is a bit more technologically sophisticated and yields more information about a particular color than the tristimulus method. The instruments that generate spectral reflectance are known as **spectrophotometers**. Though spectrophotometers can provide numerical data about the color similar to those using the tristiumlus approach, spectrometry also can provide complete spectral reflectance data for any particular image. Rather than simply characterizing a color on a three-coordinate system, color actually consists of a mixture of various wavelengths of light, varying in different proportions. A spectrometer-based colorimeter separates light reflected from an image into its component wavelengths using a **diffraction grating**. This reflected light is then presented to a large number of individual sensors, each representing a band, or portion, of the full spectrum of light. This information is collectively referred to as the spectral reflectance, or reflectance spectrum. This more accurately defines a particular color. Though hand-held colorimeters often use the less expensive tristimulus technology, some high-end colorimeters use spectrometry. Regardless, both methods of color determination are quite valuable in assessing fresh meat color and are used extensively in the industry.

Researchers at the Danish Meat Research Institute were the first to apply optical-based technology to predict pork muscle quality because of the high incidence of PSE meat at that time. Recall from previous chapters that pale, soft exudative meat, especially pork, is quite undesirable due to reduced consumer and sensory appeal and lower potential for further processing. The industry was incurring large losses because of the inferior product and there was a sense of urgency to develop systems to quickly and accurately identify PSE meat in pork carcasses and thus to discriminate against it. Thus, the Danes began working on an optical technology to predict pork quality, which is easily categorized by visual appearance. At that time, **fiber optics** was beginning to be used in a number of applications. As the name implies, fiber optics uses very thin fibers made of pure glass. These fibers are able to transmit light and thus, images. The general design of any fiber optic-based technology consists of a light source, a detector and optical fibers that transport the light to and from a probe that can be inserted into the meat (Figure 15.2). This technology was particularly useful because it allowed the placement of a fiber optic probe into the carcass or meat sample, remote from the instrument interpreting the reflected information. Though an **invasive technology**, meaning it must penetrate the surface of the product, thereby increasing the chance of microbial contamination, it was still deemed the best approach to secure information that might be useful to predict PSE. Other than being invasive,

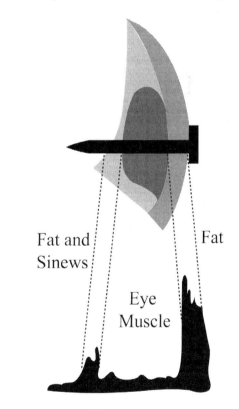

FIGURE 15.3 Illustration showing the placement of an optical probe through the subcutaneous fat and longissiumus dorsi muscle of a carcass to ascertain meat quality and composition. Source: Procedures for Estimating Pork Carcass Composition, American Meat Science Association/National Pork Producers Council Fact Sheet #04341 - 1/01.

the major limitation of the initial versions of this type of equipment was that **visible light** was used as the light source. Moreover, the size of the area sampled was quite small. Visible light, especially single wavelength light sources do not work well with pale meat, regardless of whether the pale color is caused by low myoglobin content or by the postmortem metabolism that produces PSE meat. To address this deficiency, the next generation of fiber optic probes used light in the **near-infrared** (NIR; 950 nm) spectrum, which proved more useful in detecting meat quality.

The Fat-O-Meater (Figures 15.3 and 15.4) was initially designed with the intention of detecting lean pork color (pork quality) but given the dramatic differences in the reflective nature of fat and lean, the Fat-O-Meater was subsequently equipped with a depth gauge so the self-piercing probe containing a fiber optic window could be inserted perpendicular to the long axis of the longissimus dorsi muscle between the tenth and eleventh ribs of pork carcasses and measure fat depth as well as muscle depth. This valuable information was

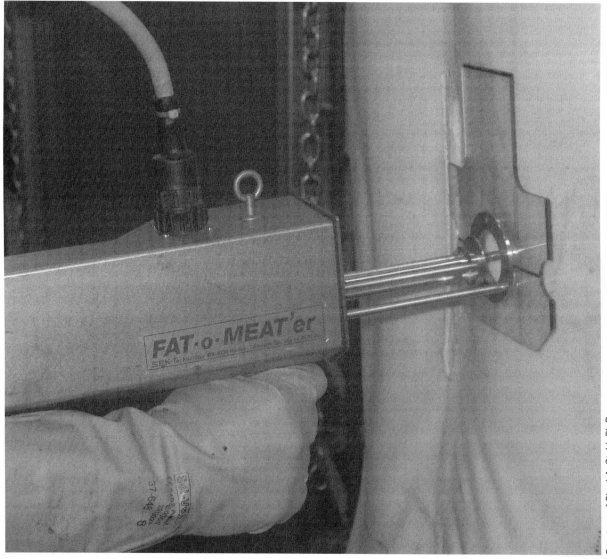

FIGURE 15.4 | Application of an optical probe to determine pork carcass composition and quality.

then used, in conjunction with hot carcass weight to predict carcass composition.

As noted previously, lean and fat have distinctly different reflective properties. Danish workers used these differences to design a fiber optic probe that could sense marbling when the probe was inserted through the *longissimus* muscle, allowing carcasses to be segregated into groups differing in marbling score. Instruments using this particular technology have been designed, tested and deployed in a number of pig processing plants around the world.

Although optical-based instruments that use colorimetry, spectrometry, or fiber optics all measure reflective properties of meat or carcasses, each has inherent limitations in its application. As mentioned previously, fiber optic probes are invasive, that is, they must be inserted into the carcass or meat sample. This increases the potential for microbial contamination in the area where the probe is inserted. Fiber optic probes also sample a very small area and they generally do not perform well in predicting fresh meat color. They have found widest application to detect PSE meat early in the postmortem period and to measure fat and muscle depth (or thickness) in carcasses.

Colorimeters and spectrophotometers that measure reflectance from meat surfaces sample substantially larger areas than fiber optic probes, although the sampled area is still quite small. Because an exposed meat surface is required, there is the potential for microbial contamination. Additionally, these instruments do not adapt well for measurements during the early postmortem period before completion of rigor mortis and thus are not useful for segregating carcasses before chilling.

Vision Systems

Vision based systems are one of the more recent electronic technologies developed to measure properties of carcasses and meat. Vision based systems capture images of large areas of the carcass or muscles, segment the images into features (elements), and use the feature data as predictors of traits that relate to product value. This technology, of course, can be piggybacked on other technologies whenever an image can be captured electronically and digitized. Recall the example of the association between beef marbling and final eating quality used earlier in this chapter. The grading of beef carcasses by federal employees is largely a subjective process and is costly because skilled personnel must perform this function. Increases in efficiency, accuracy and precision might be gained if equipment could be designed and manufactured to accomplish the grading function. Machines can consistently make repetitive measurements and are not subject to fatigue, stress and a number of other human-related shortcomings. Thus, much effort has gone into development of imaging systems that could replace human graders.

Excellent progress with vision based systems has occurred in recent years because of advances in camera technology particularly in resolution of images and the addition of true color-based information in the images. As these developments entered the market place, the application of machine vision systems increased rapidly. Using beef marbling as an example, the most important task to accomplish is to capture an accurate image of the beef rib-eye. Standardized lighting conditions are an important prerequisite for accurate imaging. The real challenge associated with machine vision grading is to segment the image into features and to isolate the portion of the image that contains the necessary information. In grading of beef carcasses, the rib-eye and the fat covering over the rib-eye must be captured. Other parts of the image that are not rib-eye, such as the lower rib lean or background, must be removed from the image using computer software. Images of the cartilage tips on the thoracic vertebrae also can be captured to determine carcass maturity characteristics. A number of vision systems are currently used to predict carcass yield and quality grades in the US and a number of vision-based technologies are used in the European Union, Australia and Canada. In addition, systems have been placed in pork, poultry, lamb and seafood plants that help replace graders and aid in the segregation of product based on quality characteristics that include but are not limited to color, marbling and textural features.

Machine Vision Systems have also been developed to yield grade carcasses on-line in European processing plants. The European beef yield grade system differs slightly from that in the US in that it contains a conformation score, or the thickness of the carcass in relation to the length. In this case grid lines are projected onto the carcass in order to allow software to determine the contour of the carcass. Then using color to determine the area of the fat on the surface of the carcass, the information is combined to indicate the predicted yield of lean from the carcass.

Further development of the technologies used in machine visioning systems, such as high resolution color video cameras and neural networks software for analyzing images will expand the possibilities to develop a wide range of applications that are now accomplished by human visual assessment. Analyzing the individual pixels in images provides the basis for assessing not only color but also shapes that are present in images. Sophisticated computer software allows these images to be analyzed and products to be sorted based on present color standards. This technology offers great potential for the on-line sorting of any meat product that currently is sorted or graded visually.

Electrochemical and Electrical Technologies

One of the most useful measurements of the physical and chemical properties of meat is that of pH. Though pH data are useful inputs at many times in meat processing, those taken during the first 24 hours postmortem are particularly useful as predictors of pork quality (see Chapter 5). Recall from courses in chemistry and biology, pH is defined as the negative log of the hydrogen ion concentration. A pH meter is a simple voltmeter that measures the voltage of an electrode sensitive to hydrogen ions relative to an internal electrode that exhibits a constant voltage. Classic pH-sensitive electrodes are manufactured of a thin glass membrane that comes in contact with a solution (in this case, meat) whose pH is being measured. Inside the electrode is a silver wire, coated with silver chloride and immersed in an HCl-based solution. When the electrode is in contact with the tissue (solution), a voltage difference between the internal standard and the sample is detected and displayed on a meter.

One of the most obvious disadvantages of early pH meters was the fragility of the glass-tipped probes. Loss of a glass tip in one of several thousand hog carcasses processed each day in some modern plants would be a serious problem. In response to this issue, more robust glass tips were designed and worked well, but muscle proteins and fat accumulated in the pores and the tips required extensive cleaning and maintenance. Finally, engineers found a way to manufacture metal pH probe tips. Now, most meat companies have a number of hand-held units that are used to predict quality on-line or simply as a means to ensure quality. Even so, measuring pH accurately is difficult and inconsistent. Moreover, pH values are temperature dependent. To overcome the later issue, many newer pH meters correct for temperature. Operator error, which is defined as that variation in measurement due to changes in probe placement and measuring, remains a significant problem for any device that relies on human-based operations (Figure 15.5). Robotics has helped reduce such errors.

Electrical **conductivity** and **resistance** (or **impedance**) have found applications for the prediction of meat and carcass characteristics. All biological tissues that contain water, including meat, are to varying degrees solutions of a number of different ions. This unique characteristic is ideally suited for technologies that measure the properties of electrical transmission to create a signal that reflects a particular characteristic of meat. As discussed in earlier chapters, meat consists of a collection of muscle fibers, each of which contains the contractile apparatus encapsulated or insulated from external surroundings by the sarcolemma. During the transformation of muscle to meat, the integrity of the sarcolemma is lost and the ions and water equilibrate across the membrane. Thus, the tissue has a low resistance or impedance to current flow, mainly because of the continuity of the moisture around the membranes and the conductivity associated with a solution of ions. Recall that PSE meat has lower water holding capacity and that more moisture has migrated from

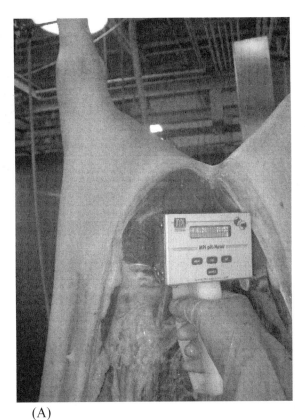
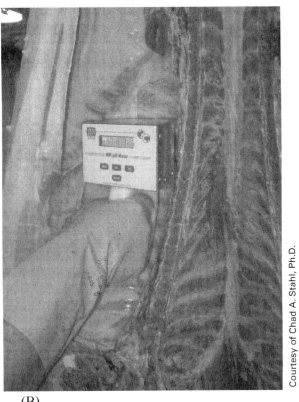

FIGURE 15.5 | Example of pH probe placement in a pork carcass. (A) pH measurement in the ham muscles. (B) pH measurement in the loin muscle.

FIGURE 15.6 | Use of a hand-held pH/impedence probe in determining meat quality.

intracellular spaces to intercellular spaces. This exudate conducts electricity better and thus, PSE meat has a lower resistance, or impedance, to electrical current flow than normal meat. Conversely, DFD (dark, firm and dry) meat retains more moisture intracellularly in the myofibrillar network, primarily because postmortem metabolism has been abbreviated by a lack of energy substrate, and has a higher ultimate pH. The higher pH in relation to the overall isoelectric point of muscle proteins collectively allows for greater repulsion between myfilaments, expansion of the myofibrils and greater retention of water within the fibers. This means the conductivity of the muscle is less, or the impedance to electrical flow is greater. Thus, impedance-based technologies are effective in objectively determining meat quality, in particular water holding capacity (Figure 15.6).

Impedance measurements also are quite effective is predicting the composition of meat, that is the proportions of lean and fat, because lean tissue has a much higher conductivity than fat. This has led to the development of procedures for prediction of carcass composition in both live animals and carcasses (Figure 15.7). However, this application has not been adopted for on-line measurement. Finally, the application of bioelectrical impedance may be useful in detecting marbling, based on the aforementioned differences in conductivity among tissue. Even so, due to inconsistencies among various conductivity-based technologies, few probes have been adopted in the industry.

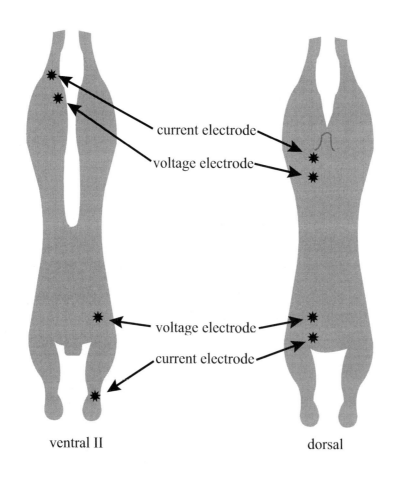

FIGURE 15.7 | Approximate placement of electrodes on a carcass necessary to ensure proper determination of carcass composition by electrical impedance.

Ultrasound Technologies

Ultrasound is defined as sound in the frequency range beyond human hearing, which is in the frequency range of 20 to 20,000 Hz. Ultrasound waves travel through different biological tissues at different speeds. Moreover, like all other sound waves, ultrasound waves are reflected back from various tissues at corresponding different speeds depending on the nature and density of the tissue. Utilization of these principles has led to the development of imaging systems that allow assessment of structures and composition of living animals and carcasses. Ultrasound equipment capable of collecting data in a meat packing facility generally utilizes sound waves with frequencies in the range of 2.5 to 15 MHz. The equipment consists of a: 1) pulser, 2) transducer, 3) receiver, 4) display and 5) memory device. The **pulser** is the source of high amplitude voltage. The **transducer** converts the electric energy to mechanical sound by applying an electrical field across an array of crystals, which vibrate and emit ultrasonic waves. The **receiver**, also assembled with the transducer, detects and amplifies signals returning to the transducer/receiver assembly. A **display** depicts a visual representative of the returning signal and a **memory device** stores the information from the display.

The simplest ultrasound imaging devices are called **A-mode** (amplitude) units. These units require an oscilloscope to interpret the returning signal because only simple signals are emitted from the transducer and collected by the detector. Essentially these transducers/receptors measure returning signal amplitude against time and determine differences in tissues by the distance between signals. A-mode units are very popular and perform nicely to determine distances to various structures within a mixed tissue. Thus, they are very useful for determining subcutaneous fat content or fat depth because the interface between adipose tissue and muscle tissue is quite pronounced.

B-mode (brightness) ultrasound units are slightly more sophisticated and generate more detailed information primarily because of the configuration of the transducers and detectors. These ultrasound units have a linear array of transducers and a corresponding set of detectors that are capable of scanning in a plane across a sample as contrasted to the small, point-source emitter and detector of an A-mode unit. Essentially, data from a

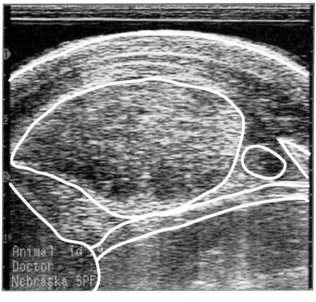

FIGURE 15.8 | Real-time ultrasound image of a pig carcass taken at, or close to the last rib. The subcutaneous fat layer and the longissimus dorsi are particularly evident with the outlining.

B-mode unit is two-dimensional and is displayed as a dot or pixel. The intensity, or brightness, of each pixel is a reflection of the amplitude of the signal returning to the detector. In much the same way as the A-mode unit, the time necessary for the signal to reflect back to the detector is calculated. This then determines its pixel location on the display. These signals are expressed on a gray scale, thus producing an image. This gray scale helps visualize tissue differences based on densities and texture. B-mode units are particularly useful because they allowed visualization of detailed structures of interest. After the signals are subjected to image segmentation and processing, image features can be used as predictors. The major drawback of these units, however, is that operators are able to collect only a single image per scan.

The limitation of B-mode ultrasound technology was eliminated with the development of **real-time ultrasound** scanning equipment. This equipment possesses real time imagining capabilities so data are collected continuously and displayed images are updated simultaneously (Figure 15.8). Regardless of the mode used, ultrasound measuring is a promising technology for predicting composition and quality (marbling) of meat products. Marbling estimation is possible due to the inherent differences in the movement of sound through lean and fat.

Several ultrasound technologies are currently used in a number of plants throughout the world and select plants in the US (CVT-2, UltraFom 300 and AutoFom). The Autofom Grading System, in particular, is part of a carcass merit buying program for pigs in the US, where producers are paid for their livestock based on carcass merit. This grading station is configured so carcasses can be passed over the transducers in such a manner that ultrasound data is collected from a number of locations on the carcass after it has been dehaired. This digital, stationary, three-dimensional scanning ultrasonic transducer array is known at the Autofom™. It does not require an operator. Some 16 ultrasonic transducers are embedded in a fixed stainless steel cradle or trough. Carcasses move across the transducer array when the carcass is still warm and pliable in order to ensure good contact with the transducers. Transducers provide an image of the carcass cross-section every 5 mm. Thus, these 16 transducers produce approximately 3200 measurements for each carcass. In addition to cross-sectional analyses, data collected from one transducer during the time a carcass is measured will provide data along the length of the carcass. After these data have been assembled into an image, image processing software is used to derive over 100 key predictors of primal cut percent. These predictors are used in an equation to generate a measurement called Primal Cut Percent, which is defined as the percentage of the carcass made up of trimmed loins, hams and shoulders. Primal Cut Percent provides a measure of carcass leanness and is the key component to determine carcass value. In addition to the Primal Cut Percent, the Autofom also provides a prediction of the individual weights of the primal cuts plus the belly. Day to day price fluctuations are fed into this equation and producers are then paid on the basis of total primal value. This type of system embodies the notion that paying producers on true value is better than simply paying for animal weight, which provides no information about lean product yield.

The aforementioned uses of ultrasound in the meat industry have been largely implemented to operate during the harvesting phase. However, ultrasound has been and is used extensively subsequent to carcass fabrication and during further processing. Ultrasound has been used to determine fat content (aside from marbling fat) and predict the underlying textural differences of various meat products. Moreover, ultrasound technology is employed to improve the functionality of meat protein and improve meat safety. A trend for additional uses will likely continue.

Electromagnetic Scanning Devices

When biological tissues are placed in a weak electromagnetic field, energy is differentially absorbed by the tissues depending upon the water content of the tissues. The fact that fat tissues are very low in water content and lean tissue is very high means that lean tissue will absorb more energy than fat. Passing carcasses through a large electromagnetic coil and measuring the energy absorbed with a detector provides a means for prediction of lean composition in the entire carcass as well as in various important primal cuts at on-line speeds. This technology has found limited application because of the necessity of removing carcasses from the rail to pass them through the coil. Numerous units consisting of a smaller coil are being utilized in the meat industry to measure the composition of hams for sorting into value classes.

X-ray Scanners

Whole body **X-ray scanners** are capable of creating cross-sectional images of live animals and carcasses that can be analyzed for the prediction of carcass composition. This technology is used for research on carcass composition; however, the high cost and low rate of speed has precluded its use in practical applications or on-line.

Direct Measurement Techniques

Lean composition is often predicted using linear and area measurements on the carcass and cut surfaces. A simple ruler can be used to measure the depth of fat at various locations on a carcass. The addition of an assessment of the area of the longissimus muscle at a standard location measured with a simple acetate grid, along with the weight of the carcass provides relatively accurate prediction of carcass composition when these measurements are used in regression equations.

16

BY-PRODUCTS OF THE MEAT INDUSTRY

OBJECTIVES: *Gain an appreciation for the role that effective by-product utilization plays in the efficient production of meat products, as well as its importance in environmental stewardship.*

By-products
Edible by-products
Inedible by-products
Variety meats
Intestines
Edible fats
Render or rendered
Shortenings
Protein residues
Cracklings
Bovine spongiform encephalitis (BSE)
Variant Creutzfeldt-Jakob disease (v-CJD)
Specified risk materials (SRMs)
Render
Transmissible spongiform encephalopathies (TSEs)
Hides

Kips
Skins
Brands
Pelts
Vegetable tanning
Chrome tanning
Lanolin
Tallows
Greases
Dry rendering
Blood meal
Meat meal
Steamed bone meal
Feather meal
Poultry by-product meal
Fish meal
Epinephrine
Estrogens
Progesterone
Insulin
Trypsin
Chymotrypsin
Glucagon

Somatostatin
Parathyroid hormone
Adrenocorticotropic hormone (ACTH)
Somatotropin
Thyroid stimulating hormone (TSH)
Hyaluronidase
Thyroxin
Calcitonin
Thymosin
Albumen
Amino acids
Fetal serum
Serum
Thrombin
Surgical ligatures
Liver extracts
Bile extracts
Cortisone
Cholesterol
Rennet
Mucin
Pepsin

Key Terms

CHAPTER 16 BY-PRODUCTS OF THE MEAT INDUSTRY

Animal **by-products** include everything of economic value, other than carcasses obtained from animals during slaughter and processing. These products are classified as either **edible** or **inedible** for humans.

The meat packing industry has long been noted for its efficiency in processing and using by-products. The ingenuity of this industry in devising new products and discovering new uses for by-products has given rise to the expression that "the packer uses everything but the squeal." Although by-product value is only a small fraction of live animal value, by-products are of considerable economic importance to the entire livestock and meat industry. Their value influences the amount livestock producers receive for animals, and the return on live animal investments by processors. For example, the by-product value from a typical slaughter steer was estimated at $23.94 per 100 kg live weight on October 13, 2008 as shown in Table 16.1. Over the past several years, value of by-products has declined with respect to value of live animals. This relative decline in value has been due largely to technological progress in producing competitive products from non-animal sources. For example, various synthetic materials have been developed to make many items that were once made of leather. Synthetic fibers, instead of wool, are used to make clothing. In many food products, vegetable fats and oils have replaced animal fats. Synthetic detergents have replaced soaps, which were once made entirely from animal fats. In fact the reduced demand for fat has driven significant genetic changes that reduce the amount of fat in animals produced for meat. Numerous pharmaceuticals that previously were obtained only from animal tissues and glands are now synthesized chemically or are produced in microorganisms through the application of molecular biology techniques. Many other examples could be cited to illustrate the competition from non-animal products.

Enormous quantities of by-products are generated at each step of the meat processing and distribution system, as shown in Tables 16.2, 16.3, 16.4 and 16.5. Yields of by-products vary greatly, depending on method of processing and weight, grade, sex, and species of animal (Tables 16.2, 16.3 and 16.4). For example, whether beef tallow is saved and processed for edible or inedible use markedly influences its value. The terms edible and inedible refer to human consumption. Value of by-products is maximized when they are processed immediately after slaughter. Otherwise, edible materials degrade quickly to inedible products. Even in inedible materials such as fats, delay in processing may cause grades to be lowered because of free fatty acid buildup and development of rancidity.

In addition to monetary value derived from processed by-products, conversion of inedible parts of animals into useful products performs very important sanitary functions. All inedible parts, unless processed, would accumulate and decompose, causing undesirable conditions in surrounding environments.

EDIBLE MEAT BY-PRODUCTS

Numerous edible meat by-products are obtained during slaughter and processing of meat animals. Some of the more common items are listed in Table 16.6. Edible organs and glands, such as tongues, brains, sweetbreads, hearts, livers, and kidneys are called **variety meats**. They are excellent sources of many essential nutrients required in human diets, and offer many interesting and appetizing variations for consumers. For example, liver is the best known and most widely served of all variety meats, partially because it ranks among the best food sources for vitamin A, riboflavin, niacin, and iron (Chapter 12).

The number of edible by-products, and the extent to which these items are processed for edible uses, varies among processors. Unless processors have potential markets for such items, it may be uneconomical to process them for edible use and therefore more profitable to process them for inedible use.

In recent years, general acceptance of variety meats has declined. This may be due to an image, held by some people, that variety meats are food for people who cannot afford more expensive meat. However, various ethnic groups traditionally have consumed large quantities of variety meats.

CHAPTER 16   BY-PRODUCTS OF THE MEAT INDUSTRY   353

TABLE 16.1 Estimated By-Product Value From A Typical Slaughter Steer (580 Kg Liveweight) On October 13, 2008

	Kg/100 kg liveweight	$/100 kg liveweight
Steer hide, butt brand/Pc	11.55	11.64
Tallow, edible	2.64	.66
Tallow, packer bleachable	9.90	2.35
Tongues, Swiss #1 0–3%, exp	.53	1.06
Cheek meat, trimmed	.70	.88
Head meat	.29	.26
Oxtail, selected	.44	.81
Hearts, regular bone out	.84	.55
Lips, unscalded	.29	.48
Livers, selected, export	2.11	1.67
Tripe, scalded edible	1.43	.90
Tripe, honeycomb bleached	.33	.51
Lungs, inedible	1.03	.04
Melts (spleen)	.31	.02
Meat and bone meal, 50% bulk/ton	8.14	1.23
Blood meal, 85% bulk/ton	1.32	.55
Totals	41.45	23.61

Source: USDA Market News, Des Moines, IA. http://www.ams.usda.gov/mnreports/nw_ls441.txt

TABLE 16.2 Annual By-Product Raw Materials Available in 2007 in the US

	Cattle	Calves	Pigs	Sheep and lambs	Chickens	Turkeys
Number of animals slaughtered, millions	34.26	.76	109.17	2.69	9,031.04	254.72
Liveweight, billion kg	19.82	.11	13.34	.17	22.65	3.41
Edible organs and trimmings at packer level, percent of liveweight	4	4	7	4	4	3
Total by-products, packers and retailers, percent of liveweight	55	49	48	57	27	21

Sources: Little, Arthur D. Inc., "Opportunities for Use of Meat By-Products in Human and Animal Foods." Report to the Iowa Development Commission (1969). US Department of Agriculture, National Agricultural Statistics Service, Livestock Slaughter 2007 Summary, Mt An 1-2-1(08) March 2008. US Department of Agriculture, National Agricultural Statistics Service, Poultry Slaughter 2007 Annual Summary, Pou 2-1(08) February 2008.

TABLE 16.3 Approximate Yield of Various Items Obtained from Meat Animals

Item	Steer	Lamb	Pig
Grade	Choice	Choice	US No. 1
Live weight, kg	455	45	115
Dressed carcass, kg	273	23	85
Retail cuts, kg	190	16	67
By-products, kg			
Hide or pelt	36	7	
Edible fats	50	4	17
Variety meats	17	1	4
Blood	18	2	5
Inedible fats, bone, and meat scrap	80	10	9
Unaccounted items (stomach contents, shrink, etc.)	64	5	13

TABLE 16.4 Approximate Yield of Items Obtained From Broiler Chickens and Turkeys

	Brioler chickens, percent of liveweight	Turkeys, percent of liveweight	
		Toms	Hens
Carcass with neck including bones	66.2	73.5	72.1
Giblets—liver, heart, gizzard	4.3	3.4	3.3
Other edible parts—kidneys, abdominal fat, gizzard fat	3.0		
Feathers and blood	9.5	9.3	8.9
Head	3.9	1.7	1.6
Feet and shanks	5.1	2.9	2.5
Organs, viscera, intestines and intestine contents	8.0	9.2	11.6

Source: Goldstrand, Richard E. An Overview of Inedible Meat, Poultry and Fishery By-products. In Advances in Meat Research: Inedible Meat By-Products, Vol 8. A.M. Pearson and T.R. Dutson, eds. Elsevier Applied Science, London and New York, 1992.

TABLE 16.5 | Approximate Yield of Items Obtained From Cod Fish

	Percent of whole body weight
Fillets	36
Skin	3
Head	22
Backbones	15
Fins and lungs	10
Roe	6 (2–7 range)
Guts removed	8 (5–8 range)

Source: Goldstand, Richard E. An Overview of Inedible Meat, Poultry and Fishery By-products. In Advances in Meat Research: Inedible Meat By-Products, Vol 8. A.M. Pearson and T.R. Dutson, eds. Elsevier Applied Science, London and New York, 1992.

Variety meats also are popular menu items in many fine restaurants. They potentially could become pleasurable sources of food for large population segments not eating them.

Almost all edible by-products, especially variety meats such as liver, are more perishable than carcass meat. Therefore, variety meats must be chilled quickly after slaughter, and processed or moved quickly into retail trade. Variety meats may be merchandised as fresh or frozen items, or used in making other processed foods.

The box appearing on page 357 illustrates sources and uses of many edible by-products. For detailed descriptions of methods used for processing these items, readers are referred to references listed in the reference section following this chapter.

Historically, animal **intestines** have been used as edible sausage casings. However, as pointed out in Chapter 8, most sausages are processed in synthetic casings made from cellulose. Edible regenerated collagen casings have characteristics of both synthetic and natural casings. Animal hides are sources of collagen used in these casings.

Fatty tissues (adipose tissue), obtained from various parts of meat animals, are processed into *edible fats*. These fats are **rendered** (separated from their supporting tissues) by a variety of heat treatments. Lard and rendered pork fat are obtained from fatty tissues of pork, and are used as **shortenings**. Fatty tissues of beef yield edible tallow used in making shortenings. Shortenings are used for frying and tenderizing foods, particularly baked foods, by interposing fat throughout the food in such a manner that protein and carbohydrates do not cook into a continuous hardened mass. If fatty tissues are rendered at low temperatures, and the fat is removed by centrifugation, remaining **protein residues** may be used as ingredients in processed luncheon meats. If fat is extracted at higher temperatures, the protein residue, known as **cracklings**, is used in animal feeds. Edible poultry fats are often used for flavoring of soups and stocks.

Since the discovery of **Bovine spongiform encephalitis (BSE)** in the United Kingdom in 1986, the linking of BSE to a **variant** form of *Creutzfeldt-Jakob Disease* (vCJD) in humans and discovery of the first case of BSE in the United States in 2003 (in a beef animal imported from Canada), stringent regulations have been implemented to safeguard against BSE-infected products entering the food supply. Tisssues that could be infective in a bovine with the disease are defined as **Specified Risk Materials (SRMs)** and are banned from the food supply. SRMs include the skull, brain, trigeminal and dorsal root ganglia, spinal cord, eyes and portions of the vertebral column of all cattle greater than 30 months of age. The tonsils and distal ileum of all cattle are defined

TABLE 16.6 Edible Usage of By-products Obtained from Meat Animals

Raw by-product	Principal use
Brains	Variety meat
Heart	Variety meat
Kidneys	Variety meat
Liver	Variety meat
Spleen (melt)	Variety meat
Sweetbreads	Variety meat
Tongue	Variety meat
Oxtails	Soup stock
Cheek and head trimmings	Sausage ingredient
Beef extract	Soup and bouillon
Blood	Sausage component
Fats	
(a) Cattle, calves, lambs and sheep	Shortening Candies, chewing gum
(b) Pork	Shortening (lard)
(c) Poultry	Soup stock, flavorings
Stomach	
(a) Suckling calves	Rennet for cheese making
(b) Pork	Sausage container, sausage ingredient
(c) Beef (1st and 2nd)	Sausage ingredient, variety meat (tripe)
Gizzard	Variety meat
Intestines, small	Sausage casings
Intestines, large (pork)	Variety meat (chitterlings)
Intestines, large (beef and pork)	Sausage casings
Esophagus (weasand)	Sausage ingredient
Bones	Gelatin for confectioneries, ice cream, and jellied food products
Pork skins	Gelatin for confectioneries, ice cream, and jellied food products; french fried pork skins
Calf skin trimmings	Gelatin for confectioneries, ice cream, and jellied food products

Beef extract	consists of concentrated cooking water from heating beef for canning. This liquid usually is concentrated by evaporation under vacuum to about 20 percent moisture and on cooling forms a pasty solid. It is a major ingredient in bouillon cubes and broth and is used for flavoring gravies. It is not produced generally in the US, but is made primarily in South America in the course of producing canned beef.
Testicles	of lambs, calves, and turkeys are sold fresh or frozen. When cooked they commonly are known as "fries" or "mountain oysters."
Blood	is used as a component of blood sausage. Special hygienic precautions must be taken when it is collected for edible use. Ultrafiltration processes have been used to recover proteins from both plasma and cellular components of blood. These blood proteins have many potential uses as binders in sausage and other food products.
Tripe	is obtained from the first (rumen) and second (reticulum) stomach compartments of cattle. It is consumed as a variety meat and used in some sausages. Pig stomachs also are processed for use as ingredients in some sausages. The Pennsylvania Dutch use cleaned pig stomachs as meat containers for delectable "filled pig stomachs" or "hog maws."
Chitterlings	are made from thoroughly cleaned and cooked intestines of pigs, and consumed as a variety meat.
Cheek and head trimmings	are commonly used in manufacture of various sausages.
Brains	which are regarded as a delicacy by some people, are distributed to consumers in either chilled, frozen, or canned form. In the US, brains from cattle more than 30 months of age cannot enter the food supply as a precaution against BSE (see discussion in text).
Beef tongues	are used fresh, cured and smoked, canned, or as ingredients in sausage. Sheep and pork tongues may be used fresh, but they are most commonly used in making other processed meat items. Pork tongues must be scalded before use because of potential contamination during hog scalding and dehairing.
Kidneys	are used fresh, or they may be frozen for meat trade. However, they are more commonly used as an ingredient in pet food.
Sweetbreads	are thymus glands obtained from the ventral side of the neck and inside the chest cavity of calves or young cattle. They are used fresh or frozen.
Hearts	are used fresh, frozen, or as an ingredient in processed luncheon meats. Chicken and turkey hearts may be included with the carcass when whole birds are merchandized.
Liver	is used fresh, frozen, or as an ingredient in liver sausages or other processed luncheon meats. Comparatively more beef liver is used fresh, whereas more pork liver is used in making sausages. Kosher style liver sausages include only beef or lamb liver. Chicken and turkey livers may be included with the carcass when whole birds are merchandized.
Gizzards	are used fresh, frozen or as an ingredient in processed meats. They may be included with the carcass when whole birds are merchandized.
Oxtails	are used mainly for making soups.

Continued...

Spleens	may be presented for retail sale as "melts," or used for manufacturing pet and mink foods.
Gelatin	is made by heating collagen-rich connective tissues, such as pork skins, calfskins, and bone, in hot water. Gelatin is used in manufacture of other processed meats; and it is used widely in gelatin desserts, consommés, marshmallows, candies, bakery products, and various dairy products including ice cream. The pharmaceuticals industry uses gelatin in making such items as capsules, ointments, cosmetics, and emulsions. Gelatin also is used in manufacture of photographic films, paper, and textiles.

as SRMs. Thus, beef small intestines are not used as sausage casings and beef brains from young cattle (less than 30 months of age) are seldom used as edible by-products.

Any discussion of potential uses of edible by-products is necessarily incomplete, because the creativity and ingenuity of meat processors continually generate more uses for these items. Unique properties and nutritional value of meat by-products should be sufficient encouragement for the meat industry to continue making concerted efforts at creating new and useful products.

INEDIBLE MEAT BY-PRODUCTS

Rendering

When the meat packing and processing industry began to centralize and grow in the late 1800's, a new service industry quietly developed to process much of what had been disposed of as waste from the nation's packing plants as well as dead animals from farms and feedlots. They **render**, i.e. reduce, convert, or melt down animal tissues (primarily fat) by heating. The modern rendering industry works hand in hand with the livestock and meat industry in highly sophisticated operations to prevent the release of many tons of highly perishable, potentially dangerous waste materials into the environment. Instead these materials are recycled into useful products. In fact the rendering industry lays claim to the title the "original recyclers" because the industry began utilizing by-products long before recycling became popular.

The rendering industry supports research and development to more efficiently utilize by-products by developing new products and processes. They also strive to develop and adopt technology that prevents the pollution of air and water near rendering operations, whether located at the packing facility or a centralized location. In fact today one can drive by a well-designed packing or rendering plant and not realize what the operation is until you see their sign or logo. Each year some 36 billion pounds of inedible animal by-products are rendered and recycled. These by-products are from slaughter and processing facilities as well as farms, supermarkets, and restaurants. In the process approximately 9 billion pounds of animal fat is reclaimed for a variety of uses annually.

A list of some of the more important inedible by-products obtained from meat animals is presented in Table 16.7. Uses of these processed by-products are varied and almost unlimited in number. New uses are found continually for these products. The discovery of BSE and the recognition of other **transmissible spongiform encephalopathies (TSE)** in other species led to a series of regulations that restrict the use of ruminant derived mammalian tissues, particularly from cattle, as inedible by-products. Restrictions for specific types of inedible by-products will be described in following sections.

HIDES, SKINS AND PELTS

Classification

Cattle hides and skins are classified on the basis of hide weight, sex, maturity, presence and location of

TABLE 16.7 Inedible By-products Obtained from Meat Animals, and Their Major Usage

Raw by-product	Processed by-product	Principal use
Hide (cattle and calves)	Leather glue	Numerous leather goods, paper boxes, sandpaper, plywood, sizing
	Hair	Felts, plaster binder, upholstery
Pork skins	Tanned skin	Leather goods
Pelts (sheep, lambs and goats)	Wool or hair	Textiles
	Skin	Leather goods
	Lanolin	Ointments
Fats		
(cattle, calves, lambs, and sheep)	Inedible tallow	Industrial oils, lubricants, soap, glycerin
(cattle, calves, lambs, sheep, swine)	Tankage	Livestock and poultry feeds
	Cracklings	
	Stick	
(swine)	Grease	Industrial oils
		Animal feeds
		Soap
(poultry)	Fats	Poultry and livestock feeds
Bones	Dry bone	Glue
		Hardening steel
		Refining sugar (bone charcoal)
	Bone meal	Animal feed
		Fertilizer
Cattle feet	Neatsfoot stock	Fine lubricants
	Neatsfoot oil	Leather preparations
Glands	Pharmaceuticals	Medicines
	Enzyme preparations	Industrial uses
Lungs (all species)		Pet foods
Blood	Blood meal	Livestock, poultry and fish feeds
	Blood albumen	Leather preparations, textile sizing
Feathers	Feather meal	Poultry feed
Viscera and meat scraps	Meat meal	Livestock and poultry feeds
Poultry necks, feet, intestines,	Poultry by-product meal	Poultry and livestock feeds

TABLE 16.8 | Classification of Cattle Hides

Origin	Weight* (kilograms)	Classification
Unborn calf		Slunk skin
Calf	Less than 4	Light calf skin
	4–7	Heavy calf skin
	7–11	Kip skin
	11–14	Overweight kip skin
Cow	14–24	Light cowhide
	Grater than 24	Heavy cowhide
Steer	Less than 22	Extra light steer hide
	22–26	Light steer hide
	Grater than 26	Heavy steer hide
Bull	27–54+	Bull hide
Stag		Accepted as steer or bull hide depending upon characteristics

*Net weights after manure and tare allowances are deducted.
Source: Price and Schweigert, The Science of Meat and Meat Products. W. H. Freeman, San Francisco, 1971.

TABLE 16.9 | Classification of Sheep Pelts

Classification	Subclass Number	Wool length (centimeters)
Shearlings	1	1.25–2.5
Shearlings	2	0.63–1.25
Shearlings	3	0.31–0.63
Shearlings	4	0.31
Fall clips		2.5–3.75
Wool pelts		3.75

brands, and method of curing. Hides and skins are classed as **hides**, **kips**, and **skins**, based on weights of clean hides (Table 16.8). Hides come from large and mature animals. Kips are skins of immature animals, and skins are from small animals, such as calves. **Brands** burned into hides affect value and consequently, branded hides are known as Colorado or Texas hides, and unbranded hides are known as natives. When hides are removed by skilled workers in packing plants, they are known as packer hides. Hides removed by unskilled workers are referred to as small packer hides.

Skins and accompanying wool from sheep are known as pelts. Sheep **pelts** are classified on the basis of wool length, as shown in Table 16.9. After wool is removed, pelts are classified as skins.

It is common practice to leave skins on pork carcasses until after chilling. Most pigskins come from the belly and backfat areas after carcass chilling and cutting.

Processing

After cattle hides and skins are removed from carcasses, the ears, lips, and tail are trimmed off. Curing with salt or brine preserves most trimmed hides. The purpose of curing is to preserve hides against bacterial decomposition until tanneries process them. The process of curing with salt consists of spreading salt over the flesh side, and stacking the hides in packs for 30 days or more.

Brine curing has become quite common in recent years. In this process, fresh hides are washed to remove dirt and manure, trimmed of excess fat and flesh, and submerged in vats of saturated brine. To facilitate curing, paddle wheels in vats keep hides moving in brine. Curing requires about 24 hours, after which hides are removed from vats, drained, and packed for shipment to tanneries.

Removal of hair from cattle hides and skins is usually done at tanneries by soaking them in lime water. Tanning processes are rather complex, and their specifics depend on ultimate use of finished products. Basically, leather is made by either of two processes: **vegetable tanning** or **chrome tanning**. Vegetable tanning consists of suspending hides and skins in vats containing solutions prepared from tannin bearing woods or barks. In chrome tanning, solutions of basic chromium sulfate are used. Tannin or chromium permeates each collagen fiber of the skin and converts it into leather. After tanning, the leather is subjected to additional operations to attain various final characteristics. These processes include dying, lubricating, filling, impregnating, flexing, surface coating, embossing, and polishing.

Sheep pelts must be cooled after removal from animals. Then they may be salted for preservation, depending on the time that will elapse before further processing. Next, fresh or salted pelts are thoroughly washed in cold water, following which excess water is removed by centrifugation. Wool is removed from skins by painting the flesh sides with paste containing sodium sulfide and lime, and allowing the pelts to hang for about 24 hours. These reagents loosen wool roots so wool may be pulled easily from skins. Pulled wool is sorted and graded as it is removed. Skins are treated with alkaline sulfide solution that dissolves residual wool. Following this treatment, chemicals are removed by thorough washing. Next, skins are treated with ammonium salts and pancreatic enzymes to remove extraneous protein material. They are washed once again, and preserved by soaking in solutions of salt and sulfuric acid. After draining, skins are graded and ready for tanning.

Most calf and pig skins are removed from carcasses after chilling. They are cured, either in dry salt or brine, to preserve them until processing by tanneries. Some pig skins are removed from carcasses before scalding to remove hair. This procedure avoids heat denaturation of superficial layers of skin by scalding water. Undenatured skins may be used for many leather products such as shoes and upholstery that scalded skins are less suitable for. Processing of these skins is similar to that described for cattle hides and skins.

Wool and Hair

Although most wool produced in the US is shorn from sheep, pulled wool is an important by-product of the packing industry. As wool is hand pulled from each sheep pelt, it is graded on the basis of cleanliness, length, and fineness of fiber. After removal and grading, wool is washed, dried, and shipped to processing mills. Pulled wool may be used in making blankets, felt, carpets, and fabrics; in other words, it is used in any product for which shorn wool is used. Fabrics may be made entirely with wool, or wool may be blended with cotton or synthetic fibers.

Wool processing results in other by-products of commercial value. One of the best known of these is **lanolin**, a refined form of wool grease, which is recovered in the washing process. Lanolin is the base of many ointments and cosmetics.

Cattle hair and hog hair were used historically for furniture upholstery, carpet padding, insulation, and brushes. However, these uses for hair have steadily declined, due to strong competition by synthetic materials. Because of this situation, disposal of waste hair is a serious environmental problem.

TALLOWS AND GREASES

Inedible fats are classified as tallow or grease mainly on the basis of their **titer**, which is the congealing or solidification point of fatty acids. Fats having titers above 40° C are classified as **tallows**. Fats with titers below 40° C are classed as **greases**. Most cattle and sheep fats are tallow; pork fat is classified as grease.

Sources of inedible tallows and greases include animals that die in transit, diseased and condemned animals and parts, and waste fat and trimmings from retail meat markets, hotels, and restaurants. However, a large portion of inedible fat produced by packinghouses comes from edible fats that had potential markets as such.

Most inedible fats are processed by **dry rendering**. Fatty tissues are ground and placed into horizontal steam-jacketed cylinders equipped with internal rotating blades. Rendering may be accomplished at atmospheric pressure, at elevated pressure, or under partial vacuum. Fat cells are ruptured, and melted fat is released from supporting tissues. When sufficient moisture has cooked out, the mixture is filtered or strained to remove cracklings from rendered tallow or grease.

Many uses have been found for products obtained from inedible tallows and greases. These fats may be split, by action of acids or bases, into glycerin and fatty acids. Glycerin is used in manufacture of pharmaceuticals, explosives, cosmetics, transparent wrapping materials, paints, and many other products. Likewise, fatty acids have many industrial uses, as in soaps, detergents, wetting agents, insecticides, herbicides, cutting oils, paints, lubricants, and asphalt. Inedible tallows, greases and oils can be converted into bio-fuels, such as bio-diesel. Such uses may be economically viable during periods of high crude oil prices.

The animal feed industry uses large quantities of stabilized inedible tallows and greases. Fats make feedstuffs less dusty, more palatable, and easier to pellet. They also add energy to feed. Fats used in animal feeds are stabilized against rancidity by addition of antioxidants such as butylated hydroxyanisole (BHA) and butylated hydroxytoluene (BHT). Current FDA regulations prohibit the use in any animal feed of tallow from BSE-positive cattle or from cattle greater than 30 months of age that are not inspected and passed for human consumption and from which the brain and spinal cord are not removed.

ANIMAL FEEDS AND FERTILIZERS

Most large packing plants have facilities for rendering inedible materials produced during slaughter operations. Independent rendering plants collect and process materials from slaughter and processing plants that do not have rendering facilities. End products of rendering are fats and proteinaceous materials. After proteinaceous materials are separated from fat, they are dried and ground. These protein concentrates are quite valuable, and are used as protein supplements in feeds for pigs, poultry, fish, and pets.

Use of ruminant derived tissues as feed ingredients is restricted as a means to protect against the spread of BSE or other TSE if they are present in the animal population. Tissues that could be infective in a bovine with BSE (SRMs) are not permitted in any animal feeds. Feeding of protein concentrates derived from mammalian tissues to ruminants is prohibited. Protein concentrates that originate entirely from porcine or equine protein can be fed to ruminants. Proteins derived from blood, milk, gelatin and processed meats, as well as poultry and fish are not restricted and may be fed to ruminants. Further research into the transmissibility of these diseases from species to species could lead to further restrictions on the use of by-products in animal feeds or result in removal of some restrictions once the safety of various products is established.

The pet food industry uses large quantities of select meat by-products. Livers, spleens, lungs, meat meal, horsemeat, and cereal products are used in making dry, semimoist, and canned cat and dog food.

Dried blood (**blood meal**) is made by coagulating fresh blood with steam, draining off liquid, and drying the coagulum. Dried blood is a rich source of protein and is used as an ingredient in animal feeds. Blood meal is used extensively in formulation of feed for commercial fish operations. **Meat meal** (proteinaceous materials from inedible rendering) and organs, such as livers, also is used for fish foods.

Steamed bone meal is made by cooking bones with steam under high pressure to remove remaining fat and meat. Dried bone is then ground and used as calcium and phosphorus supplements in animal feeds.

Feather meal is prepared by autoclaving or cooking the feathers and then drying and grinding the dried product. It is high in protein but very low in some essential amino acids, including histidine, lysine and methionine. Limited amounts of feather meal can be used in poultry diets; it may also be used in ruminant feeds. **Poultry by-product meal** consists of the rendered, clean parts of the carcass of poultry, such as necks, feet and intestines, exclusive of feathers. Its value as a protein supplement in animal feeds is similar to that of meat meal and fish meal.

Fish meal may be made from either whole fish that are caught exclusively for the purpose of producing fish meal and fish oil or from processing by-products of fish caught primarily for human consumption. Most fish meals are used in feeds for poultry, swine, fish or pets.

Use of animal by-products in fertilizers is limited almost entirely to manufacture of specialty fertilizers for home gardening use, which represents a small proportion of fertilizer production.

GLUE

Chemically and physically, glue and gelatin are very much alike. Raw materials from which glue is made include skins or hides, connective tissues, cartilage, and bones of cattle and calves. Glue is extracted from these materials by successive heating in water under specific temperature conditions. Cooking in water hydrolyzes collagen in these materials to gelatin. Extracts are concentrated, dried, and ground. Glue has many uses in woodworking, paper, and textiles. Its manufacture dates back more than 3000 years to the cabinetmakers of Egypt. Blood albumen, obtained from blood plasma after removal of red cells, is used to make adhesives, almost all of which are used in manufacturing plywood and wood veneers.

PHARMACEUTICALS AND BIOLOGICALS

Glands and internal organs have been used for their nutritional, biological and pharmaceutical properties for as long as humans have kept animals. The unique nutritional contributions of edible by-products such as the liver are discussed in Chapter 12. Before the discovery of vitamins, hormones and enzymes, preparations from animal glands and organs were often prescribed for their biological, pharmaceutical, or health-promoting properties. Often the tissues were dried, ground, and sold as powders.

With the discovery of individual vitamins, hormones and enzymes, animal organs and tissues became the sources from which these compounds were extracted, purified and manufactured for pharmaceutical or biological purposes. They have great value in treating disorders and diseases in both humans and animals. Glands and tissues used for preparation of pharmaceuticals are removed at slaughter and immediately chilled or frozen. They are shipped to pharmaceutical plants where extraction, concentration, and purification are performed to prepare the final product for medicinal use.

While animal glands and organs are still important and often the only sources for some pharmaceuticals and biologicals, in many cases a specific hormone or pharmaceutical can be synthesized using chemical or molecular biology techniques. For example, cattle and swine pancreas were the primary sources of insulin for treatment of human diabetes. Now, insulin with the same amino acid sequence and structure as human insulin can be synthesized in bacteria and then isolated, purified and concentrated for use in humans.

Discovery of BSE in cattle and recognition of the health risks for humans have led to implementation of numerous safeguards to prevent potential transmission of BSE or vCJD by biological products or pharmaceuticals derived from cattle. Beef spinal cord is an important source of cholesterol, which is then used in vitamin D or steroid synthesis. Designation of beef spinal cord as a SRM has affected that use.

Following are some animal tissues and organs that yield useful pharmaceutical products.

This list of preparations obtained from animal tissues, although incomplete, serves to illustrate the importance of animals and the meat packing industry to our health and well being.

GLANDS

Adrenal	**Epinephrine** is extracted from adrenal medullas and adrenocortical extract from adrenal cortices of cattle, swine, and sheep. Epinephrine is used to stimulate the heart, and as a remedy in treating bronchial asthma. Cortical extract is a source of several steroid hormones used to treat adrenal hormone deficiencies.
Ovaries	Bovine ovaries yield **estrogens**, which are used in treatment of menopausal syndromes, and **progesterone**, which is used to prevent abortion.
Pancreas	Pancreas glands yield **insulin**, which is used as a palliative for treatment of diabetes. **Trypsin** also is obtained from the pancreas, and is used to liquefy and remove necrotic, abscessed, or infected tissue. **Chymotrypsin** from pancreas glands also is used to remove dead tissue in wounds and lesions to promote healing. **Glucagon** and **somatostatin** are obtained for treatment of hypoglycemia and to inhibit growth, respectively.
Parathyroid	**Parathyroid hormone** extract, from beef parathyroid, is used to prevent muscular rigidity and tremors (tetany) in humans whose glands have been removed or function improperly.
Pituitary	A number of hormones, including the **adrenocorticotropic hormone (ACTH)**, is produced by, and may be obtained from, the anterior lobe of beef, sheep, and swine pituitaries. This hormone stimulates the adrenal glands and is used to treat several disorders associated with ACTH deficiency, as well as many other diseases and conditions, including arthritis, acute rheumatic fever, and numerous inflammations. Also obtained is **somatotropin** for growth stimulation, oxytocin for induction of labor, and **thyroid stimulating hormone** for diagnosis of thyroid disorders.
Testes	**Hyaluronidase**, obtained from bull testes, is used as a "spreading factor" in combination with other drugs. This enzyme has ability to easily penetrate body cells, and thus increase rate of distribution and intensity of action of administered drugs. Testosterone may be extracted for treatment of hypogonadism and for stimulation of growth.
Thyroid	Desiccated thyroid and thyroid extract, from both cattle and swine, are used to treat humans having deficiency of hormones produced by this gland. **Thyroxin** and **calcitonin** are major hormones obtained from thyroid glands for treatment of thyroid and bone disorders, respectively.
Thymus	**Thymus hormone**, known as thymosin, is used to stimulate lymphocyte development.

TISSUES AND ORGANS

Blood	In addition to industrial uses, blood yields many pharmaceutical products. Purified bovine **albumen** is used as a reagent in testing for Rh factor in human blood, a stabilizer for vaccines and other sensitive biological products, a reagent in antibiotic sensitivity tests, and a nutrient in microbiological culture media. Blood is a source of **amino acids** used in intravenous feeding of hospital patients. **Fetal serum** is used in tissue culture, **serum** is a source of growth factors, and **thrombin** is a blood coagulant.
Bone	Purified bone meal is used as a source of calcium and phosphorus in pediatric foods.
Intestines	Small intestines of sheep are made into **surgical ligatures** for suturing internal incisions or wounds. This product consists mainly of collagen, which enzymes of the body subsequently digest. Intestine is also one of the primary sources of *heparin* which is used as an anticoagulant to prevent blood clots
Liver	**Liver extracts** are used in treating pernicious anemia. However, since vitamin Bl2 was isolated from liver and subsequently synthesized, use of liver extract for treatment of this type of anemia has declined. Liver is a source of *heparin*. **Bile extract**, obtained from bile of cattle, is used to increase secretory activity of the liver. Bile extract also may be used to make **cortisone**, an adrenocortical steroid hormone with anti-inflammatory properties similar to those of ACTH. Bile is also a source of other steroid compounds with medicinal uses.
Lungs	*Heparin* is obtained from lungs.
Spinal Cord	The spinal cord is a source of *cholesterol*. The principal use of cholesterol is in preparation of vitamin D and steroid synthesis.
Stomach	**Rennet**, from calf stomachs, is used to curdle milk in cheese making. This enzyme also may be added to diets of infants to aid in digestion of milk. **Mucin** is obtained from pig stomachs, and is used in treatment of ulcers. **Pepsin** also is obtained from pig stomachs, and has been used as an aid to digestion.

REFERENCES

Chapter 1

Boorstin, Daniel J. 1973. *The Americans: The Democratic Experience*. New York: Random House.

Kinsman, D. M., E. F. Binkerd, F. W. Tauber, R. W. Bray, and D. H. Stroud. 1976. "History of Meat as a Food." Proceedings of the Reciprocal Meat Conference, National Live Stock and Meat Board, Chicago. pp. 17–85.

Mattson, H. W. 1988. "Potted, Pickled, and Ill-Preserved." *Science of Food and Agriculture*. 3, 2.

Mayer, O. 1939. *America's Meat Packing Industry*. New Jersey: Princeton University Press.

National Live Stock and Meat Board. 1980. "A Brief History of the US Meat and Livestock Industry." Chicago.

Sandburg, Carl. 1916. *Chicago Poems*. New York: Henry Holt & Co.

Sinclair, Upton. 1905. *The Jungle*. New York: The New American Library Inc.

Van Horn, H. H., T. J. Cunha, and R. H. Harms. 1972. "The Role of Livestock in Meeting Human Food Needs." *Bio Science*. 22, 710.

Chapter 2

Bailey, A. J., and N. D. Light. 1989. *Connective Tissues in Meat and Meat Products*. New York: Elsevier Applied Science.

Bechtel, P. J. 1986. "Muscle Development and Contractile Proteins." *Muscle as Food*. P. J. Bechtel, ed. New York: Academic Press. pp. 1–35.

Beermann, D. H., D. R. Campion, and R. H. Dalrymple. 1985. "Mechanisms Responsible for Partitioning Tissue Growth in Meat Animals." Chicago: Proceedings of the Reciprocal Meat Conference, National Live Stock and Meat Board.

Bendall, J. R. 1971. *Muscles, Molecules and Movement*. New York: American Elsevier Publishing Company, Inc. pp. 3–57.

Bourne, G. H. 1960. *The Structure and Function of Muscle*. New York: Academic Press. 1st ed., 3 vols.

Burkitt, H. G., B. Young and J. W. Heath. 1993. *Wheater's Functional Histology: A Text and Colour Atlas*, 3rd ed. Churchill Livingstone, New York. pp. 93-138.

Cassens, R. G. 1987. "Structure of Muscle." *The Science of Meat and Meat Products*. J. F. Price and B. S. Schweigert, eds. Westport, CT: Food and Nutrition Press. pp. 11–59.

Gerrard, D. and A. L. Grant. 2006. *Principles of Animal Growth and Development*. Kendall Hunt Publishing Company. Dubuque, IA.

Goll, D. E., R. M. Robson, and M. H. Stromer. 1984. "Muscle Proteins." *Duke's Physiology of Domestic Animals*, 10th ed., M. H. Swensen, ed. Ithaca, New York: Cornell University Press.

Lawrie, R. A. 1979. "Chemical and Biochemical Constitution of Muscle." *Meat Science*, 3rd ed. New York: Pergamon Press. pp. 75–130.

Murray, J. M., and A. Weber. 1974. "The Cooperative Action of Muscle Proteins." *Scientific American*, 230, 2. pp. 58–71.

Peachey, L. D. 1970. "Form of the Sarcoplasmic Reticulum and T System of Striated Muscle." *The Physiology and Biochemistry of Muscle as a Food, II*. E. J. Briskey, R. G. Cassens and B. B. Marsh, eds. Madison: University of Wisconsin Press. pp. 273–310.

Ross, R., and P. Borstein. 1971. "Elastic Fibers in the Body." *Scientific American*, 224, 6.

Smith, D. S. 1972. *Muscle*. New York: Academic Press. pp. 5–59.

Squire, J. 1981. *The Structural Basis of Muscular Contraction*. New York: Plenum Press.

Squire, J. M., H. A. Al-Khayat, C, Knupp, P. K. Luther. 2005. "Molecular Architecture in Muscle Contractile Assemblies." *Advances in Protein Chemistry, vol. 71, Fibrous Proteins: Muscle and Molecular Motors*. Elsevier Academic Press, New York. pp. 17-87.

Swatland, H. J. 1984. *Structure and Development of Meat Animals*. Englewood Cliffs, NJ: Prentice-Hall.

Chapter 3

Brown, J. H. and C. Cohen. 2005. "Regulation of Muscle Contraction by Tropomyosin and Troponin: How Structure Illuminates Function." *Advances in Protein Chemistry, vol. 71, Fibrous Proteins: Muscle and Molecular Motors.* Elsevier Academic Press, New York. pp. 121-159.

Guyton, A. C. 1971. *Textbook of Medical Physiology.* Philadelphia: W. B. Saunders Co.

Geeves, M. A. and K. C. Holmes. 2005. "The Molecular Mechanism of Muscle Contraction." *Advances in Protein Chemistry, vol. 71, Fibrous Proteins: Muscle and Molecular Motors.* Elsevier Academic Press, New York. pp. 161-193.

Huxley, H. E. 1965. "The Mechanism of Muscular Contraction." *Sci. Am.* 213, 18. Offprint No. 1026. San Francisco: W. H. Freeman and Company.

MacLennan, D. H., and M. S. Phillips. 1992. "Malignant Hyperthermia." *Science.* 256:789.

Murray, J. M., and A. Weber. 1974. "The Cooperative Action of Muscle Proteins." *Sci. Am.* 230:58.

Squire, J. 1981. *The Structural Basis of Muscular Contraction.* New York: Plenum Press.

Squire, J. M., H. A. Al-Khayat, C, Knupp, P. K. Luther. 2005. "Molecular Architecture in Muscle Contractile Assemblies." *Advances in Protein Chemistry, vol. 71, Fibrous Proteins: Muscle and Molecular Motors.* Elsevier Academic Press, New York. pp. 17-87.

Chapter 4

Anil, M.H., 1991. Studies on the return of physical reflexes in pigs following electrical stunning. *Meat Science.* 40:13-21.

Grandin, T. 2006. Progress and challenges in animal handling and slaughter in the US. *Applied Animal Behaviour Science.* 100:129-139.

Grandin, T. 2010. *Recommended Animal Handling guidelines & Audit Guide: A Systemic Approach to Animal Welfare.* Published by: AMI Foundation, 1150 Connecticut Ave NW, Suite 1200, Washington, D.C.

Grandin, T. 1994. Euthanasia and slaughter of livestock. *Journal of American Veterinary Medical Association.* 204:1354-1360.

Grandin, T. 1996. Factors that impede animal movement at slaughter plants. *Journal of the American Veterinary Medical Association.* 209, 4:757-759.

Grandin, T. 2001. Cattle vocalizations are associated with handling and equipment problems at beef slaughter plants. *Applied Animal Behaviour Science.* 71:191-201.

Gregory, N.G. 1994. Preslaughter handling, stunning and slaughter. *Meat Science* 36:45-56.

Gregory, N.G. 1998. *Animal Welfare and Meat Science.* CABI Publishing, Wallingford, UK.

Gregory, N.G., Lee, C.J. and Widdicombe, J.P. 2007. Depth of concussion in cattle shot by penetrating captive bolt. *Meat Science.* 77, 499-503.

Vogel, K.C., Badtram, G., Claus, J.R., Grandin, T., Turpin, S., Wegker, R.E., and Voogel, E. 2010. Head only followed by cardiac arrest electrical stunning is an effective alternative to head only electrical stunning in pigs. *Journal of Animal Science.* 89:1412-1418.

Warris, P.D., S.N. Brown, and S.J.M. Adams. 1994. Relationships between subjective and objective assessments of stress at slaughter and meat quality in pigs. *Meat Science* 38:329-340.

Chapter 5

Ammerman, G. R. 1986. "Catfish Cuisine." *Science of Food and Agriculture.* 4:21.

Bate-Smith, E. C., and J. R. Bendall. 1949. "Factors Determining the Time Course of Rigor Mortis." *J. Physiol.*, 110:47.

Bendall, J. R. 1960. "Postmortem Changes in Muscle." *In The Structure and Function of Muscle,* G. H. Bourne, ed. New York: Academic Press. Vol. 3.

Briskey, E. J., 1964. "Etiological Status and Associated Studies of Pale, Soft, Exudative Porcine Musculature," *Adv. Food Research.* 13:89.

Briskey, E. J., R. G. Cassens, and J. C. Trautman. 1966. *The Physiology and Biochemistry of Muscle as a Food.* Madison: The University of Wisconsin Press.

Forrest, J. C., J. A. Will, G. R. Schmidt, M. D. Judge, and E. J. Briskey. 1968. "Homeostasis in Animals (Sus domesticus) During Exposure to a Warm Environment." *J. Appl. Physiol.* 24:33.

Goll, D. E., Y. Otsuka, P. A. Nogainis, J. D. Shannon, S. K. Sathe, and M. Muguruma. 1983. "Role of Muscle Proteinases in Maintenance of Muscle Integrity and Mass." *J. Food Biochem.* 7:137.

Greaser, M. L. 1986. "Conversion of Muscle to Meat." *In Muscle as Food,* P. J. Bechtel, ed. New York: Academic Press. pp. 37.

Hamm, R., 1977. "Postmortem Breakdown of ATP and Glycogen in Ground Meat." *Meat Sci.* 1:15.

Hedrick, H. B. 1965. "Influence of Ante-Mortem Stress on Meat Palatability." *J. Anim. Sci.* 24:255.

Honikel, K. O., P. Roncales, and R. Hamm. 1983. "The Influence of Temperature on Shortening and Rigor Onset in Beef Muscle." *Meat Sci.* 8:221.

Judge, M. D. 1969. "Environmental Stress and Meat Quality." *J. Anim. Sci.* 28:755.

Kastner, C. L., R. L. Henrickson, L. Buchter, A. Cuthbertson, D. J. Walker and B. B. Chrystall. 1982. "Hot Processing." *International Symposium Meat Science and Technology*. K. R. Franklin and H. R. Cross, eds. (National Live Stock and Meat Board, Chicago. pp. 148.

Koohmaraie, M. 1992. "The Role of Ca2+ Dependent Proteases (Calpains) in Postmortem Proteolysis and Meat Tenderness." *Biochimie*. 74:239.

Marsh, B. B., R. L. Henrickson, J. M. Apple, H. R. Cross, and R. A. Bowling. 1981. "Fresh Meat Processing." *Proceedings of the Reciprocal Meat Conference*, National Live Stock and Meat Board, Chicago, 34:75.

Pearson, A. M., and T. R. Dutson, eds. 1985. "Electrical Stimulation." *Advances in Meat Research*, Vol. 1. Westport, CT: AVI Publishing Co.

Robson, R. M., E. Huff-Lonergan, F. C. Parrish, C.-Y. Ho., M. H. Stromer, T. W. Huiatt, R. M. Bellin and S. W. Sernett. 1997. "Postmortem Changes in the Myofibrillar and Other Cytoskeletal Proteins in Muscle." Proc. Recip. Meat Conf. 50:43.

Robson, R. M., and T. W. Huiatt. 1983. "Roles of the Cytoskeletal Proteins, Desmin, Titin and Nebulin in Muscle." Proc. Recip. Meat Conf. 36:116.

Savell, J. W. 1979. "Update: Industry Acceptance of Electrical Stimulation." Proc. Recip. Meat Conf. 32:113.

Scheffler, T.L. and D.E. Gerrard. 2007. Mechanisms controlling pork quality development: The biochemistry controlling postmortem energy metabolism. *Meat Science*. 77(1):7-16

Scheffler, T.L., Park, S.K. and D.E. Gerrard. 2011. Lessons to learn about postmortem metabolism using the AMPK gamma 3 mutation: the RN pig. *Meat Science*. 89(3):244-50.

Sebranek, J. G. 1981. "Pork Quality: A Research Review," National Pork Producers Council, Des Moines, Iowa.

Topel, D. G., E. J. Bicknell, K. S. Preston, L. L. Christian, and G. Y. Matsushima. 1968. "Porcine Stress Syndrome." *Mod. Vet. Practice*. 49:40.

Chapter 6

Dean, R. W., and C. O. Ball. 1960. "Analysis of the Myoglobin Fractions on the Surfaces of Beef Cuts." *Food Technology*. 14:271.

Hamm, R. 1960. "Biochemistry of Meat Hydration." *Adv. Food Research*. 10:355.

Goll, D. E., N. Arakawa, M. H. Stromer, W. A. Busch, and R. M. Robson. 1970. "Chemistry of Muscle Proteins as a Food." *The Physiology and Biochemistry of Muscle as a Food, II,* E J. Briskey et al., eds. Madison: University of Wisconsin Press. p. 755.

Greaser, M.L. 1997. "Postmortem Changes in Muscle Extracellular Matrix Proteins." Proceedings of the 50th Annual Reciprocal Meat Conference. American Meat Science Association. pp. 53–59.

Kropf, D. H., M. C. Hunt, and D. Piske. 1985. "Color Formation and Retention in Fresh Meat." Washington, D.C.: Proceedings of the Meat Industry Research Conference, American Meat Institute. pp 62–72.

Robson, R. M., E. Huff-Lonergan, F. C. Parrish, CY. Ho, M. H. Stromer, T. W. Huiatt, R. M. Bellin and S.W. Sernett, 1997. "Postmortem Changes in the Myofibrillar and Other Cytoskeletal Proteins in Muscle."Proceedings of the 50th Annual Reciprocal Meat Conference, American Meat Science Association. pp. 43–52.

Watts, B. M., J. Kendrick, M. W. Zipser, B. Hutchins, and B. Saleh. 1966. "Enzymatic Reducing Pathways in Meat." *J. Food Sci*. 31:855.

Chapter 7

Bittle, J. M., M. T. Younathan, and J. S. Godber. 1985. "Meat in the Food Service Industry." Chicago: Proceedings of the Reciprocal Meat Conference, National Live Stock and Meat Board. 38:70–88.

Gilbert, S. G., W. H. Gehrke, and P. Hermansen. 1983. "Meat Packaging." Chicago: Proceedings of the Reciprocal Meat Conference, National Live Stock and Meat Board. 36:47–65.

Huffman, D. L., D. H. Kropf, H. W. Walker, L. J. Ernst, and M. C. Hunt. 1980. "Meat Color." Chicago: Proceedings of the Reciprocal Meat Conference. National Live Stock and Meat Board. 33:4–46.

Johnson, H. K., B. B. Breidenstein, A. A. Taylor, D. H. Kropf, and R. T. Mansur. 1982 "Packaging Systems." *International Symposium Meat Science and Technology*. K. R. Franklin and H. R. Cross, eds. Chicago: National Live Stock and Meat Board. pp. 342–390.

Lundquist, B. R. 1987. "Protective Packaging of Meat and Meat Products." *The Science of Meat and Meat Products*, 3rd ed. J. F. Price, and B. S. Schweigert, eds. Westport, CT: Food and Nutrition Press, Inc. pp. 487–505.

Modern Packaging Encyclopedia. New York: McGraw-Hill Book Company, Inc. Vol. 40, No. 13A.

National Live Stock and Meat Board. 1973. Chicago: "Uniform Retail Meat Identity Standards."

National Live Stock and Meat Board. 1975. Chicago: "Meat in the Foodservice Industry."

National Live Stock and Meat Board. 1991. Chicago: "Lessons on Meat."

North American Meat Processors Association. 1999. "The Poultry Buyers Guide." Reston, VA.

North American Meat Processors Association. 2002. "The Meat Buyers Guide." Reston. VA.

Stern, J. S. 1986. "Fast Food: Part of a Nutritious Diet." *Science of Food and Agriculture*. 4, 2.

Chapter 8

Breidenstein, B. C. 1983. Manufacturing Guidelines for Processed Beef Products. Chicago: National Live Stock and Meat Board.

Cassens, R. G. 1990. Nitrite Cured Meat. A Food Safety Issue in Perspective. Food and Nutrition Press Inc. Trumbull, CT.

Christian, J. A. 1981. Curing Georgia Hams Country Style. Athens: Cooperative Extension Service, University of Georgia.

Cornforth, Darren. 1991. "Methods for Identification and Prevention of Pink Color in Cooked Meat." Proceedings of the Reciprocal Meat Conference. Chicago:National Live Stock and Meat Board. 44:53-56.

Pegg, R. B. and F. Shahidi. 1997. Unraveling the Chemical Identity of Meat Pigments. Crit. Rev. *Food Sci. Nutr*. 37(6)561-589.

Hamm, R. 1986. "Functional Properties of the Myofibrillar System and Their Measurements." *Muscle as Food*. P. J. Bechtel, ed. New York: Academic Press. pp. 135-199.

Killday, B. B., M. S. Tempesta, M. E. Bailey, and C. J. Metral. 1988. "Structural Characterization of Nitrosylhemochromogen of Cooked Cured Meat: Implications in the Meat-curing Reaction." *J. Agricultural and Food Chemistry*. 36:909-914.

Lanier, T. C. 1985. "Fish Protein in Processed Meats." Chicago: *Proceedings of the Reciprocal Meat Conference*, National Live Stock and Meat Board. pp. 129-132.

Saffle, R. L. 1968. Meat Emulsions. *Advances in Food Research*. C. O. Chichester, E. M. Mrak and G. F. Stewart, eds. New York: Academic Press, Vol. 16. pp. 105-160.

Sebranek, J. G. and J. N. Bacus. 2007. Cured meat products without direct addition of nitrate or nitrite: What are the issues? *Meat Science*. 77: 136–147.

Schut, J. 1978. Basic Meat Emulsion Technology. Proceedings of the Meat Industry Research Conference. Arlington, VA: American Meat Institute Foundation. pp. 1-15.

Tarte, R. 2009. *Ingredients in Meat Products: Properties, Functionality and Applications*. New York, Springer Publishing.

Xiong, Y. L. and S. P. Blanchard. 1994. Myofibrillar Protein Gelation: Viscoelastic Changes Related to Heating Procedures. *J. Food Science* 59(4):734-738

Chapter 9

Cunningham, F. E. 1987. *The Microbiology of poultry meat products*. New York, Academic Press.

Dorsa, W. J. 1997. New and Established Carcass Decontamination Procedures Commonly Used in the Beef Processing Industry. *J. Food Prot*. 60:1146.

Fung, D. 1995. Rapid Methods for Meat Microbiology. Proceedings of the Reciprocal Meat Conference, American Meat Science Association. 49:183-185.

Jay, J. M., M. J. Loessner and D.A. Golden. 2005. *Modern food microbiology*. New York: Springer Science.

Kraft, A. A. 1986. Meat Microbiology. *Muscle as Food*. P. J. Bechtel, ed. New York: Academic Press. pp. 239-278.

Niven, C. F. Jr., and E. C. Griener. 1987. "Microbiology and Parasitology of Meat."

The Microbiology of Meat and Poultry. 1998. A. Davies and R. Board, eds. Blackie Academic & Professional, London, UK.

Pearson, A. M., and T. R. Dutson. 1986. Meat and Poultry Microbiology. *Advances in Meat Research, Vol. 2*. Westport, CT: AVI Publishing Company.

Weiser, H. H., G. J. Mountney, and W. A. Gould. 1971. *Practical Food Microbiology and Technology*, 2nd ed. Westport, CT: The AVI Publishing Company, Inc.

Sofos, J.N. 2005. *Improving the safety of fresh meat*. Cambridge, UK: Woodhead Publishing.

Chapter 10

Buchalla, R., C. Shuttler, K.W. Bogl. 1993. "Effects of Ionizing Radiation on Plastic Food Packaging Materials: A Review." *J. Food Prot.* 56:998-1005.

Burke, R. F., and R. V. Decareau. 1964. "Recent Advances in the Freeze-Drying of Food Products." *Advances in Food Research*. C. O. Chichester, E. M. Mrak and G. F. Stewart, eds. New York: Academic Press. Vol. 13. pp. 1-88.

Fennema, D., and W. D. Powrie. 1964. Fundamentals of Low Temperature Food Preservation. *Advances in Food Research*. C. O. Chichester, E. M. Mrak and G. F. Stewart, eds. New York: Academic Press. pp. 219-347.

Giddings, George G. 1985. "Irradiation Utilization in the Meat Industry." Washington, D.C.: Proceeding of the Meat Industry Research Conference, American Meat Institute. pp. 187-197.

Knipe, C. L. and R.E. Rust. 2009. *Thermal Processing of Ready-to-Eat Meat Products*. Ames, IA: Wiley-Blackwell Publishing.

Luyet, B. J. 1968. "Physical Changes in Muscle During Freezing and Thawing." Chicago: *Proceedings of the Meat Industry Research Conference*, American Meat Institute Foundation. pp. 138-156.

Olson, D. G. 1998. "Meat Irradiation and Meat Safety." *Proceedings of the Reciprocal Meat Conference* 51:149-151.

Pearson, A. M. 1986. "Physical and Biochemical Changes Occurring in Muscle during Storage and Preservation." *Muscle as Food*. P. J. Bechtel, ed. New York: Academic Press. pp. 103-134.

Chapter 11

Bailey, M. E. 1992. "Meat Flavour—The Maillard Reaction and Meat Flavour Quality." *Meat Focus International*. 1:192.

Baldwin, R. E. 1977. "Microwave Cookery for Meats." Chicago: *Proceedings of the Reciprocal Meat Conference*, National Live Stock and Meat Board. 30:131–136.

Christensen, M., P. P. Purslow and L. M. Larsen. 2000. "The effect of cooking temperature on mechanical properties of whole muscle, single muscle fibers and perimysial connective tissue." *Meat Sci.* 55: 301.

Cross, H. R., P. R. Durland, and S. C. Seideman. 1986. "Sensory Qualities of Meat." *Muscle as Food*. P. J. Bechtel, ed. New York: Academic Press. pp. 279–320.

Hamm, R. 1966. "Heating of Muscle Systems." *The Physiology and Biochemistry of Muscle as a Food*. E. J. Briskey et al., eds. Madison: University of Wisconsin Press. p. 363.

Hornstein, I. and A. Wasserman. 1987. "Chemistry of Meat Flavor." *The Science of Meat and Meat Products*, 3rd ed., Eds. J. F. Price and B. S. Schweigert, Food and Nutrition Press, Westport, CT. pp. 329-347.

Leander, R. C., H. B. Hedrick, M. F. Brown, and J. A. White. 1980. "Comparison of Structural Changes in Bovine Longissimus and Semitendinosus Muscles During Cooking." *J. Food Sci.* 45:1.

Machlik, S. M. and H. N. Draudt. 1963. "The Effect of Heating Time and Temperature on the Shear of Beef Semitendinosus Muscle." *J. Food Sci.* 28:711.

Martin, A. H. 1969. "The Problem of Sex Taint in Pork in Relation to the Growth and Carcass Characteristics of Boars and Barrows: A Review." *Can. J. Anim.* Sci. 49:1.

National Live Stock and Meat Board. 1991. Chicago: "Lessons on Meat."

Paul, P. C., and H. H. Palmer. 1972. *Food Theory and Applications*. New York: John Wiley & Sons.

St. Angelo, A. J. and M. E. Bailey, eds. 1987. *Warmed-Over Flavor of Meat*. New York: Academic Press.

Starrak, G. and H. K. Johnson. 1982. "New Approaches and Methods for Microwave Cooking of Meat." Chicago: *Proceedings of the Reciprocal Meat Conference*, National Live Stock and Meat Board. pp. 86–91.

Chapter 12

Bender, A. 1992. "Meat and meat products in human nutrition in developing countries." *Food and Nutrition Paper 53*. Food and Agriculture Organization of the United Nations, Rome.

Bodwell, C. E. and B. A. Anderson. 1986. "Nutritional Composition and Value of Meat and Meat Products." *Muscle as Food.* P. J. Bechtel, ed. New York: Academic Press. pp. 321-369.

Breidenstein, B. C. 1987. "Nutrient Value of Meat." *Food and Nutrition News.* Chicago: National Livestock and Meat Board. 59. No. 2.

National Academy of Sciences. 1988. "Designing Foods: Animal Product Options in the Marketplace." Washington D.C.: National Research Council.

Smith, R. L. and E. R. Pinckney. 1988. *Diet, Blood Cholesterol and Coronary Heart Disease: A Critical Review of the Literature.* Santa Monica, CA: Vector Enterprises, Inc.

US Department of Agriculture. 2012. "Food and Nutrition, Dietary Guidance." Retrieved February 1, 2012 from http://fnic.nal.usda.gov/nal_display/index.php?info_center=4&tax_level=1&tax_subject=256.

Chapter 13

Chaudry, M.M. 1992. "Islamic Food Laws: Philosophical Basis and Practical Implications." *Food Technol.* 46: No. 10, p. 92.

Meat and Poultry Inspection Regulations. 1986. US Department of Agriculture, Food Safety Inspection Service. Washington, D.C.

National Advisory Committee on Microbiological Criteria for Foods. 1997. "Hazard Analysis and Critical Control Point Principles and Application Guidelines."

"Pathogen Reduction; Hazard Analysis and Critical Control Point Systems; Final Rule." 1996. Federal Register. Washington, D.C., Vol. 61, No. 144, pp. 38805–38989.

Pearson, A. M. and T. R. Dutson, eds. 1995. "HACCP in Meat, Poultry and Fish Processing." *Advances in Meat Research*, vol 10. Blackie Academic & Professional, New York.

"Procedures for the Safe and Sanitary Processing and Importing of Fish and Fishery Products." 1995. Federal Register. Washington, D.C., Vol. 60, No. 242, pp. 65095–65202.

"Prohibition of the Use of Specified Risk Materials for Human Food and Requirement for the Disposition of Non-Ambulatory Disabled Cattle." 2004. Federal Register, Washington, D.C. Vol. 69, No. 7, pp. 1862-1874.

"US Inspected Meat and Poultry Packing Plants: A Guide to Construction and Layout." 1984. Washington, D.C.: US Government Printing Office. Agricultural Handbook No. 570.

Chapter 14

Boggs, Donald L., Robert A. Merkel, Matthew E. Doumit and Kelly Bruns. 2006. *Livestock and Carcasses: An Integrated Approach to Evaluation, Grading and Selection,* 6th ed. Kendall/Hunt Publishing Company, Dubuque, IA.

Kiehl, E. R. and V. J. Rhodes. 1960. *Historical Development of Beef Quality and Grading Standards.* University of Missouri Agricultural Experiment Station, Columbia, Missouri, Research Bulletin 728.

American Meat Science Association. *Meat Evaluation Handbook*, 2012. P. O. Box 2187, Champaign, IL 61825.

US Department of Agriculture. 1985. "United States Standards for Grades of Pork Carcasses." Agricultural Marketing Service, Livestock and Seed Division.

US Department of Agriculture. 1998. "Poultry Grading Manual." Agricultural Marketing Service, Agriculture Handbook Number 31.

US Department of Agriculture. 1997. "United States Standards for Grades of Carcass Beef." Agricultural Marketing Service, Livestock and Seed Division.

US Department of Agriculture. 1992. "United States Standards for Grades of Lamb, Yearling Mutton, and Mutton Carcasses." Agricultural Marketing Service, Livestock and Seed Division.

Chapter 15

Berg, E. P, Forrest, J. C. and J. E. Fisher. 1994. Electromagnetic scanning of pork carcasses in an on-line industrial configuration. *Journal of Animal Sciences.* 72:2642–2652.

Damez, J. L., and S. Clerjon. 2008. "Meat quality assessment using biophysical methods related to meat structure." *Meat Science.* 80(1):132-149.

M. Altmann, U. Pliquett, R. Suess, and E. von Borell. 2005. "Prediction of carcass composition by impedance spectroscopy in lambs of similar weight." *Meat Science*, 70(2):319-327.

Prieto, N., R. Roehe, P. Lavín, G. Batten and S. Andrés. 2009. "Application of near infrared reflectance spectroscopy to predict meat and meat products quality: A review." *Meat Science* 83(2):175-186.

Warris, P. D., Brown, S. N., Lopez-Bote, C., Bevis, E. A. and S.J.M Adams. 1989. "Evaluation of lean meat quality in pigs using two electronic probes." *Meat Science.* 25:282-291.

Whitman, T. A., Forrest, J. C., Morgan, M. T. and M. R. Okos. 1996. "Electrical measurement for detecting early postmortem changes in porcine muscle." *Journal of Animal Science.* 74:80-90.

Chapter 16

Chesebro, B and Fields, B. N. 1996. *Transmissible Spongiform Encephalopathies: A brief Introduction, In: Virology*, B.N. Fields, et al., 3rd Ed., Raven Publishers, Philadelphia.

Code of Federal Regulations, Title 21, Chapter 1, Subchapter E, Part 589. "Substances Prohibited from Use in Animal Food or Feed."

Franco, D. A., and W. Swanson. 1996. *The Original Recyclers*. The Animal Protein Producers Industry, The Fats & Proteins Research Foundation, The National Renderers Association, Merrifield, VA 22116–2899.

Little, Arthur D. 1969. Opportunities for Use of Meat By-Products in Human and Animal Foods, Report to Iowa Development Corporation.

Pearson, A. M. and T. R. Dutson, eds. 1992. "Inedible Meat By-Products." *Advances in Meat Research, vol. 8.* Elsevier Applied Science, New York.

Price, J. F., and B. S. Schweigert. 1987. *The Science of Meat and Meat Products*. Westport, CT: Food and Nutrition Press.

US Department of Agriculture. 1974. Dictionary of Terms Used in the Hides, Skins, and Leather Trade. Agriculture Handbook No. 465.

US Department of Agriculture. 2011. *Livestock Slaughter 2010 Summary.* National Agricultural Statistic Service. ISSN: 0499-0544.

US Meat Export Federation. 1979. Variety Meats from the USA. Chicago: National Live Stock and Meat Board.

INDEX

A

A band, 15–20, 68
Absorption of off-flavors, deterioration of meat caused by, 241
Accelerated processing, definition of, 124–125
Acetic acid (vinegar), 257
Acetylcholine, 64–65
Acidified sodium chlorite (Sanova®), 96
Acidity, 218–219
Acid meat, 100, 118, 119
Acinetobacter, 220
Actin, 16, 66
Actin filaments, 16
Action potential, of nerve and muscle fibers
    calcium release, 66
    muscle action potentials, 65–66
    myoneural junction, 64–65
Activa™, 208
Actomyosin, 68, 261
    bonds, 108, 110
    complex, 135
Added water, 198
Adenosine diphosphate (ADP), 68, 70, 99
Adenosine monophospate (AMP), *99*, 100
Adenosine monophosphate kinase (AMPK), 119
Adenosine triphosphate (ATP), 52, 63, 68, 70, 72, 98–102, 232
    hydrolysis of, 108
    molecular structure of, *99*
    overview of reactions used to regenerate, *105*
Adenylate kinase, 100, 102
Adequate Intake (AI), 278
Adipoblasts, 31, 32–33
Adipocytes, 31
Adipose tissue, 28, 31–33, 263
Adrenaline, 108
Adrenocorticotropic hormone (ACTH), 364
Advanced Meat Recovery Systems (AMR), 194–195
Aerobic meat spoilage, 237

Aerobic (with oxygen) metabolism, 72–73, 102
    vs anaerobic metabolism, 103–106
Aerobic organisms, 218
Aeromonas, 220
Aging
    of carcass, 111
    of meat products, 211–212
Agricultural Marketing Act (1946), USA, 318
Albumen, 365
Alkaline phosphates, buffering effect of, 203
Alpha-actinin, 19
Alteromonas, 220
Amino acids, 3, 365
    complementarity of, 279, 281, 282
    nutritive value of, 279, 281
    sequence, *30*
A-mode (amplitude) units, 347
Anaerobic glycolysis, 116
Anaerobic meat spoilage, 237
Anaerobic metabolism, 73
    vs aerobic metabolism, 103–106
    products, reactants and enzymes produced during, *104*
Anaerobic organisms, 218
Androgens, 60
5-Androst-16-ene-3-one, 120
Animal aging, affect on meat quality, 119
Animal body
    chemical composition of
        carbohydrates, 51
        lipids, 47–51
        proteins, 47
        water, 47
    elemental composition of, *46*
Animal breeding, 2
Animal fat depots, fatty acid and triglyceride composition of, *48–49*
Animal feeds and fertilizers, 362–363
Animal harvesting, 3, 84–85
    animal movement and facilities, 80–82
    animal physiology and homeostasis prior to, 78–80
    animal size and densities, influence of, 83

Note: Page numbers in *italics* refer to figures and tables.

animal well-being, process and practices
associated with, 76
evisceration process, for removal of internal
organs, 94
exsanguination (bleeding), 90–92
feed withdrawal, 83–84
flight zone and point of balance, *79*, 80
food safety principles, 76
inspectors, for monitoring harvesting and ancillary
processes, 78
lairage, 78, 84
of non-ambulatory animals, 85
personnel involved in, 76
and ritual slaughter, 92
scalding and skinning, 93–94
stunning and immobilization process for, 86–90
temperature and humidity, influence of, 82–83
transportation management schemes, 76–78
Animal movement and facilities, 80–82
Animals. *See* Meat animals
Animal skins, 358–361
Animal tissues, development of
connective tissue
adipose tissue, 31–33
blood and lymph, 37
bone, 34–37
cartilage, 33–34
cells, 31
connective tissue proper, 26
ground substance, 28–31
epithelial tissues, 24–26
muscle tissue
cardiac muscle, 23–24
skeletal muscle, 9–10
skeletal muscle fiber, 11–21
smooth muscle, 21–23
nervous tissue, 26–28
Animal well-being, practices during harvesting
process, 76
Antemortem inspection, 293, 302
Antimicrobial chemicals, for beef and poultry
carcasses, 95–96
Antimicrobial peptides, 219
Antioxidants, 205, 239, 250
Aponeuroses, 29, 42
Appearance and textural properties, of cooked
meat, 260
Approved ingredients, 306
Aromatic seeds, 202
Artery pumping, 180
Articular cartilage, 34

Aspergillus oryzae, 266
Atmosphere packaging, 163
ATPase activity, 45
Autofom Grading System, 348
Autofom™, 348
Automated systems, for weighing fresh retail
meat, 164
Autoxidation, 238–239, 257
Axon, 26

B

Bacillus subtillus, 266
Backfat thickness, 331, *332*, 333
Background toughness, of meat products, 261
Bacteria, 215
corrective procedures for removing or destroying,
233–234
growth curve for a pure culture of, *216*
Bacteriocins, 206, 219, 220
B complex vitamins, 287, 288
Beef
anatomical and common terms used for bones
from, *145*
approximate yields of wholesale cuts, parts, or
edible portions from, *150*
carcass yield grade, 333
color of, 137
cooking instructions, 166
per capita consumption in selected countries, *4*
primal and subprimal cuts, 148
resistance to shear, and the tenderness rating of
selected muscles of, *121*
retail cuts from
leg, round, or ham, *157*
loin (short loin), *155*
rib, *154*
shoulder (chuck) arm, *152*
shoulder blade, *153*
sirloin, *156*
skeletal diagram of, *146*
typical cooling curves for, *242*
wholesale cuts in relation to the skeleton, *148*
Beef Carcass Data Service, 336
Beef Carcass Evaluation Service, 336
Beef Carcass Information Service, 336
3,4-Benzopyrene, 210
Beta-adrenergic agonists, 60
ß-carotene, 122
Bile extract, 365
Bind, 196

Binders, 198, 199
Biological value, of dietary proteins, 279
Bioluminescence, 232
Blast coolers, 242
Blast freezing, 247–248
Blending, 187–188
Blood and lymph, 37
Blood meal, 362, 363
Blood speckling, 92
Blood splash, 92, 122
Blood spots, 90, 92, 122
B-mode (brightness) ultrasound units, 347
Bone, 34–37
Bone darkening, 252
Bone marrow, types of, 37
Bone sour, 179, 237, 266
Botulinal toxin, 223
Botulism, 223–224
Bound water, 133
Bovine spongiform encephalopathy (BSE), 85, 95, 194, 226, 302, 355, 362
Bowl chopper, 186
Boxed meat, 142
Brahman cattle, 114
Braising, 274
Branded meat products, 143
Brands, 319, 360
Break joint, 321, 323
Brine curing, 361
Broiler meat, per capita consumption in selected countries, 4
Broiling, 273
Broken backs, 92
Bromelin, 266
Brown fat, 32
Bung, 94
Butcher shops, 144
Buttons in beef carcasses, characteristics of, 321
Butylated hydroxyanisole (BHA), 239, 362
Butylated hydroxytoluene (BHT), 239, 362
By-products, obtained from meat animals
    animal feeds and fertilizers, 362–363
    edible by-products, 352–358
    glue, 363
    hides, skins and pelts
        classification, 358–360
        processing, 361
        wool and hair, 361
    inedible by-products, 358
    pharmaceuticals and biologicals, 363–364
    tallows and greases, 362

C

Calcified bone matrix, 34
Calcitonin, 364
Calcium pump, 70
Calcium sequestering, 66, 70
Callipyge, 56–57
Callipyge gene, 119
Calpains, 113
Calpastatin, 113, 119
Campylobacter coli, 95, 96, 225
Campylobacter enterocolitis, 225
Campylobacter jejuni, 225
Canaliculi, 34
Captive bolt stunning apparatus
    general design of, *86*
    placement for an effective stun of various farm animals, 87
Cap Z, 19
Carbohydrate oxidation, 119
Carbohydrates, 51
    nutritive value of, 285–286
Carbon dioxide (CO_2)
    gondola-type CO_2 stunning apparatus for pigs, *91*
    for preservation of meat products, 257
    for stunning of farm animals, 90
Carbon monoxide (CO), 163
Carcass
    aging of, 111
    composition of
        body components, 53
        factors influencing, 54–59
        hormones and hormone-like materials, 59–60
        muscle, fat, and bone, 54
    conditioning of, 111
    cutting tests, 164
    defects, 325
    determination of carcass composition by electrical impedence, *346*
    dressing of, 53, 94
    dressing percent, 53, 336
    electrical stimulation, 125–130
    factors influencing composition of
        fat, 59
        genetics, 54
        major genes, 54–57
        nutrition, 57
        physiological age, 57
        plane of nutrition, 58–59
        protein, 59

firmness of, 324
heritability estimates of characteristics of cattle, poultry, sheep, and swine, *53*, 54
intervention treatments, 214, 234
manipulation of
   antimicrobial chemicals, 95–96
   hot water rinses, 95
   organic acid rinses, 95
   steam pasteurization, 95
   steam vacuum, 95
priority and partition of nutrients among body systems and tissues, *58*
splitting of, 95
Carcass-by-carcass visual inspection, 303
Carcass cutability and federal yield grades
   dressing percent, 336
   establishment of, 333–334
   factors used to determine, 330
      amount of fat, 331
      carcass size, 332–333
      muscle development, 331–332
Carcass grade, influence on palatability of meat, 265
Cardiac muscles, 9, 23–24
Cartilage, 33–34
   ossifies, 321
Carving skills, for serving cooked meat, 275
Case hardening, 212, 254
Casings, 206
Cathepsins, 21, 113
Cattle
   distribution of body components of, *52*
   heritability estimates of carcass characteristics of, *53*
Cattle hides, 358–361
   classification of, *360*
   processing of, 361
Caul fat, 95
Central nervous system, 26, 98
Certification of Carcasses Eligible for Approved Breed Programs, 336
"Certified" pork, 230
Cervical ligament (*ligamentum nuchae*), 31
Cesium-137 (radioactive cesium), 256
Charcoal broiling, 273–274
Chemical contamination, deterioration of meat caused by, 240
Chemical ingredients, for preservation of meat products, 257–258
Chicago meat packing industry, 5

Chicken. *See also* Poultry
   broiler parts, *158*
   skeleton of, *147*
Child Nutrition Programs, 278
Chill tunnels, 242
Chlolesterol, 284
   content of common measures of selected foods, *285*
Chlorine, 96
Cholinesterase enzyme, 65
Chondroblasts, 34
Chondrocytes, 33
Chondroitin sulfates, 28, 29
Chroma, 136
Chrome tanning, 361
Chronic Wasting Disease, 226
Chymotrypsin, 364
Cisternae, 20
Clostridium botulinum, 65, 179, 186, 223, 257
Clostridium perfringens, 224
Coagulation, of muscle proteins, 270
Cobalt-60 (radioactive cobalt), 256
Coextrusion, 207
Cold pasteurization, 256
Cold-set gelation of proteins, 195
Cold shortening, 123–124, 127, 243
Collagen, 29, 44, 196, 363
Collagen fibers, 29, 30, 34, 38, 270, 271, 361
Collagen fibril, 29–30, 33, 120, 270
Collagen shrinkage, 270–272
Color, as indicator of meat characteristics
   attributes of, 136
   chemical state, 137–139
   pigments, 136–137
   species and other influences, 137
Color balance, 167
Colorimeters, 343
Colorimetry process, for assessment of fat color, 339
Commercially sterile meat products, 252
Comminution of meat, 177, 186–187, 219
Compact bone, 34
Condemned material, control and restriction of, 306
Conditioning of carcass, 111
Conduction heating, 274
Conformation, 324
Conjugated linoleic acid (CLA), 284
Connective tissue in muscle
   adipose tissue, 31–33
   affect on the texture of meat, 140

blood and lymph, 37
bone, 34–37
cartilage, 33–34
cells, 31
ground substance, 28–31
proper, 26, 28
role in tenderness/toughness of meat, 140
Continuing Survey of Food Intake by Individuals (CSFII), 278
Continuous phase, 189
Contractile proteins, 66
Controlled atmosphere systems (CAS), 90
Convection heating, 274
Convenience stores, 144
Conversion of muscle to meat
    energy metabolism, H$^+$ production and muscle pH decline, 106–107
        heat production and dissipation, 107–108
        muscle structure, disruption of, 111–114
        physical properties of muscles, changes in, 114
        protective mechanisms, failure of, 111
        rigor mortis, phenomenon of, 108–111
    factors affecting postmortem changes and meat quality, 114
        antemortem effects, 115–122
        postmortem effects, 122–130
    pH decline curves of pork muscle during, *106*, 107
    postmortem muscle metabolism for
        adenosine triphoshpate (ATP), 98–102
        aerobic *vs* anaerobic metabolism, 103–106
        glycogen metabolism, 103
        glycolysis and mitochondrial respiration, 102–103
    postmortem temperature decline curves, *109*
    water-binding capacity during, 114
Cooked meat specialties, 177
Cooked sausages, 177
Cookery of meat
    effects of heat on meat constituents
        appearance, 267
        flavor and aroma, 272–273
        juiciness, 272
        structural changes, 267–270
        tenderness, 270–272
    methods and recommendations, for heating systems, 273–275
    and serving of cooked meat, 275
Corrective actions, for disposition of potentially unsafe products, 300

Cortisone, 365
Costameres, 20
Countable plates, 231
Cracklings, 355
Creatine kinase, 19, 100
Creutzfeldt-Jakob Disease, 226
Critical Control Point (CCP), 209, 298, 306
    decision tree to identify, *299*
Critical limit, 299
Crossbridges, between thin (actin) and thick (myosin) filaments, 66, 108
Cryogenic freezing, 244, 245, 248–249
Cure accelerators, 183
Cured meat color
    chemistry of, 182–183
    phosphates, influence of, 203
Cured meat pigment, stability of, 184–185
Curing brine, 180
Curing of meat
    chemistry of cured meat color, 182–183
    flavor stability, 184
    heat stable pink color, 185
    history and present application, 178
    ingredients for, 178–179
    mechanical methods to improve cure distribution, 181–182
    methods for, 179–181
    naturally cured meat products, 184
    public health aspects of nitrite usage, 185–186
    stability of cured meat pigment, 184–185
Cutting tests, 164
Cysteine, 240
Cytoskeletal proteins, 16
    C protein, 18, 19
    H protein, 19
    nebulin, 19
    titin, 18
Cytosolic calcium, 70

D

Daily Values (DV) for key nutrients, 278
Danish Meat Research Institute, 341
Dark cutting meat, 117
Dark, firm, dry (DFD) muscle condition, 117, 346
Death phase, of microorganisms, 216
Deboning meat, equipment for, *194*
Decomposition of meat, 236
Deep basted meat products, 177

Dehydration
- deterioration of meat caused by, 240–241
- methods for
  - freeze drying, 254–255
  - hot air drying, 254
- preservative effects on meat products, 211–212, 254

Deli departments, 144
Denaturation of muscle proteins, 107, 210, 267
Dendrites, 26
Dense connective tissue, 29
Dense irregular connective tissue, 29
Dense regular connective tissue, 29
Deoxymyoglobin, 138
Depolarization of nerve fibers, 64
Desinewing, of meat products, 207
Desmin, 19
Deterioration of meat, 236–237
- caused by
  - chemical contamination, 240
  - insects, 238
  - lipolysis and lipid oxidation, 238–240
  - microorganisms, 237–238
  - physical deteriorative changes, 240–241
  - protein oxidation, 240

Dextrose equivalent (DE), 202
Dhabiha slaughter, 92
Diaphysis, 34
Diarrhea, 225
Dietary Guidelines for Americans, 226, 229, 278, 284, 285
Dietary Reference Intakes (DRIs), 278
Diet, influence on physical properties of muscle, 121–122
Diffraction grating, 341
Dihydropyridine receptor, 66
Dinitrosylhemochromogen, 182
Direct measurement techniques, for measuring depth of fat at various locations on a carcass, 349
Dispersed phase, 189
Double muscling, 54, 57, 119
Downers. *See* Non-ambulatory animals
Dressing of carcass, 94
Dressing percentage of carcass, 53, 336
Drip, 241
Drip loss, 132
Drip pan, 241
Dry cure, 180
Dry heat cookery, 273–274
- meat products recommended for, 274

Dry rendering, 362
Dry sausages, 144, 177, 203, 205, 211, 257
Dystrophin, 19

E

Edible meat by-products, 352–358
- approximate yield of various items obtained from broiler chickens and turkeys, *354*
- cod fish, *355*
- meat animals, *354*

Elastic cartilage, 34
Elastin, 29, 31
Electrical conductivity and resistance, use in prediction of meat and carcass characteristics, 344
Electrical stimulation (ES)
- of carcass, 125–130
- differences in rate of pH decline with respect to nonstimulated beef carcasses, *127*
- effect on meat properties, *129*
- flow diagram of effects of, *130*
- meat tenderization by, 127

Electric prod, for livestock handling, 80
Electric stunning, of farm animals and poultry, 88–90
Electrochemical and electrical technologies, for measurements of the physical and chemical properties of meat, 344–346
Electrochemical process, for initiation of muscle contractions in animals, 64
Electromagnetic scanning devices, for prediction of lean composition in carcass, 348–349
Electronic assessment, of carcasses and fresh meat
- principles of
  - direct measurement techniques, 349
  - electrochemical and electrical technologies, 344–346
  - electromagnetic scanning devices, 348–349
  - optical-based technologies, 339–343
  - ultrasound technologies, 347–348
  - vision based systems, 343–344
  - X-ray scanners, 349
- requisites for successful technology adoption, 338–339

Electron transport chain, 73
Emulsification, 188–191
Emulsifying agents, 189, *190*, 210
Emulsifying capacity, 197
Emulsion mill, 186, 187
Endochondral ossification, 36
Endomysium, 38

Endotoxin, 224
Enhanced meat products, 177
Enzyme-linked immunosorbent assays (ELISA), 232
Enzyme tenderizers, 266
Epimysium, 38
Epinephrine hormone, effects of, 60, 78, 115–116, 364
Epiphyseal plate, 34
Epiphyses, 34
Epithelial tissues, 24–26
Erythrocytes, 37
Escherichia coli O157:H7, 95, 96, 166, 214, 225, 294, 302
Essential amino acids, 279
Estimated Average Requirement (EAR), 278
Estradiol-17ß, 60
Estradiol benzoate, 60
Estrogens, 60, 364
Ethylene oxide, 202
Eutectic point, 245
Evaporative cooling, 254
Evisceration process, for removal of internal organs, 94
Exchange of ions, in muscle proteins, 136
Exsanguination (bleeding), 86, 87, 90–92, 219
Extenders, 198
Extracellular fibers, 29–31
Extracellular substance, 28
Extracellular water, 139
Exudate, 107

F

F-actin (fibrous actin), 17
Facultative organisms, 218
Fair Packaging and Labeling Act, USA, 316
Fasciculi, 37
Fast freezing, 245–246
Fat caps, 192
Fat depth, 331
Fat free lean, 334
Fat-O-Meater, 342
Fat pockets, 192
Fat storage cells, 28
Fat thickness, 331
Fatty acids, 282
    metabolism, 105, 238
Fatty tissues, 355
Feathering, 324
Feather meal, 363
Feather removal in birds
    plucking procedure for, 94
    principles for, 93
    times and temperatures for, *93*
Federal Meat Grading Act (1925), USA, 318
Federal Meat Grading Service, 318–319
Federal/State National Shellfish Sanitation Program, USA, 316
Feed withdrawal, prior to animal transportation and harvesting, 83–84
Fenestrated collar, 20
Fermentation, of meat products, 211
Fermented sausages, 177
Fetal serum, 365
Fiber optic-based spectroscopy technologies, components of, *340*
Fiber optics, 341, 343
Fibroblasts, 31, 34
Fibrocartilage, 34
Ficin, 266
"Fight or flight" response, 78
Filamin, 19
Filler meats, 196
Fillers, 198
Fill, influence on carcass yield, 84
Finish, 324
Firmness, of carcass, 324
Fish, market forms of, 160
Fish meal, 363
Flaking machine, 186
Flank streaking, 324
Flavobacterium, 220
Flavor and aroma
    of cooked meat products, 264–265
    effect of heat on, 272–273
Flavorings, 203
Fleshing, 324
Flight zones, in farm animals, *79*, 80
Food allergies, 203
Food and Drug Administration (FDA), 256, 278, 316
Food-borne disease. *See* Food borne infection
Food borne infection, 95, 220
    botulism, 223–224
    campylobacter enterocolitis, 225
    Clostridium perfringens food infection, 224
    hemorrhagic colitis, 225
    listeriosis, 225
    other bacterial infections, 226
    from parasites, 226–230
    salmonellosis, 224–225
    staphylococcal food poisoning, 224
Food borne intoxication, 223

Food chain, 2
Food, Drug and Cosmetic Act, USA, 316
Food grade emulsifiers, 210
Food inspectors, 295
Food poisoning from fresh meats, 166
    characteristics of, *221–222*
Food safety, principles of animal handling and harvesting for, 76
Food service
    meat acquisition for, 169
        institutional meat purchase specifications, 170–172
        in-unit processing, 173
        meat orders, 173
        processed meat, 172
        receiving and storing meat, 173
    meat processing and distribution for, 169
    types of
        hotels, 169
        other providers, 169
        restaurants, 169
Food Stamps (food assistance program), 278
Forming film, for packaging of processed meat, 168
Formulation procedure, for meat processing
    antioxidants, 205
    food allergens, 203
    meat ingredients, 193–195
        functional properties of, 195–197
    moisture, 197–198
    other non-meat ingredients, 198–199
    phosphates, 203–205
    seasonings and flavorings, 199–203
Free radicals, 239
Free radical scavengers, 239
Free water, 133
Freeze drying, of meat products, 254–255
Freezer burn and drip, deterioration of meat caused by, 241
Freezer storage, 244
Freezing boundary, 245
Freezing methods, for storage of meat
    blast freezing, 247–248
    cryogenic freezing, 248–249
    liquid immersion and liquid sprays, 248
    plate freezer, 247
    still-air freezing, 247
Freezing rates, affect on physical and chemical properties of meat, 244–245
    fast freezing, 245–246
    slow freezing, 245

FreshBloom®, 96
Fresh meat displays. *See also* Frozen meat displays; Processed meat displays
    display lighting, 167
    display systems, 167
    display temperature, 167
    labeling
        cooking assistance, 166
        date, 166
        grade or brand, 166
        name of cut, 164–165
        name of retailer, 166
        of nutrients, 166
        package cost, 166
        price per unit of weight, 165–166
        safe food handling instructions, 166–167
        universal product code, 166
        weight, 165
    meat cutting and identification, 144–162
    packaging functions and materials
        modified atmosphere packaging, 163
        tray and overwrap, 162–163
        vacuum packaging, 163
    weighing and pricing
        automated systems, 164
        pricing policy, 164
Fresh meat, properties of
    color
        chemical state, 137–139
        pigments, 136–137
        species and other influences, 137
    meat quality, 132
    structure, firmness, and texture, 139–140
        connective tissue, 140
        intramuscular fat (marbling), 140
        rigor state, 140
        water-holding capacity, 140
    water-holding capacity, 132–133
        chemical basis of, 133–136
Fresh meat sausages, 177
Frozen meat displays. *See also* Fresh meat displays; Processed meat displays
    packaging requirements, 167–168
    temperature, 168
Frozen storage, length and conditions of, 249–251
Full scalds, 93
Fungi, 215
Fungicides, 240

G

G-actin (globular actin), 17
Game meat, 1
Gamma radiation, 255–256
Gelatin, 270
Generation interval of psychrophilic bacteria, effects of temperature on, 217
Genoa salami, 237
Genotype of meat animals, 54
Giblets, harvesting of, 94
Globin, 137
Globin hemochromes, 185
Glucagon, 364
Glucose, 74, 103
Glue, 363
Glutamic acid, 203
Glycerin, 362
Glycine, 29
Glycogen, 51, 52, 286
    general structure of, 102
Glycogenesis, 103
Glycogenolysis, 72, 103
Glycogen phosphorylase, 103
Glycolysis, 73, 99, 102–103
Glycolytic enzymes, 46
Glycolytic metabolism, 46, 52, 103
Glycosaminoglycans, 28
Golgi complex, 21
Gondola-type CO_2 stunning apparatus for pigs, *91*
Gram-negative spoilage bacteria, 220
Greases, 362
Grinding, 124
Ground substance, 28–31
Growth hormone, 59. *See also* Somatotropin
Growth medium of microorganism, 231

H

Hair removal, principles for, 93, 94, 233
Halal slaughter, 92, 316
Hal gene. *See* Halothane (Hal) gene
Halophilic (salt-loving) bacteria, 257
Halothane gas anesthesia, 118
Halothane (Hal) gene, 118
"Ham and Water Product with X percent Added Ingredients" label, 198
Hampshire breeding, 100
Ham sour, 237
"Ham with Natural Juices" label, 198
"Ham with Water Added" label, 198
Hard-scalds, 93
Hard-shelled crabs, 208
Hazard Analysis Critical Control Point (HACCP), 214, 294, 298–302, 303
    generic example of, *309–312*
        corrective actions log, *314*
        hazard analysis, *307–308*
        metal detection log, *314*
        pre-shipment review log, *314*
        room temperature log, *313*
        thermometer calibration log, 313
    generic product description for, *304*
    Plan Checklist, *300–301*
    principles applied to seafood, 316
    process flow diagram for raw ground beef for, *305*
Heart fat, 331
Heart stick, 90
Heating systems for meat cooking, categories of
    conduction heating, 274
    convection heating, 274
    dry heat, 273–274
    microwave radiation heating, 275
    moist heat, 274
    radiant heating, 274–275
Heat processing, of meat products, 208–209
Heat-set gelation of proteins, 195
Heat shock proteins, 254
Heat-stable pink color, 185
Heat transfer, 253
Heme complex of myoglobin, schematic representation of, *137*
Heme iron, nutritive value of, 286
Heme ring, 137
Hemichrome, 185
Hemoglobin, 37
Hemolytic uremic syndrome (HUS), 225
Hemorrhagic colitis, 225
Herbicides, 240
Herbs, 202
High quality protein, 3, 192, 279
High velocity air freezing. *See* Blast freezing
Homeostasis, 70, 78, *79*, 99, 110
Homestasis, 115
Horse meat, 229, 293, 296
Horse Meat Act (1919), USA, 293
Hot air drying, of meat products, 254
Hotels, food service in, 169
Hot water rinses, for cleaning of carcasses, 95

H protein, 19
Hue, 136
Humane Slaughter Act (1958), USA, 87, 293
Human foods, biological value of, 281
Hyaline cartilage, 34
Hyaluronic acid, 28
Hyaluronidase, 364
Hydrocolloids, 288
Hydrogenated vegetable fats, 284
Hydrolytic rancidity, 238
Hydrophilic emulsifying agents, 189
Hydrophobic emulsifying agents, 189
Hydroxyproline, 29, 186, 282
Hyperplasia, 33, 42
Hypertrophy, 42, 45, 54, 56–57
Hypoxanthine, 120, 264
H zone, 15–20, 68

I

I band, 15–16, 18, 20–21, 23, 68, 112, 123
Immersion curing, 180
Immersion scalding, 93
Immobilization process, affect on meat quality, 122
Immobilized water, 133
Impedance measurements, for prediction of meat and carcass characteristics, 344, 346
Inates, 163
Individual Quick Frozen (IQF) products, 248
Inedible meat by-products, 358
Inosine monophosphate (IMP), 264
Insects, deterioration of meat caused by, 238
Inspected products, identification of, 295–296
Inspection laws, application and enforcement of, 294–295
    identification of inspected products, 295–296
    meat inspectors, 295
Institutional meat purchase specifications (IMPS)
    aged beef, 172
    fat and skin limitations, 170
    identifying numbers for various species, *171*
    quality grade, 170
    refrigeration state, 170–171
    size of cut, 171–172
Insulin-like growth factors (IGFs), 59
Integrated poultry production, 5
Intercalated disks, 23
Intermolecular cross-linkages, 30, 114, 120
Intermuscular fat, 32, 39
International Commission on Illumination, 340
Intestines, 355

Intra-membranous ossification, 36
Intramuscular fat, 32, 39, 54, 140, 263, 323–324
In-unit processing of meat, 173
Invasive technology, 341
Involuntary muscles. *See* Cardiac muscles; Smooth muscles
Ionizing radiation, for preservation of meat products, 255–257
Iron, nutritive value of, 286
Irradiation odor, in fresh meat, 256
Irradiation procedure, for preservation of meat, 202, 255–257
Isoelectric point of muscle, 134

J

Juiciness in cooked meat
    effect of heat on, 272
    principal sources of, 263–264
Jungle, The (Upton Sinclair), 293

K

Kidney fat, 331
Kilogray, 256
Knock box, 88
Kosher slaughter, 92, 316
Krebs cycle. *See* Tricarboxylic acid cycle (TCA cycle)
Kuru, 226

L

Labeling, of retail meat
    cooking assistance, 166
    date, 166
    grade or brand, 166
    name of cut, 164–165
    name of retailer, 166
    nutrients, 166
    package cost, 166
    price per unit of weight, 165–166
    safe food handling instructions, 166–167
    universal product code, 166
    weight, 165
Laboratory inspection, 315
Laboratory inspectors, 295
Lactate, 106
Lactic acid, 116, 163, 179, 186, 211, 236, 257, 266, 286
Lactic acid bacteria, 211, 219–220

Lactobacillus plantarum, 211, 220
Lactoferrin, 96
Lactose, 286
Lag phase, of microorganisms, 216
Lairage, 78, 84
Lamb
    approximate yields of wholesale cuts, parts, or edible portions from, *150*
    retail cuts from
        leg, round, or ham, *157*
        loin (short loin), *155*
        rib, *154*
        shoulder (chuck) arm, *152*
        shoulder blade, *153*
        sirloin, *156*
    skeletal diagram of, *146*
    wholesale cuts in relation to the skeleton, *148*
Lanolin, 361
Lean, color and structure of, 324
Leuconostoc, 220
Leukocytes, 37
Lever arm, 68
Light fading process, for enhancing stability of cured meat pigment, 185
Lipases (lipid-hydrolyzing enzymes), 236
Lipid oxidation, 236, 238–240, 249
Lipids, 47–51, 124
    nutritive value of, 282–285
Liquid immersion or spray method, for freezing poultry, 248
Liquid smoke, 210
Listeria monocytogenes, 96, 214, 218, 225, 236, 302
Listeriosis, 225
Live hanging, 85
Liver extracts, 365
Liver glycogen, 74
Livestocks, 2
    breeding and rearing for meat production, 4
    dressing percentages of various kinds of, 334
    farm income from sale of, 4
    growing and finishing of, 5
    production of, 3, 54
Livestock Weather Safety Index, *82*, 83
Logarithmic growth, of microorganisms, 216
Loin eye, 331
Longitudinal tubules, 20
Loose connective tissue, 29
Low-Acid Canned Food program, USA, 316
Low density lipoprotein (LDL) cholesterol, 284
Lumbosacral, 147
Lysosomes, 21, 113

M

Maceration process, for improving tenderization and cure distribution in boneless smoked meat products, 181
Machine Vision Systems, 343, 344
Maillard browning reaction, 209, 257
Mammalian skeletal muscle, approximate composition of, *50–51*
Manufactured casings, 206–207
Marbling. *See* Intramuscular fat
Marking and labeling, 315
Massaging machines, basic design of, *181*
Massaging process, for improving cure distribution, 181
Matadors cape, for livestock handling, 80
Maturity in meat animal carcasses
    definition of, 321–323
    indices of, *322*
Meat. *See also* Fresh meat, properties of
    categories of, 1
    definition of, 1
    deterioration of. *See* Deterioration of meat
    heritability estimates for physical properties of, *118*
    in-unit processing, 173
    list identifying the most typical color of, 137
    maximum recommended length of storage of, *250*
    merchandising practices, 6
    muscle as, 1–2
    nutrients contained in, 3
    nutritive value of. *See* Nutritive value of animal products
    oxidation-reduction potential of, 219
    palatability of. *See* Palatability of meat
    receiving and storing of, 173
    refrigeration and transportation of, 5
    science and, 6
    storage of. *See* Refrigerated storage
    US Department of Agriculture grades for, *326–327*
Meat acquisition, for food service, 169
    institutional meat purchase, specifications for
        aged beef, 172
        fat and skin limitations, 170
        quality grade, 170
        refrigeration state, 170–171
        size of cut, 171–172
    in-unit processing, 173
    meat orders, 173

processed meat
  precooked and restructured meat, 172
  smoked meat and sausage, 172
  receiving and storing meat, 173
Meat animals
  efficiency in converting feed into meat, 58
  genotype of, 54
  harvesting of. *See* Animal harvesting
  major genes, 54–57
  muscle fibers, characteristics of, *42*
  phenotypic variations in, 54
  physiological age of, 57
  plane of nutrition, 58–59
  priority and partition of nutrients among body systems and tissues, *58*
Meat batter, 186, 187
  air in, 193
  diagrammatic illustration of, *190*
  factors affecting formation and stability of, 191–192
Meat Buyers Guide, 170
Meat constituents, effects of heat on
  appearance, 267
  flavor and aroma, 272–273
  juiciness, 272
  structural changes, 267–270
  tenderness, 270–272
Meat consumption
  and economy of the nation, 3–4
  per capita consumption in selected countries, *4*
Meat cutting and identification, 144–162
Meat flavor
  flavorings used to intensify, 203
  intensity of, 120
  stability of, 184
Meat grading and evaluation
  carcass cutability and federal yield grades
    dressing percent, 336
    establishment of, 333–334
    factors used to determine, 330–333
  federal quality grades, 320
    establishment of, 325–330
    factors used to establish, 321–325
  functions of, 319–320
  types of
    brands, 319
    commodity evaluations, 319
    federal grades, 318–319

USDA Grading Service, services offered by
  audit services, 336
  carcass data, 336
  certification and verification services, 336
US Department of Agriculture grades for meat, *326–327*
Meat grinder, 186
Meat industry
  earnings in food sector, 4
  refrigeration and transportation, role of, 5
  in United States, 5–6
Meat ingredients
  functional properties of, 195–197
  selection and preparation of, 193–195
Meat inspection
  antemortem inspection, 293, 302
  elements of
    antemortem inspection, 302
    condemned material, control and restriction of, 306
    Hazard Analysis Critical Control Points (HACCP), 298–302
    laboratory inspection, 315
    marking and labeling, 315
    postmortem inspection, 303–304
    product inspection, 306–315
    sanitation and SSOP, 297–298
  history of, 292–294
  inspection laws, application and enforcement of, 294–295
    identification of inspected products, 295–296
    meat inspectors, 295
  postmortem inspection, 293, 303–304
  product inspection, 293, 306–315
  regulations in USA, 85
  reinspection, 306
  religious inspection and certification, 316
  requirements for granting inspection service, 296–297
  seafood inspection, 316
Meat Inspection Act (1906), USA, 292, 293, 294
Meat inspectors, 295
Meat meal, 363
Meat orders, 173
Meat packaging, 4, 5
  functions of
    modified atmosphere packaging, 163
    tray and overwrap, 162–163
    vacuum packaging, 163

materials available for, 258
plants for, 5
for preservation of meat products, 258
Meat preservation. *See* Preservation of meat
Meat processing, 3–5
    basic procedures for
        comminution, blending, and emulsification, 186–193
        curing, 178–186
        fermentation, dehydration, and aging, 211–212
        forming processed meat products, 205–207
        formulation, 193–205
        restructured meat products, 207–208
        smoking and heat processing, 208–211
    history of, 178
Meat products, restructured, 207–208
Meat quality, factors affecting, 114
    antemortem effects
        age, 119–120
        diet, 121–122
        genetics and major genes, 118–119
        immobilization method, 122
        muscle location, 120
        preslaughter handling, 122
        sex, 120–121
        stress, 115–117
        stress and muscle characteristics, 117–118
    postmortem effects
        accelerated processing, 124–125
        carcass electrical stimulation, 125–130
        temperature, 122–124
Meat quality, of fresh meat, 132
Meat raw materials, binding classification of, *196*
Meat retailers, 2
Meat science, 2, 6
Meat spoilage, 237
Meat technology, 2
Meat tenderization, using electric stimulation method, 127
Mechanical deboning, 162, 194, 195
Mechanically separated (species) meat, 194
Mechanical stunning, of animals. *See* Physical stunning, of animals
Medullary cavity, 36, 37
Melting, of collagen fibrils, 270
Memory devices, 347
Mesenchymal cells, 28, 31
Mesophiles, 217
Metaphysis, 34

Methyl bromide, 238
Metmyoglobin, 138, 162, 163, 168, 183, 237
Microaerophilic organisms, 218
Microaerophilic spoilage of meat, 237
Microbial activity in meat, factors affecting, 216
    acidity, 218–219
    interactions, 220
    microbial inhibitors, 219–220
    moisture, 218
    other nutrients, 219
    oxidation-reduction potential, 219
    oxygen, 218
    physical form, 220
    temperature, 217–218
Microbial contamination of meat
    corrective procedures for removing or destroying bacteria, 233–234
    procedures for preventing, 232–233
        operational sanitation procedures, 233
        pre-operational sanitation method, 233
    sources of, 215
Microbial inhibitors, 219–220
Microbial numbers, growth, and activity in meat, methods for assessment of
    microbial testing methods, 230–231
    rapid microbiological methods, 231–232
Microbiological principles
    food borne infections and intoxications
        botulism, 223–224
        bovine spongiform encephalopathy, 226
        campylobacter enterocolitis, 225
        Clostridium perfringens food infection, 224
        hemorrhagic colitis, 225
        infection from parasites, 226–230
        listeriosis, 225
        other bacterial infections, 226
        salmonellosis, 224–225
        staphylococcal food poisoning, 224
    microbial activity in meat, factors affecting, 216
        acidity, 218–219
        interactions, 220
        microbial inhibitors, 219–220
        moisture, 218
        other nutrients, 219
        oxidation-reduction potential, 219
        oxygen, 218
        physical form, 220
        temperature, 217–218
    microbial contamination, sources of, 215

microbial numbers, growth, and activity in meat, assessment of
    microbial testing methods, 230–231
    rapid microbiological methods, 231–232
microorganisms in meat, 215–216
pathogenic bacteria, 220–223
spoilage bacteria, 220
Micrococcus, 220
Microground spices, 202
Microorganisms
    affect on palatability of meat, 266
    affects of ionizing radiation on, 255–257
    deterioration of meat caused by, 237–238
    heat resistance of, 253–254
    in meat, 215–216
Microwave radiation heating, 275
Minerals, nutritive value of, 286–287
Mitochondria, 21
Mitochondrial respiration, 99, 102–103, 105
M line, 15, 17–19
Modified atmosphere packages (MAP), 163, 238
Moist heat cookery, 274
Moisture:protein (M:P) ratio, 197
Mold growth, 206, 215
Moraxella, 220
Motendinal structure, of muscle-tendon junction, 41–42
Motor end plate, 11, 39, 64
Motor nerve fibers, 11, 62
Motor nerves, 62
M protein, 19
Mucin, 365
Multi-catalytic proteinase complex, 113
Multiple stitch injection, 180
Muscle-altering mutation, 119
Muscle cells, 11
Muscle contraction-relaxation cycle, flow diagram of, *71*
Muscle energy metabolism, 99
Muscle fibers, 11
    action potentials of, 65–66
    characteristics of, in domestic meat animals and birds, *42*
    growth of, 42–44
    size of, 44–45
    types of, 45–47
Muscle function, sources of energy for, 70–74
Muscle location, affect on meat quality, 120
Muscle proteins, categories of, 51

Muscle-tendon junction
    organization of, 41–42
    photomicrographs of, *43*
Muscle tissues
    cardiac muscles, 23–24
    dark, firm, dry (DFD) condition, 117
    isoelectric point of, 134
    lactate accumulation, 73
    locations of water that resides in, 133
    macroscopic and microscopic structure of, *10*
    as meat, 1–2
    organization of
        muscle bundles and associated connective tissues, 37–42
        muscle fiber growth, 42–44
        muscle fiber types, 45–47
        muscle size, 44–45
        muscle-tendon junction, 41–42
        nerve and vascular supply, 39–40
    pale, soft, exudative (PSE) condition, 117
    postmortem metabolism. *See* Postmortem muscle metabolism
    skeletal muscle fibers, 11–21
    skeletal muscles, 9–10
    smooth muscles, 21–23
    transformation into meat. *See* Conversion of muscle to meat
Muscling, 324
Mycotoxins, 223
Myelin sheath, 28, 39, 62
Myoblasts, 24
Myofibrillar proteins, 16–19, 44, 51, 66, 110, 112–114, 189–191, 195, 240, 252, 263, 270, 274
Myofibrils, 11, 14–15, 44, 45, 270
    and degree of tenderness of muscles, 261–263
    proteins of, 16
Myofilaments, 14, 15–16
Myogenesis, 24
Myoglobin, 46, 137, *182*, 267
Myokinase, 100
Myomesin, 19
Myoneural (neuromuscular) junction, 11, 64–65
Myosin, 15–17, 66, 240
Myosin filaments, 15, 17
Myosin isoform, 45, 192
Myostatin, 57, 118, 119
Myotubes, 24

N

National Nutrient Database, USDA, 286
Nationwide Food Consumption Survey (NFCS), 278
Natural casings, 206–207, 355
Naturally cured meat products, 184
Natural seasonings, 202
Near-infrared (NIR) spectrum, for detection of meat quality, 342
Nebulin, 16, 19, 112, 263
Nerve fibers, 26
    depolarization of, 64
Nerves and the nature of stimuli
    action potentials, 64–66
        calcium release, 66
        muscle action potentials, 65–66
        myoneural junction, 64–65
        resting membrane potential, 63
        transmembrane potentials, 62–63
Nerve trunks, 26, 28, 39
Nervous tissues, 1, 9, 26–28, 37, 228
Net charge effect, 133–134
Neuron, 26
Neuroplasm, 26
Neutral lipids, 47, 52
Nicotinamide adenine dinucleotide (NAD^+), 73, 106, 183
Nitric oxide, 182
Nitric oxide myoglobin, 182
Nitrite
    bacteriostatic properties of, 257
    burn, 185
    for curing of meat, 178, 179
    preservative properties of, 257
    public health aspects of, 185–186
Nitrosamines, 185–186
Nitrosylhemochromogen, 182–184, 209
Nodes of Ranvier, 62
Non-ambulatory animals, 85, 86, 302
Non-comminuted meat, 207, 255
Nonfat dried milk solids (NFDM), 199, 202
Non-forming film, for packaging of processed meat, 168
Non-heme iron, nutritive value of, 286
Non-ionizing radiation, germicidal effects of, 256
Nonprotein nitrogen (NPN), 12, 52
Non-sausage processed meats, 177
Norepinephrine hormone, effects of, 60, 78, 115
North American Meat Processors Association, 170
Not smoked sausages, 177
Nucleation sites, 245
Nuclei of skeletal muscle fibers, 12–14
Nutrient retention, during heating, 288
Nutritional Facts Label, *278*, 284
Nutritional labeling, of meat products, 166, 288–290
Nutritive value of animal products
    carbohydrates, 285–286
    lipids and calories, 282–285
    minerals, 286–287
    nutrient retention during heating, 288
    nutrition labeling of retail meat products, 288–290
    proteins, 279–282
    variety and processed meats, 287–288
    vitamins, 287

O

Obesity, 284
Oil-in-water emulsion, *189*
Optaflexx®, 60
Optical-based technologies, for assessment of carcasses and fresh meat, 339–343
Organic acid rinses, for cleaning of carcasses, 95
Organic acid washes, 234
Osteoblasts, 34, 36
Osteoclasts, 37
Osteocytes, 34, 36
Oxidation of iron molecule, 137
Oxidation-reduction potential of meat, 219
Oxidative metabolism, 21, 24, 46, 139
Oxidative rancidity, 163, 195, 202, 237, 238, 243, 249, 267
Oxygen, 218
    transmission rate, 162, 163
Oxymyoglobin, 138, 162, 183, 237, 244
Ozonation procedure, for destruction of bacteria, 258
Ozone, use for preservation of meat, 257

P

Palatability of meat
    characteristics of
        adipose tissue, 263
        appearance, 260
        flavor and aroma, 264–265
        juiciness, 263–264
        palatability interrelationships, 265
        tenderness, 260–263

factors associated with
    carcass grade, 265
    chemical changes, 267
    microbiological, 266
    postmortem aging, 265–266
    processing methods, 266
    tenderizers, 266
Pale, soft, exudative (PSE) muscle condition, 117, 118, 139, 343
Papain, 266
Paranemin, 19
Parasites, 215
Parathyroid hormone, 364
"Partially Hydrolyzed Soy Protein" flavoring, 203
Pasteurization, of meat products, 209, 252
Pathogenic bacteria, 214, 220–223
Pathogen reduction, regulations for, 294
Paylean®, 60
Pediococcus acidilactici, 211, 220
Peeling process, for removal of fat, 170
Pelvic fat, 54, 331, 333
Pepsin, 365
Percent retained water, 242
Perimysium, 38, 140, 261, 271
Periosteum, 34
Peripheral nervous system, 26
Perishable foods, movement of, 5
Pet food industry, 362
Phagocytic leukocytes, 37
Pharmaceuticals and biologicals, 363–364
pH decline in postmortem muscle, *106*, 107, 124
Phenotypic variations in meat animals, 54
Phosphagen system, 99
Phosphates, 179, 203–205
Phosphocreatine, 72, 99
Phospholipids, 52, 172, 236, 239, 282
Physical deteriorative changes, deterioration of meat caused by
    absorption of off-flavors, 241
    dehydration, 240–241
    freezer burn and drip, 241
Physical stunning, of animals, 87–88
    captive bolt stunning apparatus, 86–87
    general design of equipment used to restrain cattle for, 88
Physiological age of meat animals, 57, 321
Pickle, 180
Pigments in meat, 136–137
Plasmalemma, 11
Plate freezer, 247
Platelets, 37

Plucking procedure, for feather removal in birds, 94
Pluck, removal of, 94
Point of balance in farm animals, 80
Polar overdominance, 57
Polychlorinated biphenyls, 240
Polymerase Chain Reaction (PCR), 232
Polypeptides, 18, 133
Pork
    approximate yields of wholesale cuts, parts, or edible portions from, *150*
    degrees of muscle development for, *332*
    per capita consumption in selected countries, *4*
    pH probe placement in, *345*
    retail cuts from
        leg, round, or ham, *157*
        loin (short loin), *155*
        rib, *154*
        shoulder (chuck) arm, *152*
        shoulder blade, *153*
        sirloin, *156*
    salting and packing of, 5
    skeletal diagram of, *146*
    typical cooling curves for, *242*
    wholesale cuts in relation to the skeleton, *148*
Portion cuts, 170
Postmortem aging, influence on palatability of meat, 265–266
Postmortem inspection, 293, 303–304
    components of, 303
Postmortem muscle metabolism
    adenosine triphoshpate (ATP), 98–102
    aerobic *vs* anaerobic metabolism, 103–106
    flow diagram of changes in, *115*
    glycogen metabolism, 103
    glycolysis and mitochondrial respiration, 102–103
Postmortem proteolysis, 113, 114, 136
Pot roasting, 274
Poultry. *See also* Chicken
    color of meat, 137
    evisceration of, 94
    heritability estimates of carcass characteristics of, *53*
    inspection of, 293
    meat, 1
Poultry Buyers Guide, 170
Poultry by-product meal, 363
Poultry Inspection Program, USA, 293
Poultry Products Inspection Act (1957), USA, 293, 294
Power stroke, 68
Preblending, 188
Precooked and restructured meat, 172

Prerigor chilling, implications of, 124
Pre-rigor chopping, 124
Preservation of meat, 3, 5
    chemical ingredients, by, 257–258
    dehydration, by, 254–255
    irradiation procedure, by, 255–257
    packaging, by, 258
    refrigerated storage, by, 241–252
    thermal processing, by, 252–254
Preslaughter handling, affect on meat quality, 122
Pricing policies, for retail meat, 164
Primal Cut Percent, 348
Primal cuts, 148
Primary bundles of muscle fibers, 38
Prion proteins, 226
Probiotic microorganisms, 214
Processed meat displays, 168. *See also* Fresh meat displays; Frozen meat displays
Processed meat products
    formation of, 205–206
        manufactured casings, 206–207
        natural casings, 206
        non-comminuted meat, 207
    nutritive value of, 287–288
    types of
        non-sausage, 177
        sausage, 177
        whole muscle, 176–177
Processed Products Inspection Improvement Act (1986), USA, 294
Process Verification Program, USDA, 336
Product inspection, 293, 306–315
Progesterone, 60, 364
Protein fat-free (PFF), 198
Protein hardening, 252, 265, 270, 272, 273
Protein oxidation, 240
    deterioration of meat caused by, 240
Proteins
    composition in animal bodies, 47
    cytoskeletal, 16
    gelation, 209
    high quality, 279
    matrix formation, 188–191
    of myofibrils, 16
    nutritive value of, 3, 279–282
    recommended dietary allowances of, 281, 282
    residues, 355
Protein-to-DNA ratios, 57
Proteoglycans, 28, 34
Proteolysis, 113–114, 133, 135–136
Proteus, 220

Pseudo H zone, 15, 18
Pseudomonas, 219, 220, 232
Psychrobacter, 220
Psychrophiles, 217
Psychrotrophs, 231
Pulsed field gel electrophoresis (PFGE), 232
Pulsers, 347
Purge, 132, 162
Putrefaction of meat, 236
Pyruvate, 73, 103, 105, 106

Q

Quality grades, for meat products, 319
    establishment of, 325–330
    factors used to establish
        carcass defects, 325
        color and structure of lean, 324
        conformation, fleshing, and finish, 324–325
        firmness, 324
        kind and class, 321
        marbling, 323–324
        maturity, 321–323

R

Radiant heating, 274–275
Radioactive isotopes, 256
Rad units, 256
Radura symbol, 256, *256*
Rancid fat, 185
Rancidity, development of, 257
Rapid heat transfer, 244, 248
Real-time ultrasound scanning equipment, 348
Receivers, 347
Recommended Dietary Allowances (RDA), 278, 281–282, 286–287
Record keeping procedures, 300
Recrystallization, 251
Red cells (erythrocytes), 37
Red fibers, 44, *45*, 46
Red marrow, 37
Red meat, 1, 284, 286
Reductants, 179
Reduction of iron molecule, 137
Refreezing, of meat products, 251–252
Refrigerated storage, 241–242
    bone darkening, 252
    duration of, 243–244
    freezer storage, 244

freezing methods, 247–249
freezing rates, 244–246
initial chill, 242–243
length and conditions of, 249–251
thawing and refreezing, 251–252
Refrigeration process, for preservation of food products, 236
Regulatory proteins, 66
Reinspection, 306
Religious inspection and certification, 316
Rendement Napole (RN) gene, 118–119
Rendering of fats, 355, 358
Rendition, 167
Rennet, 365
Rephosphorylation of creatine, 72
Repolarization of the sarcolemma, 68
Residual nitrite, 183
Restaurant food service, 169
Restructured meat products, 207–208
Retail cutting test record form, *165*
Retail meat distribution systems
    distribution centers to retail stores, 143
    processing plants to retail stores, 142–143
    slaughter plants to retail stores, 142
Retail stores, types of
    butcher shops, 144
    convenience stores, 144
    specialty shops, catalog and Internet, 144
    supermarkets, 143–144
Rib-eye, 331, 332
Rigor mortis, phenomenon of, 108–111
    affect on meat tenderness, 110
    completion of, 110
    delay time before onset of, *110*
    isometric tension development in muscle during phases of, *110*
    resolution of, 110
Rigor state, during storage and processing of fresh meat, 140
Rigor toughening, of cooked meat products, 261
Ritual slaughter, 92
RN gene. *See* Rendement Napole (RN) gene
Roasting with dry heat, 274
Rosemary extract, 240, 257
Round sour, 237
Ryanodine receptor, 66

S

Safe food handling instructions, 166–167
Salmonella, 95, 96, 166, 214, 236, 302
Salmonellosis, 224–225
Salts, for curing of meat, 178
Salt-soluble proteins, 51, 52, 181, 192, 196, 197
Sanitation Standard Operating Procedures (SSOP), 232–233, 294, 297–298, 302
Sanova®, 96
Sarcolemma, 11–12
Sarcomere, 15
Sarcoplasm, 12, 102
Sarcoplasmic muscle proteins, 51, 282
Sarcoplasmic reticulum (SR), 20–21, 47, 66
Satellite cell, 43, 44
Saturated fatty acids, 239, 282, 284
Sausages, 172, 177
    stuffed in natural and artificial casings, *204*
Scalding and skinning of animals, 93–94
Schwann cells, 28, 39, 62
Scrapie, 226
Sea foods, 1
    inspection, 316
Seam fat, 39, 170, 194
Seasonings, 179, 199–200
Self-service stores, 144, 162
Semidry sausages, 177, 211, 254, 257
Semitendinosus muscle, 271
Serum, 365
Service counters, 144
Serving, of cooked meat, 275
Sex-associated meat quality, 120
Sex hormones, 120
Sex odor, 120
Sheep pelts, 358–361
    classification of, *360*
Shelf life, 218–219, 230
Shellfish and edible portions, *161*
Shewanella, 220
Shiga toxin producing strains (STEC), 225
Shochet (Jewish person conducting the slaughter), 92
Shortenings, use of fat tissues as, 355
Shoulder stick, 90
Shrink in animal's weight
    definition of, 83
    during storage, 132
Skelemin, 19

Skeletal muscle fibers
    cross-section of, *12*
    development of, 24
    Golgi complex, 21
    lysosomes, 21
    major contractile proteins in, 17–18
    major regulatory proteins in, 18
    mitochondria, 21
    myofibrils, 14–15
        proteins of, 16
    myofilaments, 15–16
    nuclei, 12–14
    organization of, *13*
    sarcolemma, 11–12
    sarcoplasm, 12
    sarcoplasmic reticulum and T tubules, 20–21
    structural features and their longitudinal orientation, *11*
    Z disk ultrastructure, 16
Skeletal muscles, 9–10
    changes during postmortem aging, *113*
    chemical composition of, 51–53
    contraction of, 66–68
    relaxation of, 68–70
    structural relationships among tendons, blood vessels, and nerves, *10*
Slack scald, 93
Slaughter plants, 5, 76, 94, 142, 151, 266, 292, 294, 303
Slow freezing, 168, 240, 244–245, 251
Smoked meats, 143, 144, 172, 176–177, 178, 180, 198, 206, 209, 257, 306
Smoking, of processed meats, 172, 176–177, 208, 209–211
Smooth muscles, 9, 21–23
Sodium chloride, preservative properties of, 257
Sodium-potassium pump, 63
Somatostatin, 364
Somatotropin, 59–60, 364
Soy Protein Flavoring, 203
Spatial effects, 135
Special Supplemental Nutrition Program for Women, Infants, and Children (WIC program), 278
Specialty shops, 144
Specific heat of substances, 253
Specified Risk Materials (SRMs), 304, 355
Spectral reflectance method, for assessment of fat color, 339, 341
Spectrophotometers, 341, 343

Spices commonly used in processed meats, types and origins of, *200–201*
Splitting, of carcass, 95
Spoilage bacteria, 214, 220, 237
Spongy bone, 34
Spool joint, 321, 323
Staphylococcal food poisoning, 224
Staphylococcus aureus, 224
Staphylococcus carnosus, 184
Stationary phase, of microorganisms, 216
Steamed bone meal, 363
Steam pasteurization, 95, 234
Steam vacuum systems, used in slaughter process, 95
Steric effects, 135
Sterilization of meat products, 252
Steroid hormone synthesis, 284
Sticking method, for severing the arteries and veins, 90
Still-air freezing, 247
Stitch pumping, 180
Streptococcus faecalis, 220, 226
Stress
    affect on meat quality, 115–117
    and muscle characteristics, 117–118
Stress hormones, 116, 117
    diagram of the sites of production of, *116*
Stress-resistant animals, 117
Stress syndrome, 117
Striated muscles, 9
Stromal muscle protein, 51
Stunning and immobilization process, for animal harvesting, 86–87
    carbon dioxide (CO_2) stunning, 90
    electric stunning, 88–90
    physical stunning, 87–88
    reasons for, 87
Stunning box, 88
Subcutaneous fats, 32, 54, 264, 272, 282, 347
Sublimation, 241, 254
Subprimal cuts, 132, 142, 148, 170–173
Sugar-amine browning reactions, 267, 274
Sugar, preservative properties of, 257
Super-cooling of meat surface, 245
Supermarkets, 143–144
Supportive connective tissues, 28
Surface dehydration, 241, 244, 267
Surface membrane, removal of, 170
Surgical ligatures, 365
Surimi, 195
Sweeteners, 202

Sweet Lebanon Bologna, 286
Swelling of proteins, 188
Synapses, 26
Synemin, 19
Synovial fluid, 29, 34
Synthetic estrogens, 60

T

Talin, 19
Tallows, 362
Tanking, 306
Temperature cycling, 241
Tenderizers, influence on palatability of meat, 266
Tenderness of meat
    connective tissue, 261
    contractile protein interactions and myofibril integrity, 261–263
    effect of heat on, 270–272
    perception of, 260–261
Terminal cisternae, 20
Testosterone, 120
Textured vegetable protein, 198
Texture of muscle, 323
Thabiha slaughter, 92
Thawing, of meat products, 251–252
Thaw rigor, 123
Thermal conductivity, 253
Thermal neutral zone, 83
Thermal processing of meat, 252
    principles of
        heat resistance of microorganisms, 253–254
        heat transfer, 253
Thermophiles, 217
Thrombin, 365
Thymosin. *See* Thymus hormone
Thymus hormone, 364
Thyroid stimulating hormone, 364
Thyroxin, 364
Titer, 362
Titin, 16, 18
Tolerable Upper Intake Level (UL), 278
Total microbial load, 215
Total Plate Count, 231, 298
Toxoplasma gondii, 220, 226
    life cycle of, *227*
    primary reservoir for, *228*
Toxoplasmosis, 226, 229
Transducers, 347
Trans fatty acid, 284
Trans-glutaminase enzyme, 208

Transmissible spongiform encephalopathies (TSE), 226, 358, 362
Transportation shrink, 83
Transverse tubule (T tubule) system, 11, 20–21, 47, 66
Tray and overwrap, 162–163
Triad junction, 66
Tricarboxylic acid cycle (TCA cycle), 73, *101*, 102, 103
Trichina Herd Certification program, USA, 230
Trichinella spiralis, 220, 256
    life cycle of, *228*
    storage time required, at various freezer temperatures, to destroy, *230*
    thermal death temperature for, 229
    time-temperature relationship for, *229*
Trichinosis, 226, 229
Triglycerides, 51, 52, 236, 238, 282
Trisodium phosphate, 96
Tristimulus method, for assessment of fat color, 339, 340
Tropocollagen, 28–31
Tropoelastin, 28, 31
Tropomyosin, 16, 18, 66, 68, 70
Troponin, 16, 18, 66, 70, 263
Trypsin, 364
T-system, 11
Tumbling machines, basic design of, *181*
Tumbling process, for improving cure distribution, 181
Type I muscle fibers, 45
Type IIa muscle fibers, 45
Type IIb muscle fibers, 45
Type IIx(d) muscle fibers, 45

U

Ultrasound technologies, for assessment of structures and composition of carcasses, 347–348
Ultraviolet radiation, germicidal effects of, 256–257
Uniform Retail Meat Identity Standards, 165
Universal product code, 166, 168
Unsaturated fatty acids, 184, 238–239, 273, 282, 284
US Centers for Disease Control, 226, 229, 284, 285
"US Condemned" marking, 302
USDA Dietary Guidelines. *See* Dietary Guidelines for Americans
USDA Grading Service, services offered by
    audit services, 336
    carcass data services, 336
    certification and verification services, 336
US Department of Agriculture grade, 166
US Federal Meat Grading Service, 170, 318

"US Inspected and Condemned" tags, 304
"US Inspected and Passed" tags, 304
"US Retained" tags, 303
"US Suspect" marking, 302

V

Vacuum packaging, 6, 142–143, 162, 163, 167, 168, 173, 186, 218, 237, 240, 243, 255, 256, 302
Value of a color, 136
Variant Creutzfeldt-Jakob Disease (vCJD), 226, 355
Variety meats, 193, 352
    nutritive value of, 287–288
V-belt design, for moving cattle in a processing plant, *84*, 85
Vegetable oils, 284
Vegetable tanning, 361
Ventilation systems in on-farm facilities, 82
Veterinary medical officers, 295, 303
Vinculin, 20
Viruses, 215
Viscera, 9, 94, 303
Visible light, for detection of meat quality, 342
Vision based systems, for measuring properties of carcasses and meat, 343–344
Vitamin B12, 3, 287
Vitamin E, 121, 240
Vitamins, nutritive value of, 287
Voluntary muscles. *See* Skeletal muscles

W

Warmed-over-flavor (WOF), 239, 273
Water activity, 218, 254
Water-holding capacity, of fresh meat, 132–133
    associated with each rigor stage, 140
    chemical basis of
        exchange of ions, 136
        net charge effect, 133–134
        proteolysis, 135–136
        steric effects, 135

Water vapor transmission rate, 162
Weasand, 94
Weep, 162
White cells (leukocytes), 37
White fat, 32
White fibers, *45*, 46–47, 56, 124
Whole muscle meat products. *See* Smoked meats
Wholesome Meat Act (1967), USA, 293, 294, 296, 315
Wholesome Poultry Products Act (1968), USA, 293, 294, 296, 315
Wool and hair, 361

X

X-ray scanners, for prediction of carcass composition, 349

Y

Yeasts, 215
Yellow marrow, 37
Yersinia enterocolitica, 226
Yield grades, for meat products, 319

Z

Z disk, 15, 16
    degradation of, *112*
    ultrastructure, 16
Zeranol, 60
Z filaments, 16
Zilmax®, 60
Zinc, nutritive value of, 287